Realizing Complex Integrated Systems

The creation of complex integrated systems is, in itself, complex. It requires immense planning and a large team of people with diverse backgrounds based in dispersed geographical locations (and countries) supposedly working to a coordinated schedule and cost. The systems engineering task is not new, but recent scales most definitely are. The world is now capable of designing and manufacturing systems whose complexity was not considered possible 10 years ago. While many are trained to think in terms of a complete system, where 'everything' is designed and produced by a single project team, today such systems involve integrating subsystems and components (which are also complex) that have been developed by other project teams. Inevitably, this introduces additional complexities, involving elements out of the direct control of the project, but which are essential to its overall success.

In addition to traditional systems engineering topics of hardware and software design, testability, and manufacturability, there are wider issues to be contemplated: project planning; communication language (an issue for international teams); units of measure (imperial vs. metric) used across members of the team; supply chains (pandemics, military action, and natural disasters); legal issues based on place of production and sale; the ethics associated with target use; and the threat of cyber-attack. This book is the first attempt to bring many of these issues together to highlight the complexities that need to be considered in modern system design. It is neither exhaustive nor comprehensive, but it gives pointers to the topics for the reader to follow up on in more detail.

Realizing Complex Integrated Systems

Edited by

Anthony P. Ambler and John W. Sheppard

CRC Press
Taylor & Francis Group
Boca Raton London New York

CRC Press is an imprint of the
Taylor & Francis Group, an **informa** business

Cover: www.shutterstock.com

First edition published 2025
by CRC Press
2385 NW Executive Center Drive, Suite 320, Boca Raton FL 33431

and by CRC Press
4 Park Square, Milton Park, Abingdon, Oxon, OX14 4RN

CRC Press is an imprint of Taylor & Francis Group, LLC

ISBN: 9781032036533 (hbk)
ISBN: 9781032036601 (pbk)
ISBN: 9781003188377 (ebk)

DOI: 10.1201/9781003188377

Typeset in Times
by codeMantra

Contents

PART 7 System Usage

Preface

The creation of complex integrated systems is, in itself, complex. It requires immense planning and a large number of people with very diverse backgrounds based in dispersed geographical locations (and countries) supposedly working to a coordinated schedule and cost.

Such a scenario is not new, but the scale most definitely is. The world is now capable of designing and manufacturing systems whose complexity was not considered possible 10 years ago. An individual integrated circuit can be produced with a transistor count that surpasses that found in an entire product a decade ago, but today that same integrated circuit is only a part of a complete system.

Consider the 'complete' system; it is also far easier to think in terms of a complete system where 'everything' is designed and produced by the project team. But the system will necessarily be produced using components already developed by other organizations that will need to be integrated into the new product. Inevitably, this will introduce significant perturbations in the attempt to control the creation of the new systems – parts out of the direct control of the project, but that are essential to its overall success.

Of course, any project requires planning and project management, but the scales are literally unimaginable to a casual observer and necessitate a level of hierarchical project management and carefully applied oversight.

Naturally, it depends on the application of the new system (domestic, industrial, or military), the scale/size (pocket calculator, smartphone, cellular phone network, satellite communication network, etc.), and even the laws of the producing nation, where the product is to be sold.

Communication language is an issue for the project team. International projects where teams work in different languages is an obvious case, but even the languages used by different disciplines can lead to miscommunication and misunderstanding. If care is not taken in the project specification, even colleagues two desks apart might interpret the requirements sufficiently differently to cause a problem, neglecting the case where one team was working in imperial units and the other was assuming metric… very expensive and embarrassing!

Of increasing importance is the issue of supply chain – with availability of materials, local military action, health pandemics, and critical infrastructure failures (e.g., blockage of the Suez Canal). Backup sources of components and materials are increasingly necessary.

Integration of system functions and controls into single components makes testability and reliability an increasingly critical part of the design. The use of software, while extremely useful in reducing the size of the hardware, making the system more versatile, and allowing for continuous improvements, introduces immense scope for difficult-to-identify bugs, the introduction of viruses, and the threat of other forms of cyberattack, and that entails either economic/military/life-safety issues.

This book is the first attempt to bring many of the issues together to highlight the complexities that need to be considered in the modern world. The chapter list identifies many of the required topics. It is by no means exhaustive – possible future editions (!) may need to make huge additions to the material. This work is also not intended to provide *all* the information needed to implement a project team but gives pointers to the topics for the reader to follow up on in more detail.

Note: the *creation* of complex systems is by no means purely technical – it embodies sales and marketing, creation of project definition commercial/military, division of the project into segments for design and/or further segmentation across nations/companies/design teams, definitions of maintainability and repairability and speed of return-to-service, supply chain issues critical to both creation of systems and their use, final design, contracts, project design reviews, etc. But it is important not to overlook human factors in the design process, system use, and impact of the system on the environment. It is increasingly important for a significant proportion of those involved with the system creation to be at least aware of all these issues and the importance of their role in the overall success (however measured). For this reason, many institutions of higher learning have been offering graduate programs in engineering management for some time and have gradually been pushing some of the material into undergraduate programs.

This book will serve not only as a textbook but as a reference and handbook for the practitioner. Because of this, very little "theory" will be included. It is aimed at graduate students, project managers, practitioners, and others who are embarking upon a modern complex project. It may also suggest to curriculum designers and licensing authorities to broaden the scope of education and professional requirements away from a narrow disciplinary direction.

Finally, a note of thanks. When we set out to undertake this project, we had no idea that writing such a work would, in itself, seem to follow the process of realizing a complex system design. For us, the design was in the book organization itself. As with any complex design process, this means a large number of people played a variety of roles in making this happen.

We begin by acknowledging and thanking the staff at Taylor & Francis for their guidance and support while this project unfolded, in particular Nora Konopka and Swapnil Joshi. We depended on their expertise to make sure the process would unfold as smoothly as possible, and they did not disappoint.

This work also draws on a very large number of expert authors, bringing a variety of backgrounds and perspectives to the task at hand. Neither of the editors are experts in all areas of this book, so we relied on the expertise of these authors, not only to provide the content of the chapters, but to write in a way that could be integrated into a cohesive work and lead toward the same goal. Each of the authors is identified below, with their affiliations, but here we wish to say thank you.

We also thank those who provided facility support or funding. Tony Ambler received facility support from the Provost and President of the University of Houston. John W. Sheppard received financial and facility support from Montana State University and from a variety of other sources, most notably the U.S. Navy, while completing this work. In fact, the project was initiated during a funded sabbatical from MSU.

Finally, we thank our families, and without their support, this project would never have been completed.

Anthony P. Ambler
John W. Sheppard
2025

Editors

Anthony P. Ambler is a fellow of the IEEE, elected 'For contributions to economics of testing complex digital devices and systems'. His research interests are in test economics, system test, and diagnosis. He received his B.Sc., M.Sc., and Ph.D. from the University of Manchester Institute of Science and Technology. He was appointed to a chair in Test Technology at Brunel University (UK) and then moved to the USA. He became chairman of Electrical & Computer Engineering at the University of Texas at Austin, then dean of Engineering & Computing at the University of South Carolina, and recently as dean of Technology at the University of Houston. In addition to his research work, he created the MS degree program in Engineering Management at UT Austin. He has acted as chair of the Organizing Committee of the European Design and Test Conference, general chair and program chair of IEEE International Test Conference and of IEEE International Conference on Computer Design. He has created a number of Workshops including on Test Economics, System Test and Diagnosis, and Production Test Automation.

John W. Sheppard is a Norm Asbjornson College of Engineering Distinguished Professor in the Gianforte School of Computing at Montana State University. His research interests include fault diagnosis/prognosis of complex systems, model-based and Bayesian reasoning, explainable and ethical artificial intelligence, and distributed population-based algorithms. He is a fellow of the IEEE, elected 'for contributions to system-level diagnosis and prognosis'. He received his BS in computer science from Southern Methodist University and his MS and PhD in computer science from Johns Hopkins University. Before entering academia full-time, he was a member of the industry for 20 years where his prior position was as a research fellow at ARINC Incorporated. He has been a long-time leader in the IEEE Standards Association, chairing several working groups focused on publishing standards related to complex system test and diagnosis. Previously, he also served as the "designated representative" from the IEEE Computer Society to the IEEE Standards Coordinating Committee 20 on Test and Diagnosis for Electronic Systems.

Contributors

Anthony P. Ambler
College of Technology
University of Houston
Houston, Texas

K. Suzanne Barber
Center for Identity
University of Texas at Austin
Austin, Texas

Zheyong Bian
Supply Chain and Logistics Technology
University of Texas at Austin
Austin, Texas

Jamie Blanche
James Watt School of Engineering
University of Glasgow
Glasgow, United Kingdom

Stephyn G. W. Butcher
Whiting School of Engineering
Johns Hopkins University
Baltimore, Maryland

Lila Carden
Technology Division, Cullen College of Engineering
Construction Management Department
Technology Project Management Program
University of Houston
Houston, Texas

David R. Carey
Department of Mechanical and Electrical Engineering
The School of Business and Engineering
Wilkes University
Wilkes-Barre, Pennsylvania

Wm. Arthur Conklin
Information Science and Technology
Cullen College of Engineering
University of Houston
Houston, Texas

Alessandro Ferrero
Department of Electronics
Polytechnic University of Milan
Milan, Italy

David Flynn
James Watt School of Engineering
University of Glasgow
Glasgow, United Kingdom

Ranjeetkumar Gupta
National Composites Centre
Feynman Way Central
Bristol, United Kingdom

Christian K. Hansen
Department of Mathematics
Eastern Washington University
Cheney, Washington

Sam Harper
James Watt School of Engineering
University of Glasgow
Glasgow, United Kingdom

Matthew Hu
Industrial Engineering
University of Houston
Houston, Texas

Kristen Intemann
Center for Science, Technology, Ethics, and Society
Montana State University
Bozeman, Montana

Clemente Izurieta
Computer Science | Software Engineering and Cybersecurity Lab (SECL)
Gianforte School of Computing
Montana State University
Pacific Northwest National Laboratory & Idaho National Laboratory Joint Appointments
Bozeman, Montana

Suhaib Kaissi
Supply Chain and Logistics Technology
University of Houston
Houston, Texas

Upulee Kanewala
School of Computing
University of North Florida
Jacksonville, Florida

Satyajit "Sat" Ketkar
Velentium
Richmond, Texas

Margaret Kidd
Supply Chain and Logistics Technology
University of Houston
Houston, Texas

Byeong Kil Lee
Department of Elecrical and Computer Engineering
University of Colorado
Colorado Springs, Colorado

Datong Liu
School of Electronics and Information Engineering
Harbin Institute of Technology
Harbin, China

John Marchesini
Identity Automation
Houston, Texas

Keith M. Martin
Information Security Group
Royal Holloway, University of London
Egham, United Kingdom

Mark McDermott
ECE Department
University of Texas at Austin
Austin, Texas

Shengwei Meng
School of Electronics and Information Engineering
Harbin Institute of Technology
Harbin, China

Daniel Mitchell
James Watt School of Engineering
University of Glasgow
Glasgow, United Kingdom

Dawei Pan
School of Electronics and Information Engineering
Harbin Engineering University
Harbin, China

Harjot Singh Pooni
Supply Chain and Logistics Technology
University of Houston
Houston, Texas

Vishnunarayan Girishan Prabhu
School of Modeling, Simulation and Training
University of North Carolina at Charlotte
Charlotte, North Carolina

Derek Reimanis
Gianforte School of Computing
Montana State University
Bozeman, Montana

William Ross
Department of Program Manager for Automatic Test Systems (ATS)
Naval Air Systems Command (NAVAIR) Aviation Support Equipment Program Office
Eagle Systems, Inc.
California, Maryland

Jason Rupe
CableLabs
Louisville, Colorado

Michael Seavey
Northrop Grumman
Palatine, Illinois

John W. Sheppard
Gianforte School of Computing
Montana State University
Bozeman, Montana

William R. Simpson
Institute for Defense Analyses
Alexandria, Virginia

Jason Smith
Growth & Communications Strategist
Raleigh, North Carolina

Sean Smith
Department of Computer Science
Dartmouth College
Hanover, New Hampshire

Madhusudan Srinivasan
Department of Computer Science
East Carolina University
Greenville, North Carolina

Laura Stanley
Gianforte School of Computing
Montana State University
Bozeman, Montana

Robert Taylor
College of Business and Engineering
Wilkes University
Wilkes-Barre, Pennsylvania

Kailai Wang
Supply Chain and Logistics Technology
University of Houston
Houston, Texas

Razieh Nokhbeh Zaeem
Centre for Identity
University of Texas at Austin
Austin, Texas

1 An Overview of Complex System Design

John W. Sheppard

This book is focused on the broad issues surrounding "realizing complex integrated systems." What do we mean by this? It might be helpful to break down our title, one word at a time. To do this, we start at the end. What is a "system"? We start this introductory chapter by addressing that question. We pose a number of possible definitions and perspectives, but leave open the opportunity to consider the system from the target context where it will be used.

Once we have a system in mind, we acknowledge the fact that this system needs to "integrate" a variety of pieces, components, and subsystems, in order for it to accomplish its task. Therefore, we concern ourselves with the boundaries and interfaces of different technologies and disciplines in order to determine how best to achieve that integration.

Next we raise the specter that this integrated system is going to be "complex." Complexity can be defined in a number of ways. For one, the sheer number of subsystems or components can be a measure of complexity. Alternatively, we could consider the functions being performed by the system and how those functions interact with one another. Alternatively, we could consider computational aspects such as the amount of time or the amount of memory that may be needed to accomplish one or more tasks. The extent to which new behaviors might emerge from the system can also be regarded as an element of complexity. In the end, complexity is that characteristic of a system that defines the associated challenges along the life of the system, so in this book, we are concerned with how to go about managing that complexity.

Finally, we suggest that "realization" refers to the process by which our complex integrated system moves from concept to deployment and subsequent support. It refers to the entire design, development, manufacture, deployment, operation, and support life cycle. Of particular interest here, however, is that we focus on systems that, by their very nature, are "complex." In other words, we are interested in large, complicated, interacting beasts that are intended to perform difficult tasks and meet a wide variety of end-user needs.

1.1 COMPLEX SYSTEMS

1.1.1 Systems Definition

To begin our discussion of realizing the design of complex systems, we first must talk about what constitutes a system. There are a variety of definitions one could use, and to be sure, many of these definitions will apply here. The natural place to begin is with the dictionary. Here are two dictionary definitions:

Definition 1: (Oxford English Dictionary) A system is a set of things working together as parts of a mechanism or an interconnecting network; a complex whole (OED, 2018).

Definition 2: (Merriam-Webster Dictionary) A system is a regularly interacting or interdependent group of items forming a unified whole (Merriam, 2018).

DOI: 10.1201/9781003188377-1

A unifying theme in these two definitions is the notion of parts working together as a whole. Given this, we also consider definitions drawn from different sources focusing on systems engineering. Here are three.

Definition 3: A system is the combination of elements that function together to produce the capability required to meet a need. The elements include all hardware, software, equipment, facilities, personnel, processes, and procedures needed for this purpose; that is, all things required to produce system-level results. The results include system-level qualities, properties, characteristics, functions, behavior, and performance (NASA, 2007).

Definition 4: A system is an open set of complementary interacting parts, with properties, capabilities, and behaviors emerging, both from the parts and from their interactions, to synthesize a unified whole (Hitchins, 2007).

Definition 5: A system is a set of interrelated components working together toward some common objective (Kossiakoff et al., 2011).

The theme persists; however, we are now beginning to see the idea emerge that a system accomplishes some function. If it fails to accomplish that function, then the system does not meet its purpose and is not useful. From an engineering perspective, this translates into the system failing to satisfy its requirements.

From the perspective of one of the editors of this work, we also need to consider what we mean by "system" from the perspective of managing that system's health. Thus, we need to consider how system health can be incorporated into the design process. To do so, we acknowledge that one of the added performance requirements for a system is that it needs to be able to be monitored in such a way that its health can actually be determined. However, a related issue in systems engineering arises from the fact that the systems, due to their interacting "qualities, properties, characteristics, functions, behavior, and performance" involve complexities that can make such monitoring problematic. Using this observation, Sheppard and Simpson define a system in the context of system diagnostics as follows:

Definition 6: A system is any aggregation of related elements that together form an entity of sufficient complexity for which it is impractical to treat all of the elements at the lowest level of detail (Sheppard and Simpson, 1998).

One of the key drivers in any design activity is managing complexity. As the Sheppard and Simpson definition suggests, a system by its very nature has a level of complexity that makes full-system analysis generally impractical. To support the design, analysis, and support of systems, we therefore need to employ carefully crafted design practices that draw on a variety of domains and disciplines.

1.1.2 Systems Domains

In a large number of engineering disciplines, system design is presented in a vacuum. Perhaps the three biggest exceptions would be industrial engineering, software engineering, and systems engineering. For our purposes, we approach the engineering problem from the systems engineering perspective. Doing so, we recognize and affirm that successful system design must draw from a variety of domains, including the specific technical domain(s) within which the system is being developed.

Kossiakoff et al. (2011) suggested that the domains of systems engineering can be described like faces on a Rubik's cube with each face of the cube representing a key domain (Figure 1.1). Specifically, they list the following domains as being central to systems engineering:

- **Engineering:** Applies tools to support system design, such as computer-aided design (CAD) tools, modeling and simulation tools, version control tools, etc.

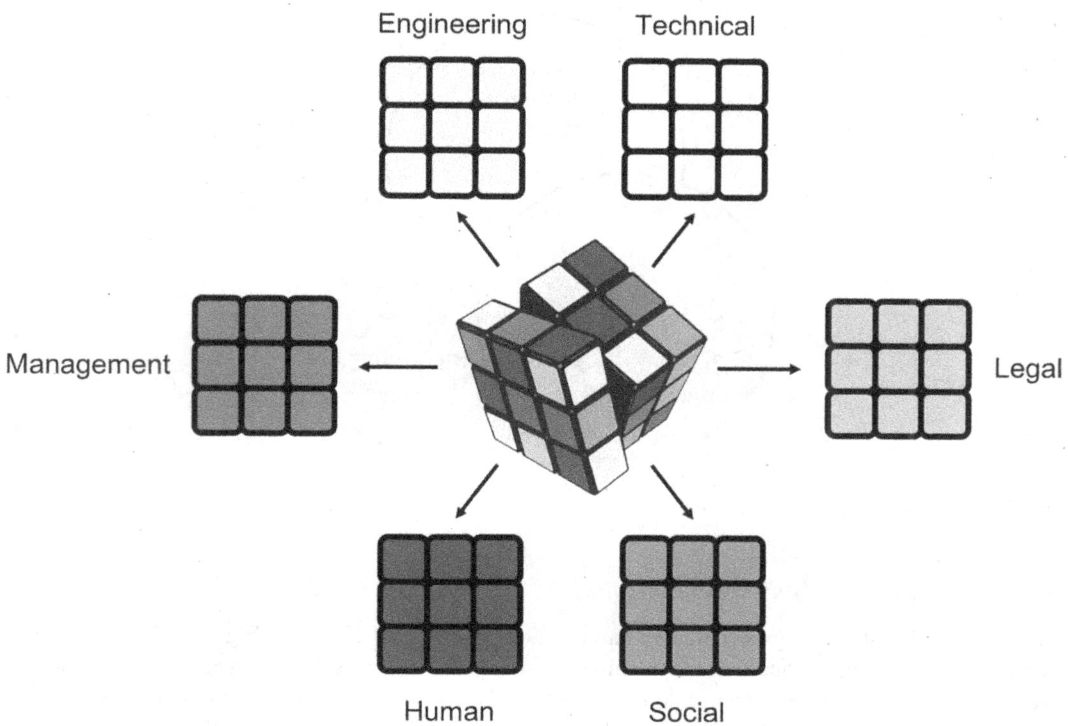

FIGURE 1.1 Systems engineering as a Rubrik's Cube.

- **Technical:** Incorporates the technical knowledge from the appropriate engineering disciplines, such as electronics, embedded software, propulsion, materials, etc.
- **Management:** Utilizes project management skills, such as scheduling, cost estimation, resource allocation, risk assessment, etc.
- **Legal:** Focuses on issues surrounding contracting, generation of intellectual property, licensing, subcontracting, etc.
- **Human:** Addresses interpersonal dynamics, team building, employment, etc.; could also be interpreted to include human factors engineering; however, this is better placed in the context of one of the technical disciplines.
- **Social:** Assesses and influences the design in ways that make positive impacts on the world around us. Evaluates those impacts on societal, economic, political, and environmental factors.

As one might expect, these "faces" are not independent. They interact in ways similar to the challenges associated with solving the Rubik's Cube. By turning one face, it can affect the neighboring faces in unanticipated ways. Thus, systems engineering needs to apply rigorous, disciplined approaches at the intersection of these domains to ensure an effective engineering process.

1.2 SYSTEMS ENGINEERING

In general, when one thinks about approaching an engineering design problem, they start by considering the specific engineering discipline central to that design. Such engineering disciplines include electrical, civil, mechanical, computer, industrial, petroleum, chemical, biological, etc. Systems engineering, however, sits above these disciplines since it is focused on how to draw the technical aspects of several of these disciplines together to create a complete, integrated system.

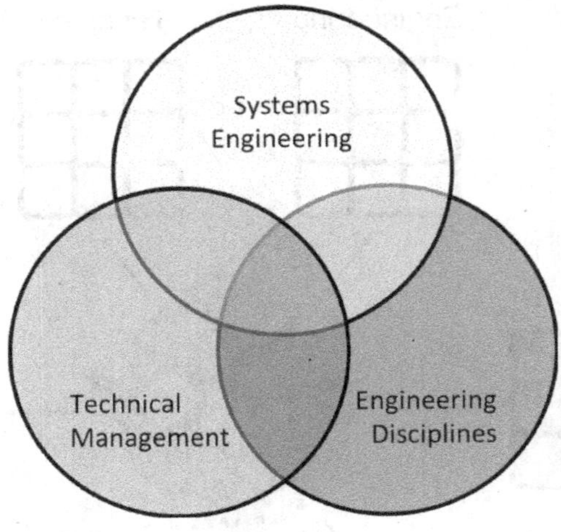

FIGURE 1.2 Systems engineering fields.

1.2.1 Disciplines in Systems Engineering

To understand systems engineering, we can consider the Venn diagram in Figure 1.2. Here we see that the system design process draws from skills specific to systems engineering as well as the relevant engineering fields and technical management. In addition to these major disciplines, we also draw on the following sub-disciplines, each of which lies at the intersection of these three areas.

- **Modeling and Simulation:** Computer-based models are developed to validate system performance and effectiveness as well as to determine whether or not specific requirements are being satisfied by the design. Modeling and simulation are usually employed by the engineering disciplines.
- **Operations Research:** Systems design and engineering is a process that involves making decisions that trade, for example, specific performance characteristics with the cost to obtaining the desired performance. Operations research is a field devoted to optimal decision-making and includes several areas relevant to system design, including financial engineering, decision analysis, and logistics and supply chain management.
- **Project Management:** Developing a complex system is, by necessity, a team-based process. To be successful, the systems engineering process draws on sound project and technical management to ensure the right resources are available to the team, proper communication occurs among team members, and schedules and deliverables are maintained.
- **Quality Assurance:** Fundamentally, system or product quality depends on system/product health. Quality assurance applies design principles to maximize system quality and incorporates process monitors to ensure that quality is maintained. Technical debt, which is a measure of system quality, is a concept that has arisen in the software engineering discipline that captures the cost of trading strategic decisions in system design for tactical decisions (Allman, 2012). The marketing slogan for FRAM oil filters of "pay me now, or pay me later" is captured formally when estimating technical debt. For example, one of the first decisions in system design that increases technical debt (by increasing life-cycle cost) is to delay addressing system test/system health requirements.

The principal challenge in systems engineering and systems design is balancing various disciplines. Designing a system that meets a vast array of requirements, including functional requirements, reliability and availability requirements, performance requirements, packaging requirements, and system maintenance requirements, requires careful management and utilization of the right technical resources.

1.2.2 System Decomposition

One of the key elements in designing a complex system is arriving at an appropriate decomposition of the system into manageable parts. A variety of approaches have been developed to accomplish this. Generally, at the core of such decomposition, is a recognition that systems are often hierarchical in nature. Systems are made up of subsystems, which are made up of sub-subsystems and so forth until we get to the constituent parts. Reversing this, we can think of each of these intermediate steps as building blocks for the next level up in the hierarchy.

Often, when considering how to decompose a system into its constituent building blocks, we need to consider both the physical decomposition and the functional decomposition. For example, consider a personal computer (PC) as our system. The PC system is decomposed physically into a monitor, keyboard, and computer, where the computer is then decomposed into a motherboard, a power supply, etc. Things start to get a bit murky in the hierarchy when we consider different types of cards and integrated circuits in the system. For example, should we regard a GPU that is plugged into the backplane at the same level as the CPU that is plugged directly into the motherboard? Each of them provides processing capabilities, but one is an IC and one is a card.

This is when we run into the issues of functional decomposition. In the same PC, a CPU and a GPU provide similar functional capabilities (albeit differently), so, functionally, they may be considered to be at the same level. However, consider a single dual in-line memory module (DIMM). We may also wish to group several memory modules together by plugging them into multiple slots on the motherboard so as to form a functional unit corresponding to the computer's random access memory (RAM). In this case, we cross the simple physical boundaries of individual DIMM cards and simply combine all of the memory ICs (and supporting circuitry) together.

In modern systems, we also begin to blur the lines between hardware and software. A given function often requires both, so pure functional decomposition breaks down. Similarly, with the increased prevalence of using field programmable gate arrays (FPGA), the distinction between hardware and software blurs. This tends to drop us into the realm of hardware/software co-design. Schaumont (2010) describes the hardware/software co-design space where the intended application for the system being designed defines the design context. We can then think of different levels of complexity for different hardware architectures. For example, ranging from the general purpose reduced instruction set computer (RISC)-based systems that provide maximum flexibility but minimum efficiency to application-specific integrated circuit (ASIC)-based systems that provide minimum flexibility but maximum efficiency, the software is integrated into the respective designs using a variety of programming languages and tools.

Once the appropriate model of system decomposition is determined, the designer must then consider the nature of the interfaces between various system building blocks. The design of the hardware interfaces must consider physical constraints, pin layouts, and routing/connectivity with other system components. Software interfaces must consider the nature and type of data being passed between the components. Once again, these two domains become blurred when we are connecting hardware to software. For example, bus systems provide the physical backbone over which signals propagate within a system, and the signals encode the data to be transmitted.

Different approaches exist for handling the hardware/software interface (Schaumont, 2010). The memory-mapped interface uses memory within a microprocessor as the means for communication via load/instructions. The coprocessor interface employs a separate coprocessor port on a microprocessor as the means for using software to control the hardware. Finally, a custom-instruction interface can be used to integrate custom, software-controlled functionality directly into the microprocessor architecture.

As a final concept to be introduced here, system design has also been influenced by the software engineering community by employing object-oriented design principles (Kumar et al., 1994; Bahill and Daniels, 2003; Drabik, 2011). Object-oriented design is based on the idea that a system can be decomposed into logical "objects" that encapsulate behavior within the system (Booch et al., 2007). The interfaces between objects are defined based on methods that implement the behaviors of those objects. Objects are generalized into "classes," which act as abstract data types and can be organized hierarchically using mechanisms such as inheritance (subclasses inherit and possibly specialize properties of superclasses) and polymorphism (the notion that the same interface can be used to interact with objects of different types) (Cardelli and Wegner, 1985).

There exists considerable debate about the appropriateness of the object-oriented design principle when designing hardware, which is somewhat surprising given the fact hardware systems are inherently object-oriented. Object-oriented principles, however, have been demonstrated in software design to lead to more maintainable code and better code reusability. With more and more software integration with hardware and hardware/software co-design, the need is becoming greater to employ the same design principles that are so successful in software; therefore, when we discuss software design issues later, the object-oriented approach will tend to stand out.

1.2.3 Managing System Complexity

Modern system design involves designing and assembling the building blocks of what will, ultimately, be highly complex. Issues related to power, behavior, packaging, health management, and support introduce competing objectives that make the design process difficult. Therefore, it is essential that the design process take steps to manage this complexity from the outset.

Complexity management begins with the requirements. Specifying requirements involves imposing constraints on the system design that influence not only the specific part of the system satisfying those requirements but also related parts of the system. Requirements interact and define what, ultimately, is feasible. In fact, we can regard the entire system design process as a large multiobjective, constrained optimization problem where we attempt to optimize competing factors (e.g., maximize performance, minimize energy, minimize manufacturing cost, maximize availability, etc.) within the constraints of the requirements (e.g., geo-locate to within one meter, fly at Mach 1.5, communicate with frequency hopping over a set frequency range, provide 100% fault isolation to three or fewer replaceable units, etc.).

Recognizing the role the requirements play, not only in specifying what needs to be done but also in determining the complexity of the resulting system, the task then becomes managing the resulting complexity. Standard approaches employ hierarchical decomposition into appropriate building blocks and careful partitioning between hardware, software, and mixed functional components. This presents us with a challenging modeling problem where, ideally, we would like to encapsulate the systems by their physical or functional boundaries, but interactions across these boundaries will remain that need to be modeled. Additional complications arise when we attempt to mix the results from multiple models. For example, attempting to combine a digital simulator with the results of finite element analysis to determine how G-forces on the structure of an on-board digital signal processor affect the DSP's functionality is difficult.

An added complication arises when we recognize that many modern systems are actually *systems of systems*. Under such conditions, the design must address the issues of unexpected emergent behavior. In general, it is impossible to be able to detect all emergent behavior, but by involving a strong team of domains that work together, some emergent behavior can be anticipated and, if deemed negative, prevented. This may require specialized models to address manageable levels of possible system interactions, usually at the interfaces to determine the extent to which encapsulation has succeeded.

1.3 A SYSTEMS ENGINEERING PROCESS

The field of systems engineering has emerged as the over-arching discipline for designing and managing complex systems. Systems engineering uses a combination of technical and management skills to provide the framework within which system design occurs. In this section, we take a look at elements in a typical systems engineering process. This includes addressing the needs for the design itself as well as the technical management of the process and the ultimate realization of the product.

1.3.1 System Design

The classical design process follows what is commonly referred to as a *waterfall* model. This model is depicted in Figure 1.3. In this model, the engineering process starts with requirements analysis and proceeds through successor stages of system design, implementation, test and verification, and maintenance. Unfortunately, this model, while logical, has been demonstrated to not be the best systems engineering approach for managing complexity.

An alternative model has emerged from the software engineering community known as the *spiral* model (Boehm, 1986). The principal idea emerging from this model is one of successive refinement. An initial set of requirements is formulated (referred to as "conceptual" requirements), from which a prototype is built. Each turn through the spiral follows the classic waterfall model; however, at the end of the validation stage, a refinement step begins. With each refinement, stronger requirements are established, and the system continues to evolve until it reaches the final version to be fielded.

Additional systems engineering models have been proposed and even standardized. In EIA 632, a top-down engineering process is defined with a considerable amount of feedback. As described in the standard, "Appropriate processes of [the systems engineering process] are applied recursively and iteratively to define the system products of the system hierarchy from the top down, and then, to implement and transition the system products from the bottom up to the user or customer (EIA, 2003)." Thus, the intent of EIA 632 is to provide a capability that combines the traditional top-down design of the waterfall model with the iterative improvement of the spiral model.

Slightly more recent than the EIA standard, the IEEE Software and Systems Engineering Standards Committee in the IEEE Computer Society published a related standard focused on interdisciplinary systems engineering (IEEE, 2005). This standard has been "dual-logoed" with both

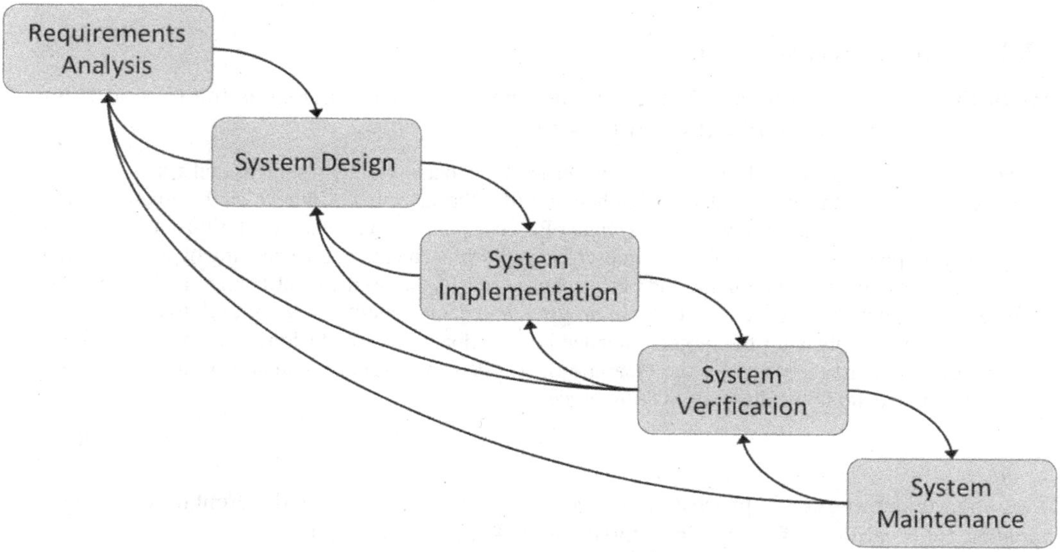

FIGURE 1.3 Waterfall model of systems engineering.

Agreement Processes

- Acquisition
- Supply

Organizational Project-Enabling Processes

- Life Cycle Model Management
- Infrastructure Management
- Portfolio Management
- Human Resource Management
- Quality Management
- Knowledge Management

Technical Management Processes

- Project Planning
- Project Assessment and Control
- Decision Management
- Risk Management
- Configuration Management
- Information Management
- Measurement
- Quality Assurance

Technical Processes

- Business or Mission Analysis
- Stakeholder Needs and Requirements Definition
- System Requirements Definition
- Architecture Definition
- Design Definition
- System Analysis
- Implementation
- Integration
- Verification
- Transition
- Validation
- Operation
- Maintenance
- Disposal

FIGURE 1.4 IEEE Standard 15288 systems engineering process.

the International Organization on Standardization (ISO) and the International Electrotechnical Commission (IEC). Similar to the EIA standard, IEEE 1220 combines a waterfall-like process in the midst of iterative refinement. This particular standard also provides detailed processes for each of the main steps that correspond to the classic waterfall process. Also of note is the strong dependence, not only on requirements analysis, decomposition, and allocation but also on conducting trade studies and assessments at each major stage to assess the impacts of various design decisions.

The last standard we consider is also published by the IEEE and also dual-logoed with ISO and IEC (IEEE, 2015). In this standard, rather than depicting a flow-type engineering model, the standard simply identifies what it sees as the major activities in the systems engineering process (Figure 1.4). Key to this approach is that any (or all) of the identified processes can be applied at any stage of development. It does not depend on a particular model of systems engineering, so it works in a waterfall environment, a spiral environment, or any other.

1.3.2 Technical Management

One of the elements in systems engineering that becomes more obvious is the role of technical management. According to ISO/IEC/IEEE 15288

> The Technical Management Processes are concerned with managing the resources and assets allocated by organization management and applying them to fulfill the agreements into which the organization or organizations enter. The Technical Management Processes relate to the technical effort of projects, in particular to planning in terms of cost, timescales, and achievements to the checking of actions to help ensure that they comply with plans and performance criteria, and to the identification and selection of corrective actions that recover shortfalls in progress and achievement. They are used to establish and perform technical plans for the project, manage information across the technical team, assess technical progress against the plans for the system products or services, control technical tasks through to completion, and aid in the decision-making process
>
> *(IEEE, 2015).*

The standard identifies eight processes that, generally speaking, capture different tasks of technical management regardless of the model implemented. These tasks include:

- **Project Planning:** This process is focused on ensuring effective and workable plans are created and maintained throughout the design process. Relevant plans include, but are not limited to, development plans, quality plans, test plans, documentation plans, build plans (for software), deployment plans, and maintenance plans.
- **Project Assessment and Control:** Given the plans that emerge from project planning, this task determines whether or not the plans are workable, feasible, and aligned with each other. It is in this context that program reviews and technical progress reviews are conducted.
- **Decision Management:** Throughout the design process, multiple design trade-offs will need to be considered. Through this process, a framework will be provided by which such decisions can be made relative to design objectives and requirements.
- **Risk Management:** Historically, this process is focused on assessing and mitigating the risks associated with the engineering process. In this case, risk is defined very broadly so it can include anything related to managing uncertainty in the engineering process, including impacts on schedule, cost, environmental impact, and safety. This process is distinct from the type of risk analysis that needs to be included within the context of operations and maintenance.
- **Configuration Management:** The critical thing in any complex engineering activity is ensuring that a clear understanding of system configuration and version is maintained. Within the software engineering community, version control is maintained through code repositories such as subversion (svn) (Pilato et al., 2010) or more recently, Git (Chacon and Straub, 2018). Similar ideas are applied in full system development in terms of maintaining versions of documentation, plans, diagrams and drawings, models, etc. An over-arching configuration management process needs to be in place to correlate all of the various versions to the appropriate system configurations.
- **Information Management:** With any systems engineering activity, a set of stakeholders drive the requirements and the development process. As the design unfolds, considerable amounts of data and information are generated. This information, which includes things like reports, plans, designs, test results, and documentation, needs to be managed in a way that is accessible to the stakeholders.
- **Measurement:** The purpose of the measurement process is to capture, analyze, and report the results of the entire design process to assess progress, quality, and overall effectiveness. This information is used to guide the decision-making processes throughout the system lifecycle.
- **Quality Assurance:** Fundamentally, the goal of any engineering activity is to design and implement/manufacture a quality product. Quality plans are produced during the project management process, and quality is assessed through the measurement process. The quality assurance process is focused on ensuring that the plans are implemented correctly and the quality indicators are analyzed to assess the quality of the process and product.

  While considered a technical process by the standard, rather than a Technical *Management* Process, we assert that Operations and Maintenance Planning falls within the purview of technical management. Ideally, the type of management activities performed here are addressed throughout the design cycle, rather than waiting until deployment or post-deployment.
- **Operations:** The operations stage of a product is the period at which the product is performing its intended function. While technically this is not part of the design phase, it is an important part of the lifecycle of the product. From a technical management perspective, operations address the process by which the product is used, performance is managed, and assessments are made to determine whether or not the product is performing as intended. From a health perspective, operational management is critical, so appropriate plans and procedures need to be put in place for system monitoring during the design phase.

- **Maintenance/Support:** Similar to the operations stage, the maintenance (or more broadly, support) stage of a product is the period in which some action needs to be executed on the product to enable that product to continue to perform its function. Maintenance involves inspections, anomaly detection, fault detection and isolation, repair, and overhaul. During the maintenance phase, the system is not necessarily available for use, unless some sort of redundancy or multi-model operational capability is provided by the product. As with the operations stage, maintenance and support management are critical, necessitating the development of proper plans and procedures during the design phase of the product.

More detail will be given on the Operations and Support aspects of a product in Section 1.4.

1.3.3 Product Realization

Once a system has been designed, it is time to realize that system – to turn it into an actual product. Typically, a number of steps are required for such product realization, ranging from implementing the system from the bottom up, integrating the components into the next level in the product hierarchy, and along the way, verifying and validating that the system meets its intended requirements. Once realized, the system then needs to be transitioned into actual use.

One way of depicting the process is described in (NASA, 2007). This process consists of three major phases that comprise five steps. The first phase focuses on design realization where the components of the system are implemented, combined, and integrated. The second phase verifies the product to verify that it functions in accordance with its requirements and validates the product to ensure that it will function as intended in its target environment. Finally, the product is transitioned into the field. More specifically, the NASA Product Realization Process consists of the following:

- **Implementation:** The first step of the product realization process is to implement the product itself. Product implementation may involve manufacturing hardware, coding software, acquiring off-the-shelf components, reusing products that have been created for some other use, incorporating simulation models, or otherwise generating an actual artifact based on the system design. It is customary that, as components are created, they are also tested to determine whether or not that portion of the system functions as intended. Such testing can include unit tests, interconnect tests, functional tests, etc.
- **Integration:** As product components or subassemblies are generated, they need to be assembled to form the next higher level in the product hierarchy (sometimes also referred to as the product breakdown structure, or PBS). As with the implementation phase, integration must include testing to ensure that the components work together as intended and do not yield any unexpected, emergent behavior. Much of the testing that occurs during integration focuses on the interfaces between the components, including both the physical and the functional interfaces.
- **Verification:** The main question to be addressed by verification is whether or not the product was realized properly. More specifically, the purpose of verification is to ensure that the realized product meets the design requirements. One can argue that, no matter how useful a product might be, if the product does not satisfy the design requirements that led to its realization, then the product is not implemented properly. Verification is performed through formal analysis, demonstration, inspection, and testing against acceptance criteria. As part of the verification process, features of the realized system are traced back to the requirements that spawned them to ensure that all of the requirements have been met.
- **Validation:** In contrast to verification, the main question to be addressed by validation is whether or not the correct product was realized. More specifically, it is the goal of validation to ensure that the system functions as intended in its target environment and in accordance with its intended use. It is customary for a complex system to have a "concept

of operations" (ConOp) defined to describe its intended use. In software development, such a ConOp often emerges as a part of "use case analysis." As part of the verification process, features of the realized system are traced back to the requirements that spawned them to ensure that only the stated requirements have been realized.

- **Transition/Deployment:** Product transition and deployment is an incremental process. The goal of product transition is to move the product into the environment within which it is intended to be used. Product transition occurs throughout the product lifecycle, including in the overall product realization process. It involves handling the move from design and development to production. In addition, as a component is realized, it is transitioned into its place within the product hierarchy; therefore, transition can be a part of product integration. It includes handling configuration management and version control. The final transition of a full system into the field corresponds to deployment and incorporates supporting products and processes such as packaging, containers, storage, shipping, and receiving.

1.4 OPERATION AND SUPPORT

Finally, we consider the operation and support (O&S) phase of the system life cycle. Typically, we consider O&S to be outside of the scope of system design and systems engineering; however, given a common lack of focus on designing for system health, we assert that the system design process must plan for how the system's health will be monitored and maintained once in the field. Therefore, it is not realistic to expect to be able to divorce the design process from the O&S process.

1.4.1 Design for Testability

Historically, system health has been addressed by designing the system with specific test capabilities. As an example, one of the most common approaches to designing testability into an integrated circuit is through the use of Boundary Scan IEEE (2013). The basic idea is to place a set of flip flops at the pins around the periphery of an integrated circuit such that signals can be passed in and out, effectively isolating the IC from the rest of the system.

A typical implementation of boundary scan is depicted in Figure 1.5. Here we see a representation of an integrated circuit with the core logic of the circuit highlighted in the middle. Around the periphery of the circuit is added a set of boundary registers/flip flops that are used to capture test inputs and outputs. These registers are connected directly to the pins of the IC and feed into the core circuitry. The registers are also connected in series to permit patterns to be passed into the registers via the Test Data In (TDI) port and read out serially through the Test Data Out (TDO) port. The IC also includes additional logic corresponding to a Test Access Port (TAP) Controller that determines the behavior when placed into test mode vs. operational model. The TAP Controller is managed through a separate Test Clock (TCK) and Test Model Select (TMS).

Another approach to designing testability into an electronic system is to incorporate Built-In Test/Built-In Test Equipment, which acts as a separate monitor on the target system to measure performance. BIT/BITE incorporates logic (software and hardware) and circuitry directly into the system to monitor the system for different error conditions. Figure 1.6 gives an example of a notional system with BITE available for doing on-board testing.

Four different types of BIT are often employed in a complex system (Butler, 2006). These include Startup BIT (SBIT), Continuous BIT (CBIT), Initiated BIT (IBIT), and Maintenance BIT (MBIT). SBIT is used to ensure the system is ready for use, prior to starting up the application on the unit. CBIT runs continuously in the background while the unit is operating and must function in a non-interfering fashion. IBIT can be triggered during operation should unexpected conditions arise, or it can be used when the system is offline to assist in fault isolation. MBIT is run purely offline when the system is in maintenance mode.

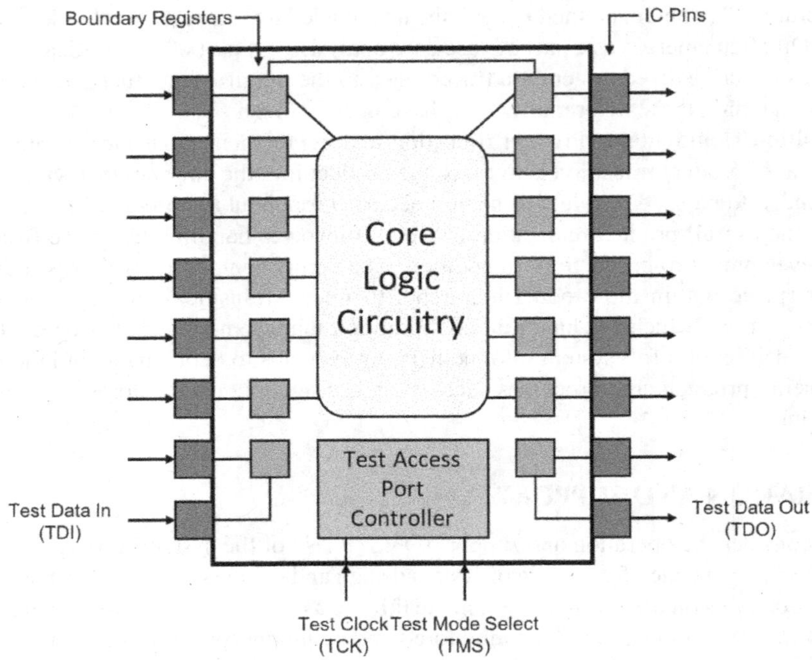

FIGURE 1.5 Boundary scan architecture for an integrated circuit.

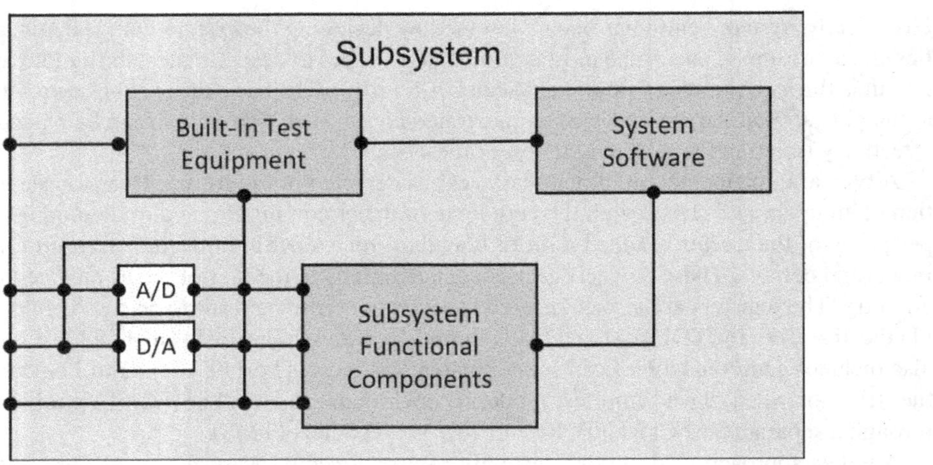

FIGURE 1.6 Design for testability with built-in test equipment.

A number of methods exist for implementing built-in test, either as part of the normal system circuitry or using separate BITE. Techniques most often applied include things like parity checks, checksums, sequence checks, memory scans for stuck-at faults, and heartbeat checks. These are all tests that check the integrity of the information being passed into and out of the components of the system to verify that this information has not been corrupted in some way. At times, additional methods based on pre-defined test patterns can be stored in read-only memory or non-volatile random access memory that can then be used to test the functionality of the components as well.

1.4.2 Maintenance Architectures

Often, a variety of maintenance architectures are employed to provide field support for a system (Simpson and Sheppard, 1994). For complex systems such as aircraft, maintenance usually occurs in various levels, the most frequently employed architecture being a three-level maintenance architecture. For the military, the three levels are often referred to as the "organizational level" (O-level), "intermediate level" (I-level), and "depot level" (D-level). Commercial airlines often employ a similar architecture for maintaining their aircraft. Of note, however, is that many maintenance architectures are being streamlined, either by eliminating a level, merging levels, or bringing the original equipment manufacturer (OEM) into the process. In settings such as the automotive industry, there is often only one level of maintenance (unless we include the automobile owner performing their own maintenance), and personal computers also often employ a single level of maintenance (unless the PC needs to be sent back to the OEM).

The process by which faults are identified so that the system can be repaired is essential to whatever maintenance architecture is employed. Key to understanding such processes is a clear understanding of what we mean by "diagnosis." For our use, we define diagnosis as follows

Definition 7: Diagnosis is a functional capability to determine that an equipment malfunction has occurred and to determine the cause of the malfunction to support operator decisions, maintainer decisions, and automated failure responses.

With this definition, we can address four aspects of diagnosis: (1) fault detection, (2) fault localization, (3) fault isolation, and (4) fault prediction. Note that we include fault prediction as part of diagnosis since we claim that prognosis is no more than applying diagnosis in a temporal context. We repeat definitions for the first three from Simpson and Sheppard (1994).

Definition 8: Detection is the ability of tests or observations to assert that a failure or malfunction in some system has occurred.

Definition 9: Localization is the ability of tests or observations to assert that a fault has been restricted to some subset of possible causes of failure or malfunction.

Definition 10: Isolation is the ability of tests or observations to assert that a specific fault is the cause of failure or malfunction. In the context of a specific maintenance level, isolation is localization to a subset of possible causes sufficiently small to lead to the repair of a single replaceable unit at the given maintenance level.

Key to any maintenance process is control of the timing over which maintenance occurs. To that end, the least desirable type of maintenance is what is referred to as "reactive" or "unscheduled" maintenance. As the name suggests, unscheduled maintenance occurs when a system must be taken offline for purposes of performing maintenance or repair as a result of some unexpected event (e.g., a failure) occurring. It should be evident that the disruptive nature of unscheduled maintenance is undesirable and is the result of poor planning in the operation and support of the system.

To prevent unscheduled maintenance, a variety of alternative models have been suggested. One of the most common strategies is to apply scheduled or periodic maintenance. In automobiles, this occurs when we change the oil every 3000 miles and run a general maintenance check perhaps once per year. Commercial airlines perform periodic checks based on number of hours flown (DOT, 2018). For example, the A- check occurs every 400–600 flight hours. This is like the car oil change where filters are replaced, control surfaces are lubricated, and all emergency equipment is inspected. A D-check, on the other hand, involves a complete inspection and overhaul of the aircraft and occurs every 6–10 years. Such inspections can lead to additional maintenance being performed.

1.4.3 Condition-Based Maintenance

Traditional approaches to viewing O&S of systems involve considering the normal product lifecycle with interruptions in system availability arising due to periodic inspections, scheduled maintenance, system upgrades, and unscheduled maintenance (failures). Recently, the US Department of Defense has turned to adapting its O&S processes to exist within the context of so-called "condition-based maintenance" (CBM) and has published a DoD instruction implementing a program referred to a CBM+ "as a principal consideration in the selection of maintenance concepts, technologies, and processes for all new weapon systems, equipment, and materiel programs based on readiness requirements, life cycle cost goals, and reliability-centered maintenance (RCM)-based functional analysis formulated in a comprehensive reliability and maintainability (R&M) engineering program (DoD 4151.22, 2012)."

CBM+ is defined to be "the application and integration of appropriate processes, technologies, and knowledge-based capabilities to improve the reliability and maintenance effectiveness of DoD systems and components (DoD, 2008)." This program emerged due to the high costs associated with O&S (taking up as much as 65%–80% of a system's life cycle cost), where much of the cost can be attributed to system maintenance. Even though CBM+ is defined within the context of DoD systems, the concepts apply to any complex system. Therefore, in the following, we approach O&S from the perspective of CBM+ with a particular focus based on Prognostics and Health Management (PHM).

1.4.4 Prognostics and Health Management

As more and more systems move into the realm of CBM and CBM+, we see maintenance decisions being made based on reliability predictions (so-called reliability-centered maintenance, or RCM) and the monitored condition of the system (thus condition-based). RCM and CBM+ have led the way to incorporating prognostics and health management (PHM) into the overall maintenance strategy whereby maintenance is performed in more of an anticipatory, "just in time" sense. A standard published recently by the IEEE focuses on defining a framework for PHM and maps the main activities to the layers of the OSA-CBM framework (Machinery Information Management Open Standards Alliance, 2017).

Specifically, OSA-CBM defines a layered framework designed to identify the information to be exchanged by the components of a CBM system. IEEE Std 1856 marries OSA-CBM and PHM capabilities into a general system support framework. The OSA-CBM standard consists of a layering of six functional blocks designed to encompass the full range of functionality required for CBM and PHM systems layered from bottom to top as follows:

1. **Data Acquisition (DA):** The DA block gathers raw data from sensors and transducers about the system being monitored.
2. **Data Manipulation (DM):** The DM block is responsible for performing signal processing, signal transformation, and feature extraction on the acquired data.
3. **State Detection (SD):** The SD block compares and analyzes the processed data with expected values or system models to discover abnormalities in the data.
4. **Health Assessment (HA):** The HA block performs fault detection and fault isolation to find the root cause of system abnormalities.
5. **Prognostics Assessment (PA):** The PA block generates predictions about the future health state of the monitored system and the progression of detected abnormalities and faults.
6. **Advisory Generation (AG):** The AG block suggests potential maintenance actions and alternative operating instructions for the duration of the mission that assist in decision-making logistics actions.

IEEE Std 1856 adds two additional layers at the extremes (IEEE, 2017).

1. **Sensors (S):** The S block as the bottom layer corresponds to the physical layer of the CBM/PHM framework where raw sensor data is collected. It also incorporates any other health-related information that might be moving through the system.
2. **Health Management (HM):** The HM block as the top layer takes the information provided from AG to determine how to respond to keep or restore the system to a healthy state. This is the layer that is ultimately responsible for implementing the recommendations from AG.

Given the OSA-CBM framework, the PHM process follows a logical sequence of actions: Sense (system condition) ⇒ Acquire (information about the system from the sensors) ⇒ Analyze (the acquired information to determine health state) ⇒ Advise (on appropriate actions to take based on health state) ⇒ Act (to carry out the recommendations and maintain system health). After taking action, the system is placed back in service, and the process continues.

Finally, the O&S process needs to be sensitive to logistics, sparing, and working with system warranties. For our purposes, logistics consists of the processes by which we manage people, goods, and services in support of a system. The objective is to maximize the availability of a system but to do so in an economically efficient manner. To that end, we can consider logistics as consisting of the following stages (Morana, 2018):

1. Identify market needs in terms of quality of service.
2. Design the logistics system in terms of product needs and constraints.
3. Develop the logistics system to meet the requirements associated with product support.
4. Implement the industrial processes necessary to put the logistics system in place.
5. Distribute the necessary items (spares, goods, consumables).
6. Support the product using the implemented system.
7. Refine the logistics process based on feedback on effectiveness, availability, and cost.

1.5 CONCLUSION

As the title of this book indicates, the endeavor we explore here is that of realizing *complex integrated systems*. Fundamentally, this book is about systems engineering but on a large scale. As laid out in this chapter, we seek to bring together and to employ the wide range of methods, tools, and techniques necessary to manage such complex systems, hopefully drawing upon the best such methods and identifying where detailed information can be found to learn more about those methods, tools, and techniques.

This book approaches system design from an *integrated* perspective, where one of the key points of integration involves keeping operations and support of the system at the forefront of the design, development, and implementation process. Historically, a figurative (and sometimes literal) wall was constructed between the design engineers and the support engineers. That approach is untenable, especially in an age of highly complex systems. Therefore, a theme runs throughout this book, reminding the reader of the importance of making O&S an integral part of the life cycle engineering process from the beginning.

REFERENCES

Allman, E. (2012). Managing technical debt. *Communications of the ACM*, *55*(5), 50–55.

Bahill, T., & Daniels, J. (2003). Using object-oriented and UML tools for hardware design: A case study. *Systems Engineering*, *6*(1), 28–48.

Boehm, B. (1986). A spiral model of software development and enhancement. *SIG-SOFT Software Engineering Notes, 11*(4), 14–24.

Booch, G., Maksimchuk, R., Engle, M., Young, B., Conallen, J., & Houston, K. (2007). *Object-Oriented Analysis and Design with Applications* (3rd ed.). Boston, CA: Addison-Wesley Professional.

Butler, J. A. (2006). Application and evaluation of built-in test (bit) techniques in building safe systems. *CrossTalk: The Journal of Defense Software Engineering, 1*(9), 15–20.

Cardelli, L., & Wegner, P. (1985). On understanding types, data abstraction, and polymorphism. *ACM Computing Surveys (CSUR), 17*(4), 471–523.

Chacon, S., & Straub, B. (2018). *Pro Git: Everything you need to know about Git.* New York: Apress.

DoD (2008). *Condition based maintenance plus DoD guidebook.* US Department of Defense, Washington, DC.

DoD 4151.22 (2012). Department of defense instruction: Condition based maintenance plus (CBM+) for materiel maintenance, Washington, DC.

DOT, U. (2018). *FAA-H-8083-30A aviation maintenance technician handbook – General.* Federal Aviation Administration.

Drabik, P. K. (2011). Object oriented approaches to hardware design. *Proceedings of SPIE, 8008*, 152–157.

EIA. (2003). *Process for engineering a system.* Electronic Industries Association, Arlington, VA.

Hitchins, D. K. (2007). *Systems engineering: A 21st century systems methodology.* Hoboken, NJ: John Wiley & Sons.

IEEE. (2005). *ISO/IEC/IEEE 26702 standard for system and software engineering—System life cycle processes.* IEEE Standards Association, Piscataway, NJ.

IEEE. (2013). *IEEE Std 1149.1–2013 standard for test access port and boundary scan architecture.* IEEE Standards Association, Piscataway, NJ.

IEEE. (2015). *ISO/IEC/IEEE 15288 standard for systems engineering—Application and management of the systems engineering process.* IEEE Standards Association, Piscataway, NJ.

IEEE. (2017). *IEEE 1856 standard framework for prognostics and health management of electronic systems.* IEEE Standards Association, Piscataway, NJ.

Kossiakoff, A., Sweet, W. N., Seymour, S. J., & Biemer, S. M. (2011). *Systems engineering principles and practice* (2nd ed). Hoboken, NJ: John Wiley & Sons.

Kumar, S., Aylor, J. H., Johnson, B., & Wulf, W. A. (1994). Object-oriented techniques in hardware design. *Computer*, 27, 64–70.

Machinery Information Management Open Standards Alliance. (2017). Open Systems Architecture for Condition Based Maintenance (OSA-CBM). Retrieved from https://www.mimosa.org/mimosa-osa-cbm

Merriam (2018). Merriam-Webster Online Dictionary. Retrieved from https://www. merriam-webster.com/

Morana, J. (2018). *Logistics.* Hoboken, NJ: John Wiley & Sons.

NASA. (2007). *NASA systems engineering handbook.* Washington, DC: NASA.

OED. (2018). Oxford English Dictionary Online. Retrieved from https://en.oxforddictionaries.com/

Pilato, C. M., Collins-Sussman, B., & Fitzpatrick, B. W. (2010). *Version control with subversion* (1.7th ed.). Sebastopol, CA: O'Reilly.

Schaumont, P. R. (2010). *A practical introduction to hardware/software codesign.* New York: Springer.

Sheppard, J. W., & Simpson, W. R. (1998). *Research perspectives and case studies in system test and diagnosis.* New York: Springer.

Simpson, W. R., & Sheppard, J. W. (1994). *System test and diagnosis.* New York: Springer.

Part 1

System Planning

2 Requirements Types – Design and Test

Satyajit "Sat" Ketkar and Jason Smith

2.1 WHAT IS A REQUIREMENT?

ISO/IEC 2007 defines a requirement as follows:

> A requirement is a statement that identifies a product or process' operational, functional, or design characteristic or constraint, which is unambiguous, testable, or measurable, and necessary for product or process acceptability.

2.2 DESIGN REQUIREMENTS

2.2.1 Types of Requirements

In common parlance, "requirement" is often an overloaded term. For the sake of clarity, we will focus on the nature of formal requirements with respect to their type or category. Requirements can be parsed in many ways, but in this section, we will consider design-specific, industry-specific, and business-specific categories.

Design-specific requirements subdivide into:

Functional Requirements: these define system functions or tasks to be performed while the system is operational.

Performance Requirements: these describe the parameters for a function or task to be executed. Any requirement expressing a limit, boundary, minimum, maximum, frequency, resolution, duration, duty cycle, etc. in relation to a system function is a performance requirement.

Interface Requirements: these govern system interactions, internal or external; the exchange and/or storage of information. External interactions may be user- or operator-facing (not necessarily the same person). External interface requirements may encompass physical controls, such as buttons, levers, knobs, or touchscreen; visual or auditory elements, such as lights (LEDs), beepers, or buzzers; and any aspect of the chassis intended to be seen or handled by the user.

Physical Requirements: these define system size or dimensions, volume, weight, material selections, shape, or even color.

Industry-specific requirements are subdivided into:

Application Requirements: for example, clinical requirements for a medical device or interoperability requirements for a vehicle-mounted GPS. Application requirements derive from the specific industry or space the system is being designed to serve primarily; they are in a sense a byproduct of the intended operating environment (physical or abstract) where the system will be applied. They can be related to or overlap design-specific requirements.

Policy and Regulatory Requirements (Including Standards Compliance and Safety): these vary by industry; a system that could serve multiple industries may have to comply

DOI: 10.1201/9781003188377-3

with a different set of similar or even overlapping requirements for each. There are process- and product-specific standards, for example, ISO, IEC, or IEEE that generate or apply regulatory requirements to a product. In these cases, the standards themselves may not be law, but compliance with those standards may be mandated by law, strongly encouraged by regulatory approval bodies, or expected and/or mandated as a matter of policy by the system's target audience. On the other hand, some regulatory requirements are law, such as (in the U.S.) FCC requirements governing radiated power limits and coexistence of wirelessly-enabled products (BLE, WiFi, Cellular). Regulatory relationships are often quite complex; for example, many legally binding FCC regulations are based on IEEE standards, and in the medical device industry, these radiated and wireless regulations are further refined in the Functional Safety Standards (IEC 60601-1, 2005) and enforced by the FDA.

Environmental (Operational) Requirements: related to the regulatory requirements, these pertain to the system's intended operating and post-distribution storage environments and may go above and beyond regulatory requirements based on industry demands. Environmental requirements define the system's ability to weather factors like shock, vibrations, operating temperatures, and humidity (extreme hot, cold, humid, dry). *Ingress* is a popular environmental requirement subtype that provides assurance that a product can handle anticipated environmental elements (e.g., air, sand, water, oil, bodily fluids) without loss of function or durability. Environmental requirements may additionally consider social environments and local laws pertaining to disposal of the product at the end of its life.

Business-specific requirements are subdivided into:

Manufacturing Requirements: these design constraints could be applied to ensure proper production, supply-chain controls, or inclusion of legacy systems and processes. (See the discussion of *manufacturing engineering* in the previous chapter).

Product Support Requirements: these are postmarket or commercialization support requirements. They define attributes of the system that will enable future troubleshooting, both at the factory (returned product support) and in the field (by users or technicians); failure investigations; issue reporting, usage data, and postmarket surveillance; and to mitigate against the possibility of a future recall.

Non-functional requirements, which further subdivide into:

Labeling Requirements: these include any markings (symbols or words) on the device, its packaging, or its accompanying documentation that are required by regulations (international or local). Examples include "single use applications", "recycle", "hazardous". In certain industries, the legal or regulatory definition of "labeling" may also include manuals (technical and user). Finally, labeling requirements may include trademark and marketing materials such as the use or positioning of logos, specific phrasing or copy, and design elements such as color schemes and shapes.

Packaging Requirements: overlapping with industry-specific requirements, these include subsystem packaging, sterilization requirements, requirements for packaging for transport. May also specify the inclusion of business information such as warranties, promotional items, and logos or copy on the box or transport packaging.

Environmental (Storage) Requirements: these generally refer to post-manufacturing, pre-distribution conditions. Overlapping with *environmental (operational) requirements*, these specify things like ambient temperature and humidity ranges for warehousing and shipping but also can refine functional and performance requirements. For example, a requirement might specify, "While in storage, the system shall have all operational functions disabled to conserve battery power". In situations where this is a concern, such as when a system relies on a non-removable, non-rechargeable power source, a corollary requirement would likely constrain how long it can acceptably be stored between its date of manufacture and date of distribution before having to be sent back to the factory to have its power source replaced.

Usability (commercialization) requirements, which further subdivide into:

Human Factors Requirements: these clarify and define the usability of the product given its intended use, user population, and specifications. They derive from researching and understanding the user population. Developing a comprehensive set of use profiles can be unexpectedly complex – for example, a pediatric hearing aid is applied to the child population with specific physical constraints, yet the parent or the caregiver is in the adult population and should be considered a potential operator with their own constraints, and both may be considered "primary" users of the product in different respects (for more, see previous chapter).

Industrial Design Requirements: these frequently overlap with functional, physical, and performance requirements as well as human factors. Industrial design requirements help refine the physical "look and feel" of the product and ensure that it meets business and market objectives.

Marketing Requirements: As discussed above, these tend to overlap across other requirement categories. They ensure the product is accessible and satisfies defined market needs. Marketing requirements extend beyond the system itself to cover associated literature, brochures, advertising, press and media engagement, and online presence and engagement.

2.2.2 Requirements Decomposition

In the previous section, we discussed type of requirements; in this section, we will describe and discuss their level of abstraction. Requirements abstraction should match and expand on the "Requirements" phase or stage on the left-hand side of the System Vee (see previous chapter): As part of the inputs analysis, initial or top-level requirements should be analyzed for further refinement and clarification, i.e., decomposed. This analysis can generate additional requirements specific to a subsystem or technical aspect of the system.

Typical requirements analysis and decomposition can generate:

System Requirements: apply to the entire product or ecosystem under development.

Subsystem Requirements: suppose, for example, you are developing an implantable defibrillator, including its leads and external accessories, such as an external controller (remote) or mobile app. The implant and leads are not interchangeable when the system is implanted and operational. A remote control may be controlled by the developer, but may be interchanged with a different remote of the same make and model if needed. The mobile app resides on a mobile device which is a system unto itself, theoretically interchangeable, and uncontrolled by the developer. Each of these *subsystems* (associated system items or elements that interface to create a system) can, and likely must, have its own set of requirements, necessitated by the nature of the subsystem and its adjacent elements, to ensure that top-level system requirements can be fulfilled.

Technical Requirements: highly coupled and detailed per technical function. Typically, these are segregated into categories of abstraction such as mechanical, electrical, firmware, software, mobile, etc.

2.2.3 Requirement Traceability

A *trace matrix* should be generated once analysis and decomposition have been executed. This matrix should be frequently reviewed and updated. In the second part of this chapter, we will discuss incorporating testing via protocol into the tracing to demonstrate the completeness of the system review.

A couple of standard keywords help clarify the association between requirements, easing the tracing process by stating outright the intended relationship between two or more requirements. These terms include:

Relates to: indicates that requirements at the same hierarchical level share common input(s).

Is Refined by or Refines: indicates that a requirement at a different hierarchical level is further specified by or further specifies this requirement.

Derived from: indicates the requirement origination. Derivation may be *direct*, such as when an applicable law, consensus standard, market analysis, use profile, or business case states outright that the system "shall" conform to a particular specification. Derivation may also be *implied*, being realized through analysis and harmonization of broad, higher-level requirements. For example, if a high-level requirement specifies that a battery-powered system has a minimum operable life of n years without needing its battery replaced, that may generate implied requirements that specify power consumption limits or conservation capabilities as well as battery-friendly environmental requirements for both operation and storage.

Example: A top-level system requirement is: "The system shall be easily and uniquely identified". The system includes subsystem requirements such as "The handheld programmer shall have a label that includes product ID numbers", and "the implant shall be able to signal its unique identifier digitally to an authorized receiver". Each of these requirements *refines* the top-level or system-level requirement as it applies to the unique needs and circumstances of those subsystems. In this case, all three of these requirements may derive from some regulation governing identifiers or labeling.

Second Example: A top-level system requirement is: "The product shall have dimensions not to exceed XYZ and shall weigh no more than n grams". It is not ideal to have a compound requirement like this, so instead, you would write each "shall" as separate requirements, each including a pointer introduced by the phrase *relates to* so that design engineers will clearly understand that changes to size or volume may impact product weight and vice versa.

2.2.4 The Good, the Bad, the Ugly: Principles for Writing a Well-Phrased Requirement

A requirement is the *what*, the *shall*, a compulsory statement that is both prescriptive and actionable. A requirement shall not prescribe the *how*! A good requirement is clear, concise, specific, and most importantly, *testable*. A bad requirement is vague, incomplete, and non-actionable. A bad requirement can also be too specific, restricting the designers from choosing the appropriate implementation and making it too difficult to test. A good requirement has tolerances and/or expectations for an acceptance criterion (testable). Good requirements provide system designers clear objectives for implementation and signposts for when to stop (answers the question "how will we know we're done?"). Good requirements also provide test engineers clear objectives for testing the design and implementation to verify it satisfies or fulfills the requirements.

Here's the Basic Structure of a Good Requirement: "The system [noun] shall [verb, actionable, the *what*], such that [specific outcome, tolerance, acceptance criteria]. Derivation: [source]".

Here's an Example of a Bad Requirement: "The system shall be small". "Small" is vague, ambiguous, and relative. No universal criteria exist for "small" and no metric is specified.

It's impossible to test whether the system fulfills this requirement and impossible to prove that it does or doesn't.

Good: "The system shall be handheld and fit in a standard lab coat pocket". This is much better! The size is specifically defined as "handheld" with additional tolerance of "lab coat pocket". Design and test engineers can research "handheld" and standard coat pocket dimensions to ascertain the form factor of the system and set an acceptance criterion for testing. (Note that if the system were intended for use by children, "handheld" may be insufficiently specific. In that case, an implied requirement could be generated to refine this requirement).

Bad: "The system shall blink a green LED every 1 second for 5 seconds when powered on". This wording is too specific. It includes implementation details ("how"), such as green LED and on-time. Requires an intricate and excessively complex test method.

Good: "The system shall provide a 5-second visual indication when powered on". At the system level, this level of abstraction is sufficient for the designer to create an appropriate visual method and the tester can have an acceptance criterion of 5 seconds' visual indication. Any further refinement of this requirement into a technical requirement or specification, if necessary, should be handled through decomposition.

Note that the body of a requirement is actionable and prescriptive, but comments or examples can be attached to the requirement to guide the designer and tester. These are not binding but may offer helpful description or clarification as to the requirement's intent or acceptable pass/fail test criteria.

2.3 TEST REQUIREMENTS

2.3.1 Test Plan and Planning

During the planning phase, even before design inputs have begun, it is important to draft a *test plan.* This activity prompts an initial strategy for testing the product or system. It also ensures and enforces the best-practice approach of "test-driven design" or "design for test" very early in the development lifecycle, which pays ample rewards by helping to keep costs and implementation time to a minimum.

A test plan describes the various types of testing applicable for the product, including test methods and approaches. This plan is a live artifact and should be reviewed and updated as-needed throughout the project lifecycle. Test methods and approaches should not be overly prescriptive or hyper-specified, but rather described to allow the designer and tester necessary flexibility when requirements are drafted. Test methods can include literature review, evaluation through modeling, analysis or inspection and finally objective testing of product with pass/fail criteria. If necessary, a competent systems engineer can draft the test plan alone, but recommended best practice is to include a test engineer in the planning process.

Tests can be classified into two general categories: *Type* (binary, pass/fail) and *Variable* (with statistical input and confidence interval). For Type tests (often used for software testing), a sample of 1 is sufficient, whereas for variable testing, a sample size justification is required based on statistical input and confidence interval.

The test plan describes in general what to test in a given phase of the project. Levels of tests can be applied per the systems engineering vee.

Design & Implementation:

Bring-Up Tests: includes low-level hardware assembly confirmation; sometimes called a "smoke test" (as in, "wire it up and see if anything starts to smoke!"). Visual inspection of mechanical assemblies and electrical PCBAs.

Unit and Module Tests: test of functional units and combined modules. This is very important for firmware and software (see IEC 62304, 2006); exact testing requirements may vary between industries.

Characterization Tests: based on the design expectation (input specification), what are the "real world" or actual observed outputs of the system or product? What are the boundary conditions – where is the system reliable and stable and at what point does it become unstable? Characterization tests can be utilized to understand and set statistical inputs for subsequent Variable testing, as well as to understand the level of confidence (or uncertainty) in the system. Examples include discovering the system's actual mechanical tolerances or performing electrical stack-up analysis (calculating how tolerances add up over a given electrical pathway in a design or model).

Integration:

Module Integration Tests: based on the integration plan, these tests ensure that the abstracted technical (Mechanical, Electrical, Firmware, Software) subsystems behave as expected when interfacing with each other and integrated into the system as a whole. A *design integration plan* can be utilized or subsystem/technical requirements can be utilized as the basis for this testing. Integration testing is sometimes referred to as a "subsystem verification dry-run".

Interface Tests: based on the system *interface control document*, these tests ensure all interfaces (internal and external) operation within parameters. Stress testing is also executed to identify limits, tolerances for error handling, and to generate potential troubleshooting guidance if an interface fails.

System Integration Tests: final level of engineering testing prior to formal Verification and Validation testing. These are per the system integration plan but can also utilize system requirements. This integration testing is sometimes referred to a "System Verification dry-run".

Verification ("did we make the thing correctly, *i.e.* according to plan?".

Validation ("did we make the correct thing, *i.e.* according to market/user needs?".

2.3.2 Test Protocols

A *protocol* is not a plan. A protocol applies to a specific level of testing, describing clear goals and objectives for testing, along with step-by-step instructions (using the test methods referenced in the plan). A protocol also defines the necessary prerequisites, test equipment and environment, acceptance criteria, and outcome (report format).

There should be one test plan for the product or system, but there will likely be multiple test protocols needed to control test activities at every level of the systems engineering vee.

2.3.3 Design for Test and Testability

Once requirements are drafted, analyzed, and sufficiently decomposed, the designer(s) and tester(s) should start reviewing them for completeness and testability. We can't see this often enough: It is important that designers and testers be involved in the analysis exercise and review cycle early.

Designers' Role: Ensure the analysis and decomposition are sufficient to proceed with Design and Implementation with testing in mind ("Design for Test").

Testers' Role: Ensure the analysis produces testable requirements and the test plan includes acceptable ways to assess the system and demonstrate that it fulfills all requirements. Start writing protocols for each level of requirement and use dry-runs to "test the tests", thereby ensuring that formal testing will satisfy the requirements. Also to ensure early that all pre-requisites and pre-conditions are well understood and met logically. The tester may need to order special equipment or reserve time at a test lab, for example.

2.3.4 Review Cycle

At all levels of testing, test results should be analyzed to ensure they are acceptable and satisfy the objectives and/or requirements. Any failure or "pass with deviations" should be subjected to further analysis and recorded. A root cause analysis can be conducted when failures are unacceptable and don't satisfy any of the requirement-specific groups, especially if the failure occurs in production or commercialization (postmarket). A systematic *impact and effects meta-analysis* should be conducted to produce recommended mitigations for future lifecycle iterations.

2.3.5 Conclusion

With referencing to the systems engineering vee, it is clear that a proper requirements analysis must include not just the stakeholders and systems engineering disciplines, but also designers and testers. This ensures that the requirements formulation and system design processes incorporate testing needs and that requirements are testable. This approach ensures a thoroughly verified and validated product.

REFERENCES

IEC 60601-1 (2005). Medical Electrical Equipment - Part 1: General Requirements for Basic Safety and Essential Performance.

IEC 62304 (2006). Medical Device Software — Software Life Cycle Processes.

3 Project Planning

Lila Carden

3.1 OVERVIEW

A project is an engagement that produces a tangible or non-tangible product. Specifically, a project is a temporary engagement that includes human and non-human resources to produce a product, service, or outcome. These human and non-human resources include project team members, software, hardware, infrastructure, and systems. Projects are diverse and cross various industries and functional areas. For example, the following are classified as projects: new product development; product and service enhancements; training programs; new software development; facility construction; software updates and releases; space flight projects; and automation of business processes.

Project management includes the skills, tools, and techniques used to complete temporary initiatives. As noted by Carden and Egan (2008, 8), project management utilizes humans, tools, and techniques to complete initiatives. This idea serves as the framework for the structure and presentation of the information, herein. Specifically, the approach in this chapter is to discuss the human and non-human resources that are integral in successful project planning with an emphasis on tools and techniques. Moreover, the tools and techniques presented assist project team members in effectively finishing projects on time, within a specified budget, and meeting the customers' requirements.

The topics that will be discussed in this chapter are related to the processes that are needed to plan a project with the focus on the tools and techniques used to manage and implement an initiative including the following: (1) project overview focusing on the project charter, the statement of work (SOW), and the work breakdown structure (WBS); (2) resource planning including human and non-human resources; and (3) management planning including business, technical, infrastructure, and system requirements; stakeholder management and communication; system development; quality management including testing and integration management; and deployment and closeout.

EXAMPLE SCENARIO

The example project scenario for this chapter, noted herein, includes the project tasks related to automating the invoice payment process using rule-based processes and machine language (ML). The technical and systems development will include robotics process automation (RPA) and ML for processing and analyzing the data and systems application and products (SAP). Some benefits related to this project's execution include 24 hours/7 days per week resources, increase in efficiency, increase in audit documentation, reduction in costs, and reduction in processing time (Carden et al. 2019).

3.2 PROJECT OVERVIEW – PROJECT CHARTER

The initiation of any project should begin with a charter (contract). The project charter officially approves the project and provides the project manager authority to utilize resources for the completion of project tasks (Project Management Institute, 2017, 66). The charter is a high-level contract that includes the terms of the agreement related to the project and the parties involved in its implementation. Specifically, these parties include the project sponsors (who provide the resources), project team members, senior leaders, and people who have direct and indirect interests in the execution

DOI: 10.1201/9781003188377-4

of the project (stakeholders). The elements of the project charter are similar no matter what type of project is implemented including new software development projects; construction projects; new project development (R&D); and in-house strategic projects such as enterprise Project Management Office (PMO). Some of the key elements that need to be included in the project charter (Table 3.1; Tool 1) are listed below. The charter needs to include an overview section that provides a summary of the project including the business need, project scope, WBS including the high-level tasks, project objectives, and sponsorship/ownership of the project (Project Management Institute, 2017, 68). The project charter also includes the internal and external organizational structure, the project management approach, technical management approach, approvals, version control, and appendices (including any additional supporting documentation). **Highlighted elements in Table 3.1: Tool 1 – Project Charter will be discussed in more detail herein this chapter.**

3.2.1 Statement of Work

The project scope includes the tasks that need to be completed to distribute a product, service, or deliverable including the detailed features and functionality (Project Management Institute, 2013). Specifically, the project scope includes the deliverables, assumptions, and constraints associated with the initiative. The project scope guides the project team's work by stating project inclusions and exclusions. The project scope also provides a baseline for determining whether requests for changes or additions fall within the original project's boundaries. The SOW is included in the project charter

TABLE 3.1
Tool 1- Project Charter

Project Charter
Project Overview
Business Need
Project Scope including Statement of Work (SOW)
Work Breakdown Structure
Project Objectives
Sponsorship & Ownership
Project Organization
Roles and Responsibilities
External and Internal Stakeholders
Project Management Approach
Methodology
Human and Non-human Resource Planning
High-level Project Schedule including Work Breakdown Structure & Budget
Stakeholder Management including Communication Plan
Risk Management
Quality Management and Testing
Deployment and Project Closeout
Technical Management Approach
Business, Technical, and Systems Requirements
Systems Development
Systems Testing
Approvals
Version Control
Appendices

Note: Highlighted elements in Table 1: Tool 1 – Project Charter will be discussed in more detail herein this chapter.

and is a part of the project scope (Project Management Institute, 2017). For example, the SOW for a RPA implementation of the invoice process will include the project tasks that need to be performed to automate the invoice process. Specifically, the SOW will include high-level tasks such as project management activities, business requirements, and technical requirements including the features and functionality needed to pay invoices as well as the system and infrastructure requirements.

3.2.1.1 Work Breakdown Structure

The WBS is a hierarchical decomposition of the scope of work also related to the project's overall scope (Project Management Institute 2017, 126). For example, the WBS can be structured by the following process groups related to project implementation: initiating, planning, executing, monitoring, and closing. The WBS is a tool that is used to assist project managers in ensuring all tasks and work deliverables are identified as well as providing a basis to control the project. The approach for completing the WBS is to first identify the project deliverables and then sub-deliverables that are required to complete the larger deliverables and thereafter other sub-deliverables until the lowest level can be managed by one or multiple project team members. For example, the project charter and the associated tasks are completed in the initiating phase of the project. The planning phase includes the development of the documents identified in the project management approach noted in Table 3.1: Tool 1 Project Charter. The executing tasks include the business, technical, infrastructure, and system development aspects of the project as well as the execution of the management approaches such as project plan tracking and control; stakeholder management and communications; risk management; quality management and testing; and deployment. The closing tasks include the activities related to delivering the product, service, or outcome as well as the project celebration, lessons learned activity, and data archiving. **See Table 3.1: Tool 1 Project Charter for some of the tasks noted in this section.**

3.3 RESOURCE PLANNING

3.3.1 Human Resources (Roles and Responsibilities)

Human resources are a major component of the project planning process. These are the individuals who will implement the tasks identified in the project charter including the SOW and the WBS. The human resources may be internal to the organization or contracted from an external vendor. One of the tools that can be used to identify and plan the human resources needed for project execution is the **RACI (Responsible, Accountable, Consulted, and Informed) Matrix** (Project Management Institute 2017, 262). The RACI matrix includes project team members and the internal project sponsor. The matrix identifies the participation level of the identified team members and project sponsor including whether the individual is responsible for executing the task, accountable for executing the task, consulting/assisting with the tasks, and being informed about the tasks. The RACI is a tool that will allow all project team members to recognize and acknowledge their responsibilities related to overall project execution. **See Table 3.2: Tool 2 RACI Matrix for more details.**

Notice in Table 3.2: Tool 2, that the technical lead, systems engineer, and RPA/ML developer are included as the individuals who are responsible for the technical functional requirements as well as the systems and architectural requirements for the automation scenario project. Additionally, note that most of the project team members (except the internal project sponsor) are responsible for the testing, risk management, and quality management components of the project. These project activities require a concerted effort to ensure system testing and health monitoring are conducted effectively to mitigate project failure. Also note, that the internal project sponsor champions the project and provides resource support including human and non-human resources. The internal project sponsor also promotes the project to the senior leadership team and is involved in the project charter documentation including the approval and sign-off.

TABLE 3.2
Tool 2 RACI Matrix

Tasks	Internal Project Sponsor	Project Manager	Business Lead	Technical Lead	Systems Engineer	RPA/ML Developer	Testing Coordinator
Project Charter	C, I	R, A	R, A	C, I	C, I	C, I	C, I
Scope Statement	C, I	R, A	R, A	C, I	C, I	C, I	C, I
WBS	C, I	R, A	R, A	C, I	C, I	C, I	R, A
Project Plan Communication Plan Stakeholder Management Deployment Management Project Closeout	C, I	R, A	R, A	C, I	C, I	C, I	C, I
RPA Business Requirements	C, I	R, A	R, A	C, I	C, I	C, I	C, I
RPA Technical Requirements including infrastructure	C, I	R, A	C, I	R, A	R, A	R, A	C, I
System Functional and Architectural Requirements	C, I	R, A	R, A	R, A	R, A	R, A	C, I
Application System Testing	C, I	R, A	R, A	R, A	R, A	R, A	R, A
Risk Management	C, I	R, A	R, A	R, A	R, A	R, A	R, A
Quality Management	C, I	R, A	R, A	R, A	R, A	R, A	R, A

A, Accountable (Task Completion); C, Consult (Assist); I, Inform (Keep Aware); R, Responsible (Executing the Task).

3.3.2 Non-Human Resources (Tools and Techniques Including Document Repository)

The non-human resources include the tools and techniques used to implement the tasks. For example, as noted in our RPA implementation of the invoice process some of the software and hardware tools include: (1) system and network for credential verification, wi-fi connectivity, report extractions, and invoice verification process; (2) Microsoft applications for communication, spreadsheets to store information, online interfaces, and work communications including the Internet and SharePoint; (3) tools to report the extraction of data; and (4) robots used for RPA development (Carden et al. 2019, 4). Another non-human resource that is necessary for project implementation is a document repository (such as SharePoint) to house all project documentation such as project management documents (project charter, project plan, meeting notes, etc.), business and technical documents, vendor contracts, project management plan documents (communication plan, risk management plan, quality management plan, testing plan), and human resource documents.

3.4 MANAGEMENT PLANNING FOR IMPLEMENTATION AND INTEGRATION

3.4.1 Functional, Technical, and System Requirements

The functional requirements are related to tasks performed to meet objectives.

The functional requirements are related to the "what" and "when" and include requirements for functional groups such as hardware, software, and human resources. These requirements consider the needs of the customer and include the functions of the overall process. **See Table 3.3: Tool 3: Functional Requirements for example scenario.**

The technical requirements are related to the "how" and the systems requirements include the needs and "like to haves" related to the functions of the systems. For example, as noted in the invoice project herein, the functional requirements will focus on the processes for robots to automate the invoicing payment process and the human intervention portion. Once the functional needs are identified then the technical and system requirements are identified to support the needs of the

TABLE 3.3
Tool 3: Example Functional Requirements for Example Scenario

Functions of the Invoicing Automation System	Pay Invoices
When functions executed	Nightly
How long to execute functions	Minutes/Hours
Frequency of functions	Nightly
Integrated systems to complete functions	Systems Application and Product (SAP), Microsoft Office Applications, Third-Party Tools, Robots
Architecture needed	Data and Security, RPA Architecture, ML Architecture

functions. Some of the considerations as the functional requirements are developed and supported by the overall system include:

- design, development, production, operations, and support (maintenance) activities including the tools and resources needed to conduct those activities;
- technical and system packaging concepts, support, interface identification, and sequencing of activities.

  For example, the project example scenario for automation includes the steps needed for configuring the robots as well as system integration including the Systems Application Product (SAP), Microsoft Outlook, and other vendor applications such as tax applications (Carden et al. 2019, 4).

3.4.2 Project Schedule

The project manager uses the project schedule including the project budget to manage the tasks and resources associated with project execution. Specifically, the project schedule is a graphic representation that can be used to capture, track, and monitor the planned dates, durations, human resources, and milestones related to any type of project. This graphic representation is one of the primary tools that can be used to manage the completion of the project on time, within the specified budget, and with the parameters for the stakeholder. Specifically, the project schedule includes estimates of the project tasks, sequencing of those tasks, resources to conduct the project tasks, and durations of those tasks (Project Management Institute 2013, 142).

The project schedule will be finalized and used as a baseline to track the project status, including project reporting. The baseline process is important in that any changes made to the project impacting time, budget or scope need to be enacted via a formal change control process. This formal change control process is related to reporting, controlling, and documenting changes to the project baseline. For example, first the change needs to be identified, then impacts to the project schedule need to be documented and proper approvals need to be obtained. Thereafter, the approved project changes need to be formally communicated and the project schedule needs to be updated accordingly. Software tools that can be used to capture project information include Microsoft Excel, Microsoft Project, and Wrike.

3.4.3 Stakeholder Management and Communication Strategy

Stakeholders are members of the project team and other internal and external interested entities including vendors, business partners, organizational groups, and functional managers (Project Management Institute 2017, 30–33). Examples of business partners include companies that support training and maintenance activities. Examples of organizational groups are internal groups

TABLE 3.4
Tool 4 – Internal/External Stakeholders for Example Scenario

Stakeholder Group	Relationship	Example Scenario Stakeholder	Power/Interest
Vendor	External Contractors	RPA Vendor, ML Vendor	Manage Closely
Organizational	Internal Employees	Finance, Accounting, Human Resources, Project Management Office (PMO)	Keep Informed
Internal Functional	Internal Functional Decision-Makers	IT, Testing, Finance, Accounting, HR	Keep Satisfied
Customers	External Purchase Power	Computer Purchaser	Monitor

that are impacted by the project activities and functional managers are business unit areas such as human resources, legal, finance, operations, manufacturing, marketing, legal, and customer service (Project Management Institute 2013, 30–33). **See Table 3.4: Tool 4 for the Internal/External Stakeholders for the example scenario.**

Stakeholder management includes the tasks used to engage, communicate, and work with stakeholders to meet their project expectations as well as balance the tasks related to stakeholder interactions. There are various stakeholder groups that need to be considered for most projects including: vendors, organizational groups, internal functional groups, and customers. These tasks are related to increasing stakeholder support and mitigating stakeholder issues. There are tools that will be discussed in this section that will help facilitate the execution of the tasks associated with managing stakeholder relationships including the Power/Interest Grid and Communication Strategy. The Power/Interest Grid focuses on categorizing the stakeholders based on their power and communicating to them based on their level in the project (Project Management Institute 2017, 396–397).

One of the approaches that can be used to manage the stakeholders is a communication strategy which includes the type of stakeholder (as noted by the Power/Interest Grid), communication method, frequency of communication, and communication artifact. The communication strategy is usually developed by the project manager with input from the project team. The communication plan not only helps with managing the stakeholders but also is a key tool in coordinating and reporting on the project status, project issues, and project action items. Specifically, some information communicated during a project includes project status reports, scope changes, document approvals, action items, risk management issues, and milestone reports.

The communication strategy needs to consider the stakeholders involved in the project communications via status reporting, such as the satisfied and closely monitored internal stakeholders. These stakeholders will be involved in interactive, push, and pull communications (Project Management Institute 2017, 294–295). Interactive communication is between two or multiple individuals that exchange information in real time. Push communication includes information that is sent to the addresses but does not ensure the message is received because a response is not required. Pull communication includes information sent to a large audience that can access the information on demand. The frequency is based on how often the information is communicated which may include daily, weekly, monthly, or as needed requirements. Communication artifacts include the ways in which information can be shared such as face-to-face meetings, phone calls, Microsoft Teams (virtual), social media, memos, reports, emails, voicemails, blogs, intranet sites, project databases, and knowledge management repositories. **See Table 3.5: Tool 5 – Communication Strategy** for the communication approach including status reporting recommendations for very satisfied and manage closely stakeholders.

TABLE 3.5
Tool 7– Communication Strategy for Example Project Scenario

Communication Stakeholder Type	Communication Method	Frequency	Communication Artifact
Very Satisfied (Status Reporting)	Interactive Communication	Weekly	Face-to-face Meeting Phone Call Microsoft Teams Social Media
Manage Closely (Status Reporting)	Interactive Communication	Monthly	Face-to-face Meeting Phone Call Microsoft Teams Social Media
Monitor	Push Communication	As Needed	Memo Report Email Voice Mail Blog
Keep Informed	Pull Communication	As Needed	Intranet Site Project Database Knowledge Repository

3.4.4 Risk Management (Both)

The risk management approach is focused on how risk management activities will be framed and conducted. During the project planning process, the project team and all internal and external stakeholders need to identify risk management processes and create a plan which will describe how risk management activities will be structured and performed. The steps for risk management discussed in our example project scenario include risk identification, risk analysis, and risk response.

The *risk identification* component of the risk management approach focuses on identifying the risks that can derail the project. Some of the tools and techniques that are used to identify risks include stakeholder meetings, expert interviews, document reviews, brainstorming sessions, strengths, weaknesses, opportunities, and threats (SWOT) analysis, and risk breakdown structure (Pritchard 2015, 321–322; Project Management Institute 2017, 344–346). The *risk analysis* component of the risk management plan includes analyzing and prioritizing the risks to understand the impact and probability of risk occurrence. In most situations, the risks are classified as high, medium, or low as they relate to the probability of occurrence and impact on the project with an emphasis on those risks that have a high probability of occurrence and high impact if occurred as well as the medium probability of occurrence and medium impact if occurred. **See Table 3.6: Tool 7 for Probability and Impact Criteria.**

Once the project team understands the implications of the risks, then they can better prepare a *risk response* for each risk. Some of the risk responses include avoid, transfer, mitigate, and accept (Project Management Institute 2013, 344–346). In risk avoidance, the project team usually changes the project plan or stops the risks entirely. To transfer a risk means to move the impact and response to the risk to a third party. If the project team chooses risk mitigation, action is taken to reduce the probability and/or impact of the risks. Examples of risk mitigation include choosing a more stable partner or including a manual process when the system is unavailable. Risk acceptance includes the organization identifying the risk but choosing not to take any action unless the risk occurs. **See Table 3.7: Tool 7 Risk Management for the example project scenario** for a list of project risks for the invoice process including the risk analysis and risk responses for each identified risk.

TABLE 3.6
Probability and Impact Criteria

Probability	Criteria
High	Likely that the risk will occur (65% or higher chance that the risk will occur).
Medium	Somewhat probable that the risk will occur (35% to 65% chance that the risk will occur).
Low	Unlikely that the risk will occur (35% or lower chance that the risk will occur).
Impact	**Criteria**
High	Significant impact to the project schedule including task completion, time frame, and cost.
Medium	Moderate impact to the project schedule including task completion, time frame, and cost.
Low	Minor impact to the project schedule including task completion, time frame, and cost.

TABLE 3.7
Tool 7: Risk Management for the Example Scenario

Risk	Probability	Impact	Mitigate	Mitigate Plan
Scope Creep	High	High	Mitigate	Document the project requirements and obtain approval from key stakeholders. Implement a change control process including signature approvals
Project Cancellation	Low	Low	Accept	
Poor Stakeholder Engagement	High	High	Mitigate	Implement a stakeholder engagement plan including the use of a power grid and communication strategy
Resistance to Change	Medium	High	Mitigate	Implement a change management plan including a discussion about the benefits of the project and how those benefits align with the business goals and objectives
Hardware/Software Interface Issue	Medium	High	Mitigate	Frequently update and test hardware and software
Schedule Overrun	Low	High	Mitigate	Track and monitor project status. The utilization of scheduling tools and charts will minimize overrun

3.4.5 Systems Development

The human resources responsible for system development are the Technical Lead and the 3A/ML Developer, as noted in the example scenario of automating the invoice process. The non-human resources, as noted in Table 3.2 herein, include system/network, Microsoft Outlook applications, Third-party Tools, and Leo Robots (RPA software). The focus of this section will include the example scenario related to the automation of the invoice process. Specifically, the Leo Robots (bot) represent the software replacement of human users in the invoice process. The bots are developed and integrated with other systems such as the Microsoft Outlook Applications and third-party tools. The system development will also include machine learning (ML) tools for data mining and data analysis such as Scikit-learn, TensorFlow, and Weka.

3.4.6 Quality Management (System Testing and Integration Management)

One of the focus areas for automation solutions is data management including data quality. The management of the data includes considerations for data preparation and exploration that are inputs to data analytics and business intelligence. In addition to quality management for the data, there is also quality management for the solution components including *informal testing* for performance and overall solution functionality that usually occurs in the RPA/ML developer environment including the systems engineer. The *formal testing* of the data, solution functionality, application integration, and systems are coordinated by the testing coordinator with assistance for the project team including the business lead, technical lead, system engineer, and the internal project sponsor's functional employees (see **Table 3.2: Tool 2 RACI Matrix.)**

The project team will test, debug, and deploy the automation solution with an emphasis on ensuring the business requirements are executed as defined in the technical documents and as per the sponsoring department's guidelines. Some of the testing will include software/hardware integration testing, security validation/verification, system functional testing, throughput and cycle verification, and certification verification requirements. Specific testing details are listed below.

- **Test Data Creation:** automation script retrieves the test data
- **Bot Execution:** automation script triggers the bots via a job and the RPA script will run through the bot execution flow related to the invoice process
- **Verification/Validation:** automation script validates the output and creates a test report; bots can be used to store the invoice detail in various locations and bots can also be used to validate return invoice details.

3.4.7 Deployment and Project Closeout (Project Management)

The deployment of the software includes tasks related to transferring the software to the department that will use the product. Specifically, as per the RPA example project, the deployment is related to the exact date on which the software will be delivered to the finance and accounting departments. The deployment includes the following activities that need to be completed prior to the product transfer: training of finance and accounting employees as well as other stakeholders; assignment of new roles and responsibilities; and transition of product to IT maintenance and support employees. Project closeout activities are conducted by the project manager with help from the entire team. The project close-out activities are related to tasks associated with project completion. These tasks are related to resources, contracts, documents, maintenance, and celebration (Project Management Institute, 2017, 104). See detailed tasks associated with project close-out in **Table 3.8: Tool 8 Project Closeout Activities.**

TABLE 3.8
Tool 9 Project Close-out Activities

Project Close-Out Activity	Date Completed
Conduct Sponsor Project Sign-Off	
Deliver Product, Service or Outcome	
Celebrate Project Completion	
Conduct Lessons Learned including positive and negative lessons	
Evaluate Project Team Members	
Release and Reassign Project Team Members	

(Continued)

TABLE 3.8 (*Continued*)
Tool 9 Project Close-out Activities

Project Close-Out Activity	Date Completed
Conduct contract closure including financial obligations	
Transfer or dispose of project assets including hardware, software, applications, and tools	
Store project documents in the data repository	
Implement the maintenance and operations plan as per the contract	
Identify and assign maintenance and operations resources	

REFERENCES

Carden, L., & Egan, T. (2008). Does our literature support sectors newer to project management? The search for quality publications relevant to nontraditional industries. *Project Management Journal*, *39*(3), 6–27.

Carden, L., Maldonado, T., Brace, C., & Marie, M. (2019). Robotics process automation at TECHSERV: An implementation case study. *Journal of Information Technology Teaching Cases*, *9*(2), 72–79.

Pritchard, C. L. (2015). *Risk management: Concepts and guidance*. Boca Raton, FL: Auerbach Publications.

Project Management Institute. (2013). *A guide to the project management body of knowledge (PMBOK Guide)* (6th ed.). Newton Square, PA: Project Management Institute.

Project Management Institute. (2017). *A guide to the project management body of knowledge (PMBOK Guide)* (6th ed.). Newton Square, PA: Project Management Institute.

4 System Decomposition

Satyajit "Sat" Ketkar and Jason Smith

4.1 SYSTEMS ENGINEERING VEE, REVISITED

In Chapter 5, we discuss the systems engineering vee, left-hand and right-hand meeting at the bottom of the vee where design-to-implementation iterations are executed. In this chapter, we will discuss the approaches and methods of taking design inputs and creating a system architecture and design (Figure 4.1). This is the Cloud (D) below (for context, refer to Chapter 5, Figure 5.2).

4.2 WHAT IS SYSTEM DECOMPOSITION?

In order to design a successful system, it is important to first understand the needs of the users who will be interacting with it. One way to accomplish this is through the process of system decomposition, which involves breaking down the system into smaller, more manageable components. By understanding the specific needs and requirements of each component, designers can create a system tailored to the needs of its users.

The process of system decomposition begins with identifying the overall goals and objectives of the system. This may include identifying the specific tasks that the system needs to perform as well as the users who will be interacting with it and their specific needs. Once these goals and objectives have been established, the system can be broken down into smaller, more manageable components.

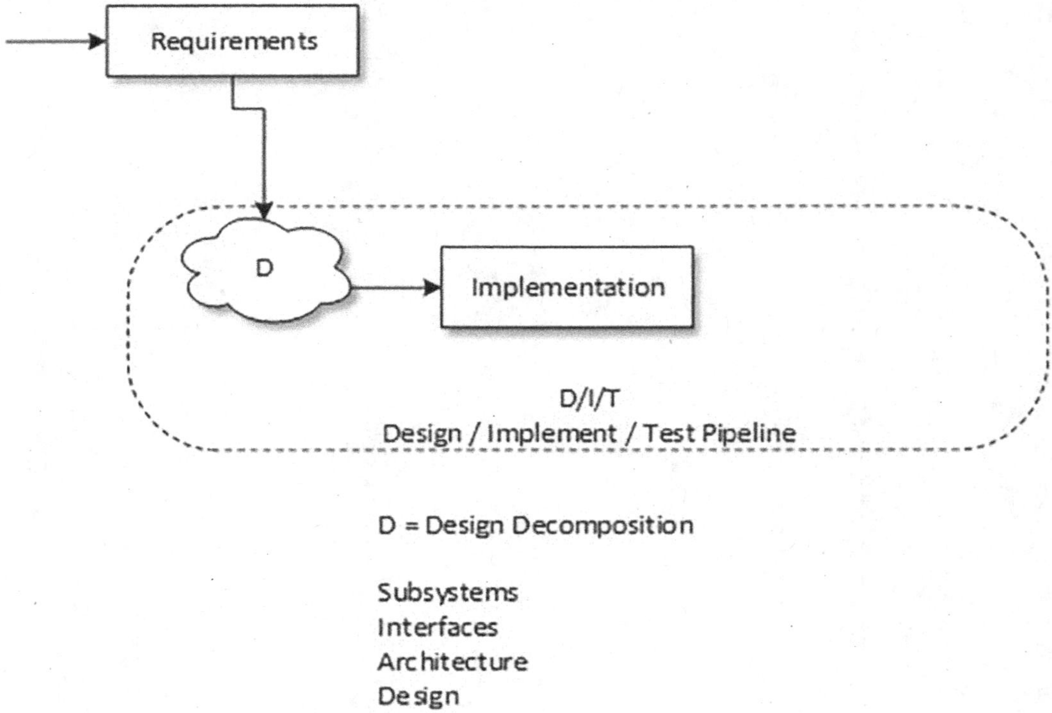

FIGURE 4.1 Close-up of systems engineering vee, bottom left.

 DOI: 10.1201/9781003188377-5

One common method of system decomposition is to use a hierarchical approach, in which the system is broken down into increasingly detailed levels. At the highest level, the system is divided into major functional areas, such as input, processing, and output. These functional areas can then be further broken down into smaller subsystems and subcomponents.

Another important aspect of system decomposition is understanding the interactions between different components of the system. By identifying the dependencies and relationships between different components, designers can create a system that is both efficient and easy to use.

Throughout the process of system decomposition, it is important to keep the needs of the users in mind. By understanding how the system will be used and what specific tasks it needs to perform, designers can create a system that is tailored to the needs of its users.

4.2.1 Approach

Decomposition should try to answer the questions generated during requirements analysis. What are we trying to achieve? What are we trying to solve?

In the real world, people often advance an answer or a solution without a clear understanding of the question or problem. During decomposition, we try to stay out of solution space. Questions and problem statements are what we use to transition from layer to layer (stage to stage of the vee). The decomposition must be systematic, answering stage-appropriate questions generated by the answers to the questions asked during the previous stage.

System decomposition should start on the left-hand side of the vee and decompose the product or system down into "primitive" elements that ultimately can be traced to testing of requirements. These primitive elements can be organized into different types based on technical discipline – mechanical, electrical, firmware/software, etc.

4.3 THE DECOMPOSITION PROCESS

Abstracted and simplified for this discussion, using the waterfall method:

System decomposition should begin during the requirements analysis and decomposition cycle. Each layer or level of requirements should generate an architectural concept and/or design concept. These can be presented in views or modes depending on the type or level of decomposition desired (Figure 4.2).

4.3.1 Decomposition Views

Logical: this view breaks the system down into its logical components or subsystems, which are organized around the system's functionality or data processing. This view is useful for understanding how the system processes information and produces output, and how different components or subsystems interact with each other to perform the system's functions. This view may include block diagrams representing power management, safety, storage, communications, and so on.

Functional: this view breaks the system down into its functional components, identifying the inputs, processes, and outputs of each component. This view is useful for understanding how the system processes information and produces output.

Physical: breaks the system down into its physical components, such as hardware, software, and data. This view is useful for understanding how the system is implemented and how its components interact. This can include internal cabling as well as user interface elements like buttons, LEDs, and switches.

Behavioral: breaks the system down into its behaviors or functions, such as user interactions, system responses, and error handling. This view is useful for understanding how the system behaves in different scenarios and how it responds to different inputs. This may

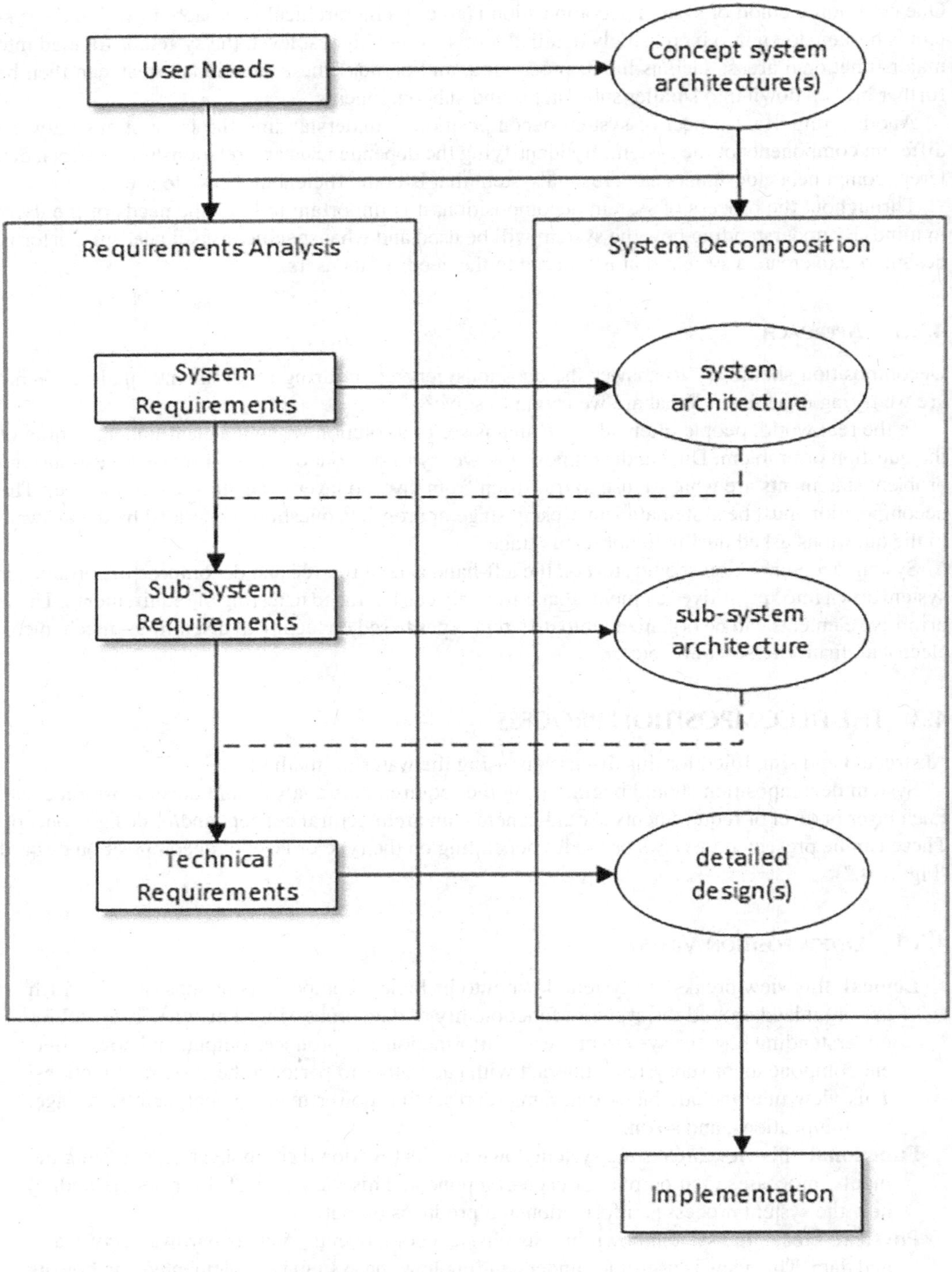

FIGURE 4.2 System decomposition process mapped to requirements analysis.

include power self-tests, troubleshooting, alarms, and failure modes, in addition to normal operational behaviors.

Data Flow: breaks the system down into the flow of data through the system, identifying the sources, destinations, and transformations of data. This view is useful for understanding how data is processed (transient) and stored (persistent) within the system.

Object-Oriented: primarily used for software products, this view breaks the system down into objects, classes, and methods, which are used in the object-oriented programming paradigm. This view is useful for understanding how the system is organized and how objects interact with each other to perform the system's functions.

Each of these views provides a different perspective on the system and can be used together to gain a more complete understanding of the system's structure, components, and interactions.

4.3.2 System States and Modes Decomposition

State: breaks down a system into different states or conditions that it can be in at any given time. The key is that a system is only in one unique state at any given time. Each state represents a specific configuration of the system, and the system's behavior and output will be different depending on the current state. For example, a product can be sleeping, idle, active, long-term storage, or end-of-life.

Mode: breaks down a system into different modes or operating conditions. Each mode represents a specific way in which the system can operate or be used. Modes can overlap with other modes as well as with states. For example, a product can be operated with low settings, normal settings, or high settings. A ceiling fan in active state (on) could have its blades spinning at slow, moderate, or high speed (mode). A printer in an idle state could have its default print quality set to draft, normal, or best; the mode is available to the user to set or change, even though the printer is not actively processing a print job.

Both states and modes are used to understand how the system behaves in different conditions and how it responds to different inputs. State decomposition focuses on understanding the different configurations of the system, whereas mode decomposition focuses on understanding the different ways in which the system can operate.

4.3.3 Decomposition Process

4.3.3.1 Requirements Analysis

The first step in the process is to identify the overall goals and objectives of the system. This may include identifying the specific tasks that the system needs to perform as well as the users who will be interacting with it and their specific needs. The requirements analysis phase is critical for ensuring that the system will meet the needs of its users.

One company that has successfully implemented a requirements analysis phase is Amazon. Amazon's e-commerce platform is a complex system that needs to perform a variety of tasks such as catalog search, product recommendation, and checkout. By understanding the specific needs of their users, Amazon has been able to create a system that is tailored to their customer's needs and is easy to use.

4.3.3.2 System Decomposition

Once the requirements have been identified, the system can be broken down into smaller, more manageable components. One common method of system decomposition is to use a hierarchical approach, in which the system is broken down into increasingly detailed levels. At the highest level, the system is divided into major functional areas, such as input, processing, and output. These functional areas can then be further broken down into smaller subsystems and subcomponents.

Companies like Tesla use a hierarchical approach to decompose their electric vehicles. They break down the vehicle into subsystems such as powertrain, battery, and infotainment, and then further decompose it into subcomponents such as motor, controller, and inverter for the powertrain subsystem.

4.3.3.3 Interaction Analysis

Another important aspect of system decomposition is understanding the interactions between different components of the system. By identifying the dependencies and relationships between different components, designers can create a system that is both efficient and easy to use.

Netflix uses an interaction analysis approach to decompose their streaming platform. They identify the interactions between different subcomponents such as the video player, recommendation engine, and content management system. By understanding these interactions, they can optimize the performance and user experience of their system.

4.3.4 Decomposition Benefits and Application

System decomposition is a critical step in the design process that helps to ensure that the device meets all of the necessary requirements and is testable. Once the requirements have been defined, the system can be decomposed into different levels, such as hardware, software, and firmware. This allows for a more modular design, where each component of the system can be developed, tested, and validated independently. Then, during the implementation stage, each component of the system is systematically integrated and tested together to form the final product. It's important to ensure that all the components are compatible and work together as expected.

The design of the system should be such that it can be easily accessed for testing and troubleshooting. This may include providing test points for measuring signals, designing the system to facilitate test automation, and ensuring that the system can be easily opened up for inspection. Additionally, the system should be designed with the ability to be tested in different environments and scenarios, such as in a laboratory or in the field.

One of the key benefits of system decomposition is that it allows for a more thorough and efficient testing process. By breaking the system down into smaller components, it becomes easier to identify and isolate any issues that may arise. This can help to improve the overall quality and reliability of the device, and ultimately ensure that it meets market, user, and business needs.

5 Systems Engineering Vee

Satyajit "Sat" Ketkar and Jason Smith

5.1 WHAT IS THE "SYSTEMS ENGINEERING VEE"?

Simply put, the "vee" is a representation of the *development lifecycle* (DLC). The DLC is the sequence of stages that every product under development goes through. Figure 5.1 shows a traditional layout of the DLC, starting from the top left and working down toward Implementation, bottom center, and up again from there to postmarket activities.

Even when a product isn't being developed formally, you can see how making something new generally follows this pattern. Let's say you have an idea for a new board game. There's a brainstorming period, hopefully informed by research and trying out several gameplay variations, followed by a period where you're designing and prototyping a version of the game, testing it to see if you've succeeded at making something that fulfills your vision and is fun to play, and then, if you want to share it with others, arranging a way to make copies of it and distribute those copies, maybe even tweaking future versions of the game based on the feedback you get from other players until, eventually, it's time to stop those activities.

A loose, casual approach to development may be okay for hobby projects, but for most engineering efforts, formal methods and controlled processes are necessary to ensure that the needs (*requirements*) of everyone who has a stake (*stakeholders*) in the development project and its final product are fulfilled. There's a lot to keep track of. Development rarely takes place in a vacuum. Most often, it is constrained by factors like

- What do customers want from a product like this one?
- How much are they willing to spend to buy one?

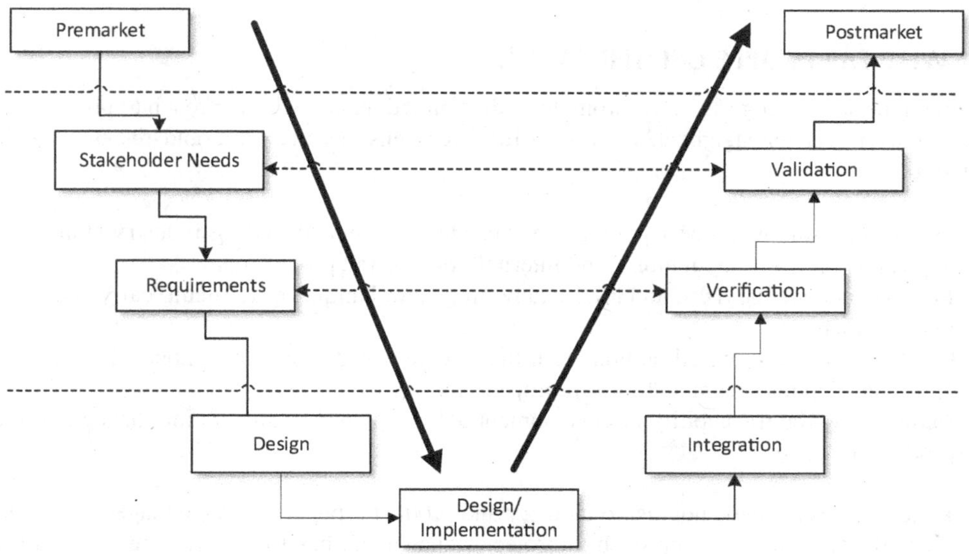

FIGURE 5.1 Industry-agnostic systems engineering "vee".

DOI: 10.1201/9781003188377-6

- What is the right balance between how quickly it operates and how much power it consumes?
- Will making, building, or selling this product create any potential legal liabilities for its creators?
- Could using (or misusing) this product cause potential harm to consumers?
- Are there any laws in effect that might apply to this product?

Those are just a few, simplified examples of things to consider when developing a product. The more complex the system being developed, the greater the need for a rigorous, methodical, well-known approach to the DLC. That's where the "vee" comes in.

The traditional "vee" isn't the only way that development can proceed, but it is the most common, reliable, tried-and-true method of taking a project through its various stages of development. The application of the formal "vee" typically consists of the stages below. These stages are defined by ISO 15288 for a generic lifecycle development scenario and are not specific to any particular type of product or industry. The generic stages are:

1. Exploratory
   Investigate new opportunities; explore technology readiness; evaluate pre-concept match with user needs
2. Concept
   Identify stakeholders' needs; evaluate alternative concepts; recommend possible solutions
3. Development
   Develop detailed planning; identify and manage risks and opportunities; implement plans and develop system
4. Production
   Produce systems; inspect and test
5. Utilization and Support
   Operate systems to satisfy user needs; provide sustained system capability
6. Retirement
   Store, archive, or dispose of system[1]

5.2 WHY DO PEOPLE USE THE "VEE"?

Organizing projects using the "vee" promotes a disciplined, responsible approach to product development and project management. Following the "vee" ensures that time and resources will be devoted to:

- Observing established best practices during development, including industry standards, applicable laws and regulations, and internal policies and procedures
- Identifying benefits, risks, and NUDs (anything New, Unique, or Difficult) early and planning for them
- Continuous reviews based on predetermined, stage-based project milestones, keeping the entire development team and overall project on track
- Maintaining the traceability of development activities and documentation throughout the product lifecycle

The classic "vee" is agnostic not just to industry and product type, but to the management model as well. Virtually every modern approach to project management has found the "vee" to be a useful method for organizing projects. In particular, some management models for software development use variants on the "vee" such as the "waterfall", rapid iterative cycles ("agile"), and others, but the "vee" remains the best available foundation for organizing DLC activities.

It's worth noting that while the "vee" is a good representation of when development activities *begin*, it does not always portray intuitively when those activities should *end*. Development is iterative. Conclusions from early-stage activities permit designers to proceed to the next stage with better understanding of the project and a manageable scope; however, those conclusions are commonly revisited and revised in light of new information resulting from subsequent activities. Controlled development practices ensure not only that we proceed in an orderly manner but also that we return to prior conclusions and iterate on them in an orderly way. Orderly iterative development maximizes project efficiency and minimizes the risk of failure.

5.3 DOES THE GENERIC "VEE" HAVE UNIVERSAL APPLICATION?

Yes and no. In this chapter, we will use more specific scenarios to move our discussion from abstract to practical. We find that it's more useful to consider specific lifecycle development cases to examine how these concepts are applied in the real world rather than to keep strictly to the generic and abstract.

There is a drawback to this approach, that is, each industry and product type often has its own terms and paradigms for the development lifecycle. This is because development needs vary from industry to industry based on how safety-critical the device under development will be, the specific regulations that apply to it, and the expectations of users and consumers. We have very different laws and expectations for an IoT-enabled coffeemaker or, a gaming platform or a car, for example! Fortunately, lifecycle stages are not radically different from industry to industry, and all follow the same basic sequence. Once you understand the principles underlying the general case, you will quickly grasp how they are applied in different industries, even when the terminology varies.

In the medical device industry, systems under development may be safety-critical in the extreme. The proper function of an implanted cardiac device or insulin pump can mean a difference between life and death for users. It's no surprise that this industry is highly regulated, and the specialized nature of its products places particular demands on the development process. Medical devices are a prime example of complex integrated systems and are more rigorous in their systems engineering aspects and the application of the "vee" than many other industries, making them an ideal lens through which to view the systems engineering "vee".

In medical device development, we use the term "phase" in place of "stage". Stage refers to the status of the system under development, whereas phase refers to the step in the development process itself. The distinction is subtle, and people often use "stage" and "phase" interchangeably. However, in many safety-critical development projects, quality controls and other regulatory measures require us to peg activities to the *process step* (phase) rather than *system state* (stage). Stages may overlap, but phases should not.

From a project management point of view, there are really just two top-level phases to a project: a feasibility phase, where you're exploring whether to make something and what that thing should be like if you choose to make it, followed by an execution phase, where you design, manufacture, sell, support, rework, and ultimately retire, the thing you chose to make. If you iterate on the product by making a Version Two, for example, all you're really doing is taking the lessons learned from working on Version One and starting again at feasibility. Lather, rinse, repeat.

But that high-level view is far too simplistic for complex systems engineering. Feasibility and Execution each get broken down into a series of discrete sub-phases, resulting in the series of steps depicted in the classic "vee" diagram. In the generic case, stages 1 and 2 fall under Feasibility, while stages 3–6 fall under Execution.

Medical device development is similar, but as you can see (Figure 5.2), its phases don't perfectly correlate to the six generic development stages:

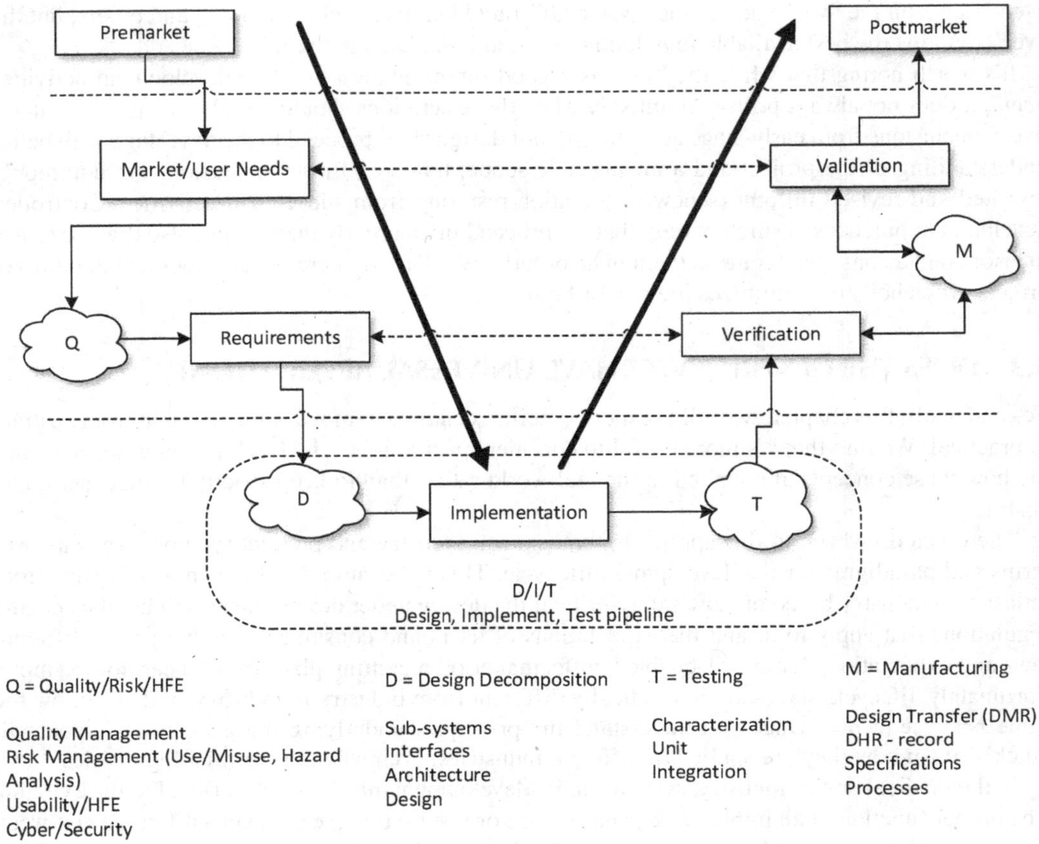

FIGURE 5.2 Systems engineering "vee" for medical device development.

TABLE 5.1
Generic Development Stages vs. Medical Device Development Phases

Industry Agnostic Development Stages	Medical Device Development Phases
Exploratory	Premarket
Concept	User requirements
	Risk profile and human factors
	Requirements
Development	Implementation
	Verification and validation
	Regulatory submissions
Production	Validation and transfer to manufacturing
	Manufacturing
Utilization and support	Postmarket
Retirement	

As you've no doubt already guessed, there is a high degree of correlation, but not 1:1 correspondence, between the industry-agnostic "vee" and any industry-specific "vee". Approximate mapping between ISO 15288's stages and medical device development phases is given in Table 5.1.

Let's look at each of these phases in more detail.

5.4 LEFT SIDE: PREMARKET

During this phase, the organization is considering its *business needs* and conducting *market analysis* to evaluate consumer needs, existing and forthcoming products in the space, and the potential for their proposed new product to represent a unique offering, thus gaining a competitive advantage in the marketplace.

Systems engineering may be involved in this phase, helping to evaluate the *feasibility* of the product concept. System engineers' general knowledge of the current technological landscape allows them to suggest various paths by which the proposed development project could achieve success. This supporting advice may include anything from a list of relevant off-the-shelf or customizable components that could be employed in the project presented with implications and pros and cons for each, to providing rough time and resource estimates for from-scratch development of hardware, software, and interfacing.

5.5 LEFT SIDE: USER NEEDS

The goal of this phase is to collate the totality of user needs. This means:

- Identifying all *users*
- Developing the *voice of user* for each type of user
- Describing all *use cases*

This process is not as straightforward as it may seem at first glance. Most products have an obvious category of user, often called the *end user*, who is intended to be the direct beneficiary of the product. In the case of a book like this one, the end user is the reader. For medical devices, it is the patient. But these are far from the only users. Medical devices may also have several other kinds of users who directly support the patient's use of the device, such as prescribers, operators, surgeons, service technicians, or other specialists who may be needed to ready the device for use. Even this book has users other than the reader, such as booksellers and instructors. And what about the organizations developing, testing, and manufacturing the product? They too can be thought of as users, with specific sets of needs corresponding to specific stages of the DLC that should be accounted for.

Each type of user has different goals for the product. While there is typically some amount of overlap between user needs, there will also be a certain amount of conflict that must be reconciled and harmonized as well. One set of market needs will focus on ensuring positive outcomes for the end user. Another set focuses on innovation and the need to establish a competitive edge over existing products. A third set speaks to the purchasing mechanisms, financial means, and buying behaviors of the target market (this is especially complicated when it comes to medical devices, where the buyer could be a healthcare delivery organization, a patient, or a patient assisted by public or private insurance – insurance that may or may not recognize the product as a valid or necessary source of therapy). A fourth set of needs ignores the market and looks the other direction along the supply chain, considering instead the manufacturer's need to access the components necessary for mass production, and to build the proposed design efficiently, safely, and cost-effectively.

During this phase, planners find it helpful to produce documentation summarizing these sets of needs as the *voice of user*. Common voice-of-users include *voice of customer* (which may need to be separated into *voice of buyer, voice of end user, and voice of operator* for products like medical devices) and *voice of business*, which considers things like schedules and costs, marketing considerations, business risks due to factors like legal exposure and fundraising, and project-agnostic development controls such as quality management and lifecycle concerns. Parsing *voice of business* into discrete units is recommended; *voice of compliance, voice of legal, voice of testing, voice of manufacturing*, and *voice of marketing* may all be worth developing in tandem. Finally, because complex integrated systems are increasingly developed for connectivity (e.g., to a network or smartphone), it is increasingly common and helpful to include a *voice of security*. Using these "voices", documentation is generated to describe and advocate for the needs and desires of each type of user as if that user type was the only one that mattered. This exercise makes it possible for systems engineering to review potentially competing or conflicting needs side-by-side and harmonize them, typically by organizing them hierarchically.

The User Needs phase also requires the compilation of user needs and premarket research into *use cases*. Each use case describes the *intended use* of the final product by an end user, operator, installer, or service technician. Use cases are critical because, along with user needs, they become the basis for all subsequent development, especially:

- requirements
- design
- risk management
- verification
- validation

User needs are collected through research, including market research, analysis of existing products, human factors assessments, user surveys, and use environment studies. Depending on the industry and type of product, it will either be necessary or advisable to circle back to these activities and conduct more nuanced and accurate research after prototyping the product (or key subsystems, such as user interfaces), sometimes with multiple iterations or by conducting A/B-type parallel testing.

5.6 LEFT SIDE: RISK PROFILE

Creating a genuinely risk-free product is impossible. Even a product as benign as this book carries, for example, the risk of confusing the reader or even imparting misinformation, when the book is used as expected. A heavy hardcover textbook could also be *misused* in a way that causes injury; a digital book in certain formats, such as PDF, could be hacked and used to spread malware. Of course, these risks are true for all books, and authors create them, publishers and retailers distribute them, and readers buy them anyway. That is because all stakeholders deem that the *benefits* of creating the product outweigh a certain amount of *acceptable risk*.

The goal of risk management is to establish a baseline of acceptable risk for the proposed product and to ensure that the final product does not exceed the acceptable limit. Risk management activities continue for the remainder of the product lifecycle, starting with the Risk Profile. The Risk Profile then becomes the basis for all subsequent risk management and the first section of the Risk Management Plan (RMP).

Developing a Risk Profile begins with reviewing applicable *standards*. Standards reviews generally proceed as follows:

1. Industry-specific *regulatory standards*. Compliance with regulatory standards is mandated by law.
2. Basic or industry-agnostic regulatory standards, typically as pointed to by industry-specific standards.
3. *Consensus standards*. Consensus standards are not legally mandated, but are best-practice recommendations agreed on by industry stakeholders. You aren't required to comply with consensus standards, but it's reasonable to assume that your potential customers, investors, and other stakeholders will expect you to, and that a lack of compliance may therefore have financial repercussions.

Complying with standards may seem burdensome or expensive in the short term, but in the long term, will benefit your budget and your schedule, and help you avoid common mistakes and known pitfalls in the development process.

One output of a standards review is the first round of *requirements* for the project. Once the set of standards that apply to the project have been identified and reviewed, each applicable section should be expressed as an individual requirement, either for the final product (product requirements) or for the development process itself (project requirements).

The next step in risk profiling is *hazard analysis.* A *hazard* is the potential harm that could come about through the *use* or *misuse* of the product. In this step, it does not matter how likely it is that any given hazard may occur, and attempting to assign a probability to any hazard would be premature. Hazard analysis bookends risk management activities during development. What's conducted during this phase is a *preliminary hazard analysis* (PHA), which, because it is based on everything that's known about the project up to this point, is necessarily incomplete. A PHA represents the development team's anticipation of any possible harm the product could cause, however unlikely that harm may be or seem to be. As the design becomes more detailed and concrete, and technical unknowns get filled in with real-world specifics, the team becomes better able to anticipate, describe, and even quantify potential hazards.

Hazards may occur as a consequence of the intended use of a product. For example, electric hairdryers are intended for use in bathrooms, where they could come into contact with water, causing electrical discharge with various undesirable outcomes. Hazards may also occur as a result of the misuse of a product; misuse can be unintentional (due to misunderstanding) or deliberate (abuse). When developing safety-critical products, regulators generally require that risk management include hazards occurring both from *intended use* and *reasonably foreseeable misuse.* While development teams aren't expected to enumerate every possible way a product could be absurdly misunderstood or creatively abused, they are expected to put thought and attention into predicting misuse. The definition of what is "reasonably foreseeable" cannot be quantified or described with certainty, making it impossible to determine whether a given list of misuse-derived hazards is complete. Risk management must instead aim for a list that is "reasonably" sufficient.

NOTE: At the time of writing, the terms "use" and "misuse" are still current. However, pending changes to ISOs will soon group these concepts together with *use cases* under the umbrella term "use specification". Once that change takes effect, it is anticipated that development teams will be expected to build a set of *Use Specifications* that describe the use cases, intended uses, and foreseeable misuses of their product. This new organization of phase activities and documentation is anticipated to improve visibility and communication for all groups involved in development.

Next, work begins on a document called the Failure Modes Effects Analysis (FMEA). As its name suggests, this document itemizes each way that the product or some element of the product could fail to perform as intended and describes what the effects of that failure could be if it were to occur. Again, we are not yet considering the likelihood of any of these failures occurring. We are simply listing any and all possible types of failure, and then describing and quantifying – typically by assigning a score based on a predetermined rubric – the *severity* of each possible failure.

Finally, advised by systems engineering, the design team itemizes *risk controls* and *mitigations.* These are laid out in a *traceability matrix,* typically using a spreadsheet, which lists out all the hazards and failure modes and quantifies their probability of occurrence, so that each hazard can be assigned a *risk classification* and a weighted *pre-mitigation score.* Risks are scored based on a predetermined rubric that takes a number of factors into consideration, but primarily weights the severity of a given hazard against the probability of it taking place.

Once each hazard has been classified as a risk, the design team describes risk controls and mitigations that shall be incorporated into the design so as to eliminate these risks or reduce their probability of occurrence and/or the severity of their impact to an acceptable level. These risk controls and mitigations then become the basis of the second round of requirements.

It's worth re-emphasizing here that all of the risk management work undertaken in this phase is preliminary to the design, and cannot be considered complete. These activities begin in this phase, but are subject to modification throughout the next several phases of the lifecycle as design and development work progresses. What is taking place now is based on the "best guesses" of the risk management team, based on research into and experience with similar products. The breadth and depth of insight brought to bear by systems engineers and quality managers is critical to the success of this phase.

5.7 LEFT SIDE: REQUIREMENTS

Having gathered the information and completed the exercises of the previous phases, it is now appropriate to formally compile the Requirements for the product. Common terminology does not align across industries; some industries interchange "Product Requirements" and "System Requirements", and if the product includes multiple elements, such as controller peripherals or a smartphone app for remote access to the main device, "Ecosystem" is sometimes used. Regardless of the terms, what's in scope for this phase stays constant. This is where all of the top-level requirements for the entire product are defined.

Table 5.2 shows the relationship between previous phase activities and their expression as requirements:

Broadly speaking,

- *Functional* requirements define what the product does
- *Performance* requirements define how it does those things (e.g., time to complete a task, minimum battery life, etc.)
- *Efficacy* requirements define benefits realized by the user when the product performs its function(s)
- *Safety* requirements ensure that using or misusing the product will not endanger users
- *Project* requirements place parameters on how the product will be developed
- *System* requirements apply to the product as a whole
- *Subsystem* requirements apply to subsystems of the whole

Project requirements are often overlooked, underemphasized, or slip through coverage gaps when multiple teams or vendors are involved in project planning and development. It's easy for teams and managers responsible for specific phases of the development lifecycle to focus on their phase without considering what enables efficient and timely work in subsequent phases. Systems Engineering can use project requirements to mitigate the *business risk* created by phase-segregated development practices. These might include product-agnostic requirements related to quality controls and

TABLE 5.2
Requirement Types Generated by Lifecycle Phase

Phase Activity	Requirement Type
User needs	Functional requirements
	Performance requirements
	Efficacy requirements
Risk profile	Safety requirements
	Project requirements
	Other compliance-based requirements
Requirements	System requirements
	Subsystem requirements

security controls that the organization or the client has determined shall apply to all development projects of a given type. They should also include product-specific requirements related to future development activities, especially *testability* and *manufacturability.*

- **Testability:** Although precisely what testing activities and when in the development lifecycle they take place may vary from product to product and industry to industry, at some point every product under development, as well as individual subsystems, must undergo testing – first to help the R&D team learn what the product requirements should be, and then to prove that the product design actually fulfills those requirements. Involving Test Engineering early in the lifecycle helps to ensure that the design will be testable, and that requirements are written in such a way that they can be objectively demonstrated. Test Engineering may recommend specific *testing requirements* to facilitate efficient and effective testing of the product design.
- **Manufacturability:** This is a closely related concept to testability. Just as designs can be developed without due concern for how they will be tested, so too can they be developed without due concern for how they will be mass-produced. The design may call for specific materials, components, or assembly procedures that work fine for prototyping, early efficacy studies or market testing, and even pilot builds, but are too costly, too scarce, or too time-consuming to be practical for mass production. Involving Manufacturing Engineering early in the lifecycle helps to ensure that the design will be optimized for mass production. Manufacturing Engineering may recommend specific *manufacturing requirements* to facilitate efficient, cost-effective production.

Delving into these considerations only comes about through decomposing *system requirements* into *subsystem requirements* entailing both physical and logical subsystems. Subsystem requirements are often highly specific to discrete subsystem aspects and may be further broken down and organized as, for example, mechanical requirements, electrical requirements, software requirements, and so on. These requirements can be very granular and are often very basic.

It is strongly urged that requirements not be written prescriptively. If the goal of a requirement is, for example, to ensure that a frequently-handled subsystem such as a remote control is able to withstand bodily contact from sweat and skin oils, it is best to state the requirement in a straightforward, solution-agnostic manner. "The remote control's performance and appearance shall not be degradable by frequent or prolonged exposure to bodily fluids" is a much better requirement than "the remote control chassis shall be made of plastic type *x* and its buttons of rubber type *z*". Although the materials specified might accomplish the same goal, codifying this solution as a requirement obscures the purpose behind the requirement and radically limits the set of possible solutions. Suppose the specified material is discontinued? Suppose stress-testing reveals that it is not capable of performing as advertised in all of your use cases? Suppose a cheaper and more environmentally responsible material comes on the market? This is a simple example, but it serves to illustrate how design and development can be needlessly complicated and delayed by moving too quickly into identifying solutions and codifying possible solutions as requirements.

5.8 LEFT SIDE: DESIGN INPUTS PHASE REVIEW

Premarket research, User Needs, Risk Profile, and Requirements are collectively termed *design inputs.* Before starting in on solutions and the real work of design, it is critical that key stakeholders meet for a comprehensive *design input phase review,* led by Systems Engineering. This review establishes that input considerations are sufficient for development to move into the next phase, and creates agreement for how the project will proceed.

5.9 LEFT SIDE: SYSTEM ARCHITECTURES

Moving now into the lower left side of the "vee", Systems Engineering begins laying out the *system architecture* for the product. System architectures are typically presented in multiple models or views:

- *physical representations* lay out the material components
- *logical representations* ignore physical layers to describe component functionality
- *data representations* describe data at rest, data in motion, communications, and interfaces

Interfaces are described in-depth in this phase through the process of *interface decomposition*. Systems Engineering maps the system's external and internal interfaces, indicating control points, communications, and data transfer nodes. An Interface Control Document (ICD) is created, defining how all communication will occur in the system. Once established, this document is particularly costly to change, as changes made here impact all design activities that follow.

5.10 LEFT SIDE: SYSTEM DESIGN

From system architecture, development moves into *system design*. Design follows the modeling layers/views established in the architecture phase, decomposing them into design concepts. These concepts are later reviewed, evaluated, and collapsed into a single, agreed-upon design at the conclusion of this phase.

Design concepts will be created for:

- Physical and mechanical elements, e.g. through CAD modeling
- Electrical elements, through circuit design and wiring diagrams
- Software elements, defining languages, design patterns, build environments, and so on
- User-facing design, including chassis, labels, status indicators like lights and sounds, user interfaces, etc.
- Packaging, including consumer packaging as well as transportation and storage

5.11 VERTEX: SYSTEM IMPLEMENTATION

After reviewing and refining all design concepts and establishing a final, agreed-upon design, it's time to begin implementing. The bulk of activities have shifted now from design activities to development activities. The team writes code, creates schematics and layouts, turns CAD models into full drawings with assembly details, commissions graphics, writes instructions such as user manuals and servicing and troubleshooting guides, and so on. Depending on the product and industry, the goal at this phase may be to develop working prototypes which could then be used to revisit and clarify user needs, optimize user experience, conduct performance and efficacy testing, or engage in market analysis. It bears repeating: Development is iterative! Engaging in this phase may be preparation for looping back to the top left of the "vee".

It is also the phase in which we look earnestly at the other side of the "vee". Activities on the right side trace back to activities on the left. We will soon be answering questions about our design – does it work as expected? Does it meet the requirements? In the course of answering those questions, DLC activities proceed through a battery of formal tests, generating evidence that the project output matches the design, and the design matches the architecture. This phase is gated by a *design output review*, which ensures that the prototype or pilot product is in fact being built as designed.

At this point, the team takes the prototype or built product and either moves back up the left side of the "vee" for further design iteration, or proceeds to the right side of the "vee" toward manufacturing and market release. In some projects, these activities may occur in tandem.

5.12 RIGHT SIDE: REGULATORY SUBMISSIONS

The timing and activities required for regulatory submissions vary greatly by industry, region, and product classification. Depending on what the development organization plans to bring to market and where they plan to market it, they may have been in touch with regulatory agencies as early as feasibility investigations. Most regulators offer support in the form of pre-submission guidance and even consultation to help product developers understand what they will need to do and to demonstrate in order to ensure their product will be approved for market. However, regardless of the level of "soft" or informal engagement, many regulations require companies to engage in a multi-round formal approval application process, which includes review and approval of test plans and protocols for the product being developed. For most controlled development projects, the team must have initiated regulatory submissions and received approval for plans and protocols before the project can move into the testing phase.

5.13 RIGHT SIDE: TEST

Although this phase is far from the only one in which testing is performed, it is the only phase in which *only testing* is performed. In this phase, Test Engineering designs and conducts:

- *Engineering characterization* or *dry run testing*, consisting of
  - *Subsystem testing*, which confirms the product's technical functions perform as expected based on high-level inputs, including across interfaces
  - *Software module testing*, which together with subsystem testing confirms that the design is complete and hasn't overlooked any intended functionality
  - *Integration testing*, which looks at how well the subsystems and their components work together, confirming that the architecture is complete and considering all component and subsystem interoperability
- *Verification testing*, which tests the system against the design requirements using pre-established protocols. Verification testing protocols are generated through requirements analysis, and are often driven by standards. The goal of verification is often described as confirming that "we made the thing correctly".

As previously stated, realizing efficient and successful testing is best assured when Test Engineering is involved in product design from early on in the DLC, particularly when collating user needs, drafting requirements, and evaluating design concepts.

5.14 RIGHT SIDE: TRANSFER TO MANUFACTURING

In this phase, Manufacturing Engineering examines the product and helps optimize its design for mass production. As previously stated, realizing efficient and successful testing is best assured when Manufacturing Engineering is involved in product design from early on in the DLC, particularly when collating user needs, drafting requirements, and evaluating design concepts.

As part of this phase, it's common for the manufacturing team to engage in low-volume pilot builds of the product. Depending on the industry, this could mean building anywhere from a few dozen to a few thousand units. During each pilot run, the manufacturing team will work through and propose improvements to assembly procedures, the design and functionality of assembly stations and manufacturing test stations (both manual and automated), and the design and functionality of the production line as a whole. Much of what they learn will be used to optimize the production process itself, but some discoveries will have implications for product design. Many product development projects, especially those that plan to engage a contract manufacturer rather than manufacture in-house, wait until this phase before showing the design to a manufacturing engineer. By this point in the DLC, it is often too late to optimize product designs for mass manufacture: Design teams are not interested in further iteration and may even be already reassigned to other projects. As a result, the company developing the product ends up leaving money on the table and/or accepting a slower manufacturing pace each time their design is built.

5.15 RIGHT SIDE: VALIDATION TESTING

Testing in this phase is focused on user needs, rather than on design and architecture verification. Generally, this kind of testing can only be performed once prototypes have been built, and in some cases may not be performed until pilot builds are available from the factory.

Types of *validation* testing include:

- *Usability* or *human factors* testing, which looks at user experience with the product in two distinct types of testing:
  - *Formative* testing, which is a learning exercise used to inform product design and is an important design input, required in some industries, recommended for all others;
  - *Summative* testing, which seeks to demonstrate the success of the design. Summative testing is totally hands-off on the part of the design team, a purely observational activity. Summative testing asks, "Can the user succeed in using the product in the absence of any instructions apart from what's 'in the box'?"
- In highly controlled industries, the product's safety and efficacy must also be validated. Does it benefit the end user? Is it safe to use? In medical device development, the answers to these questions must be definitely demonstrated, first through *preclinical testing* and then *human trials.*

The goal of validation is often described as confirming that "we made the correct thing". The activities in this phase validate that the product does what we meant it to do, without introducing new complications or unacceptable risks for the user. However, it's worth noting that although formative testing is a form of validation testing, (1) it may take place as part of a development iteration and therefore happen much "earlier" in the DLC than its introduction in this phase might suggest, and (2) unlike later validation tests, which are meant to prove that your product works, formative testing is often most valuable when it fails. Formative tests are how the design team learns what its design should be; its goal is not to make the design team look good or to rubber-stamp a fixed design so development can proceed, but to drive improvements to the product design.

5.16 RIGHT SIDE: POSTMARKET

This phase includes a host of support activities that are ongoing for the supported life of the product, which can vary widely between industries. Some regulated industries make distinctions between product *end of sale*, *end of support* (EOS), and *end of life* (EOL), with manufacturers required to engage in specific types of support for each status.

Postmarket activities generally fall into one of the following four categories:

- **Sustaining Engineering**
  Activities under this heading include providing remote and/or on-site product support and troubleshooting; conducting returned-product analysis; and assessing timing for transitioning a product design to EOS or declaring an individual product or product line to have reached EOL.
- **Surveillance**
  Activities under this heading include monitoring for complaints and field failures of the product that may not be reported to the manufacturer directly, such as by conducting user

surveys, monitoring product reviews by professional agencies or individuals, monitoring news stories and social media, and participation in industry-specific information sharing organizations. It also includes looking upstream at events and trends impacting the product's supply chain, such as material shortages, transportation interruptions, shifts in the regulatory or geopolitical landscape, and business practices of individual companies. In critical infrastructure industries, as well as an increasing number of others, surveillance also means monitoring products in the field and their material and digital supply chains for cybersecurity events, vulnerabilities, and other concerns.

- **Disposal**
  As products reach end of life or obsolescence, Sustaining Engineering is responsible for designating protocols for the safe and proper disposal of returned products, as well as unwanted, unsupported, or non-working products in the field.
- **Corrective and Preventive Action**
  Possible systemic problems relating to the product should be entered into a corrective and preventive action (CAPA) quality management process. This an investigative process which seeks to formally determine (1) whether an issue pertaining to the product is systemic, (2) assess its severity and breadth of impact, (3) determine the root cause of the problem, (4) make a recommendation for correcting the root cause of the problem or, for low-severity problems or situations where the solution would itself cause worse problems than allowing the original problem to persist, make a case for why the problem should not be corrected, (5) undertake any authorized corrections, and (6) confirm that the problem has indeed been corrected. Note that while CAPA is often formally grouped under the postmarket phase, systemic problems may be discovered in any phase of the DLC and should be entered into a CAPA process at that time.

5.17 SYSTEMS ENGINEERING AND THE "VEE"

Systems engineers (SEs) have four modes of engagement during the DLC:

- Participant
- Supervisor
- Owner (lead participant, making sure all activities are executed, reviewing output and reports)
- Leader (setting project tone, advancing the state of the art, modeling best practices)

Traditionally, SEs are heavily involved in early and late stages of the "vee". They are leaders and owners of planning and design activities, phase reviews, and verification and validation, but minimally involved as participants and supervisors during implementation. During planning and design, SEs need to have a basic understanding of the market goals and technical concepts for the product. They are critical for assessing technical feasibility and performing market analysis of predicate and competitive products. They own the collection and understanding of all stakeholder "voices" statements, developing use specifications, and the creation and review of all requirements. They guide the process of design concept review and selection. Then the leadership role is handed off to design engineers and technical leads. SEs remain involved, providing oversight and support to ensure designs meet the overall architecture and requirements.

During implementation, the SE role is even more lightweight, helping project managers maintain the schedule and project scope, reining in enthusiasm for new feature ideas, and making sure outputs are receiving timely reviews and meeting expectations. This is about keeping the project under control: Deviating from the plan mid-project, even in apparently minor ways, frequently leads to unforeseen consequences that result in missed requirements, safety or compliance concerns, and cost and schedule bloat.

Systems Engineering involvement ramps up again during testing. SEs intimate familiarity with use specifications, requirements, architecture, and design puts them in the ideal position to support Test Engineering with test planning, test protocol design, and review of test reports. In smaller organizations or teams, SEs may also need to act as stand-ins for the statistical analysis team a larger organization may have to support testing.

Lastly, the SE's role in postmarket activities is minimal. Depending on the organization, SEs may be asked to help triage incoming surveillance data and work with returned-product analysis and sustaining engineering teams to help postmarket teams understand field challenges, failures, complaints, and other incidents. SEs may be called upon to perform preliminary root cause analyses and make recommendations for escalation of issues into a formal CAPA process.

Throughout the DLC, the SE's primary function is to bridge across siloed or isolated design and development teams to maintain visibility and control of the whole picture – the 10,000-foot view as well as the critical detail. SEs should rarely or never be responsible for the day-to-day execution of development, testing, manufacturing, or other implementation engineering tasks; in fact, in larger and more traditional organizations, the SE role is more of a product development technical lead, someone who understands how to build and interface complex systems is a subject matter expert in a couple of technical functions (i.e., firmware and electrical design), and a veteran of multiple successful lifecycle product builds. In such organizations, SEs may be called away from a project entirely and retasked during implementation. By contrast, in smaller and more cutting-edge organizations, SEs may be expected to be knowledgeable enough to perform non-technical cross-function roles in areas such as industry and technical standards, regulatory and compliance assurance, quality management, risk management, and process management. This enables today's more agile SEs to maintain focus on and harmonize the interests of all stakeholders and keep the product on-target throughout its entire lifecycle.

NOTE

1 Source: https://www.sebokwiki.org/wiki/System_Life_Cycle_Process_Models:_Vee

6 Economics of Design and Test

Anthony P. Ambler

6.1 INTRODUCTION

Designing a product has enormous implications for the commercial, or military, viability of a product – over and above does it provide the functionality that the customer wants; there are economic implications associated with not just the design (e.g., the architecture of implementation, including hardware vs. software, sources and availability of the necessary materials and components), manufacturability, production testability, and the subsequent impact on field testability and maintenance.

The testing of a product *does* cost $'s to perform, but the potential cost of not testing can be enormous! It was often the case that a design team would select a simple, lower-cost test process with a lower fault coverage than a more complex, costlier one with a much higher fault coverage in the mistaken belief that *overall* design and manufacturing costs would be lower. However, the answer is still not that straightforward.

In many educational establishments, there has been a lack of appreciation for the need to have an understanding of the management-related issues with electronic design, whether it be a chip, board, or system. This is unfortunate as there are many approaches to providing a given technical solution to a specification, but which can lead to a large variation in the costs of design, manufacture, test, and maintenance. Additionally, there is the follow-on impact with warranty costs and product liability!

There has been little in the way of publications regarding analysis of design and test costs, but Di Giacomo (1989) went some way to show trends that could be used based on empirical data, comparing design productivity against complexity. Of course, a seriously complex design, spread out among multiple design teams, would show the benefit of 'divide and conquer'.

One of the earliest publications related to *test* economics was written by Brendan Davis (1994) in 1982 (now in its second edition) which had an emphasis on decision-making with the purchase of automatic test equipment (ATE). At about the same time there was beginning a wider discussion of the mechanisms for testing integrated circuits which were becoming more complex – scan path testing was being promoted, as was built-in self-test (BIST). Unfortunately, there was much argument in both academia and industry in any debate on the economics of their use, i.e., was there an increase in cost by using these new test methods?

Most early conversations on design for test techniques were oriented toward fault coverage and test time, usually pointing to the impact on implementation with the % extra circuitry required and the resultant yield degradation. Such was the production yield at that time, that this was a serious consideration.

There was little, if any, serious economic analysis of the cost and impact of testing either at chip-level, board, or system test beyond the 'rule of tens' (Figure 6.1).

The ratios state that if it costs $1 to detect and correct as fault at chip-level test, it would cost $10 to detect and correct the same fault if delayed to board-level test, etc.

This comparison will be out of date, but it does serve to show that the earlier a fault is detected, the less expensive it will be for the operation to correct. Modern system design introduces a level of complexity to extraordinary levels, and their applications (international communications/trading, safety-critical operations in aviation/nuclear reactors) and worldwide distribution creates an economic imperative to require reliability and ease of repair and diagnosis previously unheard of.

DOI: 10.1201/9781003188377-7

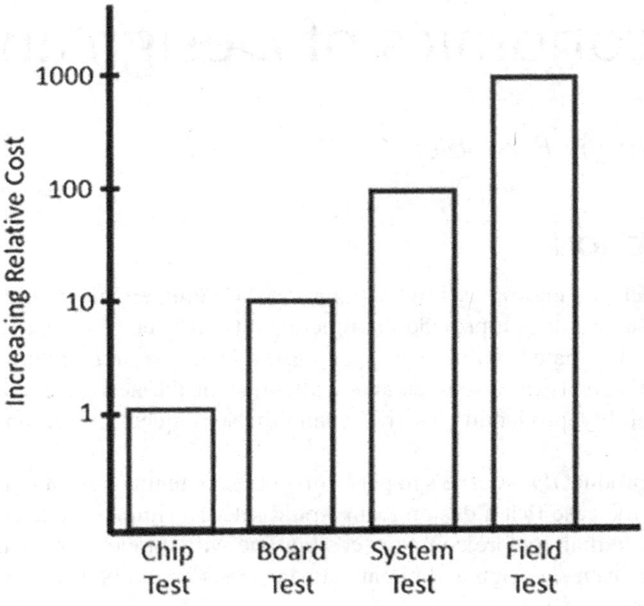

FIGURE 6.1 The rule of tens.

Further analysis of the economics of test needs to look at the total impact on design, manufacture lifecycle considerations.

One of the first works on a detailed look at design for test costs for chips, Varma et al. (1984), clearly showed that techniques, previously thought more expensive through the use of extra circuitry, yielding higher fault coverage *could* be less expensive than less invasive methods. Dislis et al. (1995) went some way to go into much more detail with the broader economics of both design and test of hardware building on the work of both Varma and Di Giacomo. Boehm (2000) looked at software cost estimation.

6.2 COST OF DESIGN AND TEST

Ultimately there is the aim to produce a product that performs the required function to specification and is manufactured to an acceptable quality. There will be a range of approaches that can be used to achieve these aims – the selected one(s) may be based on designer/corporate experience/bias, but not necessarily the optimum approach in the given application.

Then there is the conundrum of designing the cheapest or the most economic (Agrawal, 1994). And let's add other factors of possible warranty and legal costs.

6.2.1 The Cost of Test

Testing takes time, making a design more efficient to test also takes time, and can add to circuit complexity, affecting manufacturability – all of which can be measured, and compared in financial terms.

The cost of test is determined by a wide variety of factors. Design costs, DFT and test time, budget, market forces, quality/reliability, target throughput, training, packaging technology, circuit layout, real estate cost, pin out (circuit accessibility), CAD tools, designer's skills, computing equipment, capital investment, subsequent performance degradation. These can all affect the optimum selection of test strategy to be employed in any given design...and not necessarily the best choice for the given application unless sufficient investment has been given in obtaining the necessary level

TABLE 6.1
Implications for Repair Strategy

	Field	Maintenance Depot	Manufacturer Site
Equipment	Sparse quality	Medium quality	High quality
Personnel skills	Less educated	Medium educated	High educated
Test strategy	Identify, replace	Replace, repair	Replace, repair, diagnosis

of expertise and design tools. The conundrum is widespread – for example, employ highly skilled operators or purchase expensive diagnostic equipment and lower-skilled operators. The issue is wider than merely a design choice, Table 6.1 (Ambler & Schumacher, 1999; Reidel et al., 1998; 1999; Williams & Ambler, 2002).

Additionally, the investment already in place in production for automatic test equipment and operator expertise can severely impact the overall effectiveness of any test strategy chosen for the design: Purchase cost, depreciation and maintenance of equipment, operator and programming cost, setup time, test program length (test time), labor rates, program fault cover, program preparation time, test and diagnosis time, fixture cost, manufacturing yield, production volume, etc.

Then the impact on field service cannot be underestimated: Repair depot cost, logistic support, field repair cost, downtime cost, curtailing repeat visits, design for maintainability.

This is a very disparate set of parameters (and by no means exhaustive) but can be compared to each other in terms of the cost ($) impact. Dislis et al. (1995) goes into great detail of the range of parameters that can affect overall cost effectiveness; several of the references at the end of the chapter offer a further glimpse into factors impacting costs.

And do not neglect software test (Boehm, 1989; Williams & Ambler 2002)!

6.2.1.1 The Need for Test

In any product that is human designed, produced (using equipment designed by humans) and, ultimately, tested (by humans generating the tests, and using equipment used by humans!) there is plenty of scope for error in the final product. Fortunately, the quality of materials and equipment has dramatically improved so that errors and subsequent failures are greatly reduced compared to what has been the case.

Many will remember the desktop pc regularly crashing and requiring a reboot; constructing a prototype to discover that one of the memory chips had an error (a manufacturer at that time said that they expect only ~90% of those chips to be fault free). The requirement for companies designing electronic products to institute incoming goods inspection, i.e., in-house testing of devices before incorporating them in their products, was very real, time-consuming, and expensive.

Fortunately, product quality at all levels has improved dramatically, so much that incoming goods inspection is often no longer required. But errors still occur!

The way forward is not obvious and will change with each product and each application. Simplistically, some parameters will trend the cost curve downward with the increasing implementation of test-related design structures, while others will cause the curve to increase. For example, the relative weights of these parameters plotted against production volume can produce a 'U'-shaped curve leading the designer to have to determine the optimal minimum cost implementation (Figure 6.2).

Figure 6.3 shows that the overall costs of a product using either no design for test, or *with* design for test can be very different when plotted against production volume dependent on complexity/size of design (Varma et al., 1984).

How to evaluate test costs must depend on where the supplier/designer is in the supply chain! A chip vendor might only need to consider the production costs of the chip, a pc vendor may only

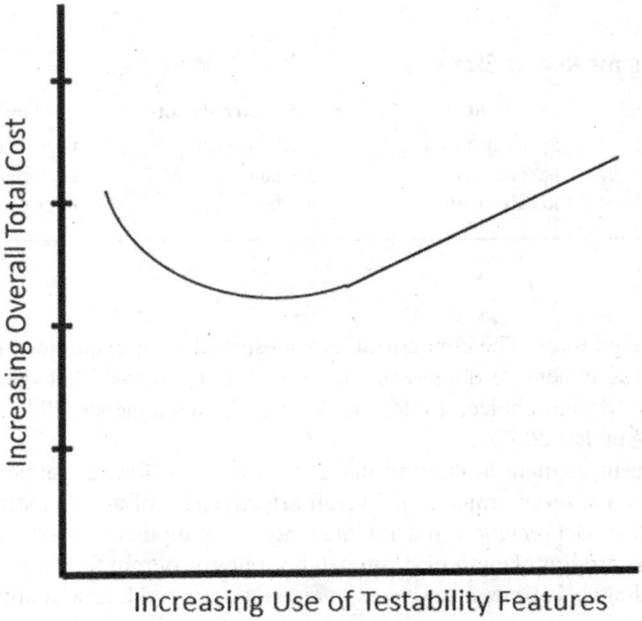

FIGURE 6.2 Problem to identify optimum cost-effective use of testability.

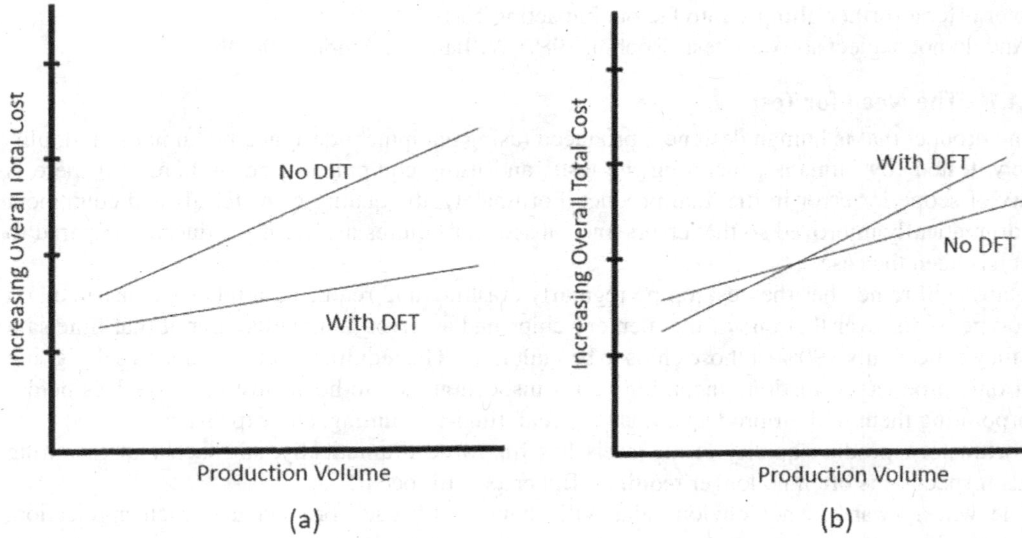

FIGURE 6.3 (a) and (b): Differing cost effectiveness of testability dependent on product type.

need to consider production costs and a short-term reliability/warranty, and a telecommunications vendor may need to consider both the reliability of a cell phone and the reliability of the complete telecommunication infrastructure, be it cellphone towers, servers, and, perhaps, satellites. A military application might either need to focus on reliability for long-term use, or not worry about reliability for a 'use once' product.

6.2.1.2 Life-Cycle Costs

When looking at life-cycle costs, the picture becomes more distinct. The life-cycle costs outweigh by a significant margin those of product design, manufacture, and purchase. It is often stated that the life-time cost of ownership has little to do with the purchase cost: 10% design, 30% manufacture, and 60% maintenance. Some sources show maintenance costs closer to 70% (Dear et al., 1991), while others show it to be even bigger. Clearly, this is dependent on application and scale of distribution. Another comparative picture of this is likened to an iceberg – the initial view of purchase cost is what is seen above water, but the associated mass of costs is submerged below the surface!

For example, improved testability has a dramatic impact on complex system life-cycle costs – one significant issue affected is the reduction of human factors in fault diagnosis which will lead to (Rosh Intelligent Systems):

- Fewer repeat calls
- Reduced training
- Faster repair time
- Fewer on-site specialists
- Fewer good components swapped (poor diagnosis by badly equipped or unskilled staff)
- Reliability degradation
- Reduced pipeline inventory

Improvements in testability will lead to a higher level of availability of these expensive complex systems.

Designers need to be aware of the downstream availability of repair and maintenance quality and take appropriate action (Blanchard et al., 1995; Brown, 1991; Pecht, 1995; Riedel et al., 1998) – they should be made aware of this and the implications for product success. To the above list, we can add (but not limited to) (Varma et al., 1984):

- Labor costs/number of staff
- ATE/field service equipment cost
- Required stock level per replaceable unit
- Purchase cost of replaceable units
- Reliability of replaceable unit
- How many times a replaceable has been repaired?
- Diagnosis time
- Repair time
- Down time

There are applications where the cost of test is not worthy of discussion – the usage *requires* that everything is functional at all times unless there is a serious failure due to unforeseen circumstances, e.g. power surge, structural failure. In such cases, a more liberal application of test-related design is required to be considered at the outset of the design process. Indeed, in safety-critical applications a form of exhaustive test (despite a high cost) may be required and justified both for production test and for online maintenance check – but still be able to ensure that the most cost-effective way of achieving this result is implemented?

6.3 CONCLUSIONS

An understanding of the economics of design and test is crucial for effective management of any project, military or commercial. The costs of current complex systems are astronomical (sometimes literally) at all stages of design, development, and in-service use.

Several of the references show that modeling of costs at different levels is possible and can be very effective especially if made specifically for the given project. Such modeling will require copious amounts of data gathering which will be time-consuming, but the end result would be potentially extraordinarily beneficial. It will be said that in given applications safety-critical concerns *demand* a high level of redundancy and testability, but as with all products, there will be cost-effective ways to achieve the same end result that might otherwise be chosen.

So, it has been shown how a detailed understanding of the overall cost of design, production, and ownership can have enormous impact on costs ($). Also, the use of a financial parameter ($'s) can be used effectively to compare the impact of a very wide range of disparate values. Some of the values will, ultimately, have little impact in the overall scheme of things, but others will have a significant impact. Unfortunately, many parts of the overall lifecycle are still not fully understood in sufficient detail.

There must also be an understanding that $'s is not *always* going to be the final arbiter in any product design – if the product is, for example, aiming to be the first one in the marketplace and time-to-market is *the* primary concern so as to corner the market, then a higher level of investment can be justified. Later versions can be analyzed to make them more efficient/cost-effective.

What has to be consistently true is that all engineers (not just 'management') need to be much more aware of the cost implications of their *engineering* choices in order to be *effective* engineers – they need to understand the 'big picture' of how their decisions will greatly impinge upon costs much further down the supply chain.

Engineering is the application of technology which, as we have seen, is much more than a purely technical debate. We have not considered the management of design teams and interdisciplinary groups etc. Therefore it is evident that engineering education requires exposure to management fundamentals in order to be truly effective (Dorf, 1998).

This topic is complex, covering a multitude of aspects that would take some time to cover in sufficient detail – the totality of product design, test, manufacture, field service, and maintenance requires detailed knowledge of each area. However, this chapter points to topic areas and references for further information – what IS clear is that an understanding of the impact of 'local' decisions on costs at each stage, and the implications for knock-on costs in all subsequent stages are essential. A comprehensive view is indispensable to understanding the overall cost effectiveness of any product – equally, a cost analysis of one product does not mean it will carry over to another (albeit similar) product. There is a danger of creating a generalized cost model as it will likely lead to inefficient decisions being made at all levels. Inevitably there will be requirements where an emphasis on cost is less of an issue, for example, safety-critical systems (Ambler et al., 1998).

Remember, we will still need to 'engineer' a solution that is 'good enough'!

Finally, let me quote Vishwani Agrawal (1994) –

> The economic laws of mass production and diminishing returns apply to all production environments. The basic ideas behind economic theories is to allow improved usage of the available resources for the good of society. In the design of a product the overall benefit/cost ration for design, test and manufacturing should be maximized. One must select the *most economic* design over the *cheapest design*. Every engineer should study economics because it teaches how to live and work intelligently.

REFERENCES

Agrawal, V. D. (1994). A tale of two designs: The cheapest and the most economic. In M. Abadir, & T. Ambler (eds.), *Economics of Electronic Design, Manufacture and Test*. Boston: Kluwer Academic Publishers.

Ambler, A. P., Zimmer, R., Hunter, A., & Dear, I. D. (1998). Design of safety-critical ASICS. In *IEEE Autotestcon*.

Ambler, T., & Schumacher, C. (1999). Cost benefits for the application of COTS ATE. In *1999 IEEE AUTOTESTCON Proceedings (Cat. No. 99CH36323)*. IEEE.

Blanchard, B. S., Verma, D., & Peterson, E. L. (1995). *Maintainability: A Key to Effective Serviceability and Maintenance Management.* New York: Wiley Interscience.

Boehm, B. (ed.) (1989). *Software Risk Management.* Washington, DC: IEEE Computer Society.

Boehm, B. W. (ed.) (2000). *Software Cost Estimation with COCOMO II.* Upper Saddle River, NJ: Prentice Hall PTR.

Brown, S. (ed.) (1991). *The Product Liability Handbook – Prevention, Risk, Consequence and Forensics of Product Failure.* New York: Van Nostrand Rainhold, 1991.

Davis, B. (1994). *The Economics of Automatic Testing*, 2nd Edition. Maidenhead: McGraw-Hill.

Dear, I. D., Dislis, C., Ambler, A. P., & Dick, J. (1991). Economic effects in design and test. *IEEE Design and Test of Computers*, 8(4), 64–77.

DiGiacomo, J. (ed.) (1989). *VLSI Handbook.* New York: McGraw-Hill.

Dislis, C., Dick, J. H., Dear, I. D., & Ambler, A. P. (1995). *Test Economics and Design for Testability.* Hemel Hempstead: Ellis Horwood.

Dorf, R. C. (ed.) (1998). *The Technology Management Handbook.* Boca Raton, CA: CRC Press.

Pecht, M. (ed.) (1995). *Product Reliability, Maintainability and Supportability Handbook.* Boca Raton, CA: CRC Press.

Reidel, T., Kaminski, J., Wahl, M., & Ambler, T. (1999). Effective testability design for the product life-cycle. In *IEEE Autotestcon.*

Riedel, T., Tiemann, N., Wahl, M. G., & Ambler, T. (1998). LCCA – Life cycle cost analysis. In *IEEE Autotestcon.*

Varma, P., Ambler, A. P., & Baker, K. (1984). An analysis of the economics of self-test. In *IEEE International Test Conference.*

Williams, D., & Ambler, A. P. (2002). System manufacturing test cost model. In *IEEE International Test Conference.*

7 Cyber-Physical Systems

Wm. Arthur Conklin

7.1 INTRODUCTION: WHAT ARE CYBER-PHYSICAL SYSTEMS?

Cyber-physical systems (CPS) are systems that integrate physical components with computer-based algorithms and systems, enabling sensing and actuating elements of the physical world using networked, computing systems. A wide variety of systems can be classified as CPS, from transportation systems, energy systems, medical devices to industrial control systems and much more. The objective of employing CPS is to improve the efficiency, safety, and reliability a system by enabling them to sense, reason, react, and adapt in real time. This allows a CPS solution to have greater utility than systems without integrated computing power.

CPS have both a physical implementation and a cyber implementation. The physical system is a set of hardware designed to do some specific set of tasks. The cyber portion is software that can monitor and control the physical system. The two sets of systems are purposefully designed to provide the value of the overall system. CPS are engineered for a specific task or purpose, and this defines much of how they are used and employed. CPS can come in a wide range of size and scale, from small singular devices such as a computerized implanted medical device, to roadway sensors reporting traffic issues to central systems, to even systems covering a significant portion of North America, such as the bulk power grid. The electric grid that serves North America has been described as one of the largest man-made systems. It is clearly a CPS comprised of multiple types of physical components, monitored, and controlled by multiple layers of computer controls to efficiently and effectively manage the flow of electrical power from production to points of usage.

An example of common places where CPS are employed include:

1. **Transportation:** CPS are used in intelligent transportation systems, such as connected and automated vehicles, traffic management systems, and smart highways. From simple signals to interconnected systems to alert motorists to road conditions ahead, the objective is to keep traffic moving.
2. **Energy:** CPS are used all across the smart grid, including advanced metering infrastructure (AMI), distribution management systems (DMS), and renewable energy systems.
3. **Manufacturing:** CPS are used in virtually all manufacturing plants, with increasing integration in efforts labeled Industry 4.0, including smart factories and the Internet of Things (IoT) for industrial control systems (ICS).
4. **Healthcare:** CPS are used in medical devices, such as implanted devices and diagnostic equipment, as well as in telemedicine and remote patient monitoring systems. Many bedside medical devices, such as infusion pumps are forms of CPS with direct connections to patients.
5. **Building automation:** CPS are used in smart buildings, including building management systems (BMS), lighting and HVAC control, and fire and life safety systems.
6. **Agriculture:** Modern precision agriculture, including soil monitoring, crop management, and livestock monitoring systems, would not be feasible without the CPS that drive them.

The list can go on and on, for CPS are used virtually everywhere, improving the efficiency of systems that make modern life affordable and sustainable.

DOI: 10.1201/9781003188377-8

There are many ways that one can characterize a CPS, by size, by industry, purpose, even criticality. One of the key aspects of examining CPS comes from the actual processes they are employed in as part of a solution. The specifics of the process, in terms of speed, criticality, accuracy, and economic aspects, are essential to the understanding of the context within which the CPS operates. The US Government has identified types of CPS employments that it considers to be critical infrastructure. The difference between these systems and other CPS is in the span of risk should things go awry. Critical infrastructure is simply a subset of CPS where the operations of the system can have implications related to national security, economic security, public health and safety, or the ability to maintain normalcy in daily societal life.

Examining the composition of a typical CPS system reveals two, or three sets of components. One is the physical hardware that is physically involved in the process. The second is the cyber element, and the third is a human element in the form of an operator. For the purpose of simplicity and focus, the remainder of the chapter will not address human elements or operator issues. Dividing the system into two components makes analysis of the system easy in separate domains, the physical and the cyber. There are a series of over-arching principles that can be applied to both sides. Techniques such as Consequence-driven Cyber-informed Engineering (CCE), Resilience, Safety, as well as OT (operational technology) cybersecurity principles can be employed in CPS to great effect. Standard IT-based cybersecurity solutions are designed for different requirements and tend to have limitations, yielding less than desired results when applied directly to CPS. This is not an indictment against well-known and respected IT security elements, but rather a result of the differing security objectives between IT and CPS systems. To understand the basis for these elements, it is important to understand how CPS are designed and built, with their specific capabilities and limitations. We will examine hardware issues first, then the cyber side of the system.

Hardware is the set of components that connect the sensors and actuators that are involved in the physical process. These can be pumps, valves, heaters, coolers, mechanical switches like those that switch railroad tracks, the list is endless. Engineers have been designing and building physical plants for centuries. There are multiple engineering disciplines dedicated to this endeavor, and in the last half century, automation has been added to improve the utility of the system. Adding the cyber side increases the effectiveness of the automation.

On the pure physical side, there is a significant effort to use good engineering practices to ensure resilience, reliability, safety, and consequence-driven designs. Resilience is the property for a system to return to a desired operation point after a disturbance from an outside force. The physical engineering of elements such as reaction vessels to hold sufficient mass to absorb additional flow without significantly disrupting the operating conditions has been practiced for eons. Having a large mixing tank, so new additions of material are inconsequential in terms of reaction can make a process more stable. This can increase resilience against a wide array of minor forces that might drive a system from its desired operational point. But these efforts come with penalties in the areas of safety and consequences-driven aspects. Larger reaction volumes can have potential safety issues. Imagine a pipeline that is 2,000 km long. If you didn't segment the pipeline into a set of smaller zones, then a problem anywhere would bring the entire pipeline contents into play. Yes, extra valves cost money, and the additional control elements increase complexity, but these are weighed against the consequences of failure. Engineering out the risk of small errors becoming large or catastrophic is one of the core principles of CCE. If you are injecting a hazardous material into a flow, where an overfeed situation could have catastrophic outcomes, then the system will be designed with pipe sizes that prohibit significant risks from ever happening. Adding extra measurement points to act as fail-safes if the primary measurement has a malfunction is another method of engineering reduced risk. In the early age of airplanes, they crashed, a lot. And then, over time, as causes of failure were identified and engineered out of the system, eliminating single points of failure, the overall system performance improved. Safety was improved. Over time the cumulative effect of engineered changes designed to reduce risk can have a significant effect. This system optimization and improvement methodology is a proven method of engineering out excess risk and has been a standard for risk reduction in hardware systems for a century or more.

The cyber side of the system is a relatively new edition into the mix of hardware. Adding a computer element to a system to allow for fully automated reactions to changes in the system. Automatic programs designed to make a system more responsive have had a tremendous impact in terms of capability and resilience of CPS. Examples of CPS like implanted medical devices that measure some critical values and then automatically react to conditions have enabled engineers to create things such as advanced pacemakers and embedded insulin pumps. But to view the cyber component as simply software is disingenuous. There is a new hardware component, the computer that runs the software. Unlike the general-purpose machines that are used in business, the computers used in most CPS applications are dedicated devices, purpose built to run a single application, and are extremely limited as general compute devices. These purpose-built computers are designed to run only a single function, and that is to act as the controller for the hardware they control. These microcontrollers can have all the functions of a computer, on a single chip, enabling significant computing capability in small and energy-efficient packages. These processors can run a variety of simple operating systems and support network connections. The smart meter on the side of a house can manage the measurement of electricity used, but it also has to manage a set of cryptography keys, as well as various network connections. For many microcontrollers in CPS, the timeliness of responses is of critical importance, so a real-time version of the operating system is deployed. This allows the designers the ability to manage how fast messages are received, processed, and sent back. Many of the processes that are being managed have time-sensitive response windows, and the systems are purposely designed with this in mind. There are no extraneous functions that one might find on a standard computer. The software side of the system is the code used to execute specific functions in response to specific inputs.

CPS can range from small embedded systems to extremely large-scale highly networked systems, and everything in between. The characteristics that make it a CPS is in the nature of its purpose. When engineered as part of a physical process, then it becomes a CPS. Small systems have smaller processors and smaller resource pools. Larger systems which have elements such as programmable logic controllers, can have computing power and resources comparable to small desktops to large server scale in terms of processing power. What is important to remember is that these compute devices, while technically computers are not at all like the PC's or servers, in that they have a dedicated OS stack and operating program to perform the mission task of the system. When discussing CPS it is important to keep the system function in context, for this is why the CPS exists, as they are purposely designed to function as part of the system and not as a general-purpose computing device. Even when composed of stock items such as PLCs, there is task-specific programming to make the CPS part of the functioning system as a whole. It is best to look at these systems as engineered solutions to physical process problems.

CPS gets frequently conflated with terms such as the Internet of Things, and other marketing terms such as Industry 4.0. While these terms represent concepts that may have CPS elements as part of them, the nature of CPS is more foundational and directly coupled to a specific process than the general terms above. CPS is the conjoining of engineered systems with cyber and physical components with the purpose of a specific problem. It is also important not to confuse the term cybersecurity and CPS. Surely CPS systems have cybersecurity concerns, but as their operational purpose and context differ from traditional IT systems, the risk model involved is different as well.

7.2 CYBERSECURITY FOR CPS

Cybersecurity is the task of managing the risk associated with a system. For CPS, this is typically focused on the cyber aspect of a CPS. Risk is defined as the cost of undesired outcomes, but unlike traditional IT systems, the risk in CPS is manifested in the less-than-optimal performance of the overall system. The context of the environment and the purpose of the CPS define the risk environment for the system, and this defines the specific needs for cybersecurity elements. One simple definition of a secure system is one that performs as it is designed to and does nothing else. For IT systems, cybersecurity is described in terms of a security policy associated with protecting the data in the system. This leads to the familiar terms of confidentiality, integrity, and availability. In CPS, the security policy has an entirely different set of outcomes and stakeholders. CPS stakeholders are concerned with efficiency, resilience, and safety. Because of the tight integration of the cyber structure and hardware elements into a cohesive system, one cannot isolate either cyber or hardware from the consequences of operating outside of desired limits. The security policy for a CPS is structured around the operation of the system in an efficient, resilient, and safe manner. This has significant implications for the elements of a CPS cybersecurity policy and operating environment. The applied implications of these elements are discussed in Section 7.3.

7.3 SECURING CPS METHODOLOGY

Cybersecurity of CPS is organized around the objective of keeping the system operating at specified setpoints. This makes specific cybersecurity elements dependent upon the process of the system. Unique to each process will be the specific measurements that tell the control elements the necessary data to facilitate the proper operation of the system. For a system with a pressure vessel, with product input and outputs, temperature, pressure, and flow measurements to mention some of the possible controls. Each of these controls, including their redundant backup systems are essential elements in the proper operation of the system. Using these devices, a computer can change the heating input if the temperature reaches a set point and turn off the heat if it is too hot.

Typical security objectives in a CPS include elements to prevent the following attacker objectives:

1. **Loss:** Loss of View, or Control
2. **Denial**: Denial of View, Control, or Safety
3. **Manipulation:** Manipulation of View, Control, Sensors and instruments, or Safety

Standard IT security controls that are reactive, waiting for already deployed preventive controls to be compromised or to fail, are at a disadvantage in a CPS, as the damage from the above attacker objectives will already have occurred to the process. This reality requires that CPS protections be designed into the process to ensure the system has visibility, control and can block attempts to degrade these critical elements.

Some controls associated with CPS are more critical than others, and the level of cybersecurity attention is directly connected to this criticality. The CPS has a series of internal functions, where control elements interact with the system to create appropriate outputs for a given set of inputs. The control elements that are associated with the critical functioning of the system need to be protected by the CPS cybersecurity function. The integration of the cyber functionality into the overall system functionality makes the system's output uniquely vulnerable to cyber threats, such as hacking and malware, that could compromise the reliability and safety of the physical systems they control. As such, cybersecurity measures must be integrated into the design and implementation of CPS to ensure the secure operation of these systems. The process to determine what controls and where is remarkably familiar to cybersecurity professionals:

1. **Identify the Critical Assets and Critical System Functions:** Determine which systems and assets need protection and prioritize based on the importance and potential impact of a breach.
2. **Conduct Risk Assessment:** Analyze the potential threats and vulnerabilities of each critical system and assess the impact of a cyber failure on the overall system output.
3. **Implement Security Measures:** Based on the risk assessment, implement technical and organizational security measures, such as firewalls, encryption, access controls, and

incident response plans. Care must be taken not to interfere with process timing elements, as the cyber system is not independent, but rather an integrated part of an engineered system.

4. **Continuously Monitor and Update Security:** Regularly monitor and assess the security measures and update them to address newly identified threats and vulnerabilities. Monitor the system communication elements as CPS have limited sets of appropriate communications.
5. **Train Employees:** Educate employees on the importance of cybersecurity and their role in maintaining the security of CPS.
6. **Test and Validate Security:** Regularly conduct penetration testing and security audits to validate the effectiveness of the security measures.
7. **Develop an Incident Response Plan:** Have a plan in place for responding to and recovering from security breaches.

One of the unique differences between CPS and IT systems is in their pattern of communications. The Internet was designed to allow any endpoint to talk with any endpoint, using virtually any protocol. One of the challenges for enterprise security managers is deciphering all of the paths of communications to find ones that are not authorized. By contrast, a CPS has a very defined set of communication pathways. When the CPS is designed, specific communication pathways are used between endpoints doing either measurements or actions and the specific control devices. A system may have a series of programmable logic controllers (PLCs) to provide some of the automated controls. There may be a control station with a human-machine interface for the operator to be involved in decision-making. There may be an engineering workstation to facilitate changes to the system's programming. Each of these unique endpoints has a defined set of elements that they are allowed to talk to, and all other communications should be blocked and treated as a security problem. This makes the monitoring of the network and the analysis of who is talking to whom a critical aspect of cybersecurity for CPS. While there may be many others who have legitimate needs for access to system data, it needs to be managed from a replica reporting system called a historian, not by accessing the sources of data directly. Any communication other than the designed and required communications is a red flag of a system not under complete control.

7.4 DESIGNING A CYBERSECURITY PROGRAM FOR CPS

To create a comprehensive cybersecurity program for a CPS is an exercise in managing risk. The following five steps can be employed to cover the risk, although some of these steps have entire processes developed to support them.[1] These are presented in the order in which they should be approached.

1. **Develop an Incident Response (IR) Plan for the CPS:** Things will go wrong, and when they do, how do you handle the failure? Having an IR plan designed around the system and its functions, rather than a generic IT-based one is critical. Many failures can be interdicted and solved before they become critical, but only when the IR effort is integrated with the operational aspects of the system. These plans should be developed by conducting exercises with operators and engineers, examining what happens if various elements fail, and how those failures are manifested in the system. This makes IR design a critical engineering design element in system design.
2. **Create a Defensible Architecture:** System architectures should make visibility into critical functions easy. Systems should be network segmented into sections based on the need to communicate, not for overall network efficiencies. It is important to identify which logs would be needed to support IR and ensure they are readily available.

3. **Establish Appropriate Network Monitoring and Visibility:** Because of the limited sets of allowed communications, having monitoring for unauthorized communications can prevent many incidents. As the known set of allowed (good) communications is relatively small, finding things outside that list is doable. For instance, any communications other than the designed set to a safety system should be considered suspect and investigated.
4. **Secure Remote Accesses:** Remote access can be a designed element that is necessary for specific functions. But these accesses should be limited to just that function and secured with elements such as multifactor authentication and isolated via jump hosts or other isolation techniques. Note that VPNs are not isolating in this regard, they actually limit many security interactions.
5. **Implement Risk-Based Vulnerability Management.** Vulnerabilities will be discovered. They are part of any system. Managing responses using a risk-based approach is essential.

While cybersecurity is renowned for complicated controls and difficult processes, these stem from attempting to do things after the fact in an IT environment. In CPS, the cybersecurity aspects can be designed as part of the purpose-driven design. Using the five guiding elements listed above as high-level guidance for the design team can go a significant way toward reducing risk in CPS systems.

NOTE

1 Adapted from SANS Whitepaper, The Five ICS Cybersecurity Critical Controls, Robert M. Lee and Tim Conway, SANS Institute, October 2022.

8 Logistics and Supply Chain Management

Matthew Hu, Margaret Kidd, Zheyong Bian,
Kailai Wang, Harjot Singh Pooni, and Suhaib Kaissi

8.1 OVERVIEW OF THE COMPLEX INTEGRATED SYSTEMS

In a global economy, everything is interconnected. Economic integration processes in recent years have led to the establishment of global supply chains, from the extraction of raw materials to manufacturing and subsequently to final consumption. Some salient trends have shaped the flows of globalization since the early 2000s in terms of the annually increased international trade, technological advances in containerized shopping, the higher level of growth in trade of Pacific Asia economies, and the well-recognized importance of international cooperation (Rodrigue et al., 2013, Chapter 5). There is a need for an in-depth understanding of the factors that influence integrated cargo shipping systems from multiple scales.

In this chapter, we present robustness thinking for robust and reliable supply chain design and models for designing supply chains that are robust (i.e., perform well with respect to uncertainties in the data, such as demand) and reliable. Broadly defined, a supply chain is robust if it performs well with respect to uncertain future conditions—for example, costs or travel times; a supply chain is reliable if it performs well when parts of the system fail—for example when a distribution center becomes unavailable because of poor weather. The goal of robust design or robust optimization, in general, is to find solutions that perform well under every realization of the uncertain parameters, though not necessarily optimally in any. Models for robust supply chain design use as an underlying model a joint location–inventory model that recently appeared in the literature (Snyder 2003); this model, called the Location Model with Risk Pooling (LMRP), chooses distribution center (DC) locations, assignments of retailers to DCs, and inventory levels at the DCs to minimize the total cost. Their models do not require parameters to be time-stationary; demand means, and variances may change over time. If we make decisions now that hedge poorly as mistake(s) against the various scenarios, we must live with the consequences as parameters change. The main risks of supply chain operations are exposed to risks of different kinds. Demand-side, supply-side, and catastrophic risks are distinguished. Demand-side risks originate in disruptions emerging from downstream supply chain operations. They can manifest in the physical distribution of products to the end customer (e.g., transportation problems or improper functioning of the warehouses), or they can come from the mismatch between the forecasted and the actual demands or from the inappropriate supply chain coordination. The well-known bullwhip effect, i.e., the amplification of the demand volatility in the upstream direction of the supply chain is such a characteristic phenomenon. The possible negative consequences of demand-side risks are costly shortages, obsolescence, and inefficient capacity utilization.

Supply-side risks include suppliers' business risks, capacity problems, technological changes, product design changes, inappropriate quality of the supply, and poor logistics performance (late delivery). Catastrophic risks involve natural hazards (e.g., tsunamis, earthquakes, hurricanes, droughts, and floods), social-political instabilities, civil unrests, economic crises, and terrorist attacks. The main drivers of supply chain vulnerability are customer dependence, supplier

 DOI: 10.1201/9781003188377-9

dependence, supplier concentration (small number of suppliers, or as an extreme, single sourcing), and global sourcing (increased uncertainty, poorer transparency, and visibility).

Reliability improvement can be approached in two critical system reliability requirements: mistake avoidance and robustness. Mistake, in this case, is defined as the error due to design decision, manufacturing, and supply chain operations. Examples of mistakes in product development include missing components, installing a component backward or interpreting a software command as being expressed in inches when it's actually in centimeters or lack of visibility, insufficient inventory or neglecting data in supply chain. Product reliability can be improved by reducing the incidence of such mistakes through a combination of knowledge-based product engineering, manufacturing engineering, supply chain, and problem-solving processes, such as Six Sigma's define, measure, analyze, improve and control (DMAIC). Robustness is the ability of a system to function (that is, insensitive to the user's environment to avoid failure) under the full range of conditions that may be experienced in the global environment. This chapter focuses on the second challenge—robustness—by proactively factoring design for reliability (DFR) in supply chain efforts through robustness improvement in the design for Six Sigma (DFSS) approach. DFSS is a method that calls on many of the fundamental design tools, including robust design. By using DFSS along with a well-defined reliability plan, you can know when to use which tool and how to integrate each together to produce a reliable product in a system in the shortest time (Hu, August 2013a).

A cargo ship is any maritime vessel that transports raw materials and manufactured goods from one port to another. Depending on the kind of cargo they carry, there can be different types of cargo ships, and these types can be categorized accordingly. A tanker, for example, is designed for transporting liquid cargo, such as petroleum products. Cargo ships that travel on a fixed prescheduled route and have a tariff for the service are called liners. Maritime vessels that are chartered by various users for shipping and do not have a predetermined schedule or route are classified as tramp ships. Maritime transportation, the most efficient and cost-effective mode of transportation, provides a reliable, cost-effective, and dependable way to transport goods globally, facilitating commerce and encouraging economic growth among nations and peoples around the world (International Maritime Organization, 2021).

As the global business environment is ever changing and competitive, the maritime shipping industry has experienced rapid growth in recent decades (Notteboom, 2004). Such growth is driven by global trade, the widespread use of Information and Communication Technology (ICT), and changes in the world's geopolitical scenarios. Figure 8.1 (World and US Port Container Traffic) illustrates the changes in port container traffic and GDP per capita from 2001 to 2019 across the globe and in the US (World Bank Group, 2021). The data used in this study shows the trends before the global outbreak of the COVID-19 pandemic. Exploding global economic growth could be seen as the major contributing factor to the increase in cargo shipping volume. The international cargo shipping industry has mushroomed into the leading means of transportation for the inter-oceanic transportation of manufactured goods. According to Kaluza et al. (2010), the global movements of cargo ships in 2007 were analyzed based on detailed arrival and departure records of 16,363 cargo ships. The complex network was constructed using actual ship journeys between ports. These cargo ships have been classified as bulk dry carriers, container ships, and oil tankers, based on their physical characteristics, mobility patterns, and network structures. They concluded that compared to simulation results of the traditional gravity model, actual cargo ship networks are not only much more complex than those generated by the traditional gravity model but also more informative for policy making.

Supply chain design models have traditionally treated the world as if we knew everything about it with certainty. In reality, however, parameter estimates may be inaccurate due to poor forecasts, measurement errors, changing demand patterns, or other factors. Moreover, even if all of the parameters of the supply chain are known with certainty, the system may face disruptions from time to time, for example, due to inclement weather, labor actions, or sabotage. Ensuring reliability and

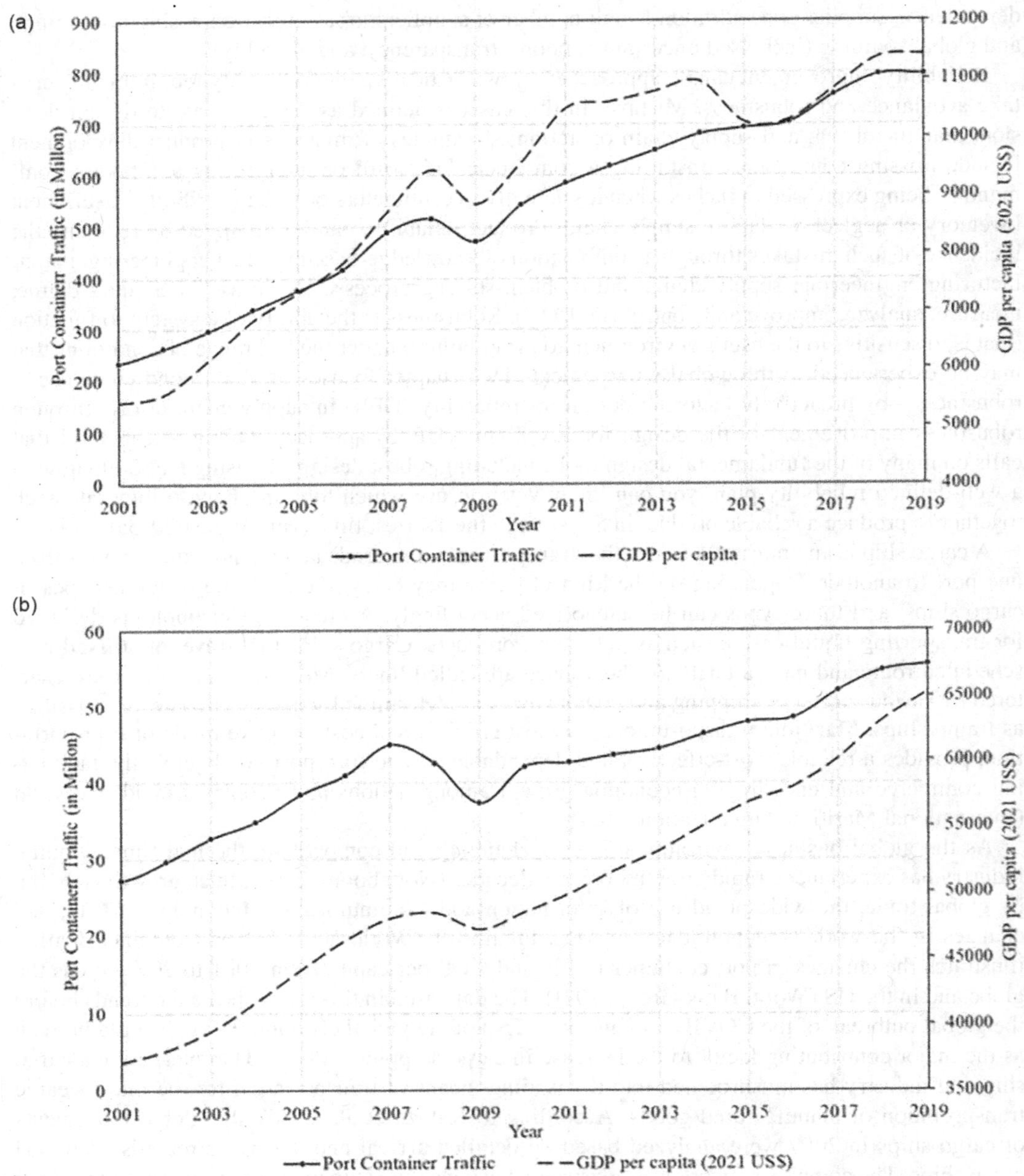

FIGURE 8.1 World and US Port Container Traffic.

robustness while considering uncertainties induced by various sources, such as transportation delay and manufacturing process variability, in individual factories and routes is a highly complex task.

The logistics and supply chain management are outlined for the cargo shipping system. Particularly, we will consider four aspects that may influence the supply chain process in terms of reliability, availability, testability, and maintainability. Building the complex integrated systems responds to a famous statement—"the whole is greater than the sum of the parts." Finally, we will discuss the importance of realizing complex integrated systems to the contemporary cargo shipping industry.

8.2 RELIABILITY

Reliability is an important component for achieving a successful supply chain management model. Shipping reliability is associated with industrial productivity and labor costs. It ensures that inventory can reach customers in a timely manner. The effectiveness of shipper-carrier collaboration has typically been evaluated based on on-time performance. In short, for the logistics of cargo ships, reliability refers to or measures the extent to which a supply chain can maintain consistent performance. Supply chain professionals place a high priority on increasing reliability, reducing inventory, and preparing for demand. Maintaining or improving transportation performance also requires strong shipper-carrier relationships. In transportation, carrier connectivity and visibility play critical roles in guaranteeing supply chain security. Good shipper-carrier relationships are also essential for ensuring transportation performance. In the twenty-first century, container shipping plays an integral role in the supply chain due to the ever-increasing competition.

Vernimmen et al. (2007) studied the causes of schedule unreliability in liner shipping. Though container shipping lines operate on fixed-day weekly schedules, Drewry Shipping Consultants reports that more than 40% of the ships deployed on worldwide liner services were delayed one or more days. Liner shipping is a mode of ocean transportation in which a shipping company operates its fleet of vessels for the carriage of goods between fixed ports regularly according to a publicly advertised schedule. Reduced schedule reliability has a significant impact on almost all rational agents across the entire supply chain including shipping lines, terminal operators, inland transportation operators, and shippers/consignees. A specific case study is provided to illustrate how schedule unreliability impacts the level of safety stock that a manufacturer with overseas spare parts should keep. The conclusion of this study is that an increase in schedule reliability can result in significant cost savings for the company under consideration.

Chun and Chiang (2011) re-examined the critical factors that are related to liner shipping schedule reliability. They conducted an expert questionnaire survey. As fuzziness and vagueness have existed for a long time in many decision processes, they adopted a fuzzy analytic hierarchy process (FAHP) approach. The findings include:

- Delays in delivery of container shipping lines had a knock-on effect on the hinterland supply chain, increasing logistics costs;
- The "well-arranged time window" was a crucial factor, providing helpful information for professionals in order to improve the reliability of container shipping lines;
- Carriers should plan their schedules with adequate buffer time in case of unexpected circumstances, such as bad weather or congestion in port. Shippers can also add more buffer time to their supply chains to prepare for unforeseen circumstances;
- Service quality will affect freight rate negotiations between shippers and carriers as well as the performance of container shipping lines.

Decò et al. (2012) provided a comprehensive examination of shipping reliability and redundancy. In this study, polar representations of reliability and redundancy are obtained for different encountered sea states, ship speeds, and headings, representing a broad spectrum of operational conditions. In international freight transport, survivability is a similar or more generalized concept to reliability. Empirical studies have given particular attention to damaged ships under different scenarios and external loading conditions (Bulian et al., 2016; Fang and Das, 2005).

For reliability improvement, it is most cost-effective and less time-consuming to make design or a supply chain system insensitive to uncontrollable user environments e.g., weather, distance, time, and others, in upfront design phase. Robustness development in supply chain design through design for reliability (DFR) process provides benefits in the improvement of reliability, availability, testability, and maintainability.

8.2.1 Essential of Robustness and Robust Design in Reliability Improvement

An important development in reliability engineering is robust parameter design pioneered by Genichi Taguchi (Taguchi Methods, 1993). For any design concept, there is a potentially large space of control factor settings that will nominally place the function at the desired target value. In robust parameter design, the engineer explores the design space seeking changes that will make the system more robust while still keeping the performance on target. Taguchi's method employs orthogonal arrays to explore the design space. At the same time, outer arrays or compounded noises are used to explore the range of possible operating conditions. Signal-to-noise ratio (S/N ratio) might be used as measures of the robustness of the system and guide the engineer to preferable levels of the control factors shown in Equation (8.1).

$$\text{S/N ratio} = \frac{\text{Energy (or power) that is transformed into intended output}}{\text{Energy (or power) that is transformed into intended output}}$$

$$\text{S/N ratio} = 10\log\frac{\bar{y}^2}{\sigma^2} \tag{8.1}$$

Compounded noise is a noise management strategy, when system uncertainty is more complicated, like in a highly dynamic system in which the behavior varies over time. Robust design methods have continued to be refined and are still an active area of systems engineering innovation. To improve reliability, improving robustness in system, parameter, and tolerance design is essential.

8.2.2 Robustness Model: P-Diagram

A P-diagram or a transfer function is a relationship between input (lower-level requirements) and output (higher-level requirements). Transfer function, including control factor (CF), noise factor (NF), input signal (M), and ideal function Y, is expressed in Equation (8.2).

$$Y = f(X_{IS}, X_{CF}, X_{NF}) = \beta M + \left[f(M, X_{CF}, X_{NF}) - \beta M\right] = \beta M \tag{8.2}$$

Y is the intended design functions cascaded from critical to satisfaction and others. ß is a slope of the ideal function. In general, a transfer function is established through an analytical or empirical approach. The process of developing or updating a transfer function is highly iterative, moving frequently between the inductive and deductive paths (Hu, 2013b). Transfer functions can be schematically represented by a P-diagram used in robust engineering design as shown in Figure 8.2: Supply Chain P-Diagram.

A product or a supply chain system can be divided into functionally oriented operating systems. Function is a keyword and basic need for describing your product or behavior. Transfer functions can enable engineers and scientists to introduce variation into the models to understand how the distribution of variation can alter the desired performance as shown in Figure 8.2: Supply Chain P-Diagram. There are three fundamental ways to improve the reliability of a product, manufacturing, and supply chain during the design stage (Hu, 2013b):

1. Reduce the sensitivity of the supply chain functions to the variation in the product parameters e.g., leveraging modern cloud applications for improved supply visibility and demand forecasting;
2. Reduce the rate of change of the supply chain and logistics parameters e.g., leveraging advanced analytics for a more proactive response.
3. Include redundancy e.g., dual sourcing or dual tooling, whereby you use two suppliers for a component, raw material, product, or service.)

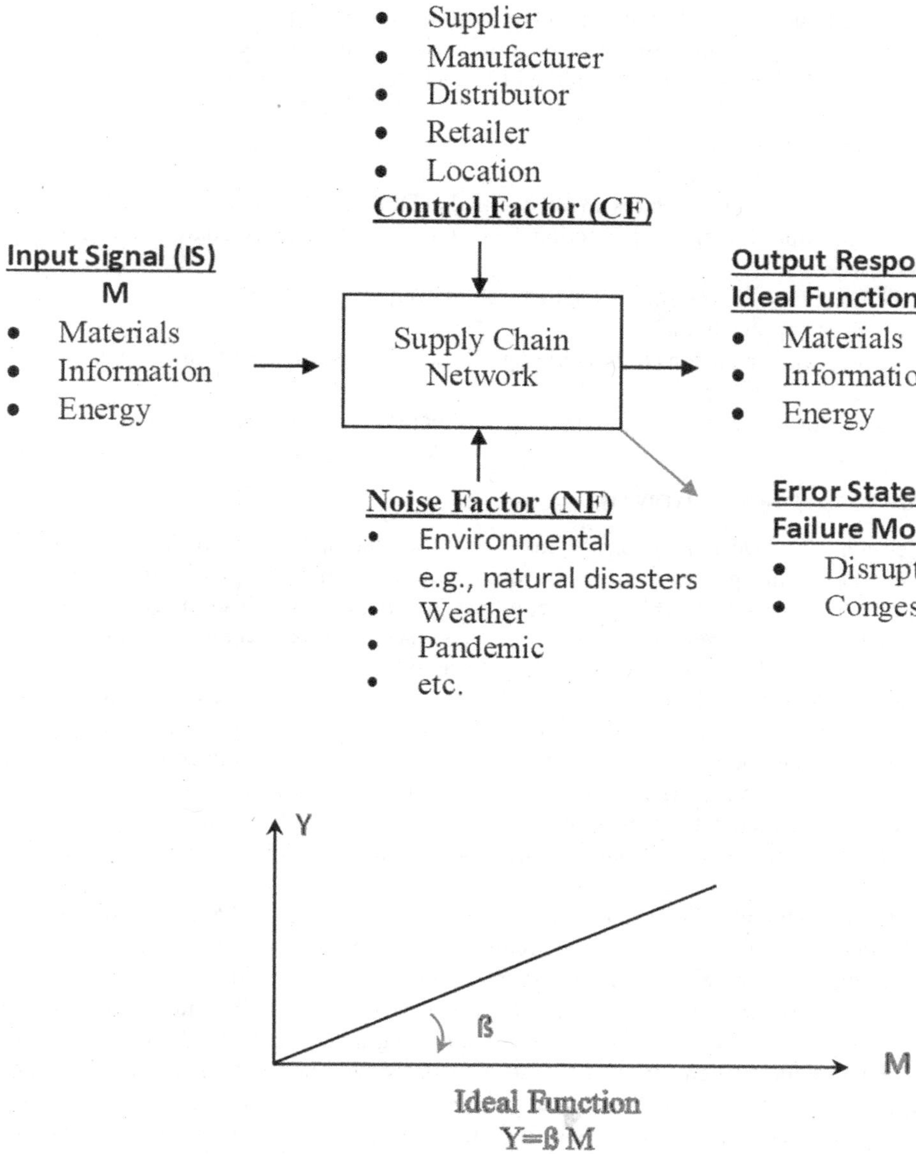

FIGURE 8.2 Supply Chain P-Diagram.

8.2.3 Design for Reliability Overview

Design for reliability (DFR) is a process. Specifically, DFR describes the entire set of tools that support product and process design (typically from early in the concept stage all the way through to product obsolescence) to ensure that customer expectations for reliability are fully met throughout the life of the product with low overall life cycle costs. The purpose of the DFR process is to provide requirements for DFR activities, which are intended to be an integral part of every product development effort to continuously improve product reliability and robustness along with earlier supply chain activities involvement in early product development phase. The reliability can be considered as part of the Design for Six Sigma (DFSS) scorecard. The DFR process can follow the DFSS roadmap—for example, the identify, design, optimize and validate (IDOV) framework. With reliability in mind, product program teams can identify the boundary and scope of system requirements and design the product. DFR activities are part of various elements in technology and product development activities during the complete product life cycle. Goals of the DFR process are:

- Integrate voice of customer (VOC) into product requirements to improve reliability and robustness of the supply chain.
- Provide requirements for activities involved in the DFR/DFSS process. Optimize the design over the time domain and functional validation of the supply chain using a test progression strategy.
- Identify methods for defining supply chain reliability requirements and activities involved in the network of supply chain.
- Provide the practitioner a means of prioritizing the reliability projects and studies that must be undertaken.
- Continuously improve supply chain reliability and robustness over time.

8.2.4 Design for Six Sigma Overview

DFSS—The acronym for "Design for Six Sigma," describes the application of Six Sigma tools to product development and process design, including supply chain. The goal is to "design in" six-sigma performance capability. DFSS is an approach to designing (re-designing) a product or service. It is equally useful in developing business processes such as supply chain or technical products. DFSS is a defined methodology – a culture and a way of viewing value creation.

The focus of DFSS begins with critical voice of the customer (VOC) analysis and rational business planning. When an understanding of the market and customer needs is obtained, design personnel work to understand and characterize critical design parameters and functionality. The time to develop a new supply chain is a critical success factor in almost any business today. DFSS facilitates development time reduction by deploying lessons learned e.g., supply chain disruption, lack of visibility throughout the development, manufacturing set-up process, and supply chain robustness development.

DFSS provides many tangible benefits to companies. For instance, the DFSS approach results in long-term cost reductions of a supply chain. There are many ways this savings is realized. Instead of debugging products, processes and or traditional supply chain that already exist, DFSS is a re-examination of the function and design parameters in supply chain. DFSS starts from scratch with the goal to design virtually error-free supply chain leveraging digital technologies. This strategy effectively replaces the trial and error or built-test-fix processes and results in product and supply chain designs that consistently meet customer requirements.

A typical Design for Six Sigma approach has four phases—Identify, Characterize, Optimize and Verify (ICOV) as follows:

1. **Identify**: Identify market needs. Define customer requirements and project goals. Identify Critical to Satisfaction (CTS) and related functional targets. Reliability is often a key CTS on the reliability aspects of a product.

   The purpose of this stage for reliability effort is to clearly and quantitatively define the reliability requirements and goals for a product as well as the end-user product environmental/usage conditions. These can be at the system level, assembly level, component level or even down to the failure mode level. Requirements can be based on contracts, benchmarks, competitive analysis, customer expectations, cost, safety, best practices, etc.

   The system reliability requirement goal can be allocated to the system e.g., vessel level, subsystem e.g., container level, assembly level, component level or even down to the failure mode level. Once the requirements have been defined, they must be translated into design requirements and then into manufacturing requirements.

2. **Characterize**: Understand the system and select design concepts. Map CTS characteristics to lower-level y factors. Relate (y) factors to CTQ or CTR (x) design factors. Determining the usage and environmental conditions is an important early step of a DFR program. Know what it is to be designed for and what types of stresses the products are supposed to withstand. The conditions can be determined based on customer surveys, environmental measurement, and sampling.

   The tendency for the potential failure mode or error state occurrence is aggravated by noise factors. Noise factors are those factors that engineers have little or no control that negatively influence designed system performance. Fundamental to designing for reliability and robustness using transfer function is the inclusion of noise factors during analysis that challenge the design and uncover potential failure modes (Clausing D., Frey D., 2005). Once uncovered, these failure modes can be avoided by developing appropriate countermeasures either in the design or manufacturing process. Inclusion of noise factors in upfront design analysis has encouraged engineers developing transfer function to consider appropriate noise factors and realistic levels as well as strategies to include them in simulations.

   It is highly important to estimate the product's reliability, even with a rough first cut estimate, early in the design phase. This can be done with estimates based on engineering judgment and expert opinion, Physics of Failure (PoF) analysis, transfer functions-based simulation models, prior warranty, and test data from similar products/components (using life data analysis techniques) or Standards-Based Reliability Prediction.
3. **Optimize**: Design for robust and reliable performance. That minimizes product, system, subsystem or process sensitivity to uncontrollable user environment to have better manufacturability and higher reliable supply chain. For example, from supply chain perspective, shifting component-production locations can result in a 20% decrease in cost, as well as reductions in logistics and trade risks (McKinsey, May 2020), Alternatively, the same parts can be manufactured domestically with only a 5% cost increase by leveraging Industry 4.0. And because supply chain risks are always changing, this stress test cannot be a one-off exercise. Optimize the supply chain in the presence of stress to reveal potential vulnerabilities in the supply chain and companies and identify appropriate mitigation steps. This calls for a combination of short- and medium- or long-term actions that are tailored to the specific issues that have been identified. Response levers can be structured across five themes: securing supply, agile manufacturing, diversifying manufacturing locations, optimizing design, and optimizing financials.

   In this stage, robust design helps further factor reliability tasks e.g., change supply location leveraging digital technologies into design process by optimizing design function in the presence of noise factors to identify important variables, to estimate their effect on a certain product characteristic and to optimize the settings of these variables to improve the design robustness.
4. **Verify**: Assess the integrated system and subsystem effects on performance. Use reliability, manufacturing, and supply chain verification to assess design performance and the ability to meet customer's requirements and avoid failure mode such as disruption. Consider what parts of the supply chain could be at high risk when major failure mode(s) e.g., disruption occurs.

   A reliable design should anticipate all that can go wrong. Design for reliability can be viewed as a means to maintain and sustain Six Sigma capabilities over time and is one of tool sets in DFSS methodology. DFSS methodology can be implemented in supply chain design to identify the voice of customer and then translate it into critical customer requirements (CCRs) such as disruption prevention showing that demand management is one of the major CCRs for a supply chain which can then be optimized and validated. Design for

> Six Sigma is not limited to tangible products but can be a useful and flexible methodology for dealing with the complexity, uncertainty, and variability of supply chain design.

As a part of supply chain reliability, building standards for visibility is a critical control factor in DFR. Digitalization within a single carrier's operation can improve the above scenario to a certain extent for shipments that are confined to a single mode and carrier, but many shipments are executed across multiple carriers and modes of transport. Currently, container shipping as a whole lacks a common approach to technology and common definitions for data and processes, making information exchange between parties difficult. Having a common data language and framework for processes based on shared requirements is the first step to ensuring that high-quality data can be communicated in a way that is understandable and usable for any stakeholder in the supply chain regardless of carrier, choice of technology platforms, or even the mode of transportation.

Digital Container Shipping Association (DCSA) works in close conjunction with our ocean carrier members, industry stakeholders, solutions providers, industry regulators, and other standards bodies to develop open-source standards that can be freely adopted by all to provide seamless, uninterrupted, and automatic (if desired) communication of shipping data. Specifically in the areas of Track & Trace and Internet of Things (IoT), our standards drive maximal visibility from end to end, enabling shippers to receive updates regarding the whereabouts of containers and the status of their contents at any point along the container journey.

Digital transformation in highly standardized industries such as banking and telecommunications didn't happen overnight, but it did happen decades ago. This contributed to their ability to weather the COVID pandemic without disrupting global operations. While disruptions along the physical supply chain due to COVID (such as factory closures) were unavoidable, the impact on container shipping could have been diminished if the industry had been more digitalized and standardized. Container transportation can be more resilient if (API-based) digital standards such as those published by DCSA are widely implemented by carriers, solution providers, and other participants in the supply chain.

Shippers who wish to have more reliable shipping services must start by working with their carriers and service providers to adopt standards that will create a foundation for timely exception handling, driven by reliable, real-time data that is consistent across carriers. With these standards in place, shippers can have instant access to the information they need to make informed, data-driven decisions about inventory management, helping establish trust between all stakeholders. Once trust through transparency is established, real innovation can occur—innovation that will enable transparent, reliable, easy to use, secure, and environmentally friendly container transportation services.

A single carrier's digitalization of operations can benefit this scenario to some extent for shipments confined to a single mode and carrier, but many shipments are executed across multiple carriers and modes of transportation. Container shipping lacks a common approach to technology and standard definitions for data and processes, making information exchange a challenge. Improving visibility can be an efficient way to help container shipping become reliable. A common data language and framework should be adopted for the supply chain processes. This is the first step to support a high-quality and efficient dialogue environment. The information will be understandable and usable for all stakeholders within the supply chain, regardless of carriers, selections of technology-enabled platforms, or modes of transportation. The timeliness of the data or information across the entire supply is a crucial component of visibility. More research attention is needed to the linkage between real-time communication and global industry demands.

Product development has a huge impact on revenue stream and reliability. Enhancing product development process with design for six sigma disciplines will improve the product development process to develop customized design for supply chain with reliability focus process with required tools to support specific reliability tasks. It is most cost-effective and less time-consuming to make design insensitive to uncontrollable user environments using transfer function. Design for reliability or design for supply chain tasks can be best accomplished through DFSS roadmap.

8.3 AVAILABILITY

In today's uncertain environment, it has become increasingly important for businesses to guarantee that when a customer wants to make a purchase, their supply chain is efficient enough to ensure that the item they want is available. If a business can't fulfill its customers' requirements, they are likely to turn to the nearest competitor who can sell the item immediately. The threat of not being able to meet demand is therefore a key concern within the supply chain.

The current economic environment is forcing many firms to drastically cut stock levels. Some businesses have taken this philosophy too far and have stripped away strategic buffers which protect the supply chain from risk factors. Companies that take stock-cutting measures to extreme levels are leaving themselves at risk of not being able to fulfill normal orders if there is a disruption in the supply chain. Ever-changing environmental factors such as natural disasters, the upheaval of financial markets, and commodity prices are difficult to avoid. The ensuing impact on availability cannot go unnoticed. Therefore, businesses must take a far more comprehensive view of the supply chain in order to better understand, anticipate, and plan for a changing environment.

However, many organizations continue to rely on legacy systems or 'gut feeling' to identify trends and make supply decisions. It is the challenge that this 'gut feeling' method will begin to be phased out in 2013, as forward-thinking organizations realize that the only way to achieve a fully holistic view of the supply chain and to manage it more effectively is to adopt intelligent process management systems. Businesses can use this intelligence to take immediate action to ensure optimal stock levels are reached and availability is maximized. By using these intelligent tools to effectively optimize inventory, firms can ensure they maintain availability and service levels while also protecting liquidity as optimal inventory levels can be reached, even in volatile markets.

Utilizing the benefits of technology for inventory optimization becomes even more imperative when working within environments that can change at speed, which may leave businesses struggling to meet availability requirements. Supply chain management will become an ever-pressing focus for businesses across the board. Proactive organizations will respond with more coordinated approaches to supply chain management, adopting technology that enables them to quickly and effectively manage operations and gain a greater perspective over the entire supply chain.

Inventory management tools will aid this process, allowing for reduced planning efforts, a better overview of consumer buying patterns and coordinated management of entire the supply chain network. Organizations will utilize these technologies to obtain vital information that will enable them to adjust quickly to reflect market changes, ensure availability, and enable customers' expectations to be met—adding value in the process. The sophistication of supply chain management will continue to grow as businesses face increasingly difficult conditions and look to balance lean supply chains while maintaining availability and high service levels.

Companies will also be forced to look at all activities through the lens of liquidity; exploiting technology to reduce stock and procurement expenses while freeing up additional liquidity at the same time. Although this is a careful balancing act, it is vital to guarantee availability and maintain optimal customer service levels. By ensuring that the optimal inventory is held in the right place at the right time—customer satisfaction can be guaranteed. Now is the right time for businesses to re-evaluate their approach to supply chain management. Those who successfully implement advanced intelligent inventory planning tools will begin to make the supply chain work for them and will reap the rewards.

As an example of global shipping container shortage, monitoring the availability of containers is for better container availability and trading decisions. The global shortage of shipping containers, primarily caused by the COVID-19 pandemic, has led to drastic inflation in shipping and container prices and increased delay times for companies. Freight shipping has found itself in a unique situation where unforeseen events have left a global shortage of containers, which has had a domino effect down the supply chain, disrupting global trade. Why is the industry facing this shortage? Where have the containers vanished to? A significant number of containers have found themselves in inland depots while others have been stacking up at cargo ports. As Asia slowly began to recover, other countries were still faced with national lockdown restrictions meaning containers could not be sent back to Asia (where they were really needed) to continue the trade partnership. The pairing of lockdown regulations alongside other factors, such as a staff deficit, meant a backlog of containers began to develop.

The COVID-19 pandemic has resulted in a shift in customer spending away from services to goods due to restrictions imposed by national lockdowns. This resulted in the issue of container shortage becoming even more prominent as desirability increased, with many companies being unable to secure containers. Aside from the impacts of COVID-19 disturbing the usual flow of trade and container availability, another major factor was the Chinese New Year. Celebrated on 12 February, the holiday played a part more than ever when it comes to container shortage.

Global industry giants are thrown off balance. Many industry shipping giants, such as Maersk and Hapag-Lloyd, have been hit hard by the global container shortage, which has resulted in them adapting their methods and facing shipping setbacks. Hapag-Lloyd has increased container refilling and emptying times to 25% faster than usual to ensure optimal container use. Hapag-Lloyd has thought outside the box when it comes to adapting their containers for alternative uses. Turning the reefer containers off allows dry goods such as shoes, electronics, or textiles to be sent to reefer demand locations. Once there, they can be emptied and switched back on to continue their journey. With the situation being unprecedented and ever shifting, it is stressed that the importance of close industry collaboration is needed to prevent situations like this from occurring again. A strategic closer cooperation between customers, shipping lines, and port terminals is needed more than ever. The shipping industry is very volatile and needs a closer cooperation between customers, shipping lines, and port terminals so they can be better prepared for situations like this in the future.

As the value of goods traded has increased, as well as the integration of the global economy, many opportunities have been presented for companies to expand their market reach. It can be concluded that the factors that are considered as the most important factors in the selection of carriers for land, sea, and air cargo transportation have more to do with the mode of transportation rather than with the quality of transportation service. The interests of the carrier are influenced by a variety of factors, such as the quantity, characteristics, and the circumstances of the company, which means that shippers who export small quantities of cargo abroad are frequently faced with changes in transportation modes and transportation service providers. In context with the above, Jung et al. (2019) argued that the availability of transportation service is one of the most important factors that determine shippers' selection on carriers. It consists of three dimensions, in terms of connection with other transportation modes for inland transport (intermodal service), number of international transportations per day (frequency of transportation), and the possibility of cargo loading space to handle during peak season (guaranteed space during peak season).

Container terminals play a critical role in linking the various elements of the intermodal transport system. They must provide sufficient facilities for transferring containers from ships to trucks or trains, or to transship containers between ships, to smooth the flow of containers in the system. The capacity of container terminal capacity influences the availability of an integrated cargo shipping system. Chu and Huang (2005) reported a comparison of various container handling systems regarding the facility's capacity when considering different methods of container handling. In this approach, the intention is to support decisions regarding terminal design and handling equipment, thereby setting the basis for decisions concerning terminal planning. This study proposed a general equation for container terminal planning, accounting for the dimensional characteristics of equipment, the transshipment ratio, and the average container dwell times. This study also ranked the annual handling capacities for three commonly adopted container handling systems in Taiwan ports and suggested that selecting appropriate equipment at container terminals requires a careful evaluation.

Factors that influence freight or carrier service choice for shippers have received attention since the late 1960s (Cook, 1967; McGinnis, 1989). A consensus is that the important factors to shippers are the results of a balance between cost and service issues. Matear and Gray (1993) executed principal components analysis to group the important factors in freight modal choice. Reliability and service availability (i.e., space capacity) were found to be the most important ones. A more recent study argued that one of the reasons that makes Eurasian rail freight a cost-efficient option for shippers interested in building more sustainable supply chains between China and Eurasian countries is service availability. Other reasons are transit times, transport costs, and reliability (Zhang and Schramm, 2020).

Meng et al. (2019) conducted a critical review on revenue management for container liner shipping services. The availability of container slots is one of two major issues that influence the revenue management of container liner shipping services. The other one is the price of service. Meng et al. (2019) argued that the state-of-the-art on network capacity control studies could be improved by considering more current real-life conditions in the industry, like dividing booking classes based on the characteristics of containerized cargo (e.g., reefer and dry cargoes), and differentiating the spot market and the long-term market. This study also pointed out that a better understanding of customer behaviors, such as the number of slots they book, the kinds of freight contracts they sign, and whether they accept the offered freight rates, can be useful for container shipping companies to develop optimal strategies in the future.

The level of product availability is measured using the cycle service level or the fill rate, which are metrics for the amount of customer demand satisfied from available inventory. The level of product availability, also referred to as the customer service level, is one of the primary measures of a supply chain's responsiveness. A supply chain can use a high level of product availability to improve its responsiveness and attract customers, thus increasing revenue for the supply chain. However, a high level of product availability requires large inventories, which raise supply chain costs. Therefore, a supply chain must achieve a balance between the level of availability and the cost of inventory. The optimal level of product availability is one that maximizes supply chain profits.

In the fourth quarter of 2008, U.S. inventories shot up by $6.2 billion because of the rapid decline in demand that hit retailers and manufacturers. For some manufacturers, the situation was exaggerated because of the excess inventory of raw materials, such as steel and plastics, that they had built up while anticipating price increases. Retailers were also hit hard, with some, such as Saks Fifth Avenue, slashing prices by 70% during the holiday season to spur demand. The excess inventories and the drop in demand led to several retailers, such as Steve & Barry's and Circuit City, declaring bankruptcy during this period. In contrast, Nintendo missed out on an estimated $1.3 billion in sales during the 2007 holiday season because of a failure to meet soaring global demand for its Wii video game console. These examples make clear that having too high or too low a level of product availability has a significant impact on supply chain profits.

Contrary to common belief, product availability does not require businesses to keep all items on hand at all times throughout all locations. Instead, product availability refers to having access to inventory whenever the customer needs it. This requires having connections to suppliers and external partners, as well as excellent inventory visibility. Product availability includes the cost of designing, manufacturing, storing, and delivering different item variances. With high product availability, customers can visit a business knowing all of their shopping needs can be fulfilled. Alternatively, low availability can result in lost sales and low customer retention. By increasing product reliability and then availability, companies can ensure they meet customer demands.

There are several ways businesses can enhance their availability, such as utilizing sales & operations planning. Sales and operations planning (S&OP) requires complete cooperation and transparency between a company's departments in order to supply products and meet revenue targets. S&OP involves sourcing inventory, logistics planning, calculating stock quantities, and meeting cost objectives. With an inventory management system, businesses can automate cycle counts, product transferals, and analytic reports to improve their stock control. Management systems make locating an item as simple as clicking a button, improving companies' visibility and efficiency.

8.4 TESTABILITY

Testability is a way of thinking wherein the designer possesses an awareness of the importance of testing. Testability is designed at the beginning of design process. Before starting any design, the designer should ask himself, "How will I test it, how will production test it, how will it be tested in supply chain and in the field?"

The major benefit of enhanced testability is lower product cost. Operation, supply chain, and support costs can in turn be lowered by improving product testability, to isolate faults more rapidly and without ambiguity to the field and consumer replaceable items in deployed systems. Testability objectives include: (1) maximizing operational capability through the use of built-in-test (BIT), which includes user-friendly controls and displays; (2) optimizing equipment availability by rapid fault detection and unambiguous fault isolation, (3) reducing test complexity by providing standardized hardware/software, (4) exciting and uncovering potential failure modes in early design phase, and others.

Testability is also a property that applies to an empirical hypothesis. The meaning of testable hypothesis is when there is a possibility of deciding whether a hypothesis is true or false based on experimentation. As a result, we can determine whether, or not a theory can be supported or refuted by data. The interpretation of experimental data can also be uncertain or inconclusive as the results may be inconclusive. Simply put, testability refers to which critical components or functions can be tested when operational.

As supply chains and logistics operations become more globalized, companies and industries are tasked with matching up to the increasing demands. Newer and more complex methods are being developed to reach these goals. In response to these developments, supply chain and logistics management have come to play an increasingly important role for companies. Through numerous research and development, tests are performed regularly to gauge the effectiveness and efficiency of the three modes of moving goods and products, which are by land, sea, and air.

With land transportation, trucks have become the predominant mode of transportation of goods and materials and within last mile delivery. They are, by far, the most common way to move goods. Trucks can get to places that other transportation methods cannot get to. Rail does not travel to all destinations, and air freight is expensive and limited by weight and available airports. Because of this, in the US trucks are often the first option for moving large amounts of cargo versus railway, water, air freight, or pipelines (Blume Global, 2021). With advancements in technology, companies are researching and testing self-driving trucks, also known as an autonomous truck or the robo-truck. Research is persistently being conducted on ways of improving autonomous trucks to maximize its potential, eliminate or minimize dangers on the road, and ultimately cut down on the delivery time of products and goods. Another reason why companies are funding these robo-trucks is the promise of becoming sustainable companies and impacting the environment positively by eliminating trucks that emit hazardous pollutants or products (Paddeu and Denby, 2021).

Ocean or sea freight is the method of transporting large quantities of goods via maritime vessels. Products are packed into typically 20- or 40-foot standardized containers, which are then loaded onto vessels and then sailed to their final destinations. Numerous tests and research have been carried out to help facilitate effective and efficient maritime transportation. By doing so, companies can detect faulty containers, find ways to combat cybersecurity and the ever-changing technology landscape, and optimize routes for ocean freight. To add to the above issues, others such as fuel also play a role when it comes to maritime transportation. With the fluctuation in fuel prices, companies adjust their rate or fees accordingly. Other issues are the changing seasonal demands of goods as well as the currency fluctuation. Just like with land transportation, more companies are looking into sustainable means of moving products effectively across major sea lanes while reducing the emissions produced from the vessels and avoiding major accidents or spills that might cause harm to marine life.

8.5 MAINTAINABILITY

Maintainability is defined as the relative ease of time and resources within which the maintenance can be performed on any equipment. An item's maintenance abilities determine its ability to maintain or restore its specified condition when it is performed by people with specific skills and resources, using specified procedures. Maintenance is performed at various levels of maintenance and repair; each will require a specific skill level. Every company with the goal of producing goods and quality products strives to implement a smooth system of operations. The objective of maintainability is to sustain or return the equipment or supply chain products to its intended condition when the maintenance is performed. From a design engineering perspective, there is also a probability the equipment can be restored to its intended state once maintenance is performed after failure. Maintainability is an important factor in the total cost of assets. An increase in maintainability can lead to reduction in operation and support costs. For example, a more maintainable product lowers maintenance time and operating costs. Furthermore, more efficient maintenance means a faster return to operation or service, decreasing downtime. For industry, an important issue is how to reduce maintenance cost e.g., supply chain maintenance cost and manage business risks while increasing asset reliability, availability, and safety.

Recent studies show that, for the average factory, inefficient maintenance policies are responsible for costs ranging from 5% to 20% of the plant's entire productive capacity (Biggio and Kastanis, 2020). Furthermore, according to the International Society of Automation (ISA), the overall burden of unplanned downtime on industrial manufacturers across all industry segments is estimated to touch the impressive figure of $647 billion per year. According to a recent trend study gathering interviews with more than 230 senior European businesses, roughly 93% of them deem their maintenance policy inefficient.

Prognostics and Health Management (PHM) is a technology to enhance effective reliability and availability of a product in its life cycle conditions by detection of current and approaching failures by providing for mitigation of the system risks and is a maintenance policy aimed at predicting the occurrence of a failure in components and consequently minimizing unexpected downtimes of complex systems and have been studied by many researchers from many different engineering fields to increase system reliability, availability, safety and to reduce the maintenance cost of engineering assets. Many works conducted in PHM research concentrate on designing robust and accurate models to assess the health state of components for particular applications to support decision-making. Prior knowledge to implement PHM in complex systems is crucial to building highly reliable systems. PHM implementation steps consist of:

- critical component analysis,
- appropriate sensor selection for condition monitoring (CM),
- prognostics feature evaluation under data analysis and,
- prognostics methodology and tool evaluation matrices.

PHM systems are some of the main protagonists of the Industry 4.0 revolution (Biggio and Kastanis, 2020). Efficiently detecting whether an industrial component has deviated from its normal operating condition or predicting when a fault will occur are the main challenges these systems aim at addressing. Furthermore, they could potentially drastically reduce the often-conspicuous costs associated with scheduled maintenance operations. The increasing availability of data and the stunning progress of machine learning (ML) and deep learning (DL) techniques over the last decade represent two strong motivating factors for the development of data-driven PHM systems.

The current most popular approaches to maintenance are divided into two categories, namely reactive maintenance and scheduled maintenance. In general, the first implements maintenance operations immediately after a system failure occurs, whereas the second is based on scheduling maintenance operations at regular time intervals. These strategies naturally introduce significant extra costs due to machine or system e.g., vehicle downtime, component replacement, or unnecessary maintenance interventions.

On the other hand, predictive maintenance (PM) represents a completely different paradigm that holds the promise of overcoming the inefficiencies of the aforementioned methods. PM is one of the hallmarks of the Industry 4.0 revolution, i.e., the process of modernization of the industrial world induced by the advent of the digitalization era. The goal of PM systems is to implement a smarter and more dynamical approach to maintenance leveraging recent advances in sensor engineering and data analysis. The health state of a machine is now constantly monitored by a network of sensors and future maintenance operations are based on the analysis of the resulting data. An increasing number of organizations, motivated by their need for reducing costs and by the potential of PM, are starting to invest significant amounts of resources on the modernization of their current maintenance strategies.

Prognostic and Health Management (PHM) is an engineering field whose goal is to provide users with a thorough analysis of the health condition of a system and its components. To this extent, PHM employs tools from data science, statistics, and physics in order to detect an eventual fault (anomaly detection) in the system, classify it according to its specific type (diagnostic) and forecast how long a system e.g., machine will be able to work in presence of this fault (prognostic). Different maintenance approaches are compared and summarized as follows:

- Reactive maintenance
- Scheduled maintenance
- Condition based and predictive maintenance (CBM)

The predictive maintenance approach inevitably results in lower maintenance costs compared to CBM, since each component can be fully exploited without sacrificing safety and efficiency (Fink, 2020). PM, as a key player in the Industry 4.0 paradigm, strongly relies on some of the most recent advances in hardware technology, communication systems, and data science. Among them, DL techniques have gained popularity over the last few years in light of their excellent performances in processing complex data in an end-to-end fashion.

Digitalization is a fundamental driver of Industry 4.0 and Supply Chain 4.0, a novel paradigm that enhances production efficiency through information and communication technologies. These technologies also provide the foundation for predictive PM for industrial components and systems, whereby CM data is employed to perform three tasks:

1. detection of abnormal states, by identifying deviations from normal operating conditions in production processes, manufacturing equipment, and products;
2. diagnostics, by classifying abnormal states;
3. prognostics, by predicting the evolution of abnormal states up to failure.

These tasks help implement efficient, just-in-time and just-right maintenance strategies by selecting the right action for the right component at the right time, thus maximizing production revenues and minimizing costs and losses, including assets. Furthermore, PHM performance metrics have been introduced to characterize errors in detection, diagnostics, and prognostics. Health Management Systems are responsible for fault detection, and often diagnosis of fault. Recently, advanced sensors and improved on-board computing have made damage prediction a capability that is rapidly gaining attention as a viable means to dramatically reduce the cost of operations for large-scale fleet maintainers.

The maintainability costs for the logistics infrastructure necessary to conduct the repairs can be modeled as a supply chain from which the necessary resources must be brought in time to make a scheduled repair. Based on an a priori estimate of the average rate of repairs, and the variance from this average rate, the maintenance infrastructure is designed to facilitate on-time repairs with a predetermined quality-of-service level. A classical model for safety stock in serial supply chains can be extended to address the cost implications of the introduction of PHM technologies in logistics. This model parameterizes the PHM estimate using a log-normal distribution of the predicted failure time, but can handle an arbitrary time dependency of the variance of this distribution, capturing the fact that the estimate improves as failure approaches. The supply chain model can be extended in a number of ways and does not account for capacity constraints, or nonlinearities in the costs and time lags as a function of capacity.

Taking advantage of robust design, as we discussed in the section of reliability above, to identify key product or system health characteristics and determine appropriate PHM based on the findings from risk assessment e.g., design failure mode and effect analysis (DFMEA). Document the decision regarding PHM i.e., why it needs to be implemented and to what degree or the reason why it is not required. From maintainability perspective, we need to gather and analyze maintainability requirements including, customer maintainability expectations, return of field experience, compared to existing or previous generation products, compared to competition, document results as summary of comparisons and highlight risks, will built-in-test or PHM be required.

As companies and supply chains move automation and into using autonomous trucks, many factors are considered. Not only does the move increase safety on the roads or reduce the emissions and thus positively impacting the environment, but repairs and maintainability become a factor as well. Regardless of a traditional or autonomous truck, no company wants to spend money on a product that constantly breaks down. PHM discipline provides the overall health state of logistics or complex supply chain systems and assists in making correct decisions on maintenance. For practitioners, to select and implement PHM technology is based on risk assessment and their ability and knowledge about PHM approaches, tools, etc. PHM can be well applied to the service and maintenance of self-driving vehicles including autonomous driving vehicles including trucks.

There are different maintenance plans that every maritime or sea freight organization must follow to ensure the machinery on their ships maintain a certain level of functionality and safety. It is important to note that one of the objectives of a maintenance plan is to provide a quick and efficient means of completing the necessary maintenance and repair work. To conduct a successful maintenance plan, some strategic steps should be taken with the objective of identifying the problem and determining its cause, proposing, and evaluating the solution, implementing of the solution and its effectiveness, and finally resolving the issue (Marine Insight, 2021a). The shipping company can and should take into consideration the emerging maintenance approach such as PHM guidelines and specifications based on a particular manufacturer of the ship, the history of the ship and its history of failure, defects, damages, and remedial actions during the development of a maintenance procedure for that ship. Others are the guidelines mentioned in the ISM (International Safety Management) code, the age of the ship, third-party inspections, consequences of failure of equipment on safe operation of the ship, critical equipment and systems and the intervals of maintenance, which should be incorporated with PHM methodology. As a result of the points mentioned above, a systemic approach to maintenance has been developed. When performing the maintenance procedure, it is necessary to make sure that a maintenance plan is well planned, executed, and monitored. The steps included in the conventional approach for maintenance procedure along with maintenance intervals, methods and frequency of inspections, specification of the type of inspection, type of measuring equipment to be used, the establishment of appropriate acceptance criteria, assignment of responsibility for inspection activities to appropriately qualified personnel and a clear definition of reporting requirements and mechanisms should be reviewed, updated and deployed for effective and efficient maintenance approach at the lowest possible cost to achieve operational excellence and business success.

8.6 CONCLUSION AND NEXT STEPS

In this chapter, we have outlined the characteristics of reliability, availability, testability, and maintainability and the factors that influence integrated cargo shipping systems in terms of reliability, availability, testability, and maintainability, an essential part of today's global supply chains, and the primary link between manufacturing sources in developing economies and the markets in the global market. We have introduced robustness thinking, robust design, and prognostic health management as emerging best practices to improve supply chain reliability, robustness, and resilience to prevent disruption as much as possible. Illustrative of this sector and the exponential growth of container shipping are companies, such as Mediterranean Shipping (MSC) which was founded in 1970, currently owns a fleet of 570 vessels and calls on 500 ports on over 230 trade routes. Container shipping has become the most important means of transport for inter-oceanic transportation of manufactured products, as evidenced by the explosive growth of the container shipping industry. As it is pointed out, the supply chain impact of ocean container shipping of transport is significantly understudied, and there are four areas of research that can contribute to a better understanding and improved performance of the worldwide container supply chains leveraging digital technologies. The four identified research areas are:

1. Improving visibility and automation of container shipments across the container supply chain leveraging digital technologies e.g., IOT, AI/ML/DL, autonomous solutions e.g., autonomous driving trucks and others
2. Pricing and risk management in the container supply chain
3. Reliability, robustness, resilience improvement through robustness thinking based prognostic health management in the container supply chain
4. Competition between ports, carriers, and container terminals

Shipping reliability has a direct connection to industrial productivity and labor costs. A supply chain's reliability refers to its ability to maintain consistent performance while trying to achieve as many objectives as possible. Future research and practices could focus more on how to improve the visibility and real-time communication of the supply chain process. Availability refers to a percentage of time or services that a system is available for use once it has been assigned a responsibility. For shippers, service availability is a critical factor that influences the selection of carriers. There are two major factors to consider, in terms of connectivity to other modes of transportation such as rail, road, and air, the number of international transportations per day, and the amount of loading space to handle during peak season. Undoubtedly, a comprehensive evaluation is needed when selecting the right equipment for container terminals. Container shipping companies could benefit from a better understanding of customer behavior to formulate effective strategies for revenue management.

It has become evident in the past several decades that supply chain and logistics management has become more complex, globalized, and integrated, which means companies and industries are faced with increasing demands. The testability must be interrelated with all other objectives by the system planners in order to achieve a balanced design, a product whose testability characteristics meet system and supply chain and lifecycle management requirements at the lowest life cycle cost. In addition, the outcomes related to environmental sustainability should be taken into consideration in the contemporary integrated cargo shipping systems. Gauging the effectiveness and efficiency of the three modes of moving goods and products, which are by land, sea, and air, requires numerous research and development efforts at regular intervals. Last but not least, investments in automation, digital technologies, and technology-enabled devices are useful for developing an effective maintenance protocol. It is a fact that improved performance in maintainability provides companies with more flexibility and options when it comes to how to proceed with transporting their products.

REFERENCES

Biggio, L., & Kastanis I. (2020). Prognostics and health management of industrial assets: Current progress and road ahead. *Frontiers in Artificial Intelligence*, 3, 578613.

Blume Global (2021). Modes of transportation in the global supply chain. https://www.blumeglobal.com/learning/supply-chain-modes-of-transportation/.

Bulian, G., Lindroth, D., Ruponen, P., & Zaraphonitis, G. (2016). Probabilistic assessment of damaged ship survivability in case of grounding: Development and testing of a direct non-zonal approach. *Ocean Engineering*, 120, 331–338.

Chakraborty, S. (2021). Understanding Design of Container Ships. https://www.marineinsight.com/naval-architecture/understanding-design-of-container-ships/.

Chu, C. Y., & Huang, W. C. (2005). Determining container terminal capacity on the basis of an adopted yard handling system. *Transport Reviews*, 25(2), 181–199.

Chun, C. C., & Chiang, C. H. (2011). *Critical Factors in Schedule Reliability of Container Shipping Carriers for 2011*. Institute of Traffic and Transportation National Chiao Tung University, Taiwan, ROC.

Chung, C. C., & Chiang, C. H. (2011). The critical factors: An evaluation of schedule reliability in liner shipping. *International Journal of Operations Research*, 8(4), 3–9.

Clausing, D., & Frey, D. D. (2005). Improving system reliability by failure-mode avoidance including four concept design strategies. *Systems Engineering*, 8, 245–261.

Cook, W. R. (1967). Transport decisions of certain firms in the Black Country. *Journal of Transport Economics and Policy*, 1(3), 325–344.

Decò, A., Frangopol, D. M., & Zhu, B. (2012). Reliability and redundancy assessment of ships under different operational conditions. *Engineering Structures*, 42, 457–471.

Fang, C., & Das, P. K. (2005). Survivability and reliability of damaged ships after collision and grounding. *Ocean Engineering*, 32(3–4), 293–307.

Fink (2020). *Data-Driven Intelligent Predictive Maintenance of Industrial Assets*, January 2020, Women in Industrial and Systems Engineering (pp.589-605).

Gosling, J., Purvis, L., & Naim, M. M. (2010). Supply chain flexibility as a determinant of supplier selection. *International Journal of Production Economics*, 128(1), 11–21.

Hu, M. (2013a). *Robustness Plan*, Six Sigma Forum Magazine, American Society for Quality (ASQ), Milwaukee, WI.

Hu, M. (2013b). *Readiness Plan Put Into Action*, Six Sigma Forum Magazine, American Society for Quality (ASQ), Milwaukee, WI.

International Maritime Organization (2021). Introduction to IMO. https://www.imo.org/en/About/Pages/Default.aspx.

Jung, H., Kim, J., & Shin, K.S. (March 2019). Importance analysis of decision making factors for selecting international freight transportation mode. *The Asian Journal of Shipping and Logistics*, 35(1), 55–62.

Kaluza, P., Kolzsch, A., Gastner, M. T., & Blasius, B. (2010). The complex network of global cargo ship movements. *Journal of the Royal Society Interface*, 7(48), 1093–1103. doi:10.1098/rsif.2009.0495

Matear, S., & Gray, R. (1993). Factors influencing freight service choice for shippers and freight suppliers. *International Journal of Physical Distribution & Logistics Management*, 23(2), 25–35.

McGinnis, M. A. (1989). A comparative evaluation of freight transportation choice models. *Transportation Journal*, 29(2) (Winter), 36–46.

McKinsey (2020). Why Now is the Time to Stress-Test Your Industrial Supply Chain. https://www.mckinsey.com/business-functions/operations/our-insights/why-now-is-the-time-to-stress-test-your-industrial-supply-chain.

Meng, Q., Zhao, H., & Wang, Y. (2019). Revenue management for container liner shipping services: Critical review and future research directions. *Transportation Research Part E: Logistics and Transportation Review*, 128, 280–292.

Notteboom, T. E. (2004). Container shipping and ports: An overview. *Review of Network Economics*, 3(2), 86–106.

Paddeu, D., & Denby, J. (2021). Decarbonizing road freight: Is truck automation and platooning an opportunity? *Clean Technologies and Environmental Policy*, 24, 1021–1035.

Rodrigue (2013). *The Geography of Transport Systems*, 3rd Edition. London: ISBN9780203371183.

Snyder, L.V. (2003). *Supply Chain Robustness and Reliability: Models & Algorithms*. Evanston: Northwestern University.

Taguchi, G., & Yokoyama, Y. (1993). *Taguchi Methods: Design of Experiments*. American Supplier Institute.

Vernimmen, B., Dullaert, W., & Engelen, S. (2007). Schedule unreliability in liner shipping: Origins and consequences for the hinterland supply chain. *Maritime Economics & Logistics*, 9(3), 193–213.

World Bank Group (2021). World Bank Open Data. https://data.worldbank.org/.

Zhang, X., & Schramm, H. J. (2020). Assessing the market niche of Eurasian rail freight in the belt and road era. *The International Journal of Logistics Management*, 31(4), 729–751.

Part 2

System Design

9 Chip-Level Design and Test

Mark McDermott

9.1 INTRODUCTION

This chapter focuses on the design, verification, and testing of ultra large scale integrated (ULSI) circuits. These ultra-deep submicron (UDSM) integrated circuits (ICs) would not be possible without the enormous amount of computer-aided design (CAD) tool development effort that has been done over the last 50 years. These tools, coupled with specialized multi-parameter design convergence flows allow designers to produce integrated circuits with 40+ billion transistors.

While the IC designers are busy, the software programmers are busy coding millions of lines of code for the system. This is made possible by having a development environment that supports HW/SW Co-Design as shown in Figure 9.1. This development environment contains a virtual prototyping simulation and emulation system that allows a seamless transition from a software simulation to hardware emulation. The SW can be loaded onto the production system as soon as the IC is fabricated, packaged, and tested.

9.2 DESIGN STAGES AND PHASES

There are three main stages in the IC design process: front-end (logical) design, back-end (physical) design, and the design planning phases as shown in Figure 9.2. The front-end design stage focuses on describing the functionality of the IC using various levels of abstraction ranging from highest to lowest: behavioral (architectural), register transfer (RTL), logic gate, and switch (transistor) level. Synthesis tools are used to map from a higher-level abstraction to a lower level. At each level of abstraction, the functionality of the IC is verified using a logical equivalency checker (LEC) tool. The back-end stage focuses on converting the logical level data to physical data that is used for manufacturing and is verified using a layout vs. schematic (LVS) tool (see Figure 9.2). The design planning stage is used to plan the placement of physical objects that occur during the back-end phase.

It should be noted that the design and verification of state-of-the-art ULSI designs will not converge without feedback loops in the design flow. There are typically hundreds of feedback loops in a typical flow. This is due in part to the large number of design and manufacturing parameters

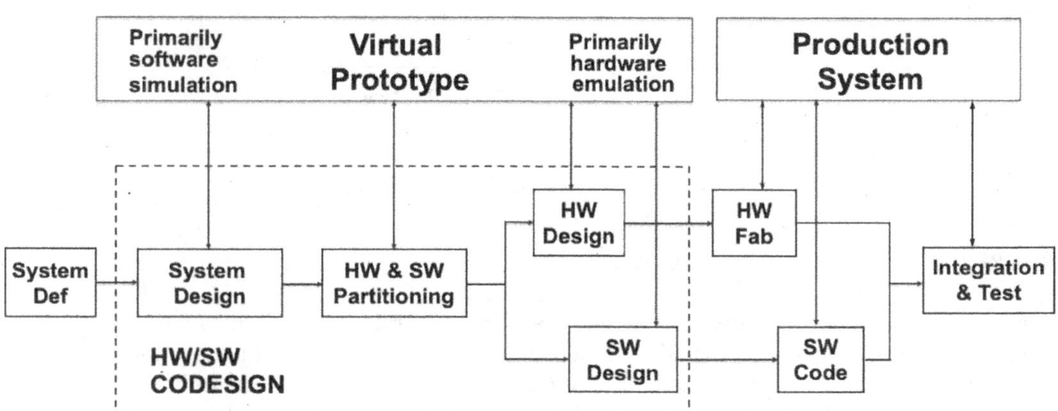

FIGURE 9.1 HW/SW system co-design environment.

DOI: 10.1201/9781003188377-11

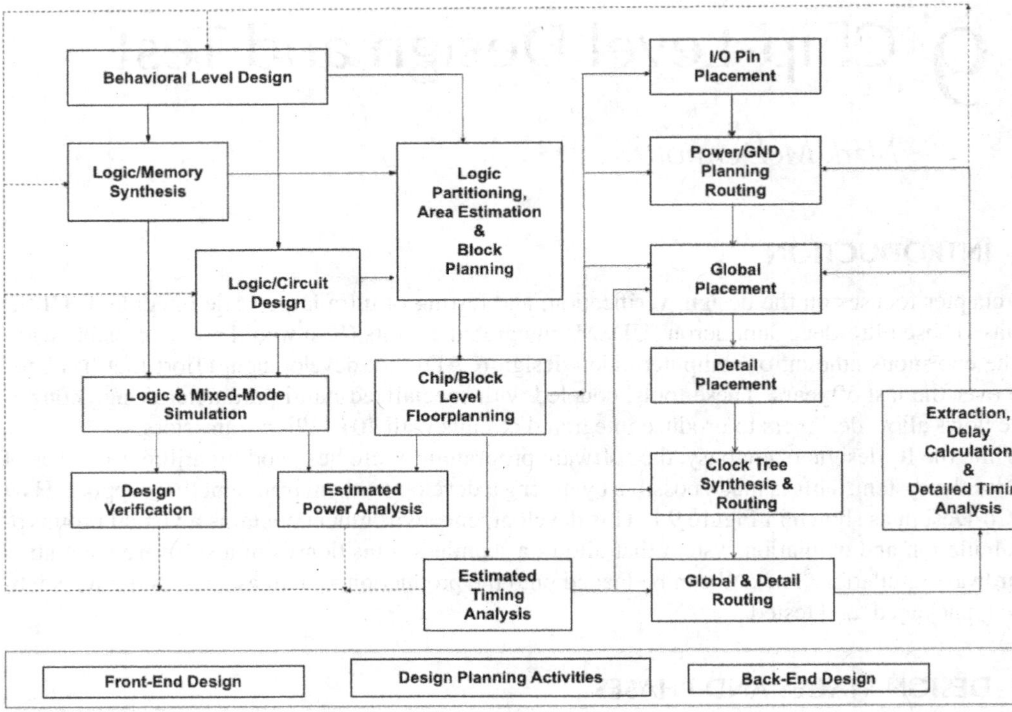

FIGURE 9.2 Basic integrated circuit design flow.

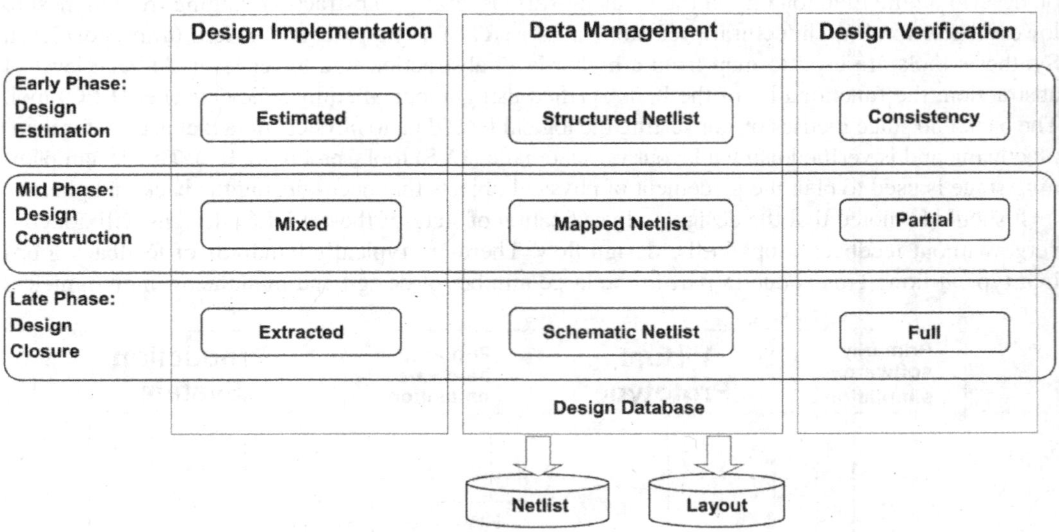

FIGURE 9.3 Integrated-circuit design phases.

(constraints) that need to be met before tape out. Once several similar designs have been implemented using a design convergence flow, it becomes easier to predict when a new design will converge, and the IC can be manufactured. This topic will be covered in more detail in Section 9.2.2.

The front-end and back-end design stages have three phases: design estimation, design construction, and design closure as shown in Figure 9.3. The design estimation phase uses estimated data to produce a structured netlist that can be used for consistency checks. The estimated data is generated from previous similar designs or from senior engineers who can extrapolate the data based on experience. The design construction phase uses the design estimation data to begin the construction of the integrated circuit. This phase continues until it is possible to extract all the physical features from the layout database. The design closure phase can run concurrently with the design construction phase working on sub-blocks that have fully extracted netlists.

There are three key areas associated with each phase: design implementation, design verification, and design data management. Design implementation utilizes architectural and hardware description languages (Verilog, VHDL) and graphical representations (schematics and layout) to fully describe the integrated circuit. The design is complete when a fully extracted layout netlist is available. Design verification confirms that the design implementation meets the design specification at each level of abstraction. The IC can only tape out once the verification steps have been completed. Refer to Sections 9.5 and 9.6 for more detail.

The design collateral is stored in a data management system that is accessed by the implementation and verification tools. The design database is continuously monitored for coherency between levels of abstraction. Additionally, various types of data mining are performed on the design database to predict potential issues that may arise during the design convergence process.

9.2.1 Design Planning Overview

The design planning step starts once the system requirements are solidified. The system requirements include hardware and software, packaging, power, price, etc. These requirements go through a partitioning process where the outcome is a starting point for both software and hardware as shown below in Figure 9.4.

The ideal ULSI circuit design flow would optimize hardware and software concurrently. There are billions of transistors and millions of lines of software in a typical embedded system-on-chip (SOC). Serializing the development of the hardware and then the software will result in long design convergence times and subsequently loss of market share.

Figure 9.5 illustrates a feedback design flow that drives constraints from a high level of abstraction to a lower level during the planning phase and annotates results during the analysis phase. A wide range of design constraints would need to be manually or auto generated for each level of abstraction. Initially, the constraints are estimated based on previous designs or from experienced designers. The more analysis that is performed the more realistic the constraints become.

The feedback loop is generated by analyzing the design and back annotating the results upwards. The constraints are modified and fed-forward again. If the design is unable to converge, various specification parameters may need to be changed. The sooner this can be determined the faster the design team can tape out the IC.

9.2.2 Design Convergence

As mentioned above, it would be virtually impossible to tape out a multi-billion transistor integrated circuit without aggressive design convergence techniques. There are several key parametrics that must be met before taping out. These include frequency, operating power, standby power, die size, test coverage, manufacturability, resiliency, redundancy, and repairability. The convergence techniques must be capable of simultaneously optimizing ALL parameters during the design flow.

Figure 9.6 below shows a "typical" convergence flow starting early in the front-end phase and continuing to tape out. The *Y*-axis is the number of possible starting points. The time between snapshots of starting points is determined by the amount of detailed data that is available. Recall that during the front-end design phase (also referred to as rapid exploration) estimated data is used until the design construction phase begins to generate detailed data.

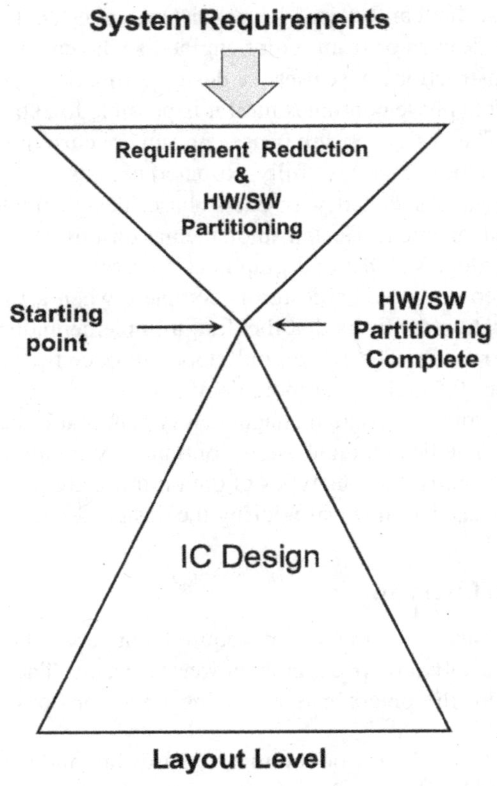

FIGURE 9.4 HW/SW starting point.

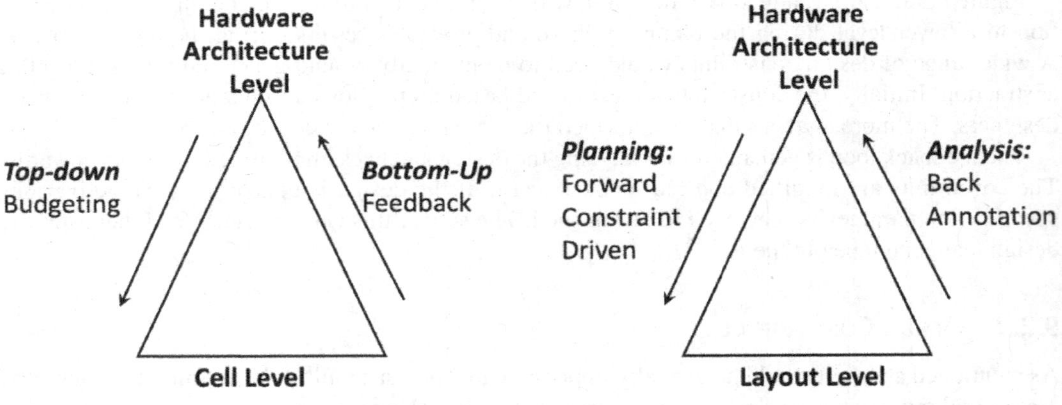

FIGURE 9.5 Hardware design planning and analysis feedback loops.

Once the back-end design phase starts there is a transition from the rapid exploration activities to the rapid traversal activities. This is denoted by the reduction in possible starting points and an increase in convergence time. Periodically there will be an increase in starting points due to repartitioning and retiming efforts. This will result in a rapid reduction of starting points as the optimal starting point(s) become more obvious to the physical design team. Once the convergence times decrease and the starting points decrease, it becomes easier to predict when the IC can be taped out (Wikipedia, n.d.a).

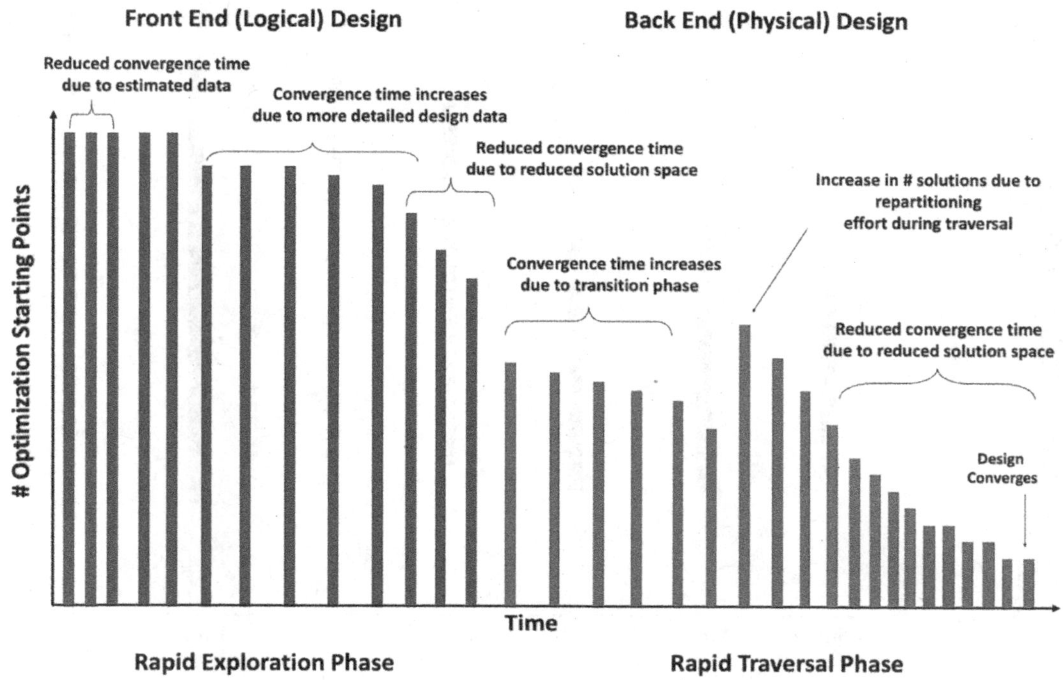

FIGURE 9.6 IC design convergence.

9.2.3 Statistical Design

Ultra-deep-submicron (UDSM) technologies have introduced several new issues that must be carefully analyzed to confirm that the IC will function over a wide range of voltage, temperature, and processing corners. This has introduced the need to statistically analyze both digital and analog circuits to confirm that they will function correctly on the same die and on the same wafer. There are three sources of variation in the design and manufacturing of ULSI circuits: process variations, layout, and CAD tools.

Process variations include ion implantation, photolithography, gate etch, thermal processing, tolerance in the size of the mask image, mask alignment, angle of the etched polysilicon, temperature, polishing, and planarization. These variations cause "across chip" and "across wafer" differences which affect performance and yield. UDSM manufacturing requires that the layout database be modified to enhance the printability of the shapes. This is called optical proximity correction (OPC) and is used to compensate for image errors due to process effects or optical diffraction during manufacturing.

Process variations and process scaling affect device mis-tracking, metallization, effective channel length, thickness of the oxide, doping irregularities, transistor fatigue, and many other issues. To analyze these effects using classical process, voltage, and temperature (PVT) techniques takes an incredible number of computational cycles to determine the parametric yield from the design slack distribution for a target clock frequency. Figure 9.7 below shows Min/Max performance variation for various transistor gate lengths. Note: transistor gate lengths 22 nm and below are Fin-FET devices (Wikipedia, n.d.b) or GAA-FET devices (Hu and Li, 2021).

9.2.4 Design Analysis

Design analysis is a very broad and tedious effort in IC design. Table 9.1 shows some key design analysis activities and key metrics. The complexity metric is an indication of the level of human

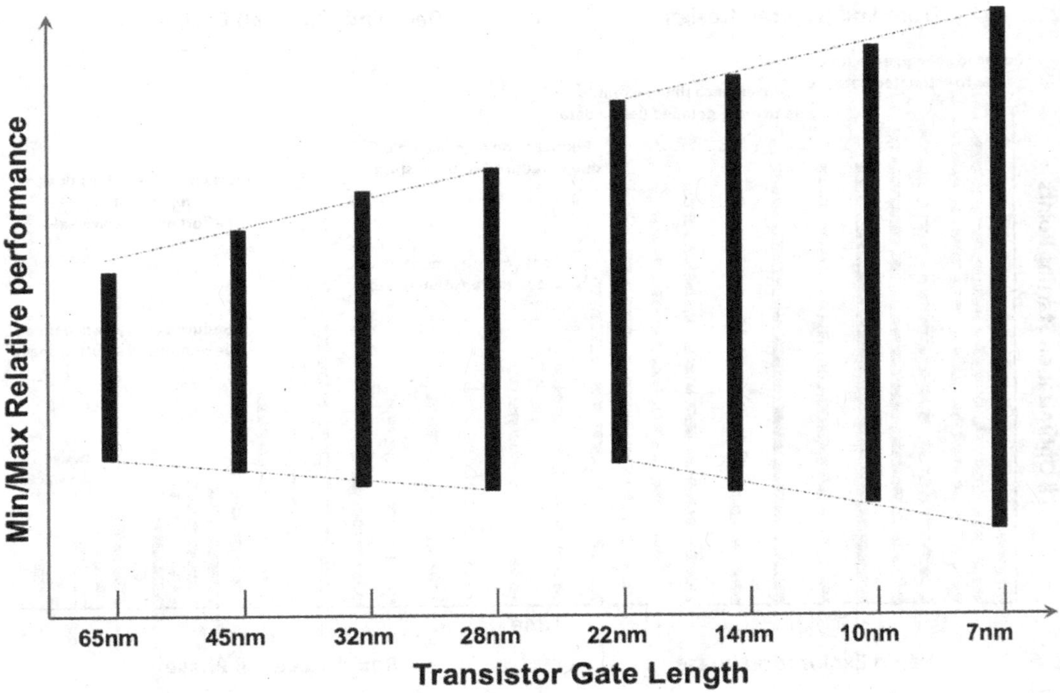

FIGURE 9.7 Min/Max transistor performance variation.

TABLE 9.1
Design Analysis Activities and Key Metrics

Design Analysis Activity	Complexity	Data Generated	Computational Requirements
Parasitic Extraction	Very High	Very Large	Very High
Logic Simulation	Medium	Large	High
Mixed Mode Circuit Simulation	High	Very Large	High
Static Timing Analysis	Very High	Large	Very High
Statistical Timing Analysis	High	Medium	High
Dynamic Power Analysis	High	Large	High
Noise and Signal Integrity	Very High	Large	Very High
Design Rule Checking	High	Medium	High
Electrical Rule Checking	High	Medium	Medium
Layout vs. Schematic Checking	Medium	Medium	Medium
Design for Manufacturability	Medium	Medium	Medium
Optical Proximity Correction	High	High	High

involvement that is needed to perform the activity. The amount of data generated during a particular design analysis activity is a proxy for the time it takes to execute the design analysis tool. There is a tradeoff between time and computational requirements. Many of the latest IC-CAD tools can split a design analysis activity across multiple CPUs and/or server nodes.

One of the key activities is parasitic extraction. This involves doing 3D analysis of the IC layout and extracting ALL the transistors, resistors, and capacitors. There are over 40 layers in a typical ULSI device which results in hundreds of gigabytes of extraction data being generated depending on how accurate the results need to be. This requires hundreds of computational nodes to complete in a reasonable amount of time. Note: the only activity in this list above that does not use the extraction results is logic simulation, which is an indication of the importance of parasitic extraction. Once the extracted data is obtained, it is further modulated to mimic the effects of OPC and the manufacturing process. For example, crosstalk between two signal wires is a function of the distance between the two wires. This distance varies depending on how well the manufacturing process replicates what was in the layout database. The minimum and maximum distances are modeled as a Gaussian distribution for statistical design analysis but not for static design analysis. The modulation of the extracted data for static design analysis results in more than 100 different extraction databases that must be analyzed individually.

9.3 DETAILED IC DESIGN FLOW

Figure 9.8 illustrates a detailed view of the IC design "cockpit". The design partitioning phase and the design planning phase focus on the rapid exploration of potential starting points for the implementation phase consisting of synthesis and integration phases. Each one of these areas will be discussed in more detail below.

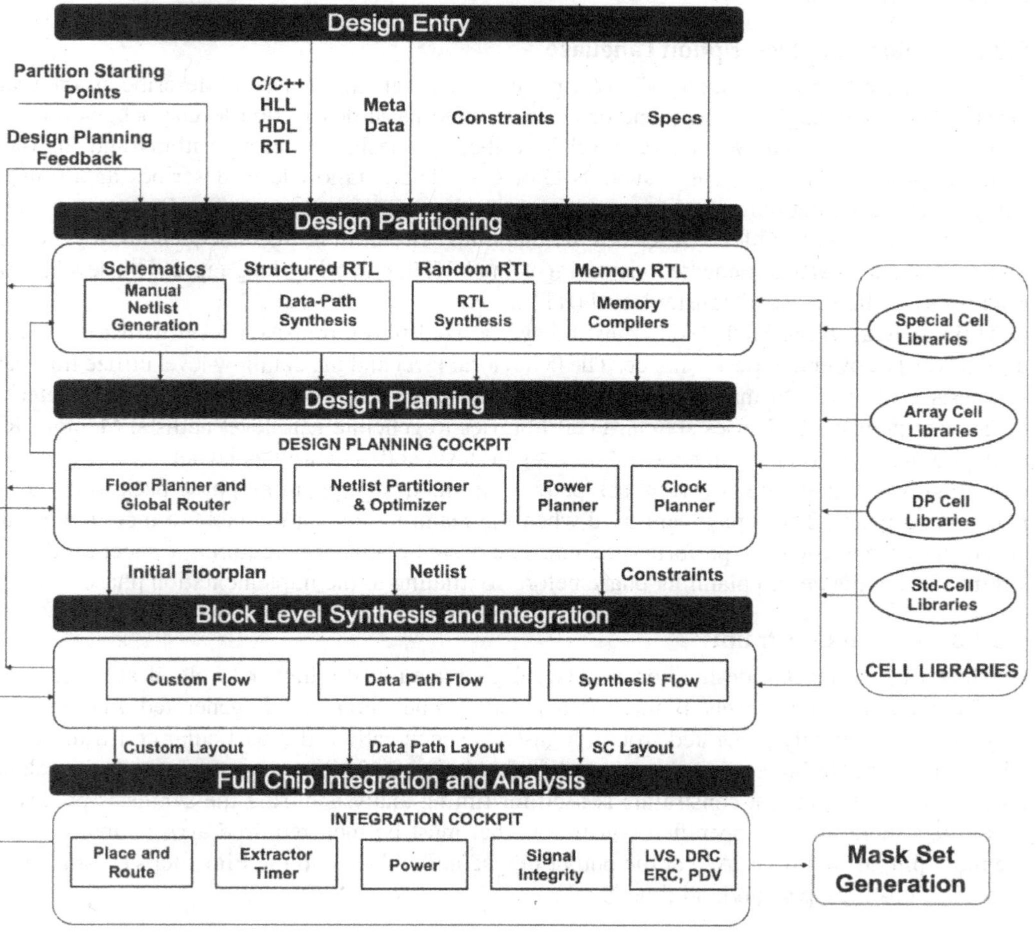

FIGURE 9.8 Detailed IC design flow.

9.3.1 Design Entry

The design entry phase consists of a number of different activities that generate the data and meta-data needed for the design partitioning phase. These include the architectural and logical description(s) of the integrated circuit, design specifications, design constraints, and potential starting points. Ideally, much of this information can be derived from previous designs. World-class design organizations have automated data collection systems that capture critical data and meta-data and store the information in tagged repositories for use on derivative IC designs. Manufacturing data is also captured and placed in these repositories for analysis and correlation to model data used during the implementation phase.

9.3.1.1 Schematics

In the early days of IC design, schematics were the only form of design entry. They are a very tedious method for describing an IC. As the complexity of ICs grew, many hardware description languages (HDL) were developed to speed up the design entry process. However, there are still many functional units on an IC that must be described using schematics. For example, a Digital-to-Analog (DAC) converter may utilize a resistor ladder tree to convert digital values to an analog voltage value. A switched capacitor filter may have 50 unique capacitor structures that need to be described. These functional units cannot be automatically synthesized and require custom schematic and layout implementations.

9.3.1.2 Hardware Description Language

There are more than 30 "languages" (Wikipedia, n.d.c) that can be used to describe an integrated circuit. These languages fall into one or more of three basic description levels: behavioral, dataflow, and structural. The behavioral level describes the design using algorithms and high-level software programming languages such as C or C++. The dataflow level describes data transfers using synchronous dataflow (SDF) event-driven simulators such as Matlab/Sim-Events, Ptolemy, or Lab-View. The structural level describes the integrated circuit in terms of sequential (register) and logical elements that are connected via wiring using VHDL or Verilog languages. This level is often referred to as the Register Transfer Level (RTL).

Ultimately all three HDL levels must be synthesized into a transistor level netlist that can be used to verify power and performance. The behavioral level and the dataflow level utilize high-level synthesis CAD tools to map the language constructs into known elements using parameterized macro-components. RTL uses standard cell libraries to generate gate-level netlists. Memory RTL utilizes a memory compiler to generate both a gate-level netlist as well as layout.

In Figure 9.8, there are two feedback paths from the design planning phase back to the design partitioning phase. These paths are used when the planning tools are unable to meet a particular constraint such as die size, performance, etc. The design engineer is required to meet all the constraints during the design planning phase before continuing to the implementation phase.

9.3.1.3 Design Constraints

As shown in Figure 9.5, design constraints are generated at the higher levels of abstraction and "fed" down to the lower levels. In many cases, they can be automatically generated. In other cases, they must be manually generated from a high-level specification. If a particular constraint cannot be met, then the design engineer must modify the design or petition to get the constraint relaxed. One of the most common constraints is used for timing analysis during the synthesis process. In Figure 9.9, there are four specified constraints that must be met: required arrival time (RAT) at the input pin, arrival time (AT) at the output pin, setup/hold time to the Flip-Flops (Flop), and the internal Flop-to-Flop setup/hold time.

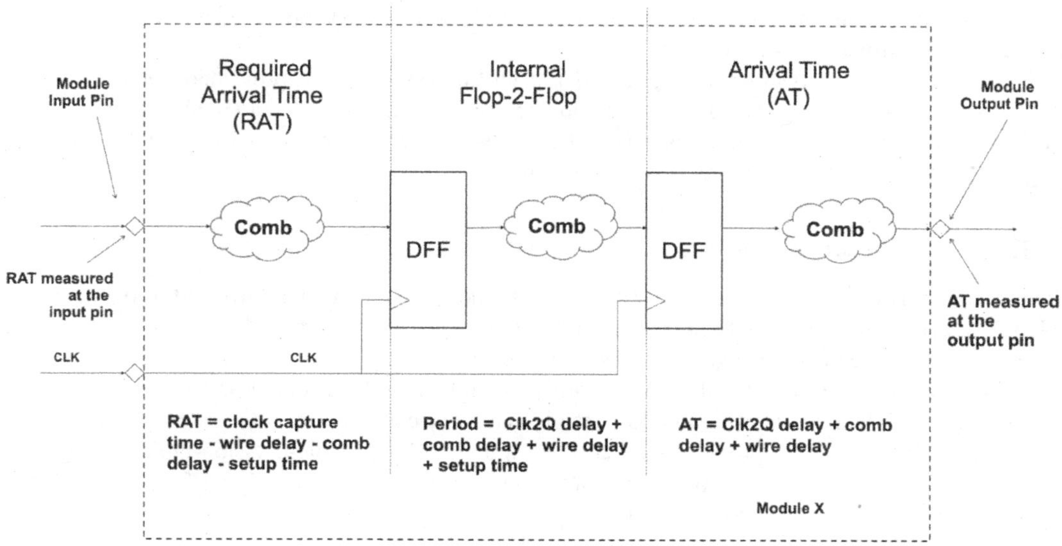

FIGURE 9.9 Timing constraints.

Constraint generation is one of the most critical activities in integrated circuit design. It is very easy to make a simple mistake that will cause an IC to miss performance goals when it is manufactured. Strict adherence to a robust design flow is required.

9.3.1.4 Meta-Data

Meta-data is the output of human capital and is NOT expressed in the form of specifications, user manuals, etc. It is considered the next generation institutional learning and often called Knowledge Meta-Data (KM). In Figure 9.10 we can see three potential sources of data that would be used for seeding a new IC design. In many cases, there is an overlap between the data implying that there is more data than just the data itself (meta-data).

Constraints are covered in the section above. Decisions and associated context are useful information when considering similar contextual situations on a new design. There are many types of assertions (truths) that can be captured. Many of these assertions are specific to design verification and are covered in Section 9.5.1.3. Other assertions having to do with design specifications and non-verification activities are very useful in generating viable starting points on future designs.

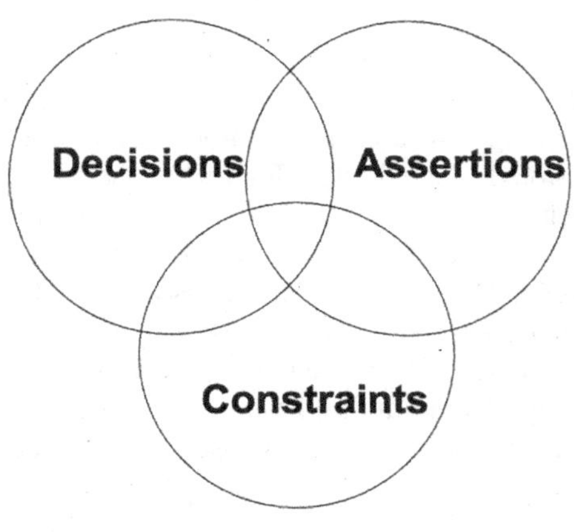

FIGURE 9.10 Knowledge meta-data.

Assertions and decisions along with related context provide an insightful view into the thought processes that drove previous designs.

With the advent of machine learning (ML), it will be possible to capture additional meta-data from the terabytes of data that is generated during the implementation of a ULSI circuit. As mentioned above constraint generation is very critical. Research is underway to use ML to check existing constraints as well as generate new constraints.

9.3.2 Design Partitioning

The design partitioning phase is very critical. It sets the basis for the planning and implementation phases. During the specification and HLL modeling stage, the system architects will specify the major blocks of the design as well as the width of critical data buses that connect the blocks. This specification must be rationalized with the ability to implement the design. For example, specifying a 1,000-bit wide data-bus is generally not feasible due to the excessive area and power requirements. The micro-architects may propose a much narrower bus that is time-division-multiplexed using low-voltage differential bussing to save area and power at the expense of a higher frequency bus and more complicated transmit and receive logic.

Three types of building blocks are used in a VLSI chip:

- **SC: Standard Cell Block**
  - Typically synthesized using standard cell library.
  - Layout is done using Automatic Place & Route (APR) tools.
- **SDP: Structured Data Paths**
  - Typically designed using datapath libraries.
  - Layout is generated by tiling engines using relative placement constraints.
  - Routing can be done manually (for busses, clocks) or with automated routers or a mix of both.
- **Customs Macros:** Memory arrays, Register Files, CAMs, PLLs, Thermal sensors, off-chip IO buffers, voltage regulators, etc.
  - Memory generators can be used; High-performance arrays are typically done manually. Memory generators will produce layout.
  - Semicustom design is also used: leaf cells are pure custom, but block can be built with AP&R tools.

The output of the design partitioning phase is a set of high-level netlists that are forwarded to the design planning phase.

9.3.3 Design Planning

The inputs to the design planning flow are the partitioned high-level netlists. Each partition goes through most of the steps outlined in the design planning phase flow is shown in Figure 9.11.

The design planning "cockpit" allows the architect, micro-architect, logic designer, and physical designer to work together on generating the optimal layout. Wire delays are the main source of timing issues at 22 nm and below. Wire capacitance is also one of the key contributors to dynamic power. Optimization of wires is only possible if the design can be rapidly re-partitioned using feedback from the design planning phase. There are typically 20–30 high-level floorplans concurrently being developed during the rapid exploration phase. Once three or more of these floorplans start to converge the focus shifts to the rapid traversal phase where the detailed block-level design and integration activities begin.

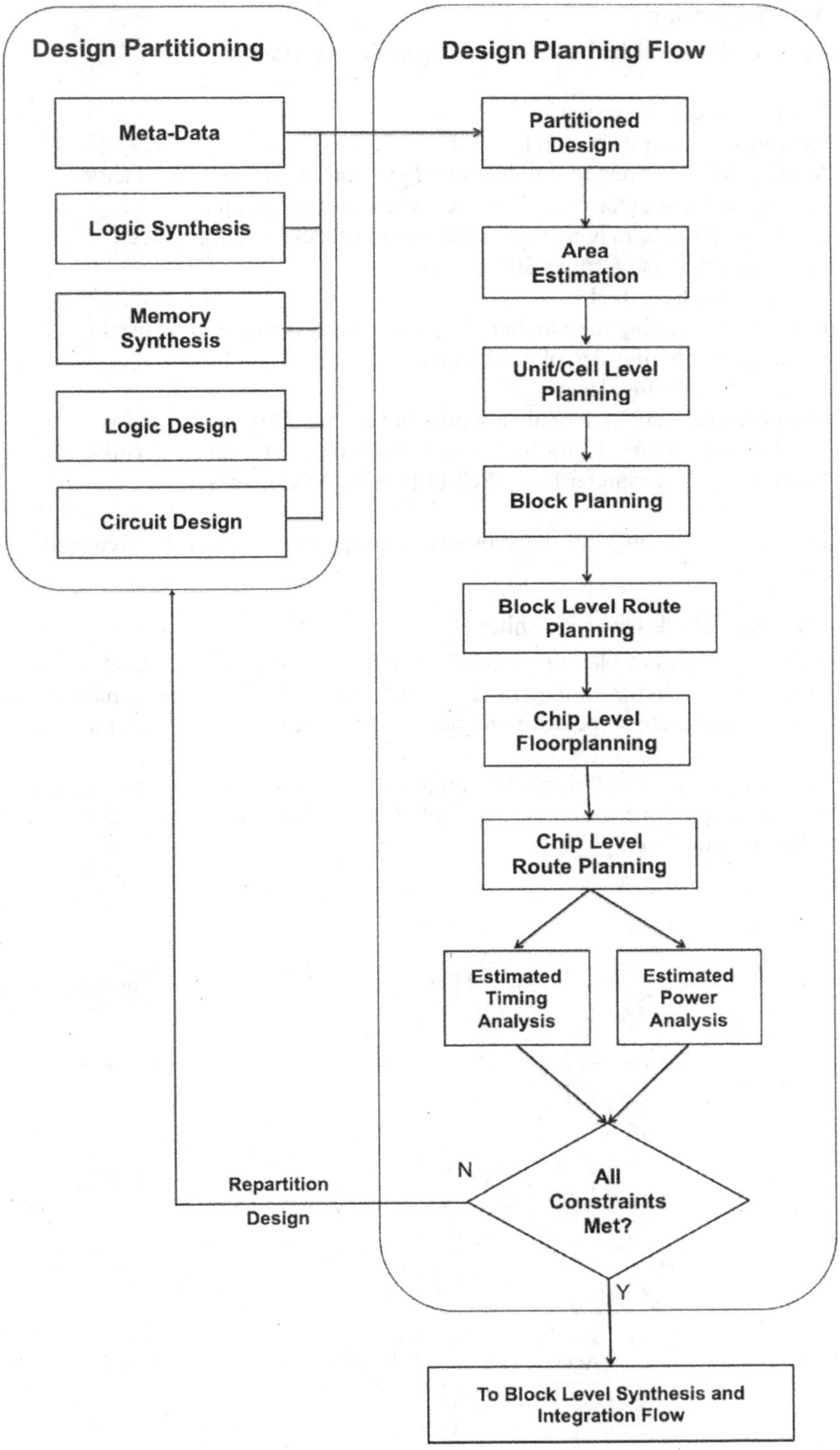

FIGURE 9.11 Design planning flow chart.

9.3.3.1 Area Estimation

Area estimation is accomplished using one of these three methods:

- Scaling from previous design
  - Reasonably accurate method
  - Modify area by direct multiplication of previous area by scaling factor
  - Scaling factor is determined by process technology group
    - Scaling factor may be non-linear due to process scaling issues.
- Manual estimation using spreadsheets
  - Least accurate method
  - Requires estimating the number of logical elements that will be used
  - Requires estimating size of hard macros
- Synthesis of an existing design
  - Most reliable method. Not always possible during early design phase
  - Need to add "fudge" factor to accommodate future growth (or shrinkage)
  - Need to estimate standard cell (SC) utilization percentages

This activity continues throughout the subsequent design phases until the area estimations for all blocks are finalized.

9.3.3.2 Chip and Block-Level Planning

This activity focuses on floor planning a block or many blocks. A block would be a module that is one or more levels down in the chip hierarchy. While it is not optimal, it is sometimes more practical for the physical hierarchy to be different than the logical hierarchy. An example of this is shown in Figure 9.12.

In this example, the physical hierarchy collapses units and sub-units into larger layout cells to help with timing, pin placement, and over-the-cell (OTC) cell routing. It does cause issues with LVS checking at the sub-unit level.

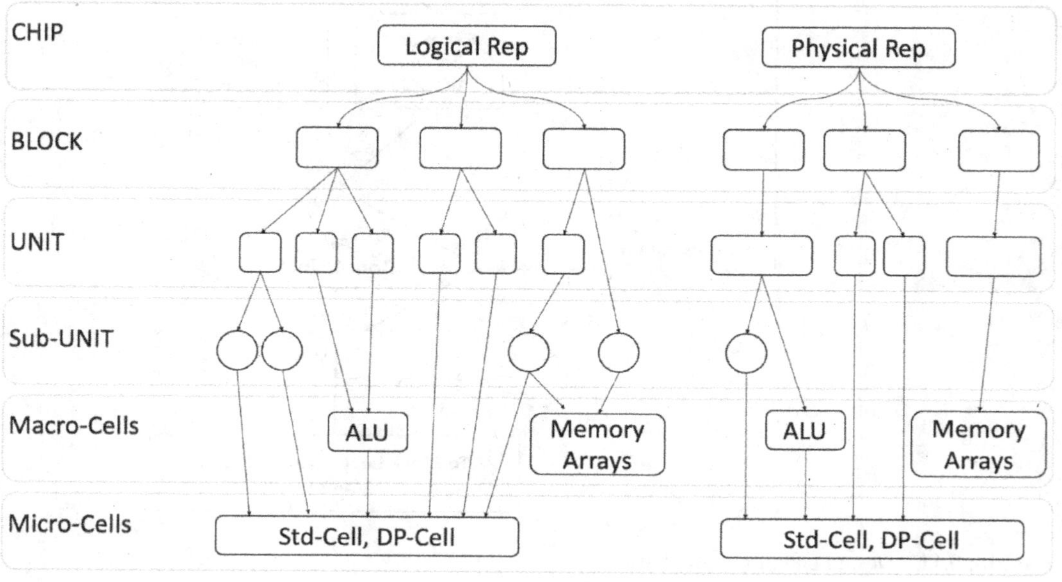

FIGURE 9.12 Logical vs. physical hierarchy.

The following activities are performed during Chip- and Block-level planning:

- Block pin placement
- Power grid planning
- Clock route and buffer planning
- Signal Routing constraint generation including pre-routes & shielding requirements
- Structured Datapath planning
- Random logic area planning
- Identification of hard macros (custom-designed sub-blocks)
- White space
- Over-the-Block routing
- Identify clock-domain 'islands'
- Identify power 'islands'

One of the key outcomes of the block planning activity is pin placement. This helps with chip-level routing by eliminating "scenic" routes from pins on one block to pins on another block.

9.3.3.3 Chip-Level Floorplanning

Chip-level floorplanning and route planning starts almost at the same time as block-level planning. Typically, there are five or more chip-level floorplans in development. This may require multiple block-level floorplans with different pin placements, different hierarchies, different OTC routes, etc.

A chip-level floorplan of the Motorola MC68332 (circa 1989) (McDermott, 2021) is shown below in Figure 9.13. The diagram on the left shows the top-level block-level placement and one routing channel. The die photo on the right shows the manufactured die and how closely it matched the final chip-level floorplan.

The *X* and *Y* dimensions of the die are determined during this planning phase. The manufacturing team will use this floorplan to refine the design of the package that the IC will be placed in. The die size (in sq mm) also determines the yield and the subsequent cost of the die. Figure 9.14 shows two wafer level die placements. The wafer on the left has four die per. The wafer on the right has 44 die per. The dots indicate where a manufacturing defect occurred. The wafer on the left has one good die and three bad die (25% yield) while the wafer on the right has 36 good die and eight bad die (81% yield). The number of dies per wafers are determined by the following equation:

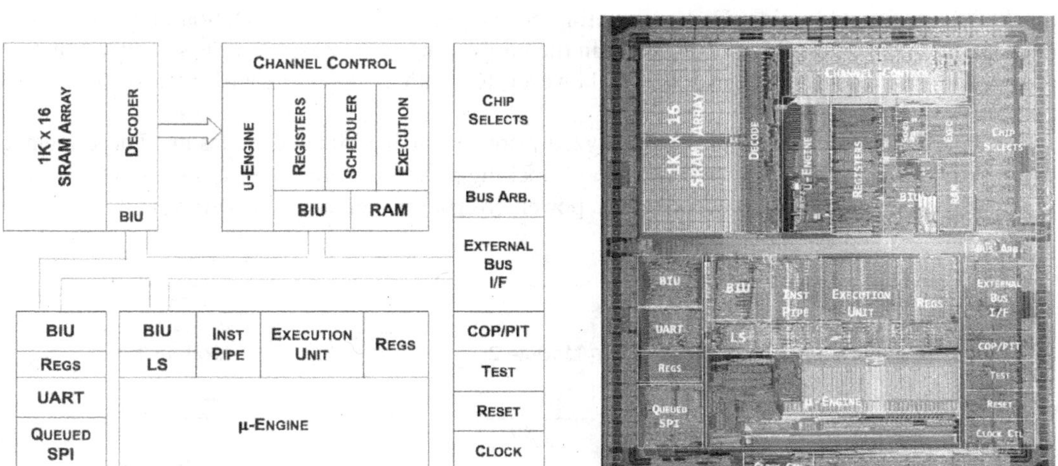

FIGURE 9.13 Chip-level floorplan and die photo for the MC68332.

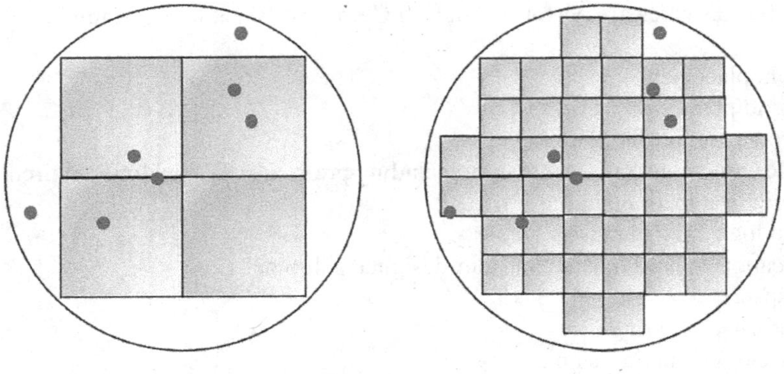

FIGURE 9.14 Impact of die size and wafer level defects.

9.3.3.4 Chip-Level Estimated Timing Analysis

Estimated chip-level timing analysis can start once the first chip-level floorplan is available and the pin placement on each block has been finalized. The blocks are represented as "black boxes" with simple timing constraints (RAT/AT) on each pin as shown below in Figure 9.15. The top-level routing is then performed, and the wire delays are extracted. The timing analysis tool will determine if the top-level routing operates at the required speed. If not, there are several options to remedy the problem:

- Modify the constraints on the offending pins
- Redo the floorplan to reduce wire delays
- Clone logic to remove global wiring
- Add buffers to the wires to restore signal rise/fall times
- Move critical signals to less capacitive metal layers and pre-route

This activity continues until top-level timing is met and the block-level constraints are determined to be implementable.

9.3.3.5 Chip-Level Estimated Power Analysis

One of the most critical estimation exercises that must be done prior to committing to the design of an IC is power estimation. Underestimating power can lead to the cancelation of an IC project (Wikipedia, n.d.d). This is best done early in the project. In most cases, power is strictly a function of charging and discharging capacitances; however, ICs with analog units will have resistive paths between VDD and VSS.

There are eight areas that must be analyzed prior to starting block-level design. These include block activity factor, signal switching factors, clocking power, memory power, logic power, glitching power, wire capacitance power, leakage power, crowbar power, and I/O power.

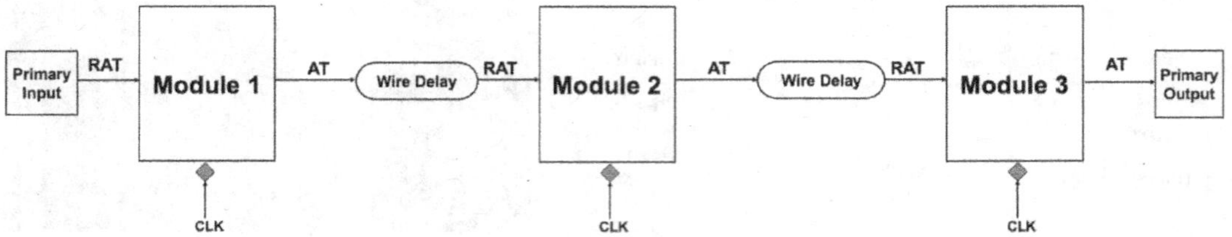

FIGURE 9.15 Chip-level timing analysis using black boxes.

1. **Block Activity Factor**: Each block/unit on an IC will be active some percentage of a time. For example, an instruction fetch unit is active 80%–90% of the time, whereas a debug unit would be active 0.1% of the time. The system architects will determine the block activity factors from the architectural simulations running specific workloads.
2. **Signal Switching Factor**: This factor is dependent on the type of circuitry being analyzed. It is also represented as a percentage and indicates how often the internal nodes of a specific block/unit toggle. For example, inverters switch all the time, four-input NAND gates switch considerably less and complex gates have even lower switching factors.
3. **Clocking Power**: There are numerous clocks on a typical integrated circuit. A small number of clocks are free running which means that they have signal switching factor of 200% (since clocks switch twice per cycle). The remaining clocks are architecturally gated or have clock gating circuitry inserted into the gate-level netlist during the logic synthesis stage.
4. **Memory Power**: Most power dissipation in a memory array occurs in the bit-lines and sense amplifiers. The bit-lines are high capacitance and continuously precharged and discharged on every memory access. The sense-amps are enabled for every memory access. For high-performance memories, the sense-amps are usually biased into a very high gain mode which uses excessive current.
5. **Logic Power**: This power is a function of the number of transistors used in the control logic and datapath logic and can dominate the power used in an integrated circuit if you include glitching power (see below). The logic is synthesized from HDL and the number of resulting transistors is directly influenced by the level of abstraction used in the HDL.
6. **Interconnect Wiring Capacitance**: This has become one of the largest contributors UDSM power due in part to the increased number of routing layers that have been added the new process nodes. Careful attention to the placement of blocks/units will help reduce wiring induced power.
7. **Glitching (False Switching) Power**: This is caused by several different effects:
   a. Spurious transitions (hazards) before an output approaches steady state
   b. Uneven propagation delays of input signals to a logic gate. UDSM technology nodes make interconnect delays more dominant, so delays are more uneven
   c. Multiplies as the signals propagate through many combinational logic blocks

   Figure 9.16 shows the generation and propagation of spurious transitions in a typical logic circuit.

   Glitching can account for 10%–60% of the total power in some logic elements like arithmetic units, multipliers, etc. As mentioned above, complex gates have low switching factor which helps prevent excessive glitching power. Logic synthesis can be directed to use more complex gates when synthesizing a unit if timing constraints permit.
8. **Leakage Power**: There are three main sources of leakage power on an integrated circuit as shown below in Figure 9.17.
   a. Subthreshold leakage is due to the inability to turn off transistors when the gate voltage is equal to the source voltage. FinFET technology (Wikipedia, n.d.b) has helped to

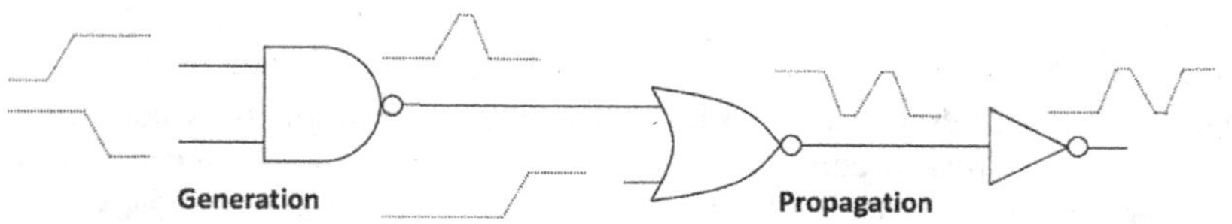

FIGURE 9.16 Generation and propagation of spurious signals.

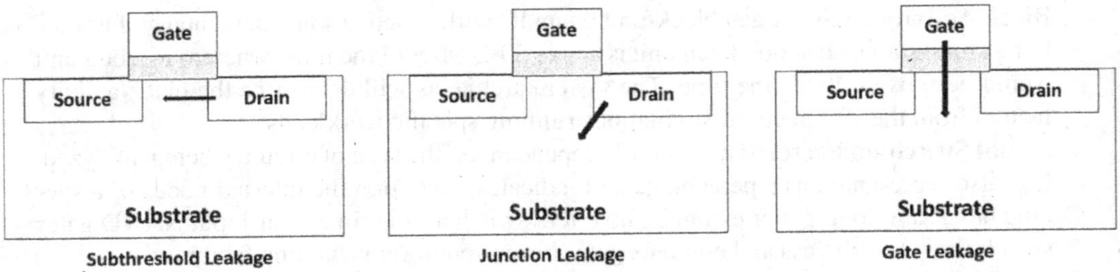

FIGURE 9.17 Sources of leakage power.

minimize the problem. It is anticipated that the Gate-All-Around (GAA) (Hu and Li, 2021) technology will further minimize subthreshold leakage power.

b. Junction leakage is due to the relative doping concentrations of the source or drain with respect to the substrate which determines the width of the depletion region. As the doping concentrations increase and the depletion region shrinks the leakage increases.

c. Gate leakage has been increasing rapidly as the native gate oxide (SiO_2) approached four monolayers. SiO_2 was replaced with Hafnium Dioxide in 2007 (Wikipedia, n.d.e) which reduced the effects of gate-to-substrate leakage by reducing tunneling between the gate and substrate.

9. **Crowbar (Shoot-through) Power**: This is typically caused by slow rise-fall times on the inputs to a CMOS logic element. As the input signal transitions through the switch-point, both the N-channel and P-channel are on resulting in a "resistive short" between VDD and VSS. This effect can be mitigated by speeding up the rise-fall times on the inputs.
10. **Input and Output Power**: As we build more complex integrated circuits, the number of Input and Output (I/O) pins increases. There can be many pico-farads of load on a typical I/O pin which can be equivalent to many thousands of on-die interconnect wires.

9.3.4 Custom-Designed Blocks

In many cases, synthesis tools are unable to produce circuits and layouts that meet the specifications of the design. In these cases, we must design custom circuitry and layout. There are six major custom areas: Digital, Analog, Mixed Signal, Memory, Library Elements, and Input/Output elements.

9.3.4.1 Digital Logic

There are a few key areas where custom digital logic is required including adders, multipliers, Delay-Lock-Loops (DLLs), Phase Lock Loops (PLLs), clock recovery circuits, subthreshold logic circuits, asynchronous logic circuits, etc.

9.3.4.2 Analog

Analog circuits include operational amplifiers, sense amplifiers, analog-to-digital converters (ADC), voltage-controlled-oscillators (VCO), comparators, etc. Analog circuits require extensive simulation before going into production. In some cases, the circuits are tested on actual silicon before being committed to production status.

9.3.4.3 Mixed Signal

Mixed signal design is generally referred to as digitally assisted analog where the best features of digital circuits and analog circuits are merged to produce a component with superior power-performance characteristics. An example of this is a biosensor pixel used for DNA and RNA testing (Wikipedia, n.d.f) where the digital logic provides precise clock timing for controlling the analog circuits.

9.3.4.4 Memory

Memory elements are critical components of most integrated circuits and are usually custom designed for each application. There are two basic forms of memory - volatile and non-volatile, as shown in Figure 9.18. Volatile indicates that the memory is lost when the supply voltage is removed from the integrated circuit compared to non-volatile memory which will retain its memory state.

Volatile Memory: There are two types of volatile random-access memory (RAM): Dynamic (DRAM) and Static (SRAM). The data is stored in capacitors in the DRAM and must be refreshed periodically, or it will be lost. DRAMs can store 6–10 times more data than an SRAM. However, SRAMs do not need to be refreshed and can retain data at very low supply voltages.

Non-Volatile Memory: There are different storage mechanisms for this form of memory including fuse, electric charge, ferroelectric, magneto-resistive, and phase change. The list of memories using these storage mechanisms are:

- ROM – Read-Only Memory: Fuse or diffusion
- OTP – One time programmable: Fuse or floating gate
- EEPROM – Electrically Erasable Programmable ROM: Floating gate
- FLASH memory: Floating gate/nitride
- FRAM/FeRAM: Ferroelectric RAM
- MRAM: Magneto-resistive RAM
- ReRAM/RRAM: Resistive random-access memory
- PCM: Phase Change Memory

The dominant storage mechanism is currently electric charge on a floating gate or floating nitride. This mechanism has reached its end-of-life due to the physical size of the floating gate which limits the amount of charge that can be stored.

9.3.4.5 Standard Cell Library Elements

The logic synthesis process requires many library components of varying functionality and drive strength to meet the timing constraints of the block/unit. The functionality ranges from simple logic elements like inverters, NAND, NOR gates to sequential elements like JK-Flip-Flops and D-Flip-Flops. Some commercially available libraries will also contain macro-elements such as Adders, Multipliers, DSP Slices, etc. Schematics are generated for each element in the library and then manually converted to layout. The layout is converted to an extracted schematic (one with parasitic devices) and simulated using a circuit simulator. The results of the circuit simulation are

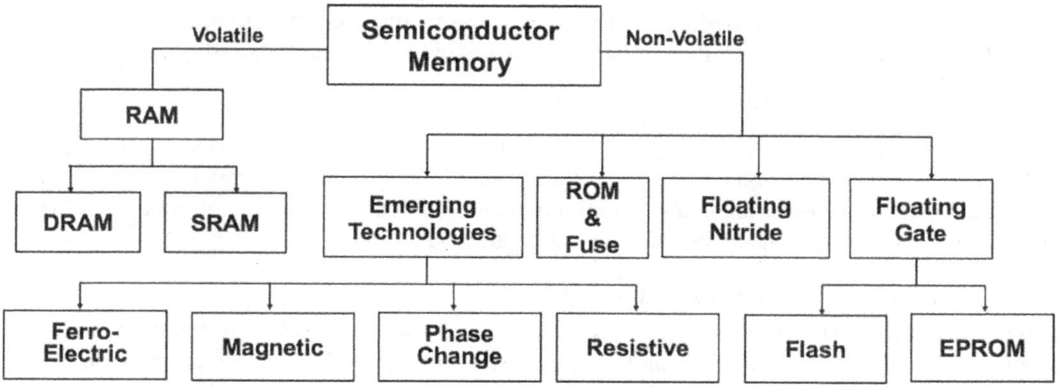

FIGURE 9.18 Types of semiconductor memories.

converted to a lookup table that is used by the logic synthesis tool during the block or unit level implementation phase.

9.3.4.6 Input/Output Elements

I/O elements are used to connect the IC to the external world. In many cases, they are also level shifters that convert the internal voltage levels to a standard external voltage level. The standard external voltage levels guarantee that ICs can communicate with each other when they are placed on a printed circuit board (PCB). There are several different types of I/O elements including:

- General purpose I/O
- Low-voltage differential
- Analog
- Hysteresis inputs
- Open drain or open source

9.3.5 Block/Unit Level Implementation

A well-planned block/unit level implementation is critical for the chip-level integration process. There are hundreds of constraints that are generated in the planning phase including:

- Well characterized library elements
- Standard cells
- Hard macro cells
- Placement locations for the hard macros
- Block pin placements
- Timing constraints on all pins
- Power constraints for the block
- Maximum Block/Unit area (mm^2) and aspect ratio (if applicable)
- Over-the-Block routing channels
- Percentage of spare gates
- Percentage of bypass capacitors
- Testability constraints

9.3.5.1 Logic Synthesis

The process of converting HDL code to schematics is done using logic synthesis and mapping. The first step is to generate a generic schematic of the logic function described in the HDL. Next the generic logic gates are mapped to the available logic elements in the standard cell library. Lastly, the standard cells are sized based on the timing constraints that were generated by the block/unit planning phase. The final schematic netlist is now ready to be placed and routed.

9.3.5.2 Memory Compilation

This step uses parameterized building blocks to compile (synthesize) ROMs, RAMs, and Register Files. The compilation process generates layout, gate-level logic simulation model, a timing simulation model, built-in self-test logic, and power simulation model. The layout is considered a "hard" macro that is used during the block/unit level place and route phase.

9.3.5.3 Block/Unit Level Place and Route

The place and route (PNR) phase generates the final layout for the block or unit level component. The first step is to place the pins using the information provided by the planning phase described above. The next step is to place the hard macros. These are components that have had the layout generated during a previous phase such as compiled memories, datapaths, etc. The final step is to place the standard cells from the logic synthesis phase. This step can take considerable amount of time and computational resources. Once the placement step is complete, there are three critical routing steps that must completed before the final route.

1. **Over-the-Cell Routes**: These are simply routes that need to pass over the top of the current block to connect signals between other adjacent blocks. These routes are routed early to preserve the timing constraints of signals.
2. **Clock Tree Synthesis**: The next set of critical routes involves the clock signals in the block. The specialized pre-routing of these clock signals attempts to balance the arrival times to every critical flip-flop in block.
3. **Pre-Routes**: These are the remaining critical signals that need to be routed early. These are determined during the planning phase and from earlier block routing attempts.

Once the critical signals are pre-routed the remaining signals are routed. This process continues until the timing analysis engine in the PNR tool determines that it can meet the pre-planned time constraints. The last step is to route the scan chain. The scan chain is used for manufacturing test and signal timing is not considered critical.

9.3.5.4 Block/Unit Level Timing Closure

Figure 9.19 shows a typical block/unit level timing closure flow. The input to this flow is a synthesized netlist that has gone through global placement, detailed placement, clock tree synthesis, and detailed routing. The timing engine used in the logic synthesis and in PNR approximates timing delays based on a wire load model.

A signoff Static Timing Analysis (STA) tool must be used for detailed timing analysis and closure. This requires a complete parasitic extraction of all wires, diffusions, and transistor gates for the STA tool. The output of the STA tool will highlight setup/hold time violations and high fanout nets with slow rise/fall times. The designer uses this information to determine what modifications

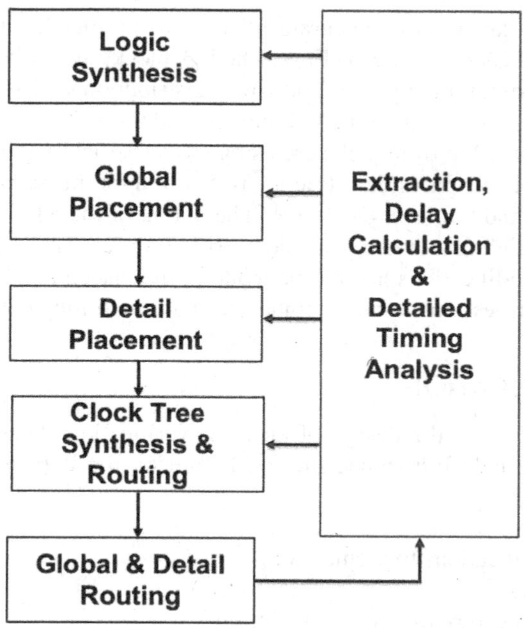

FIGURE 9.19 Block/unit level timing closure flow.

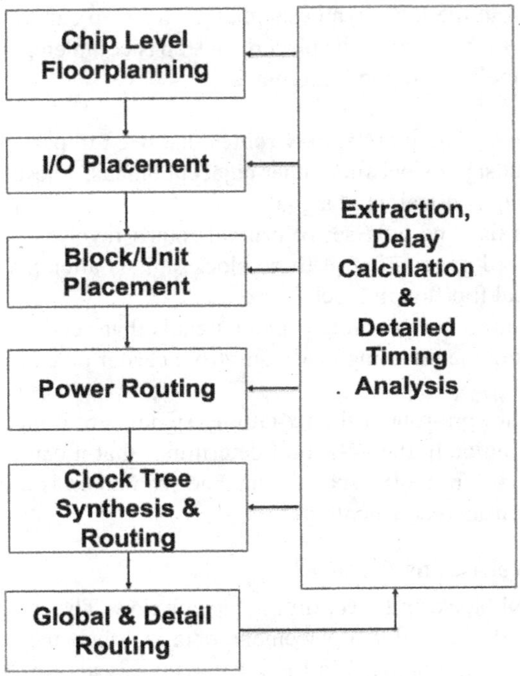

FIGURE 9.20 Full chip integration and timing closure flow.

need to be made in either the logic synthesis flow or in the PNR flow. This loop continues until all the timing constraints are met.

9.4 FULL CHIP INTEGRATION

The last step in building an IC is to integrate all the blocks together to form the whole chip. Figure 9.20 shows the Full Chip Integration Flow. The I/O, blocks and units are placed in their areas determined during the floorplanning phase. The power distribution grid and the top-level clock tree are routed to all the blocks. At this time the I/O, blocks, and units are wired together.

The last step is to do full chip timing closure using the STA tool that was used in the block/unit level timing closure. A full chip parasitic extraction is required for this step to account for the effects of adding the power grid and top-level clock tree. The output of the STA tool will highlight setup/hold time violations and high fanout nets with slow rise/fall times. The designer uses this information to determine what modifications need to be made in the placement of the I/O's, blocks, units, power routing, and clock trees. This loop continues until all the timing constraints are met.

9.5 LOGICAL VERIFICATION

The most critical phase during the design of an IC is verification. There are numerous sources of bugs (SoftwareQuotes, n.d.) introduced into an IC during the implementation phase: The top sources include:

- Mapping from specifications to architecture
- Miscommunication
- Mapping from architecture to microarchitecture
- Logic changes
- Corner cases

- Clock domains and synchronization
- Power domains and level shifters
- Complexity
- Initialization (reset)
- Incorrect RTL assertions
- Design mistakes

It should be noted that verification is not testing nor is it validation. Verification confirms that the implemented design matches the product specification, i.e., *"Are we designing the IC correctly?"*. Validation confirms that the product meets the expectation of the customer(s) i.e., *"Are we designing the correct IC?"*. Testing confirms that the manufactured ICs are free of defects and work as expected i.e., *"Can we sell the IC to a customer?"*. This section covers the logical verification of the IC, and the next section covers the physical verification of the IC. The last section covers the testing of the IC after it has been manufactured.

Over the last 25 years, the ability to design (and manufacture) more complex ICs has outpaced the ability to completely verify that the ICs meet their product specifications. This has helped to generate several CAD tools and techniques to reduce the gap. These include:

- Functional Verification
  - Random
  - Directed
  - Assertions
  - Coverage
  - Emulation
  - Formal property checking
  - Equivalence checking

9.5.1 Functional Simulation

A typical functional verification cockpit is shown below in Figure 9.21. The scoreboard provides a real-time view of the progress of a verification run. The assertion checker and the coverage checker are used to validate the quality of the verification patterns. Based on the quality of the patterns the

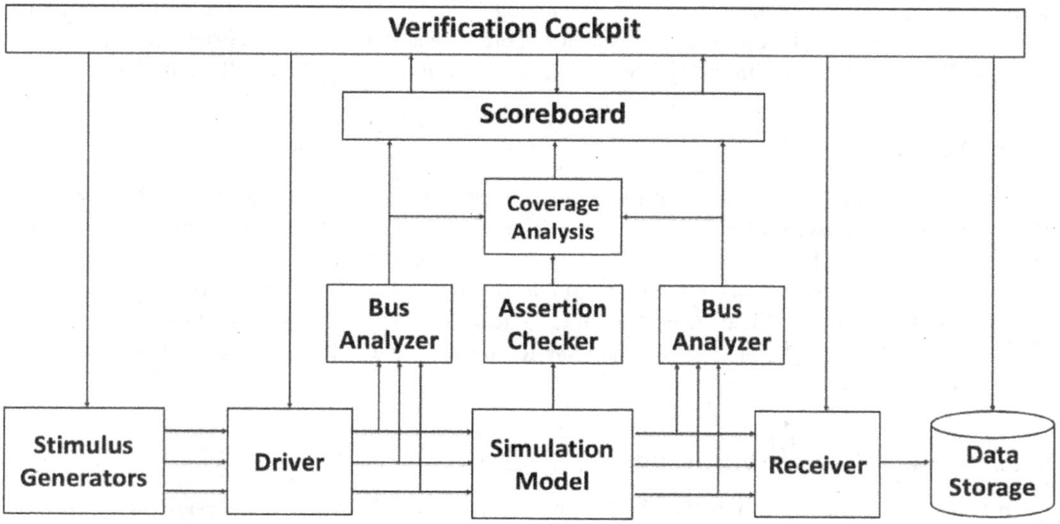

FIGURE 9.21 Functional verification environment.

stimulus generator can be dynamically reprogrammed to provide more or different directed patterns to the simulation model.

9.5.1.1 Random Patterns

As the name implies, the stimulus generator generates random patterns and sends them to the simulation model. The intent is to randomly run patterns until the simulation model fails. There are a couple of issues with this approach: (1) The patterns can generate many verification failures that need to be debugged. In many cases, the test patterns are wrong. (2) It may take many months of simulation to happen upon a sequence of stimulus patterns the detect a real bug.

9.5.1.2 Directed Patterns

The stimulus generator can be "directed" to produce stimulus patterns that stimulate a particular block(s) or unit(s) in the IC. The intent is to use the knowledge of the designer to find the bugs quicker by directing the stimulus generator to produce targeted patterns that verify a specific behavior of the IC. A combination of random patterns and directed patterns are used to improve the coverage of functional simulation.

9.5.1.3 Assertions

Functional simulation assertions capture designer knowledge about how an IC should behave. The assertions provide additional observability into a design and as such provide information on the controllability of the design. An assertion specifies legal and illegal behavior of a section of logic in the design. An example of some assertions include:

- "The state-machine should not go to an unknown state"
- "If this signal is valid, then this (some activity) should happen"
- "Register<123> should never be accessed in user-mode"

Note: A more formal language is used to describe an assertion such as System Verilog Assertions (SVA) (Wikipedia, n.d.g).

9.5.1.4 Coverage Analysis

The goal of coverage analysis is to confirm that specified coverage signals have been stimulated across a wide range of conditions such that there is a high confidence level that the functionality has been adequately verified. Each signal to be tested is described in terms of the number of events, sequences, and values. The coverage analysis tool determines if all the edge/corner cases have been explored. If not, additional directed patterns can be generated and applied to the simulation model.

9.5.2 Emulation

An emulator is a hardware accelerator that emulates the gate-level models of the IC. As mentioned above emulators are used in the HW/SW Co-design (Embedded Website, n.d.a) prototyping phase. Figure 9.22 shows a typical flow where the virtual prototype progresses from being primarily software based to primarily hardware (emulation) based. Emulators are considerably faster than functional simulation and use Field Programmable Gate Arrays (FPGA) or Application Specific ICs (ASICs) to speed up the execution time. The SW team uses the emulator to boot operating systems, debug code, run regression tests, etc.

9.5.3 Formal Property Verification

"Formal property verification (FPV) is the process of checking whether a design satisfies some requirements (properties)" (Embedded Website, n.d.b). The goal of FPV is to determine properties

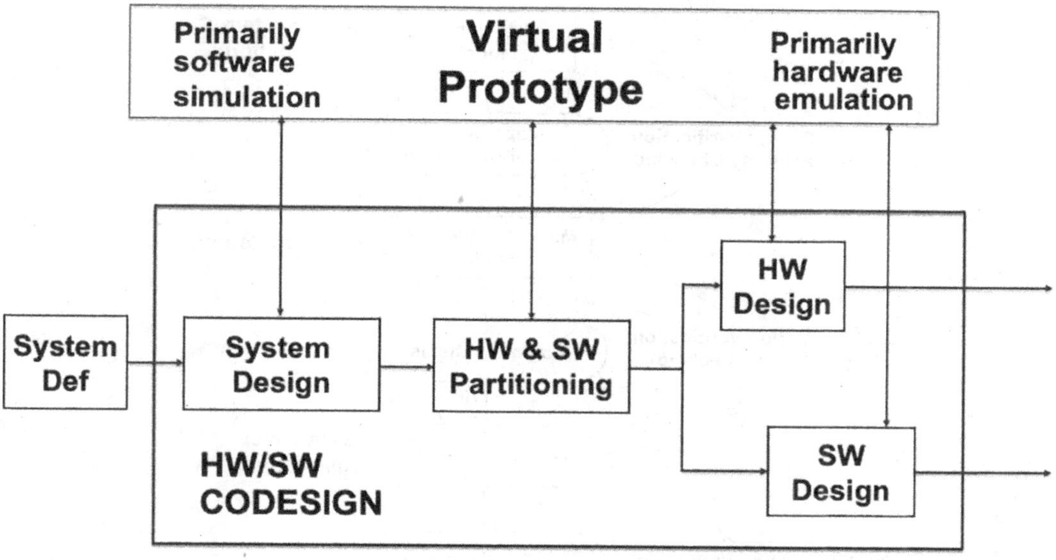

FIGURE 9.22 Virtual prototyping environment.

of the design that are independent of all inputs and then prove the correctness of the design. This is the opposite of functional simulation which attempts to prove correctness using verification patterns, which is not possible for complex IC designs. For example, an IC design with 300 state variables (flip-flops) has more states than the number of protons in the universe (1080). It would take hundreds of years to functionally verify the microprocessor in a cell phone. It should be noted that FPV is not capable of proving the correctness of an entire microprocessor. Instead, it is used to verify critical sections of the design such as state machines, arbiters, bus controllers, etc. FPV is also used to prove the validity of assertions and guiding functional simulation.

9.5.4 Equivalence Checking

Equivalence checking is used to check whether two simulation models are equivalent. Figure 9.23 shows the various simulation models that are generated during the implementation of an IC. There are four primary models and three primary synthesis steps. Each simulation model that is generated is compared to the previous model to confirm that the mapping/synthesis step did not introduce an error. Equivalence checking is a formal verification technique that proves correctness between the two models. This step occurs when the RTL model is synthesized into the gate-level model. NOTE: do not use functional simulation test patterns to prove correctness between the two models because as mentioned above it will take hundreds of years to prove equivalency between the two models.

9.6 PHYSICAL DESIGN VERIFICATION

The physical design verification (PDV) step confirms that the mapping from the logical model to the physical model is correct and can be manufactured correctly. The physical model is referred to as "layout". It is simply polygons placed on different layers with respect to each other. The layers are converted to photographic masks and used to pattern the transistors, contacts, metal routing, and doping regions. There are four types of PDV:

- Layout Versus Schematic (LVS)
- Design Rule Checking (DRC)

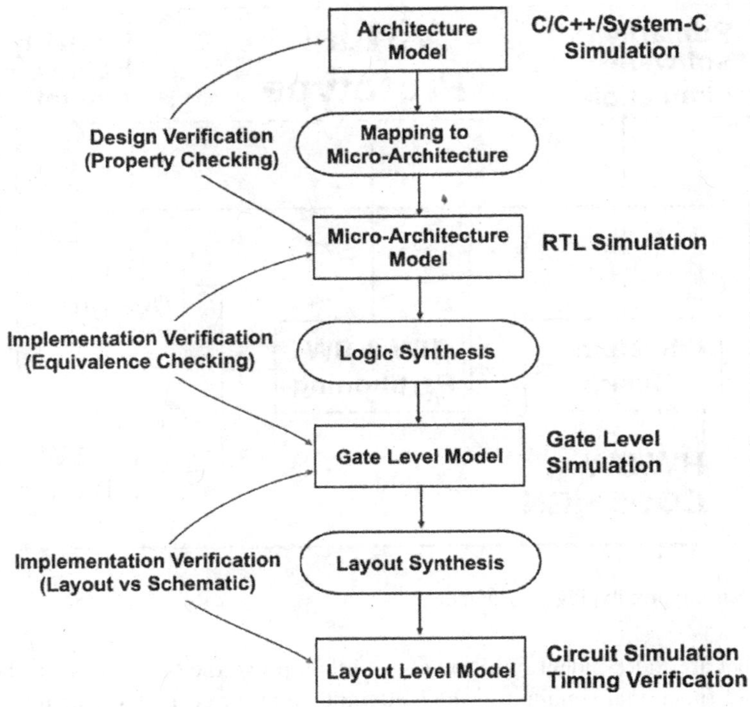

FIGURE 9.23 Comparisons of synthesized models.

- Design for Manufacturability (DFM)
- Electrical Rules Checking (ERC)

9.6.1 Layout versus Schematic

As the name implies LVS confirms that polygons are in the correct pattern to generate a Physical model that represents the Logical model. In Figure 9.24 the layout (a) represents a three input AND gate and (b) represents a three input AND gate in schematic form.

The LVS process involves extracting features from layout and generating an equivalent schematic. For example, transistors are formed when Gate (cross hatched shape) crosses a Diffusion (dithered shape) as shown above. Metal makes a connection to either a Gate or Diffusion when it is on top of a Contact (square cross hatched). The equivalent layout schematic is compared to the logical schematic to determine if there are connection issues, sizing issues, etc.

9.6.2 Design Rule Checking

To confirm that the layout polygons can be manufactured correctly the DRC tool checks the spacing and width of the polygons on a single layer and between layers (Wikipedia, n.d.h). Figure 9.25 shows an example of eight rules. Note: there are hundreds of rules that must be checked before sending the IC to be manufactured.

1. Metal width
2. Metal spacing
3. Metal overlap of contact
4. Diffusion overlap of contact

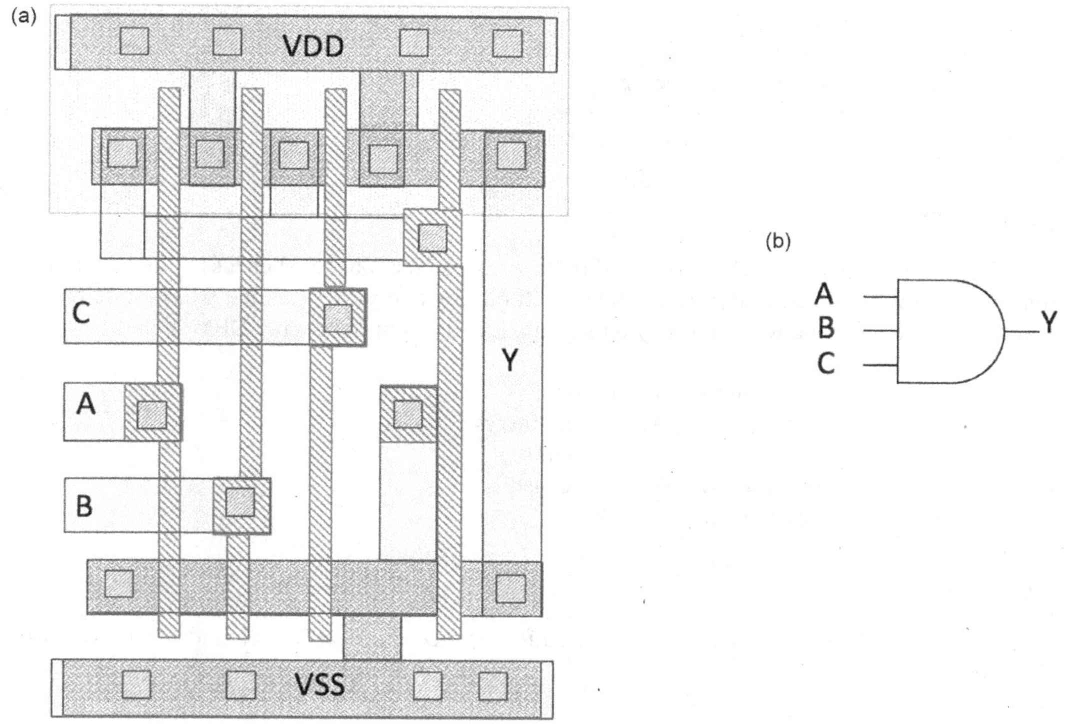

FIGURE 9.24 (a) Layout and (b) schematic.

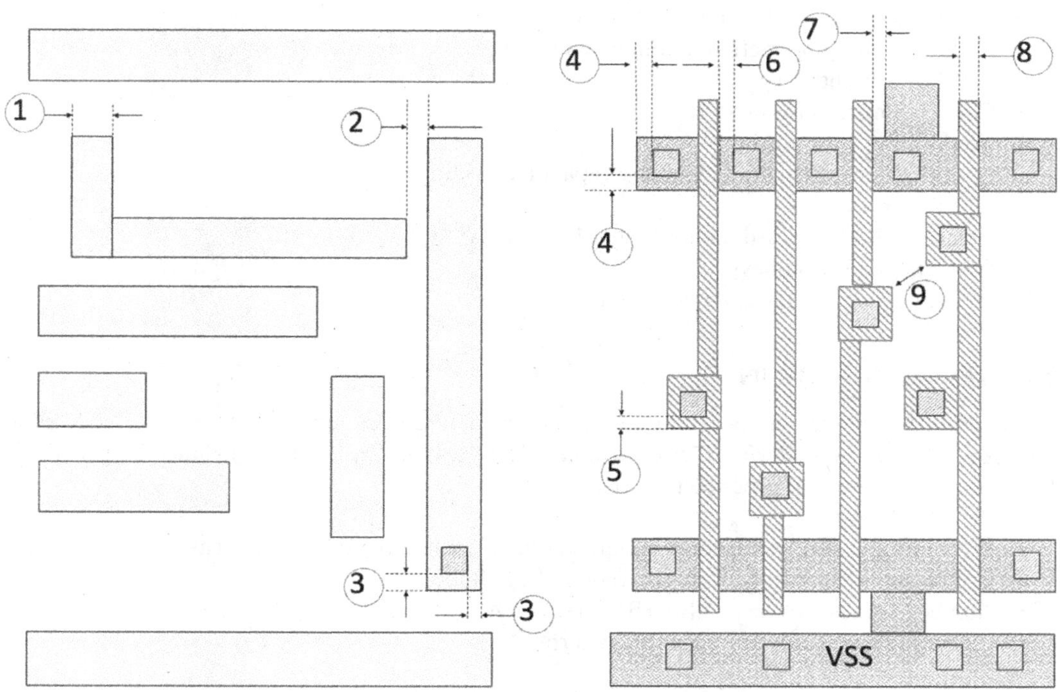

FIGURE 9.25 Design rules.

5. Gate overlap of contact
6. Gate to contact spacing
7. Gate extension to diffusion spacing
8. Minimum gate length
9. Gate to gate corner spacing

9.6.3 DFM

DFM is an extension to the DRCs discussed in the previous section. DFM checking looks for layout areas where it is possible to improve the manufacturability and yield of an IC. DFM guidelines are generally supplied by the wafer fabrication facility. Examples of improving DFM include:

1. Increasing metal spacing to reduce crosstalk
2. Adding redundant contacts between metal layers
3. Adding metal fill and slotting to wide metal
4. Adding redundant standard cells and transistors
5. Adding redundancy to memory blocks

9.6.4 ERC

There are several ERC techniques that are used to enhance the electrical characteristics of an IC. These include:

1. Adding antenna diodes
2. Adding bypass capacitors
3. Adding tie-high or tie-low cells are on all pull-high/pull-low inputs
4. Adding tie-high or tie-low cells to redundant standard cells and transistors
5. Adding additional substrate and well ties
6. Adding latch-up protection (dummy collectors)
7. Level shifter checking
8. Clock boundary crossing checking
9. Configuring reset correctly
10. Checking for maximum fanout and capacitive loading
11. Signal slew rates
12. Detecting Unconnected inputs or shorted outputs
13. Widening power busses
14. Shielding clock signals and other critical signals

9.7 TEST GENERATION

As mentioned above testing confirms that the manufactured ICs are free of defects and work as expected i.e., "*Can we sell the IC to a customer?*". To test an IC, test patterns must be generated. There are several ways to generate patterns:

- Convert functional verification patterns or hand-write directed test patterns
- Use an automatic test pattern generation (ATPG) tool
- Add built-in-self-test capability (BIST) to the IC
- Reuse and/or modify previous test patterns

The test patterns must be able to detect a high percentage of manufacturing defects, typically 99.999% or better. Detecting a defect requires activating the resulting fault and then propagating the fault to the tester pins. The exact techniques of how this accomplished are beyond the scope of this overview of IC testing.

The source of defects in an IC include:

- Particulate matter in the wafer processing facility
- Defects in the reticle that is used to print a layer
- Faulty reticle alignment
- Under/over-etching of diffusion, gate, contacts, vias, and metal regions
- Under/over depositing of gate oxide, gate, dielectric, via plug and metal material
- Across die doping variations (see statistical design above)
- Design timing faults: setup, hold, transition times, power droop, noise, and crosstalk

The faults manifest themselves as:

- Signals that are Stuck-at-1 or Stuck-at-0
- Transistor defects that appear as Stuck-at-1, Stuck-at-0, or something in between
- Signals that are shorted to each other due to bridging, these faults can generate analog voltages which cause digital circuits to become metastable
- Signals that are open due to incomplete metallization, unfilled contacts/vias
- Slow signals due to excessive wire resistance caused by over-etching metal

9.7.1 Pattern Generation

9.7.1.1 Using Functional Patterns

As mentioned above, using (only) functional patterns for testing (or verifying) an IC is discouraged. However, using a subset of functional patterns to generate probe patterns can be very useful in detecting that some minimal amount of functionality is present in the IC before doing additional testing.

9.7.1.2 Automatic Test Pattern Generation

The preferred mechanism for testing complex ICs is to use an automated tool to generate scan patterns. Scan patterns are serially shifted into the IC using a JTAG (Wikipedia, n.d.i) controller and a prewired scan chain through all the flop-flops in the IC. Figure 9.26 shows a subset of the IC scan chain. Note: there can be a single scan chain or multiple scan chains in a block. The flip-flops are loaded with the ATPG patterns via the scan-in and scan-out signals. The output of the flip-flops in scan chain #1 propagates through the combinational logic between scan chain #1 and scan chain #2. The flip-flops in scan chain #2 sample the output of the combinational logic and the scan chains are shifted out of the IC. The tester evaluates the results and determines if a fault has been detected. This repeats until all the ATPG patterns have been shifted into the IC and the results have been shifted out.

9.7.1.3 Built-In Self-Test

In many cases, it is useful to add additional logic to the IC to test standalone components of the IC. These generally include memory units which return the same values that were asserted to the memory. For read-only memories, a hash value can be determined a-priori that the BIST controller uses to confirm that the ROM data is correct. Figure 9.27 shows a block diagram of a BIST Controller. During normal operation, the inputs are applied to the unit. In BIST mode a test pattern generator (TPG) applies patterns to the unit. The output of the unit under test is compared to the expected results.

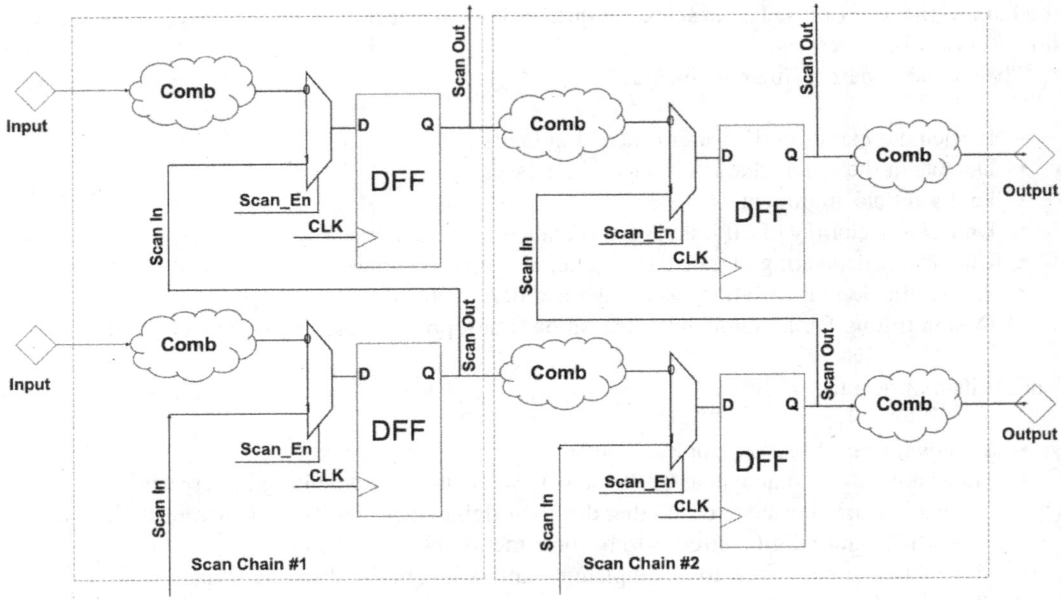

FIGURE 9.26 Block scan chain configuration.

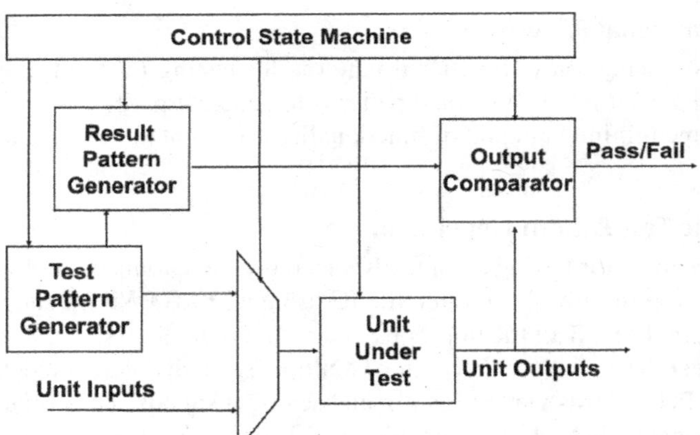

FIGURE 9.27 Block diagram of a BIST controller.

REFERENCES

Embedded Website (n.d.a). https://ptolemy.berkeley.edu/projects/embedded/Research/hsc/abstract.html.

Embedded Website (n.d.b). https://ptolemy.berkeley.edu/projects/embedded/research/vis/doc/VisUser/vis_user/node4.html.

Hu, W., & Li, F. (2021). Scaling beyond 7nm node: An overview of gate-all-around FETs. In *2021 9th International Symposium on Next Generation Electronics (ISNE)* (pp. 1–6). doi: 10.1109/ISNE48910.2021.9493305.

McDermott, M. (2021). Motorola MC68332: One of the first true SoCs. *IEEE Micro*, 41(6), 105–106. doi: 10.1109/MM.2021.3114897.

SoftwareQuotes (n.d.). https://softwarequotes.com/blog/the-origin-of-the-world-s-first-computer-bug.

The University of Texas at Austin (n.d.). https://repositories.lib.utexas.edu/handle/2152/24851.

Wikipedia (n.d.a). https://en.wikipedia.org/wiki/Tape-out.
Wikipedia (n.d.b). https://en.wikipedia.org/wiki/FinFET.
Wikipedia (n.d.c). https://en.wikipedia.org/wiki/Hardware_description_language.
Wikipedia (n.d.d). https://en.wikipedia.org/wiki/Tejas_and_Jayhawk
Wikipedia (n.d.e). https://en.wikipedia.org/wiki/Hafnium(IV)_oxide.
Wikipedia (n.d.f). https://en.wikipedia.org/wiki/Mixed-signal_integrated_circuit.
Wikipedia (n.d.g). https://en.wikipedia.org/wiki/SystemVerilog#Assertions.
Wikipedia (n.d.h). https://en.wikipedia.org/wiki/Design_rule_checking.
Wikipedia (n.d.i). https://en.wikipedia.org/wiki/JTAG.

10 Board-Level Design and Test

Datong Liu, Dawei Pan, and Shengwei Meng

10.1 INTRODUCTION: BACKGROUND AND DRIVING FORCES

The quality of the circuit board directly determines the stability, life, and application range of electronic products. Circuit board design is a complex project that requires designers to have a theoretical foundation and practical experience. The purpose of the board-level test is to verify whether the components, integrated devices, plates, topological structure, trace length, etc., on the circuit board have been properly utilized, whether the circuit can normally achieve the expected functions, and whether the board reaches its normal performance. For researchers and designers working in the electronics industry, it is with great importance to master PCB design methods and test skills, for effectively circuit design and development cost reduction. The following part of this chapter introduces the design and testing method and protocols of circuit boards. At the same time, taking a USB-HUB card soldered in the author's laboratory as an example, the steps and precautions required for board-level testing are introduced in detail.

10.2 BOARD-LEVEL DESIGN

10.2.1 Conceptual Design

PCB circuit board design, correct or not, is one of the most important parts to improve the reliability of electronic products. The board-level design first needs to be programmed. Program design should focus on the following aspects.

10.2.1.1 Circuit Board Size and Device Placement

The size of the printed circuit board should be moderate. If it is too large, the printed lines will be long and the impedance will increase, which will not only increase the signal transmission noise but also increase the cost; if it is too small, the heat dissipation will not be good, and it will be easily interfered by adjacent lines. In terms of device layout, the devices related to each other should be placed as close as possible, so that a better anti-noise effect can be obtained. Noise-prone devices must be kept away from logic circuits as much as possible, and if possible, separate circuit boards should be made. Draw the PCB in the PCB design environment according to the determined board size and various mechanical positioning, and place the required connectors, buttons, switches, digital tubes, indicators, inputs, outputs, screw holes, mounting holes, etc. Fully consider and determine the wiring area and non-wiring area. In the actual design, special attention should be paid to the actual size of the components (occupied area and height), the relative position and space between the components, and the surface on which the device is placed to ensure the electrical performance of the circuit board.

10.2.1.2 Grounding Design

In electronic equipment, grounding is an important part to control the signal noise. If the grounding and shielding can be properly combined and used, most of the interference problems can be solved. The ground structure of electronic equipment roughly includes system ground, enclosure ground (shield ground), digital ground (logical ground), and analog ground. The following points should be paid attention to in the ground wire design.

DOI: 10.1201/9781003188377-12

1. Correctly choose single-point grounding and multipoint grounding in low-frequency circuits. The operating frequency of the signal is less than 1MHz, and the inductance between its wiring and devices has little effect, and the circulating current formed by the grounding circuit has a greater impact on interference, so single-point grounding should be used. When the signal operating frequency is greater than 10MHz, the ground wire impedance becomes very large. At this time, the ground wire impedance should be reduced as much as possible, and multipoint grounding should be used;
2. Separate the digital circuit from the analog circuit. There are both high-speed logic circuits and linear circuits on the PCB circuit board. They should be separated as much as possible, and the ground wires of the two should not be mixed, and they should be connected to the ground wire of the power supply terminal. Try to increase the grounding area of the linear circuit;
3. Thicken the grounding wire as much as possible. If the grounding wire is very thin, the ground potential will change with the change of current, resulting in an unstable timing signal level of electronic equipment and deterioration of anti-noise performance. Therefore, the ground wire should be as thick as possible so that it can pass the allowable current of the printed circuit board.

10.2.1.3 Heat Dissipation Design

From the perspective of conducive to heat dissipation, the printed plate is best installed upright, the distance between the board and the board should not be less than 50px, and the arrangement of the devices on the printed plate should follow certain rules.

1. The components on the same printed board should be arranged as far as possible according to their calorific value and degree of heat dissipation. The components with small calorific value or poor heat resistance should be placed at the most upstream (entrance) of the cooling airflow. Devices with good thermal properties are placed at the most downstream of the cooling airflow.
2. For equipment that adopts free convection air cooling, it is best to arrange integrated circuits (or other devices) vertically; for equipment that uses forced air cooling, it is best to arrange integrated circuits (or other devices) horizontally.

10.2.1.4 EMC Design

The purpose of electromagnetic compatibility design is to enable electronic equipment to suppress all kinds of external interference so that the electronic equipment can work normally in a specific electromagnetic environment, and at the same time to reduce the electromagnetic interference of the electronic equipment itself to other electronic equipment.

1. **Use the Correct Wiring Strategy:** Parallel wiring can reduce the wire inductance, but the mutual inductance and distributed capacitance between the wires increase. If the layout permits, it is best to use a tic-tac-toe mesh wiring structure. The specific method is that one side of the printed board is wired horizontally, and the other side is wired vertically, and then metalized holes are connected at the cross holes. To suppress the crosstalk between the conductors of the printed circuit board, long-distance equal routing should be avoided as much as possible when designing the wiring.
2. **Choose a Reasonable Wire Width:** Since the impact interference caused by the transient current on the printed lines is mainly caused by the inductance of the printed wires, the inductance of the printed wires should be minimized. The inductance of the printed wire is proportional to its length and inversely proportional to its width, so short wires are beneficial to suppress interference. For integrated circuits, the printed wire width should be selected between 0.2 and 1.0mm.

10.2.2 Schematic Design

To carry out the design of circuit engineering and make these circuit boards in the scheme design, we need to draw the schematic diagram of the circuit, that is, the design in which devices are connected to finally constitute the pathway to achieve a certain function. A circuit schematic diagram is a circuit symbol connected to reflect the electrical connection of the components of the drawing, that is, a circuit diagram composed of simple diagrams. The specific design steps of a schematic diagram are as follows.

10.2.2.1 Determine the Needs

Understand the design requirements in detail, and sort out the circuit function modules and performance index requirements from the requirements.

10.2.2.2 Determine the Core Processor

According to the function and performance requirements, the overall design plan is formulated, and the processor is selected. The core processor selection has the following requirements:

1. Cost-effective; 2. Easy development; 3. Good scalability.

10.2.2.3 Peripheral Device Selection and Design

The selection of peripheral components means the selection of components for peripheral function modules according to requirements. The selection of components should comply with the following principles.

1. **The Principle of Universality:** the selected components must be widely used and verified, and use as few unpopular chips as possible to reduce risks.
2. **The Principle of Cost-Effective:** in the case of similar functions, performance, and utilization, try to choose components with better prices to reduce costs.
3. **The Principle of Convenient Procurement:** try to choose components that are easy to buy and have a short lead time.
4. **The Principle of Sustainable Development:** try to choose components that will not stop production within a foreseeable time.
5. Substitution principle: try to choose more pin-to-pin compatible components.
6. **The Principle of Upward Compatibility:** try to choose components that have been used in old products before.
7. **The Principle of Resource Conservation:** try to use all the functions and pins of the components.

10.2.2.4 Design Principles

Hardware schematic design should abide by some basic principles. These principles should be implemented throughout the design process. Although these principles are also reflected in a successful reference design, the schematics may be "pieced together", so we still have to follow these principles at any time. In the schematic diagram of the design review, these principles include the following.

1. Digital power and analog power are separated, and digital ground and analog ground are separated.
2. The component parameters must be accurately identified. Pay special attention to the power resistance, which must be marked with the power value, and the high withstand voltage filter capacitor, which must be marked with the withstand voltage value.

3. For power consumption and product heat dissipation issues, heat sinks or fans can be added near the chips with higher power consumption and heat generation. The product chassis should also consider this issue. The chassis cannot be made into an insulation box; also consider the placement of the product. It is best to place it in a place with a relatively large space and unobstructed airflow, which is conducive to the dissipation of heat.
4. The corresponding functions such as adjustable components (such as potentiometers) and switches must be marked.
5. Important signal lines need to indicate the flow direction and use text to indicate the function.
6. Try to ensure the convenience of software development without increasing the difficulty of hardware design.
7. Ensure that the module resources of the system cannot conflict, for example, the device addresses on the same I2C bus cannot be the same, etc.
8. The layout of each functional module must be reasonable, and the entire schematic diagram must be well-balanced. Avoid some very crowded places and some places are loose, which is the same as PCB design.

10.2.2.5 Schematic Review

After the hardware schematic design is completed, you should first conduct a self-review according to the above steps and requirements and then submit it to others for review. Other reviewers also strictly review the schematic following the above requirements. If problems are found, they must promptly discuss and analyze, and analyze the resolution process. Also, follow the above principles and steps.

10.2.3 PCB Layout

After completing the schematic design and schematic design, a PCB layout design is required. The following describes the PCB layout specifications, layout techniques, and how to perform layout inspections.

10.2.3.1 PCB Layout Rules

1. Under the premise of ensuring electrical performance, the components should be placed on the grid and are arranged in parallel or vertical, pursue neat, beautiful, in general, not allowed to overlap.
2. Generally, all components should be arranged on the same side of the circuit board. Only when the top components are too dense, can some devices with limited height and low heat generation be placed on the lower layer, such as chip resistors, chip capacitors, SMD ICs, etc.
3. The minimum spacing between adjacent pad patterns of different components on the circuit board should be at least 1 mm.
4. The distance between the components and the edge of the circuit board is generally not less than 2 mm. The best shape of the circuit board is rectangular, the aspect ratio is 3:2 or 4:3, and the circuit board size is larger than 200 mm × 150 mm. The mechanical strength that the circuit board can withstand should be considered.

10.2.3.2 PCB Layout Tips

In the layout design of the PCB, the units of the circuit board should be analyzed, and the layout design should be carried out according to the function. When laying out all the components of the circuit, the following principles should be met.

1. Take the core components of each functional unit as the center and layout around it. The components should be uniformly, integrally, and compactly arranged on the PCB to minimize the leads between the components and shorten the connection.
2. Arrange the position of each functional circuit unit according to the circuit flow, so that the layout is convenient for signal circulation, and the signal is kept in the same direction as possible.
3. For circuits operating at high frequencies, the distributed parameters between components must be considered. This tip is the same as Point 1 in 10.2.3.1.

10.2.3.3 PCB Layout Check

1. Whether all the components have been laid out, whether the layout is balanced and neatly arranged.
2. Whether the commonly used components are convenient to use. Such as switches, plug-in equipment, components that must be replaced frequently, etc.
3. Whether there are conflicts at all levels. Such as components, frame.
4. Whether the heat dissipation is good, whether the distance between the thermal components and the heating components is reasonable.
5. Whether the size of the circuit board and the processing size required by the drawing are consistent.
6. Line interference problem.

10.3 BOARD-LEVEL TEST

In the production process and use of circuit boards, external factors will inevitably cause short-circuit, open circuit, and leakage defects. In addition, the circuit boards continue to evolve toward high density, fine pitch, and multiple levels. If the defective boards are not screened in time. If it is detected and allowed to flow into the manufacturing process, it will inevitably cause more cost waste. Therefore, it is necessary to improve the process control and the technical level of the test. At the same time, the necessary tests can be taken for the circuit board at different stages of the work. Manufacturers provide support to reduce scrap rates and increase product yield. The circuit board-level test is divided into the pre-soldering test, passive test, and active test, as shown in Figure 10.1; it will be introduced in detail below.

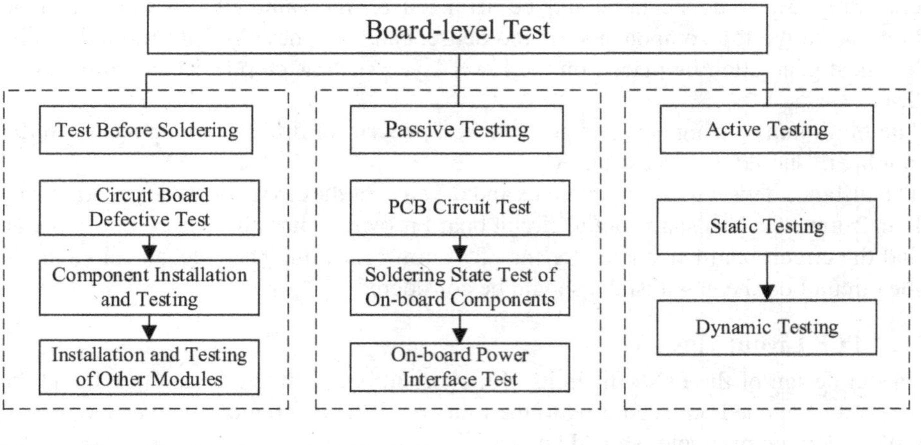

FIGURE 10.1 Board-level design content block diagram.

10.3.1 Test before Soldering

In the production process of electronic products, the estimation of the cost loss caused by the failure is different at each stage. The sooner it is discovered, the lower the cost of salvation. When the blank board is finished, it needs to be tested before soldering.

10.3.1.1 Board-Level Test Steps

1. Circuit board defective test

   When you get the circuit board, you should first carefully observe whether the surface of the circuit board is smooth and tidy, and there are no burrs. Sometimes the burrs will be short-circuited to the nearby signal line. Then check the schematic diagram to see if the board design is consistent with the circuit principle, and then check the circuit board against the designed board diagram to see if there are obvious defects such as cracks, short circuits, few connections, and open circuits. Use a multimeter to measure the power supply and ground to ensure that there is no short circuit. Use a magnifying glass to check that the circuit board is not open. You can check whether the resistance between the power supply and the ground is large enough. The most important thing is to use a multimeter to check whether the power supply used is short-circuited to the ground before soldering. At least it can ensure that the device will not be burned out after power-on.

   The common types of test defects are as follows:

   1. Whether the circuit board is short-circuited: In the original design, two unconnected conductors caused an unexpected energization situation, as shown in Figure 10.2.
   2. Whether the circuit board open circuit: In the original design, any two points of the same circuit should be energized, but a power outage occurred, as shown in Figure 10.3.
   3. Whether the circuit board conduction is poor: the conduction effect between two points of the same circuit is not good, which is a kind of open circuit, as shown in Figure 10.4.

2. Component installation and testing

   As the most important components on the circuit board, the quality of electronic components determines the quality of the circuit board. When installing components, it is better not to install all independent modules, but to install parts of them if you are not sure that they work properly (for smaller circuits, you can install them all at once). In this way, it is easy to determine the scope of the failure. Generally, you can install the power supply part first and then check whether the power module can output normally after power-on. If you

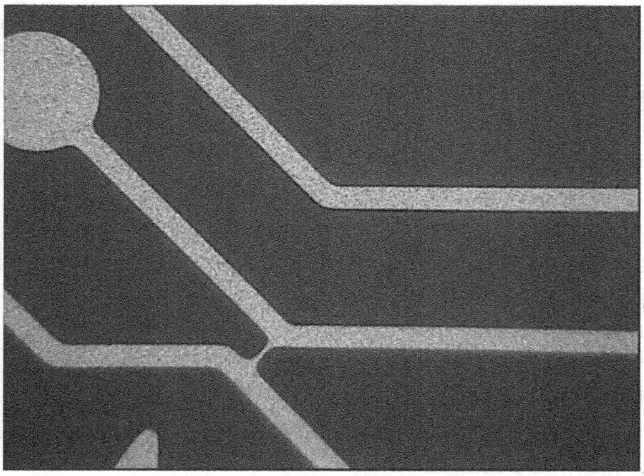

FIGURE 10.2 Circuit board short circuit schematic.

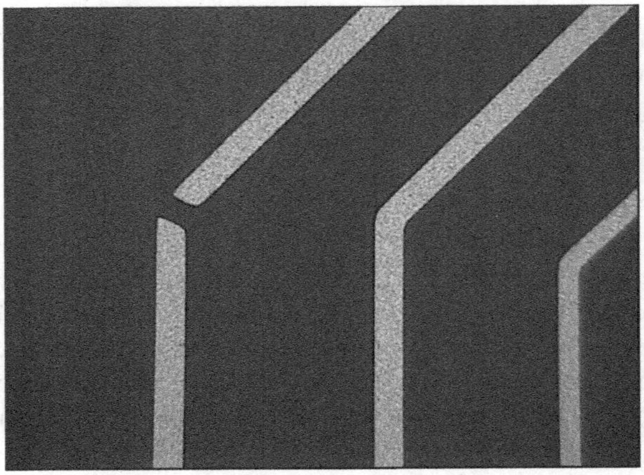

FIGURE 10.3 Circuit board open circuit schematic.

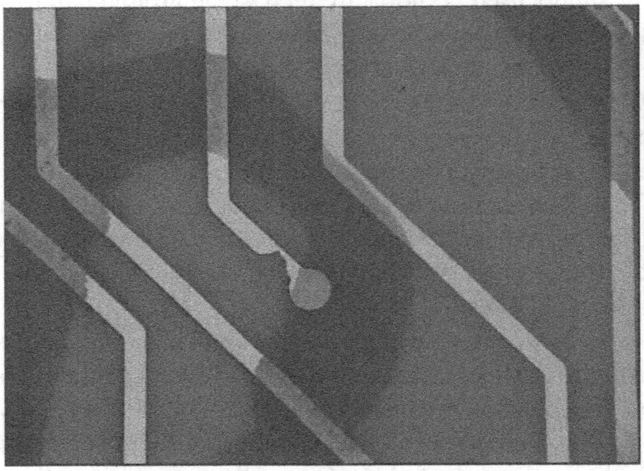

FIGURE 10.4 Circuit board poor conduction schematic.

are worried about a failure during the power-on process (the risk of failure can be reduced by adding a fuse), you can use an adjustable regulated power supply with a current limiting function. You can preset the over-current protection current first and then slowly increase the voltage value of the regulated power supply and monitor the input current, input voltage, and output voltage simultaneously. If there are no problems such as over-current protection during the upward adjustment process, and the output voltage is normal, it means that there is no problem with the power supply part. Otherwise, you should disconnect the power supply, then find the fault point, and repeat the above steps until the power supply is normal.

3. Installation and testing of other modules

   Install other functional modules on the circuit board step by step, such as power interface modules. Whenever a module is installed, power-on and test it. To avoid over-current or burnout of components due to errors from design or installation, you should follow the steps above when powering on.

10.3.1.2 Main Steps of Soldering

Surface mounted technology (SMT) is the most popular technology and process in the electronic assembly industry. It is a kind of circuit assembly technology that installs the pinless or short lead surface assembly components on the surface of PCB or other substrates and is soldered and assembled by reflow soldering or dip soldering. Although SMT automation equipment has been very advanced and has replaced most of the manual work, manual soldering is still needed due to the particularity of some components. Concretely, soldering iron can be divided into five steps, which must be strictly operated to obtain high soldering quality.

1. Preparation before soldering

   After getting the circuit board, you need to observe whether the circuit board circuit is damaged, whether the pad is damaged, and whether the green oil is damaged. Use a multimeter to measure whether there is a short circuit between the power supply and the ground, different power supplies, and different grounds. Prepare solder wire and soldering iron. It should be noted that the tip of the soldering iron must be clean before the solder can be applied.
2. Heating the weldment

   Touch the soldering iron to the soldering point. Keep the soldering iron heating all the weldment firstly, such as the leads and pads on the PCB, and secondly, let the flat part (larger part) of the soldering iron tip contact the soldering parts with larger heat capacity. The side or edge of the soldering iron tip contacts the weldment with a smaller heat capacity to keep it heated evenly.
3. Molten the solder

   When the solder is heated to a temperature that can melt the solder wire is placed on the soldering point, the solder begins to melt and wet the soldering point.
4. Removing the solder

   Remove the solder wire after melting a certain amount of solder.
5. Removing the soldering iron

   When the solder wets the solder joints completely, remove the soldering iron. Note that the direction of removing the soldering iron should be approximately 45°.
6. Precautions for circuit board soldering

   Following these steps is one of the keys to a good solder joint. In the production process, one of the most prone ways to violate the operation steps is that the soldering iron tip does not contact the welded part firstly, but the solder wire. The molten solder drips on the part to be soldered that has not been preheated, which is easy to produce false soldering, so the soldering iron tip must be in contact with the soldered part. Preheating the soldered part is an important means to prevent false soldering.

   Solder the power supply at first, and then solder the power supply according to the voltage and its relationship. Next, solder the core processor and peripheral resistors and capacitors, crystal oscillator and reset circuit, and retest the power and ground after soldering to avoid short circuit and thermal breakdown during soldering.

   Test after soldering the core processor:

   1. It is necessary to confirm that the crystal oscillator can work normally: Measure the waveform of the crystal oscillator pin with an oscilloscope. If the waveform is sinusoidal and the frequency fluctuates with that of the crystal oscillator, the crystal oscillator is good and can be started normally;
   2. Check whether the reset is normal;
   3. Check whether it can download the program and simulation normally and then check whether the downloaded part of the circuit is normal, and when debugging, make sure

that the corresponding IO port can be set, and the IO port can be set directly to turn on the LED.

After that, the serial port is soldered and connected to the computer for serial port input and output debugging. Finally, other peripheral devices are soldered, and the corresponding hardware debugging is carried out in combination with the software.

10.3.1.3 Tips for Board-Level Testing

1. It is strictly forbidden to use grounded test equipment to touch the live TV, audio, video, and other equipment of the bottom plate to test the PCB without an isolation transformer. It is strictly forbidden to directly test TV, audio, video, and other equipment without a power isolation transformer using instruments with grounded shells. First of all, it is necessary to make clear whether the chassis of this machine is electrified; otherwise, it is easy to cause short circuit of power supply with TV, audio and other equipment with electrified base plate, affecting integrated circuit and causing worse failure;
2. Before testing PCB, it is necessary to understand the working principle of the integrated circuit and related circuit. Before checking and repairing integrated circuits, we should be familiar with the functions of integrated circuits, internal circuits, main electrical parameters, the role of each pin and the normal voltage, waveform of the pin, and the working principle of the circuit composed of peripheral components. If these conditions are met, analysis and inspection are much easier;
3. Note the heat dissipation of power integrated circuit when testing PCB;
4. The internal resistance of the test instrument used to test the PCB should be large. When measuring the DC voltage of the IC pins, you should choose a multimeter with the internal resistance of the meter head greater than 20 KΩ/V, otherwise, there will be a large measurement error for some pin voltages;
5. Avoid short circuits between pins when testing PCB. When measuring voltage or using an oscilloscope probe to test the waveform, do not cause a short circuit between the pins of the integrated circuit due to sliding of the test leads or probes. It is best to measure on the peripheral printed circuit directly connected to the pins. Any short circuit can easily damage the integrated circuit;
6. Do not easily determine the damage of integrated circuit when testing PCB;
7. Note the insulation performance of electric soldering iron when testing PCB. It is not allowed to use a soldering iron for soldering with electricity. Make sure that the soldering iron is not charged. It is best to ground the shell of the soldering iron.

10.3.1.4 Test Case

This section takes a USB-HUB card as an example to introduce the test procedures and precautions before soldering. The USB-HUB blank board is shown in Figure 10.5.

First, you need to check whether the surface of the circuit board is smooth and tidy and then perform a circuit board defect test, as shown in Figure 10.6. Set the multimeter to the buzzer position, and test whether the circuit board has short circuits, short circuits, and poor conduction conditions. If there is no problem, solder the prepared components and test them one by one. Refer to 10.3.1.2 and 10.3.1.3 for the soldering steps and precautions. The circuit board with components is shown in Figure 10.7.

10.3.2 Passive Testing

The board-level testing of the PCB after the component is soldered and before the power-on test is called passive testing. The main object of passive testing is circuits, components, and some main interfaces such as power interfaces.

FIGURE 10.5 USB-422 circuit board empty board picture.

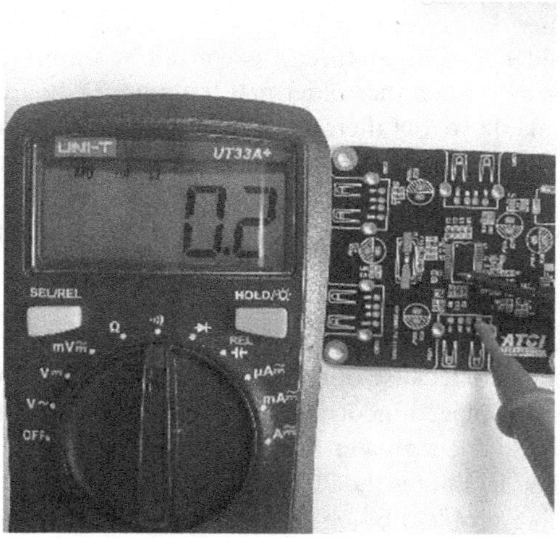

FIGURE 10.6 Board-level pre-soldering testing picture.

FIGURE 10.7 Soldered USB-422 circuit board.

10.3.2.1 PCB Circuit Testing

In terms of the manufacturing process, a certain proportion of PCB has defects in connectivity, which cannot be completely avoided. Therefore, it is necessary to conduct circuit testing on PCBs. The main basis of the circuit test is the PCB schematic diagram during the design process. For the PCB circuit test, you can test through the buzzer setting of the multimeter to determine whether the circuit is connected normally and whether there is an open circuit fault. The test sequence can be based on one or several key components on the PCB, starting with checking the loop where they are located, and then testing the entire PCB circuit. In addition to the loops around the components, the power lines and ground lines on the PCB also need special attention and focus on inspection. On the premise that the designed PCB schematic diagram is correct, the circuit test is the basic guarantee for the subsequent power-on test.

10.3.2.2 Soldering State Test of On-Board Components

The test of the components on the PCB refers to the test of the components after soldering on the PCB in the state of no power, which is different from the device test in the active test. Because the PCB soldering process also has certain individual differences, it is necessary to check whether the components are correctly soldered to the PCB.

The test is mainly divided into two parts: The first part is to detect whether there is a false soldering problem in the components. False soldering is generally caused by oxidation or impurities at the soldering point and poor soldering temperature. It is caused by improper methods. The essence is that there is an isolation layer between the solder and the pins. They are not completely contacted and cannot be seen by the naked eye, but their electrical characteristics are not conducted or poorly conducted, which affects the circuit characteristics. In addition to the traditional manual inspection methods, there are automated optical inspection (AOI) inspection methods, electromagnetic vibration inspection methods, and X-ray inspection methods for the detection methods of virtual soldering.

1. AOI is based on optical principles to detect common defects encountered in the soldering production process, but also because of its working principle, when the inspection position of the PCB to be tested is placed incorrectly, or in the case of oil stains on the board, the AOI detection result will have a greater false detection rate.
2. X-ray inspection method irradiates the PCB with X-rays, you can intuitively judge whether there is a false soldering problem by observing whether there are soldering voids. At the same time, the software operation can adapt to multi-position and multi-size PCB, which means relatively simple and easy to operate. After comprehensively weighing the inspection efficiency, inspection cost, and inspection accuracy, the X-ray inspection method is generally selected to detect the false soldering problem.
3. The manual inspection method, as the name suggests, is to manually inspect each component pin on the PCB. When performing false solder detection on a highly integrated PCB, manual detection will consume a lot of time, and due to a certain error in human judgment, it may cause problems such as missed detection or false detection.
4. The electromagnetic vibration detection method uses the principle of resonance to detect the soldering condition of the PCB. Although this method can detect the problems caused by poor soldering to a large extent, the operation of the electromagnetic vibration test bench is very complicated and the details are cumbersome, such as the test bench needs to be calibrated before each experiment; otherwise, the test results will be meaningless.

In addition to false soldering, it is also necessary to check whether the components are installed correctly. The components mentioned here generally refer to components and chips with polarity. Common polar components include polar capacitors, diodes, etc. For this kind of relatively simple component with polarity, you can generally check the polarity mark on the package and compare with the PCB schematic diagram whether the polarity is installed correctly. For chips, you need to check the chip packaging information in the data manual provided by the chip manufacturer. Especially for the center-symmetric chip (such as square), you need to check whether the pin sequence is correct. The general manufacturer will put the PIN1 on the position with some marks, such as a groove or dot, and other pins are arranged in a counterclockwise direction usually. You can also judge whether the chip is installed correctly by observing whether the chip model printed on the chip is correct.

Take the AD9527S of Analog Devices as an example. The layout of its pins is shown in Figure 10.8. If a problem is found, it should be removed and re-soldered immediately to avoid failure when the power-on test.

10.3.2.3 On-Board Power Interface Test

After testing the circuits, polar components, and chips on the PCB, the power interface of the PCB needs to be checked. Since the power supply voltage or reference voltage required by the chips on the PCB board is different, in addition to the power supply for the entire PCB board, several level conversion modules need to be designed to supply power to each part of the board. This involves the isolation between the power supplies and the power-on sequence. In the passive test, it is mainly necessary to check whether the power supplies are properly isolated.

The power interface detection is mainly divided into two parts, namely the power interface detection of a single power supply module and the power interface detection between power supply modules. For a single power supply module, the focus of detection is power and ground. Under normal circumstances, there will never be a short circuit between the two. You can check whether there is a short circuit fault on the interface through the buzzer of the multimeter. Once there is a

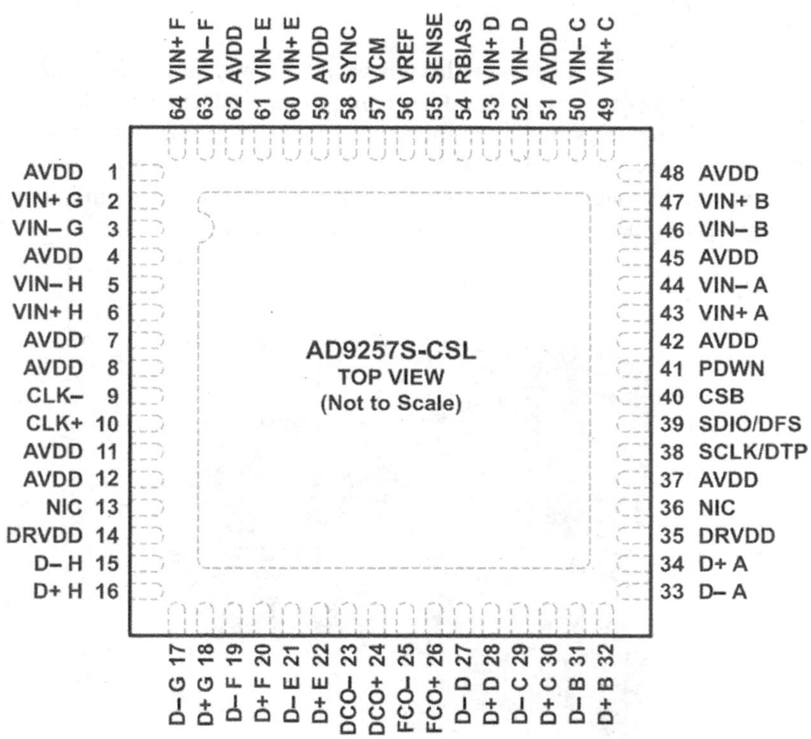

FIGURE 10.8 The pin distribution diagram of AD9257S. (Source: ADI.)

short-circuit fault problem, you should first compare the design drawing of the PCB board to highlight the part where the short-circuit fault occurs, analyze the parts with a higher probability of failure, and perform troubleshooting one by one. If there is no problem in the above inspection, it may be related to the short circuit caused by the internal failure of the chip (such as the power chip). In this case, the chip needs to be replaced in time, and then the short circuit failure test is performed.

After inspecting each power supply module, it is necessary to detect the interface between the power supply modules, and then conduct a joint inspection between the power supply modules. The focus of the inspection still falls on the power supply and the ground, but the inspection at this stage also includes whether there is a short-circuit fault between the power supply levels. The order of detection can be based on the power supply topology diagram formulated at the beginning of the PCB design, and the short-circuit detection of the board power supply can be carried out in an orderly manner. If necessary, a protection circuit, such as a self-recovery fuse, can be added to the circuit design to avoid failures or aging failures that are not found in the detection link and cause harm to the entire PCB.

10.3.2.4 Test Case

This section will take the soldered USB-HUB as an example to introduce the steps and precautions of passive testing. Mainly test whether there is a problem with the circuit on the board, test the soldering status of the components on the board, and test the soldering status of the power interface on the board. It should be noted that after inspecting each power module, the interface between the power modules needs to be tested. As shown in Figure 10.9, the USB-422 circuit board power interface test is in progress.

10.3.3 Active Testing

10.3.3.1 Static Testing

1. The Concept of Static Testing

   Static testing is a kind of DC test with a power supply only. Multimeters can be used for measuring the voltage of every on-board test point. Compared with the value which is estimated with theory and the structure of the circuit, the DC working state of the circuit can be determined. Furthermore, components that have been damaged or are in critical working conditions can be found. By replacing components or adjusting circuit parameters, the DC working state of the circuit can meet the design requirements.

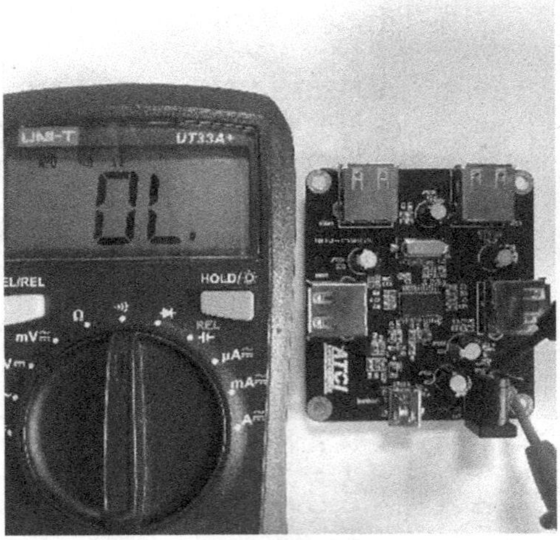

FIGURE 10.9 Board-level passive testing: on-board power interface test picture.

2. Static Testing Process
   1. Supply the circuit board with DC according to the design, and the power supply level needs to refer to the input level of the circuit during design.
   2. Observe whether the circuit board has abnormal phenomena, such as whether there is smoke or abnormal odor. After confirming that you are wearing anti-static equipment or ensuring that there is no static electricity, touch the outer package of the integrated circuit to see whether there is abnormal hotness, etc. If there is an abnormal phenomenon, turn off the power immediately, and then turn on the power for inspection after troubleshooting.
   3. According to the circuit schematic diagram, check whether the power supply level reaches the expectation step by step. If there is an abnormality, the location of the abnormality can be inferred based on this checking and the abnormality can be excluded.

10.3.3.2 Dynamic Testing

Dynamic testing is carried out based on static testing. Appropriate input signals are added to the input of the circuit, and the output signals of each test point are checked in sequence according to the flow of the signals. If abnormal phenomena are found, the reasons should be analyzed, and solve the fault until the requirements are met.

10.3.3.3 The Method of Active Testing

Grasping some analytical ideas and methods of circuit board inspection can help quickly find the abnormalities and failures and test rapidly and accurately. Some methods of active testing will be introduced as follows:

1. **Signal Injection Method:** This method uses different output signals of external signal sources for testing and uses the external device to indicate the test results of the tested board. During the inspection, select the corresponding signal source according to the specific requirements to obtain the signal with different indicators. This checking should be executed from the last level to the first level. In the maintenance of industrial circuit boards, inject the known and different test signals into the circuits at all levels, observe whether the output of the board is normal, and determine where and why the faults happen.
2. **Substitution and Comparison:** Use an intact circuit board of the same model and same specification with the same input signal, and compare the output of the same level of two circuit boards to evaluate the state of the board to be tested.
3. **Observe Parameters:** It is a method of electronic equipment fault inspection in which instruments (such as online maintenance testers) are used for testing the voltage and current of circuits in the power-on state.
4. **Capacitor Bypass:** It is a method of bypassing a certain part of the circuit to be inspected by using a capacitor of appropriate capacity and withstand voltage. This is a quick and easy fault detection method. It is suitable for judging the modules that produce power interference and parasitic oscillations in circuits.
5. **Separation Test:** also known as the circuit segmentation method, it is a fault detection method to reasonably separate the circuits related to the fault, to clarify the range of the circuit where the fault is located. This method is to narrow down the range of the circuit where the fault may occur step by step through multiple separate inspections until the fault location is found.
6. **Observation:** This method is to check the fault of the circuit to be repaired only through observation (including vision, smell, hearing, touch, etc.) without using any equipment or soldering any circuit components.

10.3.3.4 Precautions

Whether the test result is correct is largely affected by the setting of the experiment and the test accuracy. To ensure the accuracy of the test result, it is necessary to reduce the test error and improve the test accuracy. For this, the following points need to be paid attention to:

1. Use the ground of the test instrument correctly. Use an instrument with a ground connected to the chassis for testing. The ground should be connected to the ground of the amplifier. Otherwise, the interference introduced by the instrument chassis will not only change the working state of the amplifier but also cause errors in the test results. For example, when debugging the emitter bias circuit, if you need to test Vce, you should not directly connect the two ends of the instrument to the collector and emitter, but measure Vc and Ve to the ground respectively, and then subtract the two; when using a multimeter powered by a dry battery, since the two input terminals of the electric meter are floating, it is allowed to connect it with the two test points directly.
2. The input impedance of the instrument used to measure the voltage must be much greater than the equivalent impedance being measured. If the input impedance of the test instrument is small, it will cause shunting during the measurement and bring great errors to the test results.
3. The bandwidth of the test instrument must be greater than the circuit under test.
4. Select the test point correctly. When measuring with the same test instrument, the error caused by the internal resistance of the instrument will be very different if the measuring point is different.
5. The method of measurement should be convenient and feasible. When it is necessary to measure the current of a circuit, it is generally possible to measure the voltage because it is not necessary to destroy the circuit which is needed to measure the current. If you need to know the current value of a branch, you can get it by measuring the voltage at both ends of the resistor on the branch and then converting it.
6. Recording during debugging. The recorded content includes experimental conditions, observed phenomena, measured data, waveforms and phase relationships, etc. Only by comparing a large number of reliable experimental records with theoretical results, can the circuit design problems be discovered and the design plan can be improved.

10.3.3.5 Test Case

This section introduces the steps and precautions of active testing with examples. Mainly include static test and dynamic test. As shown in Figure 10.10, the USB-422 circuit board is being tested after power-on. When the active test of the circuit board is okay, it can work normally. Figure 10.11 shows the USB-422 circuit board that can work normally.

10.4 THE REASON AND METHOD OF DETECTION FOR THE BOARD-LEVEL FAULTS

For a complex system, it is not easy to accurately find faults in many components and circuits. The general fault diagnosis process starts from the fault phenomenon, makes analysis and judgments through repeated tests, and gradually finds out the fault. It is necessary to carefully find the cause of the failure, and must not remove the component and reinstall it once the failure cannot be solved. Because if it is a problem in principle, even reinstalling the component will not solve the problem. Common causes of board-level failure and detection methods are listed as follows.

FIGURE 10.10 Board-level active testing picture.

FIGURE 10.11 Picture of USB-422 that has been tested and works normally.

10.4.1 The Reason for Faults

When the finalized product fails after using it for some time, it may be due to component damage, short circuit or open circuit in the connection, changes in conditions, etc. Common phenomena of failure include: the amplifying circuit has no input signal, but has an output waveform. The amplifier circuit has an input signal, but no output waveform or the waveform is abnormal. The series regulated power supply has no voltage output or the output voltage is too high to be adjusted, or the output voltage regulation performance deteriorates, the output voltage is unstable, etc. The oscillation circuit does not produce oscillation, and the counter waveform is unstable.

10.4.2 Fault Detection Methods

Except for the methods which are introduced in 10.3.3, the faults can also be detected with short-circuit and open-circuit.

Temporarily short-circuiting a part of the circuit to find the fault. The short-circuit method is the most effective for checking open-circuit faults. But it should be noted that the short-circuit method cannot be used for the power supply (circuit).

The open-circuit method is the most effective for checking short-circuit faults. The open-circuit method is also a method to gradually narrow the scope of the suspected fault point. For example, because a certain regulated power supply is connected to a circuit with a fault, the output current is too large, we can disconnect a branch of the circuit in turn to check the fault. If the current returns to normal after the branch is disconnected, the fault has occurred in this branch.

In actual debugging, there are many ways to find the cause of the fault, and only a few commonly used methods are listed above. These methods can find the place where the simple fault happens, but for more complex faults, it is necessary to adopt a variety of methods to complement and cooperate to find the fault point.

11 Power Distribution

Michael Seavey

11.1 WHAT IS POWER DISTRIBUTION?

Automatic test equipment (ATE) typically consists of various sources, sensors, loads, control mechanisms, and active subsystems. These pieces of test equipment each require electrical power. The forms of electricity for each piece of test equipment may or may not be the same (e.g., there may be single-phase 120VAC 15A power required for one instrument, and 208VAC 30A/phase three-phase power for another).

Additionally, to facilitate not having to plug each item into an electrical outlet separately; as well as having to turn each one on and off, ATE designs incorporate a power distribution subsystem.

An ATE power distribution subsystem is the final stage in the delivery of electrical power; it carries electricity from the facility in which the ATE is located, to the individual ATE internal sources, sensors, loads, control mechanisms, and active subsystems.

11.2 WHY DOES AN ATE NEED POWER DISTRIBUTION?

- To convert the facility power form to the power form of the various sources, sensors, loads, control mechanisms, and active subsystems within the ATE.
- To protect the ATE internal equipment from noisy facility power using EMI filters, surge suppressors, and other techniques.
- To comply with general and industry-specific standards for electrical power handling equipment.
- To save space by integrating electricity conditioning and conversion within the ATE instead of housing them as external boxes.

  a. **Standards and Compliance**

  Power distribution-related requirements are contained in the following standards:
- ISO 9001:2015 Quality management systems – Requirements (with AS9100 is a widely adopted and standardized quality management system for the aerospace industry).
- UL/IEC 62368-1 Audio/video, information and communication technology equipment – Part 1: Safety requirements and 61010-1 (others when required).
- UL and CE certification (others when required).
- MIL-STDs in support of development for land, air, and shipboard applications.

11.3 WHAT MAKES UP AN ATE POWER DISTRIBUTION SYSTEM?

a. **Entry Panels**

The Power Entry Panel provides an input/output pass through connector power interface between the facilities electrical single or three-phase source and the ATE station.

b. **Power Distribution Units (PDU's)**

Power Distribution Units (PDUs) ensure sufficient voltage and current is provided to each ATE component while also protecting this equipment from surges, sequencing power provision, and may include various safety or functionality-based controls. These may be single-phase or three-phase systems; networked or locally controlled.

Some of the features of PDUs include the following:

DOI: 10.1201/9781003188377-13

- **Remote On/Off:** Remote on/off is a function that gives operators easy access to full control of the power system. The remote capabilities allow operators to change power states or use the Emergency Power Off (EPO) mechanism if an emergency arises.
- **Circuit Protection:** This comes in the form of fuses or circuit breakers, which can protect equipment from unexpected power surges.
- **Power Sequencing:** Power sequencing capabilities ensure that ATE equipment turns on in a specific order. For example, certain components may need to start first to allow other pieces of equipment to function normally. The PDU can be programmed to start these primary elements first in the power-on sequence.
- **Power Balancing:** PDUs can be designed to have multiple banks of power. This characteristic allows engineers to balance power loading on the PDU, preventing overcurrent conditions that can potentially result in faulty tests or even damage to system equipment. When paired with a UPS to power critical components and an EMI filter to protect from noise, PDUs provide excellent power infrastructure for ATE stations.
- **Addressing Noisy Facility Power:** Noisy facility power signals occur in factories, test labs, service depots, and other environments due to the varying loads of nearby high-powered equipment like motors, switching power supplies, office air conditioning, lighting, etc. Whether the noise comes from the next bench over, or the lab down the hall, or just building infrastructure, it can adversely affect an ATE. Surges of voltage and current in electrical lines cause signal abnormalities, which result in vulnerable components in an ATE possibly failing. Where clean power is essential, a PDU should integrate surge suppression, EMI filters, isolation transformers, and other tools to mitigate noise.
- **Power Forms:** Many test environments need a mixture of different power forms. A power form is the combination of specifications for voltage type, voltage rating, and current rating. For example, 120 VAC 60Hz. vs. 240 VAC 50Hz. are two different power forms.

  A building's facility power will usually be wired for two or three forms, but each of the forms will not be available in all work areas. In industrial or commercial buildings, these forms are typically three-phase AC power for some areas in addition to common 120Volt wall outlets. In various manufacturing or laboratories, there may be a need for multiple three-phase and single-phase power forms. Different voltages and currents may be needed to serve the larger and smaller pieces of equipment. DC power may also be needed.

c. Uninterruptable Power

Test Program Set (TPS) execution can be affected by power losses, brownouts, and malfunctions during normal TPS execution.

Uninterruptible power supplies (UPSs) ensure components in the ATE continue to receive continuous power when the ATE's primary power source encounters problems. For example, it may be critical that the ATE's cooling, control, and monitoring systems remain operational during brownouts, as failure may damage the ATE or Unit Under Test (UUT).

d. Emergency Power Off

Emergency power off (EPO) is a safety mechanism when there is a risk of injury to the ATE or TPS operator(s), or the execution of tests requires to be stopped immediately when there is a risk of damage to the ATE, TPS hardware or the UUT; by immediately removing electrical power and initiating ATE and TPS shut-down procedures.

EPO is defined by ISO 13850:1996 Safety of machinery – Emergency stop – Principles for design.

An emergency off panel typically consists of an emergency stop switch mounted in a panel which is in turn mounted on the front of the ATE; easily accessible to the operator(s).

OSHA 29 CFR 1910.179(a)(59) defines an emergency stop switch as a manually or automatically operated electric switch to cut off electric power independently of the regular operating controls. Emergency stop switches are wired in series with the facility electrical power input control circuit of the ATE. Pushing the mushroom head of emergency stop button will break the circuit and remove electrical power.

The release of emergency stop button requires pulling, rotating, or using a key, under strict authorized operation to ensure people and equipment safety to avoid injury again.

e. EMI and EMC

EMC regulations (there are several different IEC, IEEE, and Mil standards that are to be used depending on the application, such as Military, Industrial, Commercial, and Medical) to limit the electromagnetic noise a product can inject onto the local electrical power line. Just as the components and functions of ATEs vary, so does the EMC noise it generates. To ensure EMC compliance and to protect the ATE from any noise already on the electrical line, EMI Filters are standard in test equipment system architecture, these are typically located near the electrical power input.

There are a few basic rules that apply to the implementation of EMI and EMC regulations. First, identify the noise source; and second, place the filter of the noise as close as possible to the source (assuming the noise cannot be avoided). Unfortunately, the power distribution network is a good candidate for distributing conducted noise within the ATE system as well as radiating noise around the ATE system racks. It is therefore extremely important to implement proper filtering.

f. Grounding

Grounding is a crucial aspect of ATE design, both for safety and measurement quality. Grounding secures the safety of operators and equipment by ensuring that all system equipment has a proper path for current to flow to the ground. Quality grounding also includes assurances that electrical connections to the ground are direct, which minimizes the production of radio frequency emissions that can interfere with measurement equipment.

g. Reliability and Redundancy

The lifespan of an ATE may be decades, therefore, identifying and controlling critical components' mean time between failure (MTBF) is necessary.

This includes electrical and thermal de-rating (to reduce thermal stresses) and redundancy of the ATE's power distribution system. Critical components whose failure jeopardizes ATE power distribution systems should implement redundancy and provisions should be made for field replacement.

h. Serviceability and Diagnostics

Enable an elevated level of modularity and provide online diagnostics. Implement flexibility by allowing hot swapping non-critical components (e.g., fans) that can be field replaced while the ATE is operating. Additionally, provide for easy access to the components of the ATE's power system.

Implementing diagnostics permits the identification and isolation of power distribution components or subsystem issues quickly.

11.3.1 Example ATE Power Distribution System

Figure 11.1 depicts a generic power distribution system for a generic ATE, employing the elements described in this chapter.

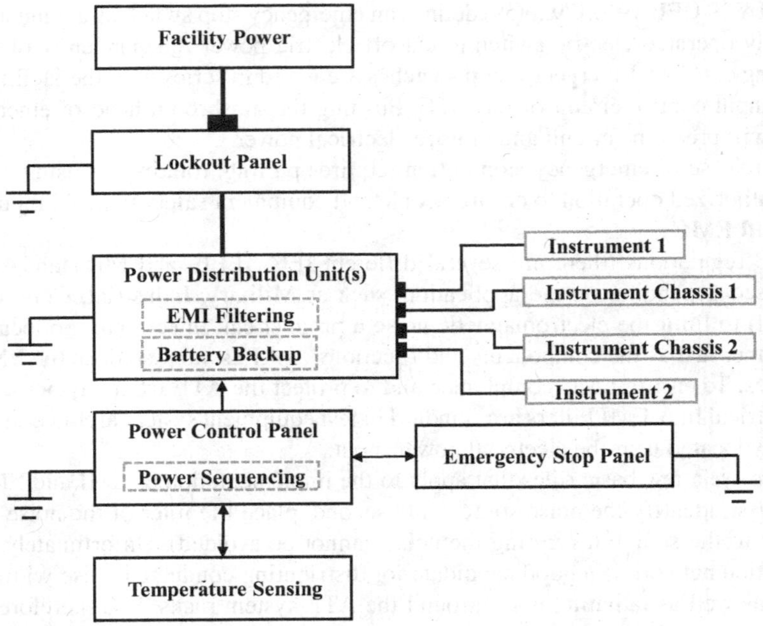

FIGURE 11.1 Generic power distribution system.

12 Design Patterns and Reusability

Clemente Izurieta

12.1 INTRODUCTION

Every successful software system undergoes multiple evolution cycles. The evolution of systems (Lehman, 1997; 1980; Lehman et al., 1998) occurs in response to demands for new functional changes or to attenuate technical debt through refactoring of designs and code. The complexity of such systems has demanded that solutions need to occur during the design stages of its evolution, and design patterns represent an approach to help characterize aspects of a system that facilitate ease of maintainability, reusability, and reduction of defects (Gueheneuc and Albin-Amiot, 2001) as a system evolves. Many proponents of design patterns have argued that their use improves architecture.

Design patterns are agreed-upon solutions to complex problems. *"Instead of code reuse, with patterns you get experience reuse"* (Freeman and Freeman, 2004). Patterns cannot be invented; however, if enough evidence of similar solutions to problems can be collected, then an abstraction can be constructed to help organize these experiences, and these abstractions form the basis for design patterns. Characterizations of experiences can be represented either in visual or in textual form; however, the most common approaches use the Unified Modeling Language (UML) (The Object Management Group, n.d.) or UML-style-like languages. The ever-increasing complexity of systems demands that said abstractions not only encapsulate the constraints necessary to characterize a solution but also provide the flexibility necessary to allow for meaningful and practical extensions. We capture such abstractions using declarative modeling techniques.

The balance between capturing flexibility to allow for extensions and the constraints to flag violations of a pattern is one that requires careful balance. Relaxing the constraints too much allows any instance of a design pattern to be conformant to its model, whereas tightening the constraints too much does not allow a pattern instance to be extensible, which defeats the purpose of patterns from a practitioner's perspective.

This chapter takes a look at design patterns from the perspective of software evolution. This is an approach I decided to take because the purpose of design patterns is to ease the maintainability of a system and to ease the reusability of existing solutions as the system matures and changes over time. As a practitioner of software engineering, the approaches to the engineering of systems, their maintenance and operation require that we experiment with techniques that eventually help the practice, and hence the study of design patterns from a practitioner's approach requires that we describe their usefulness from an evolutionary perspective.

This chapter is organized by first providing the necessary background and reviewing the classic set of original design patterns by the Gang of Four (GoF) (Gamma et al., 1995). We then describe approaches to characterize instances of design patterns using role-based techniques. We describe the specification and validation of design pattern instances. As systems evolve, so do the design patterns embedded in its design, and evolution can have negative effects if patterns do not adhere to their intended designs. We describe the tracking and the associated challenges that design pattern instances exhibit as they evolve. These challenges reveal different ways that can impact the quality of design patterns, and we describe a taxonomy that helps categorize the different types of negative impacts observed. They affect both maintainability and reusability of the system as a whole. We describe a selection of design patterns (i.e., Singleton, Adapter, Observer and Factory)

DOI: 10.1201/9781003188377-14

and their evolution by showing how grime accumulation of their respective instances increases over time. Design pattern decay is also a good prelude to a discussion about design anti-patterns. We tackle this topic in the last section of this chapter, discuss some of the most commonly encountered anti-patterns (i.e., The Blob, Cut and Paste, Lava Flow, and Functional Decomposition), and explain the relationships they have with traditional design patterns through a diagram that contrasts them in an abstract way.

12.2 BACKGROUND

Design patterns are pervasive constructs in software designs. They are widely used in object-oriented specifications and provide an agreed-upon characterization of a solution, a micro architecture, that serves as a reference to practitioners during the implementation phases of the software development lifecycle. Characterizations of solutions in object-oriented designs are typically generalized to show abstract interactions between classifiers and their respective relationships. An ideal design pattern language that can characterize solutions should be able to provide comprehensive support for their specification, modeling, and detection. Many languages (visual and textual) have been proposed that capture these relationships and include among others UML (The Object Management Group, n.d.), the Meta Role-Based Modeling Language (RBML) (France et al., 2002), SPINE (Blewitt et al., 2005), Fujaba (Fujaba Software Engineering Group, 2005), Pinot and Muscat (Shi, 2007), Layom (Bosch, 1998), LePus (Eden, 2006), and DeMIMA (Gueheneuc and Antoniol, 2008). An abridged description of the capabilities of each of these languages can be found in the comprehensive work by Izurieta (Izurieta, 2009).

Although design patterns in the context of software can be traced back to the work of Beck and Cunningham (Beck and Cunningham, 1987) in the late 80s, they never truly gained popularity until the *Gang of Four* (GoF) published a book describing the application of patterns in software (Metsker and Wake, 2006). The book uses UML to describe 23 classic software design patterns, mostly written in C++. To date, this book continues to serve as a reference that is used by academics and practitioners alike.

Gamma et al. (1995) categorized the selection of design patterns into three sub-classifications that are tied to the types of problems that the patterns address. Behavioral patterns (e.g., Strategy and State) characterize common communication patterns observed between objects, creational patterns (e.g., Factory and Singleton) characterize object creation, and structural patterns (e.g., Decorator and Adapter) characterize the organization of classes and objects. The full list of specifications can be found in GoF's publication. Although the publication of this reference book served a large community, it exemplified all pattern implementations using the C++ language, thus limiting its utility for the Java practitioner community. Interface constructs in Java are a popular method to capture abstractions and establish a contract for classifiers that choose to realize said interface. *"A class's interface, speaking abstractly, is the collection of methods and fields that the class permits objects of other classes to access"* (Metsker and Wake, 2006). In Java, there is a distinct separation between the interface (i.e., the signatures of the methods that a client can use) and the implementation of the methods. In C++ you can use abstract classes and the concepts of pure virtual functions to achieve the same semantics as interfaces. In this chapter, I make no distinction between interfaces and abstract classes. This is an implementation detail that is left for practitioners to address. Instead, I use the notion of abstraction, as a concept in modeling that is well understood.

The largest corpus of patterns assembled continues to be Booch's *Handbook of Software Architecture* (Handbook of Software Architecture, n.d.) with more than two thousand patterns spanning many domains, and the taxonomy of pattern specifications has matured to encompass derivations into domain-specific rubrics such as security (Fernandez-Buglioni, 2013; Schumacher et al., 2006; Steel et al., 2005), distributed systems (Buschmann et al., 2007), architectural (Buschman et al., 1996), pedagogical, human–computer interaction and web design to name a few. Of notable interest in modern day cybersecurity systems, security by design has been the subject of renewed interest. The work by Fernandez-Buglioni (Fernandez-Buglioni, 2013) provides

TABLE 12.1
Metadata Associated with Design Pattern Descriptions

Metadata	Description
Pattern Name and Classification	Name and a *creational, structural, or behavioral* classification
Intent	The rationale and intent of the pattern
Also Known As	Other well-known names
Motivation	A scenario that exemplifies how the pattern solves a problem
Applicability	The situation when the design pattern can be used
Structure	A graphical representation
Participants	The classes and objects (with responsibilities) that participate in the design
Collaboration	How classifiers collaborate to carry out responsibilities
Consequences	Tradeoffs and results of using the pattern
Implementation	Language-specific issues associated with instantiating the pattern in a given language
Sample Code	Code fragment exemplifying use
Known Uses	Examples used in real systems
Related Patterns	Closely related patterns

insights into the pragmatic applications of security design patterns in software designs. The focus of Fernandez-Buglioni is to build secure systems, and he focuses on the modeling approach to provide abstractions of the solutions. Other work (Schumacher et al., 2006; Steel et al., 2005) also exemplifies solutions for well-known problems in the triad *Confidentiality, Integrity* and *Authentication* (CIA) of cybersecurity. Additionally, many publications also provide descriptions of various patterns that demonstrate solutions in the product lifecycle of software with many architectural, test, and build patterns. At their core, however, GoF's patterns remain the main contribution.

In practice, design patterns are the subject of mixed success. UML has been adopted as a *de facto* standard to communicate these micro architectures, and although highly utilized during the design phase of software components and architecture, once the occurrences of a design pattern are implemented, they tend to drift from their original intent. The lack of alignment between the evolution of specifications and its complementary source code is well understood, and in large part, it is due to the intentional technical debt (Cunningham, 1992; Schloss Dagstuhl, 2016; Izurieta et al., 2016) tradeoffs necessary to meet the market demands of products.

The communication and documentation of design patterns are not strictly limited to their structural (UML class diagrams) and behavioral (UML sequence diagrams). Other significant and important information must accompany a definition. The metadata associated with a design pattern (shown in Table 12.1) is well known and comes from the GoF (Gamma et al., 1995):

12.3 CHARACTERIZATION OF DESIGN PATTERNS

In order to capture experience from design patterns, we first have to characterize it. UML diagrams of design patterns can vary significantly depending on their context and embedding, and although generic UML diagrams can be generated, a more formal approach is necessary. In this section, we discuss the specification of design patterns at higher levels of abstraction that allow us to capture experience by focusing on the roles of each element. We then discuss the different approaches necessary to validate these diagrams.

12.3.1 SPECIFICATION

In order to characterize design patterns, their informal definitions using UML, as used by GoF in their book, are not sufficient, and a more formal approach is necessary. RBML (France et al., 2002; 2004)

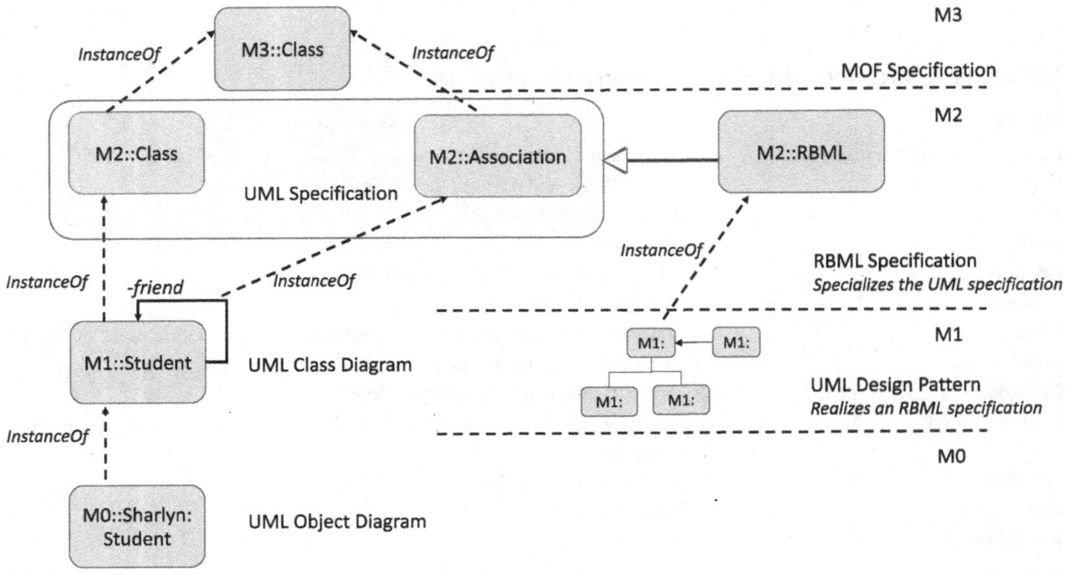

FIGURE 12.1 Visual Representation of the relationship between UML and RBML.

has been shown to be a promising approach that meets this task. RBML is defined as a specialization of UML and is a visual language implemented as an extension of UML that can be used to model both structural as well as behavioral aspects of design patterns. *"The pattern specifications created by the technique are metamodels that characterize UML design models of pattern solutions. A pattern's metamodel is obtained by specializing the UML metamodel"* (France et al., 2004). The structural components of a pattern specification are captured using Structural Pattern Specification (SPS) diagrams, and the behavioral components are captured using Interaction Pattern Specification (IPS) diagrams. An SPS and an IPS diagram are analogous to the corresponding UML class and sequence diagrams at the M2 level of design modeling specifications. In the UML, there are four levels of specifications where M0 corresponds to an instantiation of a class diagram (i.e., an Object diagram), an M1 level diagram corresponds to an instantiation of an M2 level, and so on. For example, an M1 diagram can represent a class diagram, which is an instantiation of the UML specification (M2 level). The M3 metamodel is OMG's Meta-Object Facility (MOF) (Meta Object Facility (MOF), n.d.). We generally refer to a parent level as a metamodel. Additional descriptions of SPS and IPS diagrams can be found in the work performed by Kim (2004) and further GoF examples in RBML form in the work by Izurieta (2009). Figure 12.1 provides a visual representation of the relationship between UML and RBML from a structural perspective, and there is a corresponding behavioral perspective view as well. A behavioral representation can be constructed by replacing class diagrams with sequence diagrams. Together with RBML, additional constraints can be used to decorate the model specification using the Object Constraint Language (OCL) (Warmer and Kleppe, 1998). The OCL can be specialized to refer to elements of the RBML at the M2 level. Other representations that use matrices have been proposed and are further explained in Section 12.3.2.

12.3.2 Validation

Few automated approaches to validating instances exist. Instead, most rely on manual techniques to find patterns in a design and rely on experienced developers and architects to assert if said instance maintains its structure and behavior. One approach to automating the validation of patterns is through the use of M2 modeling languages. RBML allows a modeler with the ability to precisely specify a design pattern. The precise roles that classes play and their behavioral expectations exemplified through an M1 UML model need to be validated against their M2 level metamodel. Thus, a design pattern (e.g., a strategy pattern) has to be checked for conformance against its RBML specification in order to establish compliance. Failure to establish structural and behavioral compliance signifies a deviation from the intended role of the pattern. A potential drawback with using RBML to characterize design patterns, however; is that it may not always be simple to check whether a UML implementation conforms to its RBML specification. If the M1 UML model of a pattern does not conform to its M2 RBML specification, a developer will want to know the exact classifiers of the M1 model that do not conform to the intended M2 RMBL specification. In their work, Kim et al. (Kim, 2005; Kim and Shen, 2008) proposed an algorithm to check for conformance. The algorithm was implemented by the RBML Conformance Checker (RBMLCC) tool on top of IBM's Rational Rose software suite of products. RBMLCC takes two input parameters, the RBML specification and the actual design pattern in UML. The RBMLCC tool is not easy to use; however, it allows to the inspection of traces that allow a practitioner to look for potential violations.

In another tool developed by Strasser et al. (2011), and based on the work by Kim et al., the goal was to ease the entry point for novice developers, yet maintain a similar level of rigor when checking for compliance. The algorithm used by this tool checks instances of design patterns for potential violations in classifiers, relationships, and multiplicities against their RBML specification. The algorithm takes the same two parameters as the RBMLCC tool and provides a normalized score between 0 and 100 that signifies that the instance is not a pattern, or that the instance is 100% compliant. Figure 12.2 exemplifies the tool RBML Visualizer in the process of providing a score for an instance of an observer pattern. The tool (Google Code, n.d.) checks a number of rules and provides a score. To compare an RBML and a UML diagram, the approach uses the divide-and-conquer algorithm developed by Kim and Shen (2008).

Another approach to validating and detecting patterns was performed by Tsantalis and Chatzigeorgiou (2006). They correctly describe the difficulty in identifying patterns *in-situ,* as the search space for said patterns can be very large. The identification of modified patterns that have been extended to novel patterns can also pose problems. The analogy to searching for patterns in a design is that of a graph isomorphism problem and it is in the class of NP problems, but we suspect it is harder because there can exist multiple relationships (i.e., edges) types and multiple classifier types in UML diagrams, thus making their graph representation more complex. Tsantalis and Chatzigeorgiou (2006) engineered a detection methodology that is based on the similarity scoring between graph instances with the ability to recognize patterns that are modified from their standard representation. This is a typical scenario practitioners will encounter as instances of patterns evolve over time. Evaluations of their approach have shown that their algorithm is accurate and allows for practitioners to define their own structures of interest to search. The representations of the graphs are stored as matrices. Rather than using a metamodel like RBML to represent patterns, Tsantalis et al. use square matrices because of their ease of use and manipulation. Their method uses two graphs. A similarity algorithm compares a graph representing a generic pattern (similar to an M2 metamodel) and a graph representing an instance of a pattern in a system. A score between 0 and 1 is produced.

12.4 DESIGN PATTERN EVOLUTION

Patterns evolve, and their evolution can significantly help with the maintainability of any design or architecture if the pattern is extended properly. Unfortunately, this is not typical, and many challenges exist. We describe the tracking of pattern instances and the potential challenges as patterns evolve.

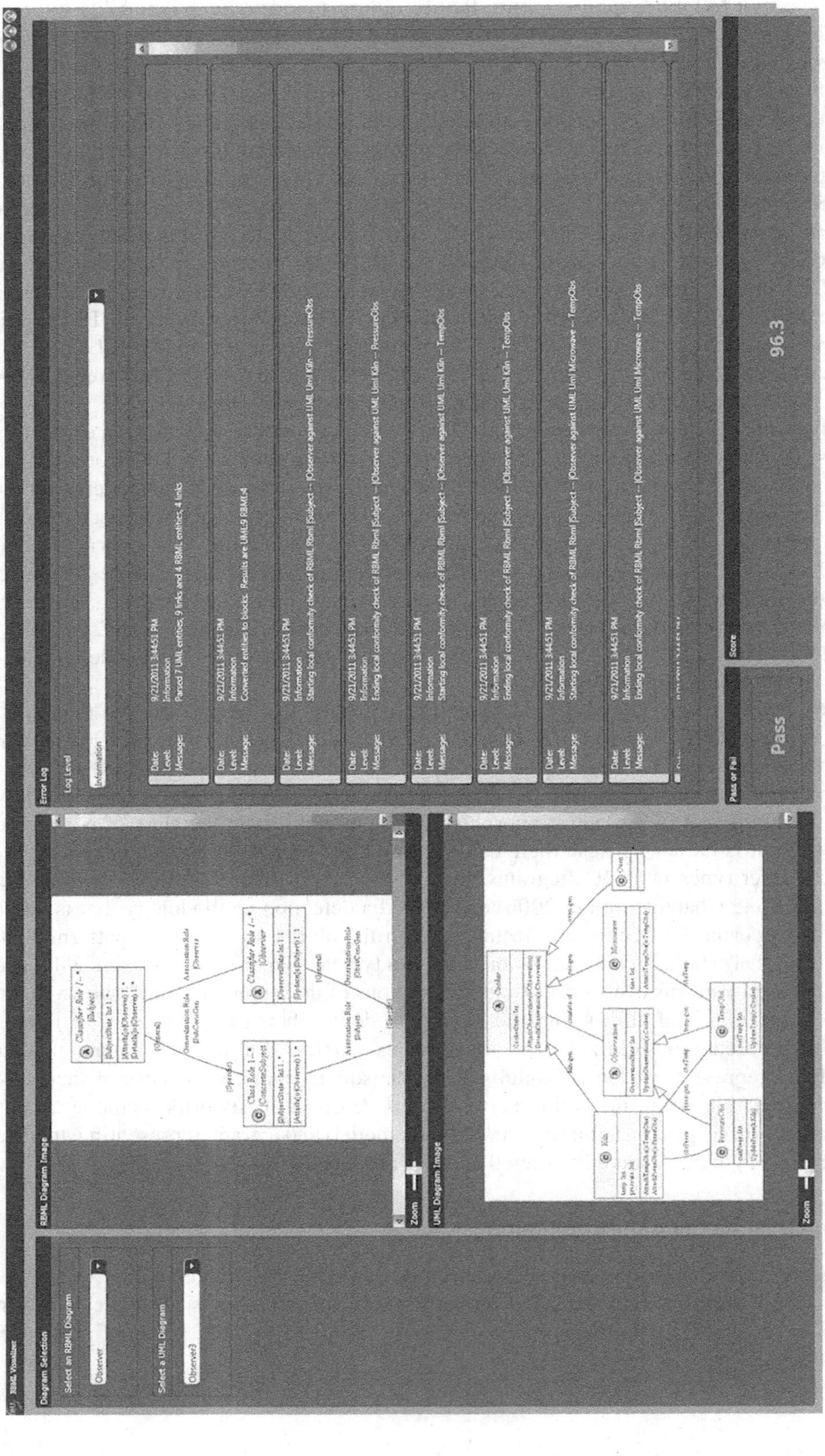

FIGURE 12.2 The RBML Visualizer allows users to check the conformance of a pattern instance.

12.4.1 Instance Tracking

Longitudinal tracking of the evolution of a design pattern instance allows designers to detect deviations from intent early. The microarchitectures of design patterns capture experience in a formal design specification. This, in turn, promotes the reusability of the experience to any problem. The experience is specified and implemented in (typically) an object-oriented language. However, although these micro architectures promote reusability of solutions, they tend to break down once a system begins to evolve.

As instances of design patterns evolve Izurieta et al. (Izurieta and Bieman, 2007; 2008) experimented with understanding the consequences associated with the deterioration of design pattern instances. Consequences were observed in testability as well as adaptability of design patterns. However, in order to measure the deterioration observed, informal definitions of design patterns are not enough, and a formal definition was required. By contrasting a pattern instance against its RBML definition, the authors coined the terms design pattern decay, grime, and rot.

To quantify decay, rot, and grime, it needs to be characterized first, and prior definitions of decay (Eick et al., 2001; Parnas, 1994) did not quite describe what was observed in design pattern realizations. Izurieta and Bieman (Izurieta, 2009) defined decay as *"as the deterioration of the internal structure of system designs. Internal structural breakdown of a design is caused by changes that do not conform to intended architectural methodologies,"* which served as an overarching definition from which two subtypes of decay are formalized: (1) design pattern rot –the structural degradation of a pattern through the negative evolution of its parent system; which can be captured by systematically checking all classifiers and relationships in a pattern instance against its RBML specification, and (2) design pattern grime –the buildup of unnecessary or unwanted extensions to a design pattern instance that are not admissible when compared against its formal RBML specification. Design pattern rot occurs when there is evidence of the removal of key classifiers or relationships from the pattern such that the pattern instance no longer retains its core elements. A pattern instance that has relented to rot is no longer representative of the pattern captured by the specification, and they are rare and hard to find because of the difficulty in identifying them. Grime identification is easier to detect because the pattern instance remains hidden underneath all the potential grime buildup, and although hidden, the pattern can still be found. Izurieta (2009) found that grime increases over time, regardless of the type of design pattern in use. Feitosa et al.(2018) found that grime has a tendency to accumulate linearly, providing further evidence that the quality of a pattern worsens as the grime of that pattern increases (Feitosa et al., 2017).

It is also worth mentioning the existence of near instances of design patterns. Near instances are found when a developer appears to have had the intention of developing a pattern, but critical components of the pattern are left out, thus rendering said pattern a near instance (i.e., not quite a match). There is no need for evolution to occur to find near misses in a system and they can be identified because developers name some classifiers with strings that provide hints as to the pattern they intended to use. In many cases near instances can also occur as a result of recent design pattern rot. However, the latter case cannot be identified via pattern detection techniques that focus on structural matches alone (since the instance has critical components missing, it does not match any design) and requires that techniques also use other heuristics like searching for hints in the naming of the classifiers. In Section 12.4.2.3, we discuss near instances in depth. Figure 12.3 exemplifies the process of the evolution of a design pattern.

12.4.2 Tracking Challenges

Many challenges associated with tracking the lifecycle of software design patterns exist. The lifecycle of a design pattern instance is generally tracked across many versions of the software. Each instance undergoes changes as the design of the software evolves. However, it is unlikely that one can find multiple versions of a design, as many changes are typically only incurred in the code. Thus, the process of tracking changes is difficult because we first have to find techniques to detect a pattern instance

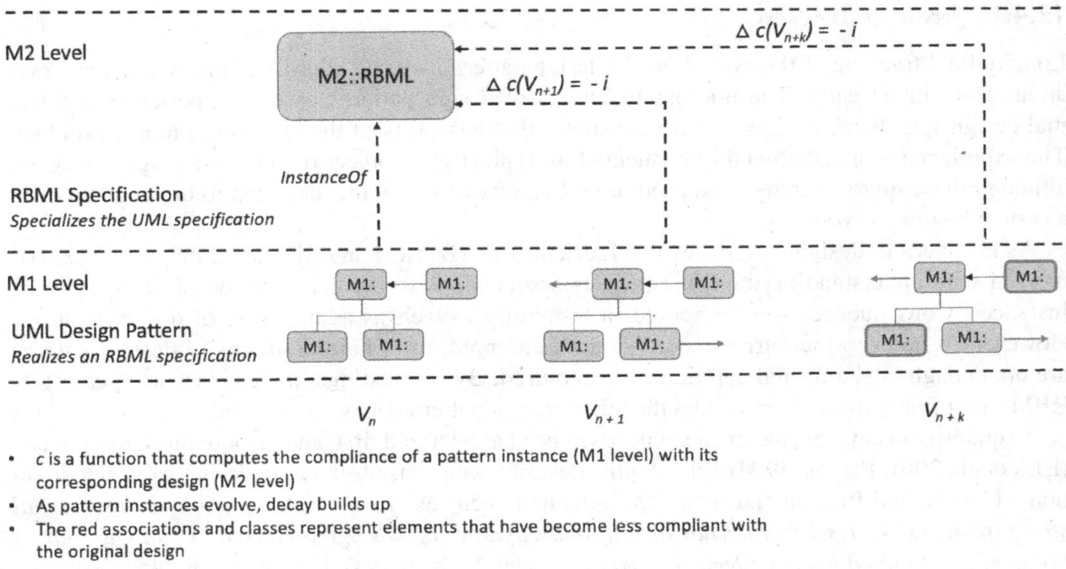

FIGURE 12.3 Evolution of a design pattern instance.

in its setting (i.e., the source code), and then we have to derive (i.e., reverse engineer) the UML diagrams that represent the design pattern. Once a UML diagram is available and validated, then we can analyze the potential forms of grime that affect the instance. First, we discuss the classification of the various forms of grime that pattern instances can be subject to, then we discuss the common problem of pattern coupling (Wang and Bieman, 2006) which can make their detection difficult because elements in the design pattern can be shared across multiple patterns, and finally we discuss the notion of near instances (not quite a design pattern), that can also make tracking difficult because detection can be challenging.

12.4.2.1 A Taxonomy of Design Pattern Decay

As design patterns evolve, there exist many ways in which said pattern can drift from its intended design. It is important to understand the different forms of deviations that can occur. Izurieta et al. (2019) present a taxonomy that describes the structural and behavioral deviations that a pattern instance can undergo as it builds grime. The evolution of a design pattern instance from its original intent (i.e., drift) can be described as rot or grime:

> Pattern rot is the breakdown of a design pattern such that a critical element in the pattern ceases to exist. Thus, a realization of a pattern that experiences rot is no longer a representation of its microarchitecture. Rot is difficult to find because as the realization evolves, it becomes harder to identify. Design pattern grime is the buildup of unrelated artifacts in classes that play roles in a design pattern realization.
>
> *(Izurieta et al., 2019).*

The taxonomy divides grime into two major categories: structural and behavioral. The structural category has three major sub-categories: class, modular, and organizational. The behavioral category is divided into Package (i.e., organizational) and modular sub-categories. In Figures 12.4–12.6, we depict the corresponding structural taxonomies of class, modularity, and organization. The left-hand column in all these figures describes the dimensions and categories that are used to classify the different forms of grime. *Strength, Scope and Direction* were used to classify grime at its highest level and provide an intuitive way to separate the different forms of grime. In general, when discussing relationships (i.e., associations) between classifiers, *strength* refers to the difficulty of removing a coupling relationship. Persistent and temporary coupling are the most common forms in

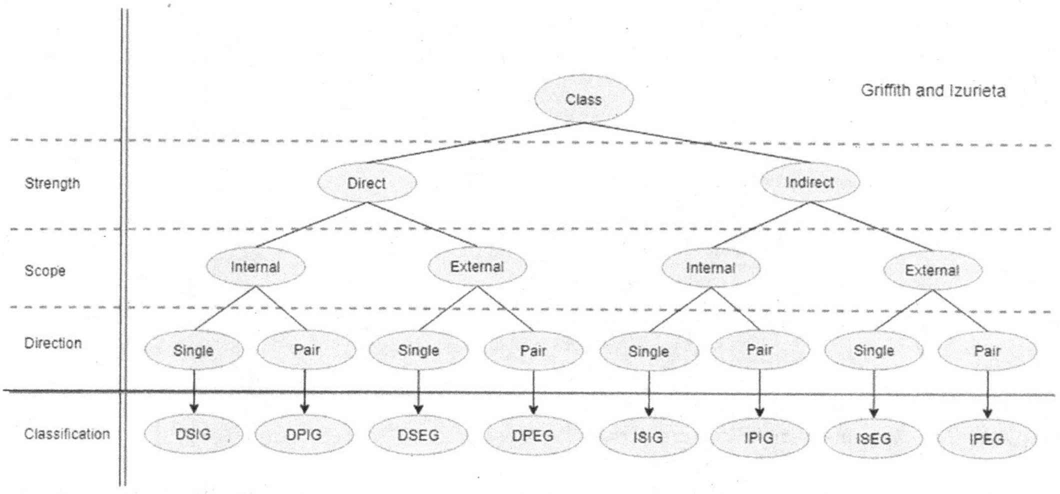

FIGURE 12.4 Class Structural Grime Taxonomy.

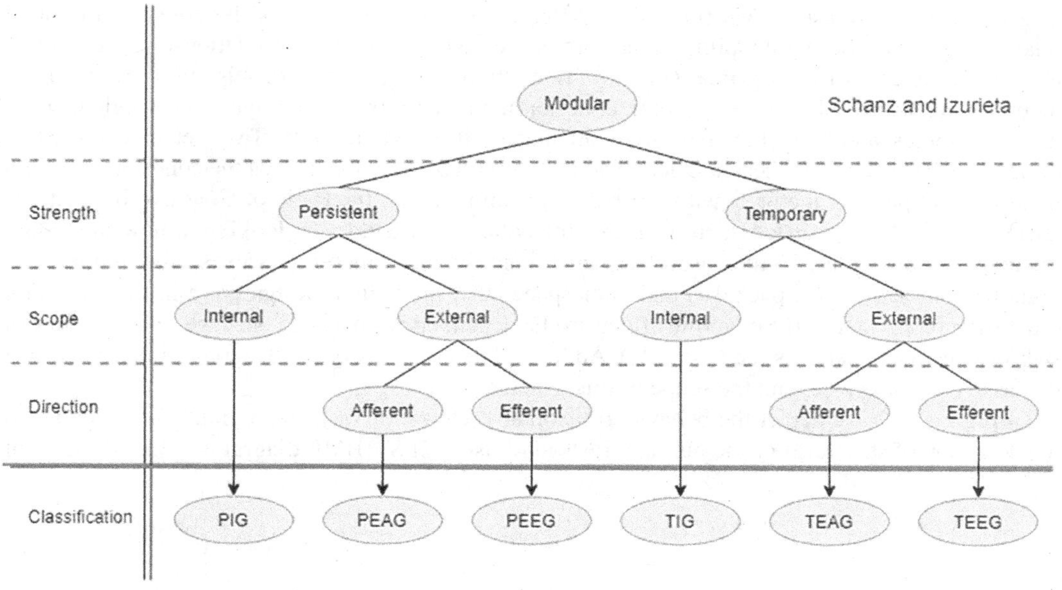

FIGURE 12.5 Modular Structural Grime Taxonomy.

object-oriented systems. Persistent forms of coupling tend to occur when associations are present, and temporary coupling occurs when transient relationships (i.e., dependencies) are present. The latter are much easier to refactor or remove than the former. When discussing classifiers (i.e. classes and interfaces), *Strength* is indicated by the approach in which attributes are locally accessed by the methods of a class. The method to reference attributes can be either direct (attributes are directly accessed by methods) or indirect (attributes are accessed through the use of an accessor or mutator method). Accessing attributes directly allows for an intimate relationship between a method and its attributes, but not considered a good practice. Indirect attribute access is more amenable to refactoring because it is also considered good use of software design techniques.

The *Scope* dimension refers to whether the grime observed is constrained to the internal members (i.e., classifiers and relationships) of the design pattern instance or whether the grime observed

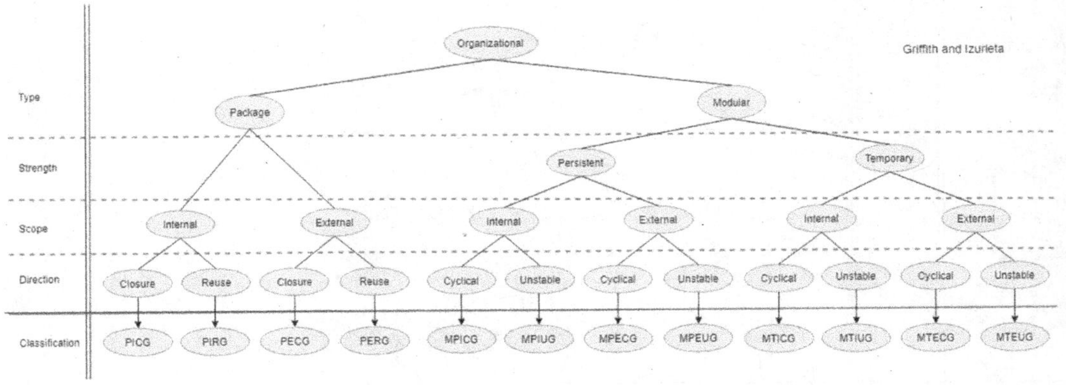

FIGURE 12.6 Organizational Structural Grime Taxonomy.

refers to a relationship to an unwanted external classifier. Scope is more subjective because pattern instances must interact with external components in a system, and in many cases, those external components provide additional functionality that is necessary for the system to perform its function.

In the case of modular and behavioral grime, the *Direction* dimension refers to the direction of a coupling relationship (Martin, 1994). Afferent coupling or fan-in is the count of in-bound relationships and efferent coupling or fan-out is the count of out-bound relationships of a set of classes. In the case of class grime, *Direction* refers to the types of relationships used by surrogate metrics to measure cohesion. In general, cohesion metrics are one of two types: (1) single method use of attributes, and (2) method pair use of attributes (Briand et al., 1998). Two metrics capture this dimension: Tight Class Cohesion (Bieman and Kang, 1995) which measures the cohesion of a class by looking at pairs of methods with attributes in common, and the Ratio of Cohesive Interactions (Briand et al., 1993) metric which measures the cohesion of a class by looking at how individual methods use attributes. In the case of organizational grime, which refers to the distribution and organization of the files, packages and namespaces that make up a design pattern, the *Direction* dimension refers to how the coupling affects cyclic dependencies between packages, and the flow of stability between packages (i.e., stability). Additional descriptions of the different dimensions used to classify grime are beyond the scope of this chapter.

In Figure 12.7, we depict the behavioral taxonomy. Detection of grime buildup is analogous to the detection of structural grime buildup. Instead of using SPS RBML diagrams at the M2 level of

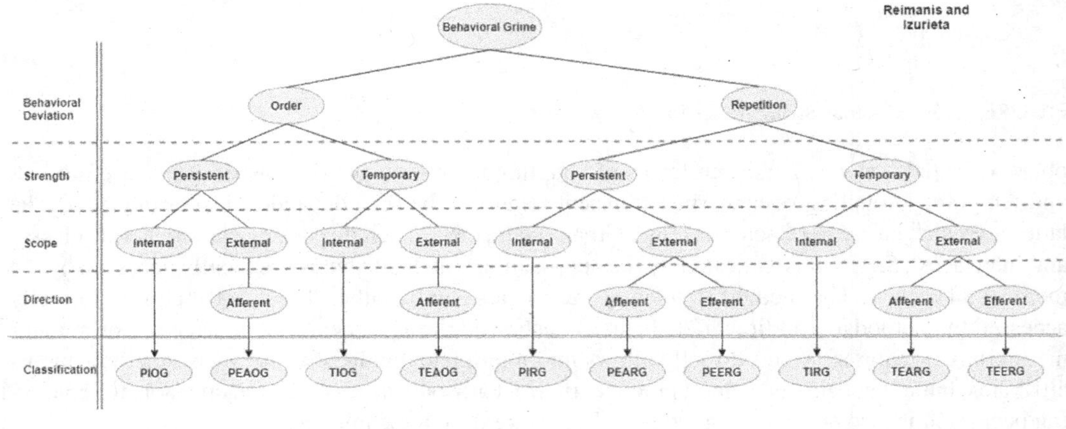

FIGURE 12.7 Behavioral Grime Taxonomy.

modeling, researchers use IPS RBML diagrams. UML sequence diagrams of a specific pattern are checked for conformance against their RBML specifications.

12.4.2.2 Pattern Coupling

A design pattern can be described as a cluster of relationships between classifiers, but in real world systems, we find that many times some of the classifiers are shared between instances of patterns. This occurs either intentionally or inadvertently. In 2006, Wang and Bieman wrote a report (Wang and Bieman, 2006) that identified ways in which patterns can be coupled and they measured the relative tightness of the connections. Further, although patterns can be coupled via classifiers or relationships, they found that patterns that are coupled via relationships (i.e., associations, dependencies, etc.) are more prone to change. Figure 12.8 shows an example of two pattern instances that are coupled together via a common element, and another example where both patterns are coupled via an association.

Authors Buschmann et al. (2007) further discuss the most common types of pattern relationships. They identify three common types: (1) *pattern complements*, where one pattern provides the functionality needed by another pattern. This results in a sort of cooperation between two patterns where both solutions have a natural symbiotic relationship that makes sense and provides for good extensible design, (2) *pattern compounds*, where a solution to a problem is addressed by a common set of patterns that work well together. These pattern collaborations are so well understood that they could be treated as one compound pattern, and (3) *pattern sequences*, that string together in a daisy chain-like approach, a series of patterns – *"by joining together predecessor patterns to form part of the context of each successive pattern"* (Buschmann et al., 2007).

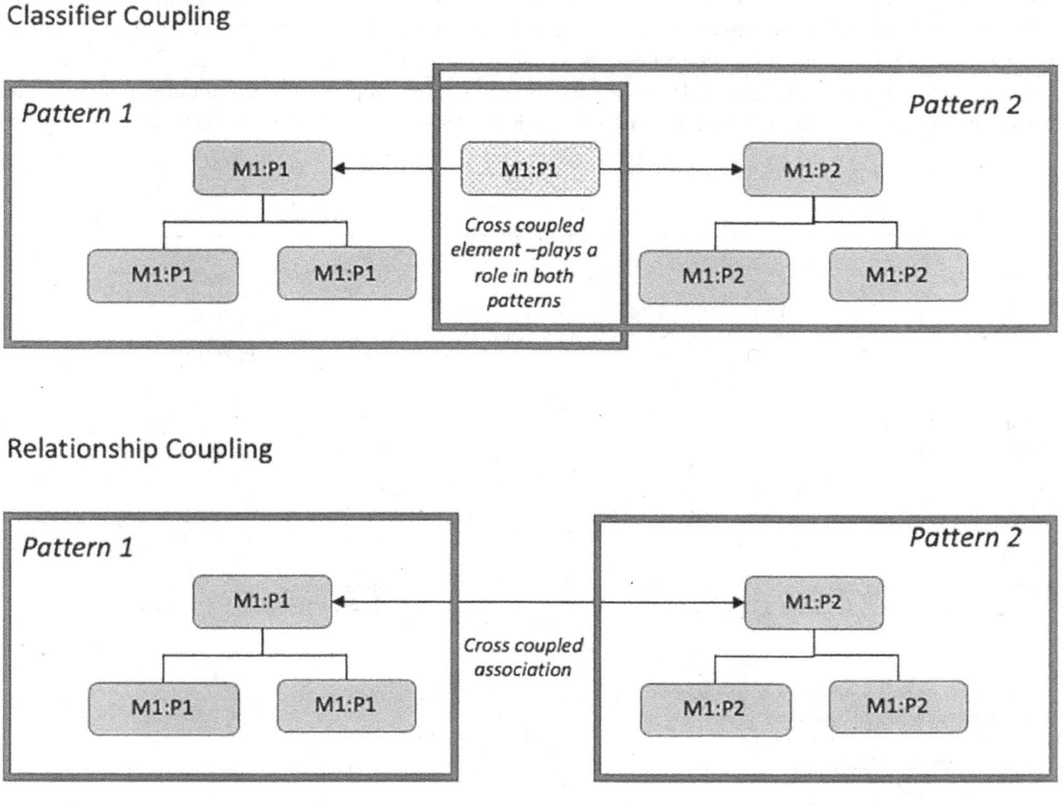

FIGURE 12.8 Cross coupling of design pattern instances.

As pattern instances evolve and decay sets in (i.e., grime buildup) the removal of disharmonies through refactoring mechanisms becomes harder to do because the pattern instances are hard to detect, and depending on the type of coupling after detection, the removal can incur a significant effort to the developer. In general, coupling relationships like associations (including composition and aggregations) tend to be the most difficult to refactor, and dependencies tend to be the easiest to refactor given their transient lifecycles.

The effects of pattern coupling on external quality factors have been studied and although the coupling of patterns increases impacts on the quality of a system across all ISO (ISO/IEC 9126-1:2001, 2001; ISO/IEC 25010:2011, 2011) characteristics, an important aspect to note is the *ripple effect* (Yau et al., 1978). The latter refers to the spread of grime across the rest of a design. Wang and Bieman found that patterns coupled via associations, the tightest form of relationship coupling, had the greatest tendency to be changed together, thus causing ripple effects across the design.

12.4.2.3 Near Instances

A near instance of a design pattern occurs when there exists a close match to a pattern RBML specification that violates some requirements. The violations create an instance that is almost a match, but force the instance to miss some critical requirements necessary to deem the instance a match. In many cases, this occurs because of staff changes, where a new developer without design pattern training may not identify the solution and change it in ways that drift the implementation from the intended design. In other cases, the naming of the classifiers may not be intuitive and are thus not recognized by a developer. If a developer does not know that he/she is dealing with a design pattern, then the evolution of the pattern is also likely to drift away from the intended design. For example, a near instance of the Singleton pattern can occur when all elements are bound to an RBML classifier, but the static private variable that holds the real instance may be named something that is unexpected, thus forcing the algorithm that checks for compliance to flag the variable. A change that turns a pattern into a near instance represents a case of pattern rot. Figure 12.9 shows an example of a near instance. On the left-hand side of the picture we display a representation of pattern instance at the M1 level and its relationship with its design (in RBML) at the M2 level. On the right-hand side of

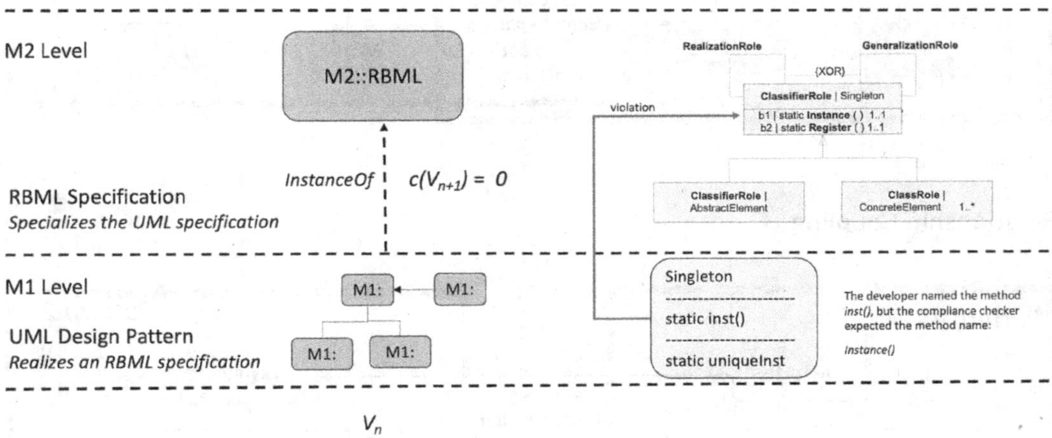

- c is a function that computes the compliance of a pattern instance (M1 level) with its corresponding design (M2 level)
- A score of 0 indicates no compliance, thus not an instance. This means that a mandatory requirement was violated

FIGURE 12.9 Compliance of a Near Instance.

the picture, we depict a real Singleton design pattern at the M1 level that is checked for conformance against its real RBML characterization at the M2 level. The solid arrow line that crosses from the M1 to the M2 level indicates a violation, but one that is mandatory for the pattern instance to claim being a Singleton. This example brings up an interesting discussion regarding the level of compliance necessary to match a design. The compliance algorithms can be parameterized to be lenient, but the drawback would be that anything could be compliant with the RBML design, so we *tighten* the algorithm to be strict. If the algorithm is too restrictive, as is the case in our example, then a violation that finds a disconnect in the naming of a method can trigger a violation. Thus, careful balancing is necessary to find a parameter that produces high accuracy and prediction.

12.4.3 Design Pattern Instance Tracking

To exemplify the evolution of pattern instance decay we show an example from the Vuze system (Vuze, n.d.) where we track four types of patterns. Data was gathered from eight versions of Vuze across 38 months of development. In Figure 12.10, we display the evolution of all the Singleton, Adapter, Observer, and Factory pattern instances from 2007 to 2008. Before an instance is tracked, the UML class diagram is checked for compliance against its RBML specification. The graphs show the increase in modular grime as the system evolves. The *y*-axis shows the count of grime instances according to the taxonomy for modular structural grime, and the *x*-axis shows the various versions of the software. For this example, we focus on only four types of grime: Persistent External Afferent Grime (PEAG), Persistent External Efferent Grime (PEEG), Temporary External Afferent Grime (TEAG), and Temporary External Efferent Grime (TEEG).

Each type of grime can be weighted differently. In the case of this example, all four types of grime were equally weighted at 1 (scale [0, 1]), however, a practitioner that is not interested in temporary grime can set these weights to 0. The dips we observe in some of the curves correspond to potential refactorings of the source code.

12.5 ANTI-PATTERNS

A discussion of design patterns is not complete without a discussion of their less desirable yet equally important counterpart –anti-patterns. According to Brown et al. (1998), *"Anti-Patterns identify those bad design concepts, technical approaches, and development practices that lead to poor quality software and project failure."* Rather than abstracting experience into a potential solution that can be instantiated to help architects produce better designs, an anti-pattern characterizes a commonly made mistake. These mistakes seem to occur systemically across domains and solutions, and abstractions can be built for them. In most cases, the evolution of a software system is the main cause of anti-patterns. When systems evolve without mirroring designs, or by incurring technical debt (intentionally or unintentionally) they accumulate disharmonies. Another way in which anti-patterns can appear is when a practitioner uses a design pattern to solve the wrong problem. Some of these decidedly negative evolutions or wrong designs are so common that the community has generated names for them. Some of the most common anti-patterns include:

i. **The Blob or the God Class:** A solution where the majority of the responsibilities for a given component or system are assumed by a single class. The coherence of the class is usually very low, and the refactoring solution requires the redistribution of responsibilities among other classes.
ii. **Cut and Paste:** The same source code can be detected in multiple classes or parts of the system. This usually occurs because the easiest solution is to just "cut and paste" code into a specific area of the code. The refactoring solution requires that the code be encapsulated and parameterized in a separate entity.

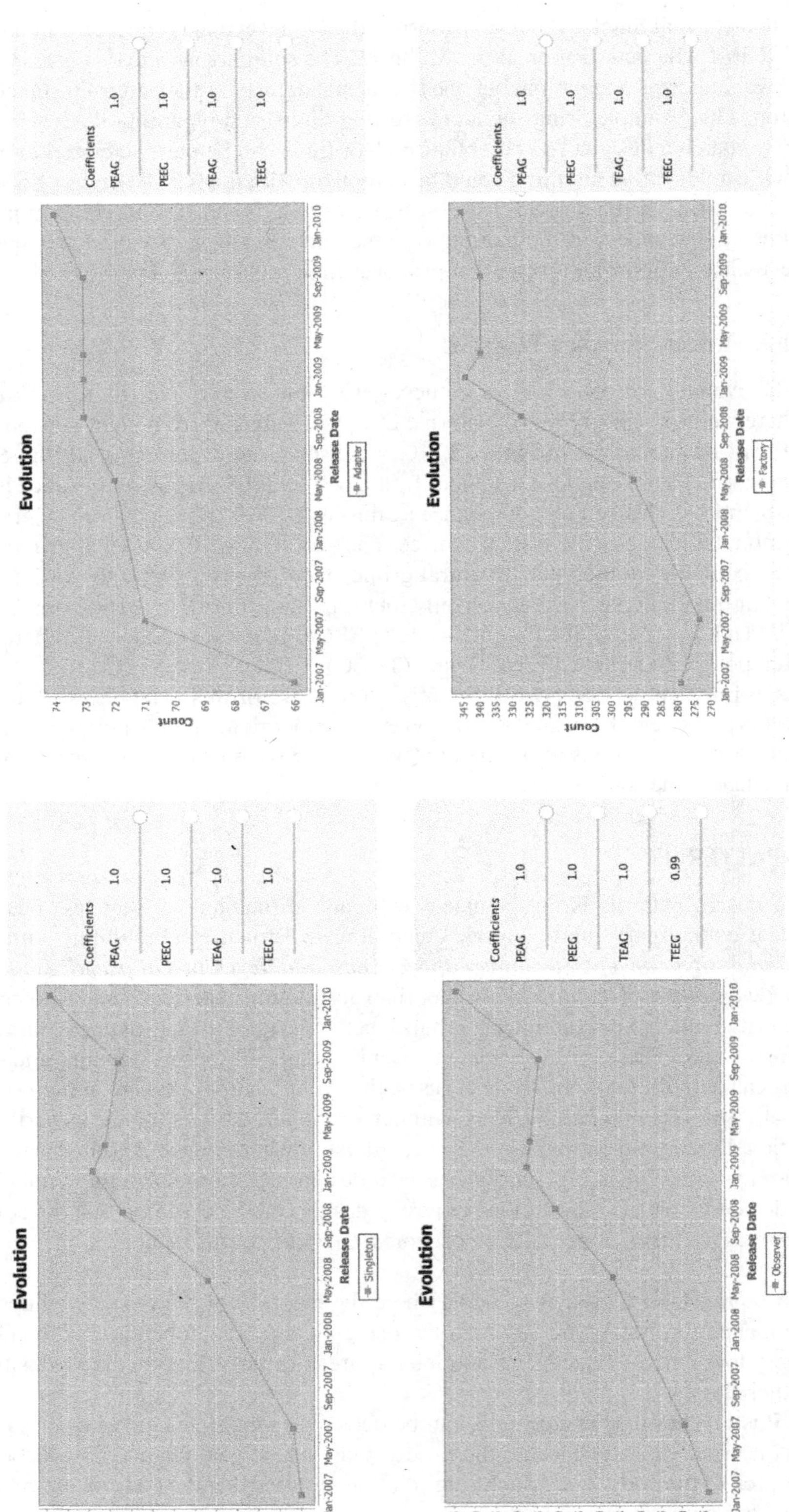

FIGURE 12.10 Evolution of grime counts of four design patterns.

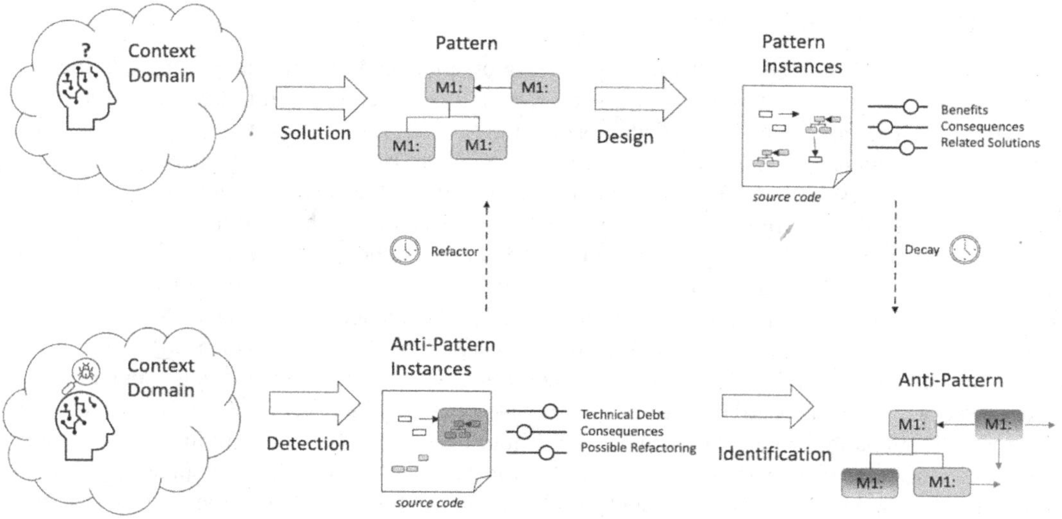

FIGURE 12.11 The relationship between design patterns and anti-patterns.

iii. **Lava Flow:** This is more common than anticipated. If you have ever heard the phrase *"if it ain't broken don't fix it,"* then you are probably witnessing the causes of the formation of this anti-pattern. Not fixing or removing obsolete code leads to dead code that seems to linger until a system runs its time. Refactoring requires quality assurance techniques to identify branch coverage and actively managing the source code.

iv. **Functional Decomposition:** This is typically a situation where a procedural developer wraps code with object-oriented constructs without understanding object-oriented principles. Solutions require active refactoring by experienced object-oriented developers.

Although many more anti-patterns have been documented, there are many books that offer complete references. Anti-patterns, like proper design patterns, are also described with templates that decorate anti-patterns with meta data that helps with their classification and documentation.

Figure 12.11 provides a side-by-side comparison of the relationship between design patterns and anti-patterns. This figure is based on work by Brown et al. (1998) and shows that in the context of a domain space, a practitioner can either propose a solution by using a design pattern or may detect the presence of an anti-pattern. The presence of an anti-pattern can be detected either through automated means and quality assurance software or manually through code inspections and reviews. If a solution to a problem is possible by using a design pattern, then the design is instantiated in the source code of the system, and multiple instances of the design pattern may be observed in the system. If the detection of an anti-pattern matches with well-known descriptions of anti-patterns, then a possible refactoring may already exist. Design patterns embedded in the source code (i.e., instantiated) bring benefits associated with many of the ISO metrics for internal and external quality as well as related solutions. Anti-patterns that emerge in the source code carry consequences such as technical debt (i.e., principal and interest). Over time, anti-pattern instances will continue to decay (e.g., the accumulation of design pattern grime) which needs to be refactored. Similarly, if care is not taken to properly extend good solutions to problems (i.e., design pattern instances), then over time these instances can turn into anti-pattern instances.

12.6 CONCLUSION

Design patterns promote good software engineering practice, and the knowledge surrounding their specification and implementation continues to grow. This can be observed in both, the practitioner as well as the academic community, with most having a good grasp on how to make effective use of patterns. Countless papers and books have been published on the subject which provides further evidence of their success and adoption. In general, design patterns have been adopted in every domain, yet significant challenges remain and continue to plague the practitioner community. As described in this chapter, the evolution of systems can have demonstrable negative effects on the overall quality of software, and although design patterns are meant to help increase the internal quality of a system (both structural and behavioral), we have observed evidence of their decay. Not immune are design pattern instances that are embedded in the designs. We now have evidence that design pattern instances do not parallel the evolution of their designs. This is not unsurprising, and as pattern instances evolve, they tend to drift from their intended designs, and there are two important forms of decay: design pattern rot, and design pattern grime. The former is difficult to find because detection techniques that focus strictly on structural comparisons cannot find an adequate instantiation that matches a specification, so heuristics such as name hints embedded in the strings of pattern components are used. The latter is present in all systems and can be classified according to known taxonomies. Grime obscures the presence of a design pattern; however, the principal components of the pattern are still present, thus allowing algorithms to successfully mine them.

As we look to the not-so-distant future, the theory of design pattern work will continue to evolve and aid our understanding of their usage, consequences, and best practices. Emerging areas in technology such as cloud computing, mobile computing devices, secure computing, industrial control systems, and stream·processing architectures will undoubtedly have new developments to report. Further, and especially relevant today, our understanding of the cybersecurity computing space will also help generate patterns that will help aid in defending software systems. We have already seen emerging and similar techniques (i.e., patterns) used in ransomware attacks, and threats to critical infrastructure (e.g., the power grid, transportation, water resources, and telecommunications).

On a final note, and given the perspective of a practitioner that was taken to write this chapter, the evolution of better tools is still necessary. Refactoring tools remain behind. This is understandable given the complexity of characterizing experience –which makes it difficult to detect and specify solutions in the real world, but progress in this space is critical if we are to maintain the overall quality of systems at healthy levels.

REFERENCES

Beck, K., & Cunningham, W. (1987). Using pattern languages for object-oriented program. In *OOPSLA '87 Workshop on Specification and Design for Object-Oriented Programming*.

Bieman, J. M., & Kang, B. K. (1995). Cohesion and reuse in an object-oriented system. In *Proceedings of the ACM 2nd Symposium on Software Reusability*, Seattle, Washington, USA (pp. 259–262). New York, NY: Association for Computing Machinery.

Blewitt, A., Bundy, A., & Stark, I. (2005). Automatic verification of design patterns in Java. In *Proceedings of the 20th IEEE/ACM International Conference on Automated Software Engineering*, Long Beach, CA.

Bosch, J. (1998). Design patterns as language constructs. *Journal of Object-Oriented Programming*, 11(2), 18–32.

Briand, L. C., Daly, J. W., & Wust, J. K. (1998). A unified framework for cohesion measurement in object-oriented systems. *Empirical Software Engineering*, 3(1), 65–117.

Briand, L., Morasca, S., & Basili, V. (1993). Measuring and assessing maintainability at the end of high level design. In *Proceedings of IEEE Conference on Software Maintenance*, Montreal, Canada (pp. 88–87).

Brown, W., Malveau, R., McCormick, III H., & Mowbray, T. (1998). Anti-Patterns. *Refactoring Software, Architectures, and Projects in Crisis*. New York: Wiley & Sons.

Buschman, F., Meunier, R., Rohnert, H., Sommerland, P., & Stal, M. (1996). *Pattern-Oriented Software Architecture.* West Sussex: Wiley, Various Volumes.

Buschmann, F., Henney, K., & Schmidt, D. (2007). Past, present and future trends in software patterns. *IEEE Software*. doi: 10.1109/MS.2007.115.

Buschmann, F., Henney, K., & Schmidt, D. (2007). *Pattern-Oriented Software Architecture, Volume 4: A Pattern Language for Distributed Computing*. Wiley.

Cunningham, W. (1992). The WyCash portfolio management system. In *Addendum to the Proceedings on Object-Oriented Programming Systems, Languages, and Applications* (pp. 29–30).

Eden, A. (2006). *Object-Oriented Modeling in LePUS3 and Class-Z.* Spring. https://www.lepus.org.uk/ref/lepus3-tutorial.pdf.

Eick, S. G., Graves, T. L., Karr, A. F., Marron, J. S., & Mockus, A. (2001). Does code decay? Assessing the evidence from change management data. *IEEE Transactions on Software Engineering*, 27(1), 1–12.

Feitosa, D., Ampatzoglou, A., Avgeriou, P., & Nakagawa, E. Y. (2018). Correlating pattern grime and quality attributes. *IEEE Access*, 6, 23065–23078.

Feitosa, D., Avgeriou, P., Ampatzoglou, A., & Nakagawa, E. Y. (2017). The evolution of design pattern grime: An industrial case study. In *International Conference on Product-Focused Software Process Improvement* (pp. 165–181). Springer.

Fernandez-Buglioni, E. (2013). *Security Patterns in Practice, Designing Secure Architectures Using Software Patterns*. West Sussex: John Wiley & Sons.

France, R. B., Kim, D. K, Song, E., & Ghosh, S. (2004). A UML-based pattern specification technique. *IEEE Transactions on Software Engineering*, 30(3), 193–206.

France, R., Kim, D., Song, E., & Ghosh, S. (2002). Metarole-Based Modeling Language (RBML) Specification v1.0, Technical Report 02-106. Computer Science Department, Colorado State University.

Freeman, E., & Freeman, E. (2004). *Head First Design Patterns*. Sebastopol CA: O'Reilly Media.

Fujaba Software Engineering Group (2005). *FUJABA Tool Suite RE*. University of Paderborn. https://web.cs.upb.de/archive/fujaba/no_cache/projects/older-projects/fujaba-tool-suite-re.html

Gamma, E., Helm, R., Johnson, R., & Vlissides, J. (1995). *Design Patterns: Elements of Reusable Object-Oriented Software*. Upper Saddle River, NJ: Addison-Wesley.

Google Code (n.d.). https://code.google.com/archive/p/rbml-uml-visualizer/

Gueheneuc, Y. G., & Albin-Amiot, H. (2001). Using design patterns and constraints to automate the detection and correction of inter-class design defects. In *Proc 39th International Conference and Exhibition Technology of Object-Oriented Languages and Systems* (pp. 296–305).

Gueheneuc, Y. G., & Antoniol, G. (2008). DeMIMA: A multilayered approach for design pattern identification. *IEEE Transactions on Software Engineering*, 34(5), 667–684.

Handbook of Software Architecture (n.d.). https://handbookofsoftwarearchitecture.com/

ISO/IEC 25010:2011 (2011). Systems and Software Engineering – Systems and Software Quality Requirements and Evaluation (SQuaRE) – System and Software Quality Models.

ISO/IEC 9126-1:2001 (2001). Software Engineering – Product Quality – Part 1: Quality Model.

Izurieta, C. (2009). Decay and Grime Buildup in Evolving Object-Oriented Design Patterns. PhD. Dissertation, Colorado State University, Fort Collins, CO.

Izurieta, C., & Bieman, J. M. (2007). How software designs decay: A pilot study of pattern evolution. In *1st ACM-IEEE International Symposium on Empirical Software Engineering and Measurement, ESEM 2007*, Madrid, Spain.

Izurieta, C., & Bieman, J. M. (2008). Testing consequences of grime buildup in object oriented design patterns. In *1st ACM-IEEE International Conference on Software Testing, ICST 2008*, Lillehammer, Norway.

Izurieta, C., Ozkaya, I., Seaman, C., Kruchten, P., Nord, R., Snipes, W., & Avgeriou, P. (2016). Perspectives on managing technical debt. A transition point and roadmap from Dagstuhl. In *1st Int. Workshop on Technical Debt Analytics (TDA)*, December 6–9, Hamilton, New Zealand.

Izurieta, C., Reimanis, D., Griffith, I., & Schanz T. (2019). Structural and behavioral taxonomies of design pattern grime. In *12th Seminar on Advanced Techniques & Tools for Software Evolution, SATToSE 2019*, Bolzano, Italy, July 8–10 (vol. 2510). https://ceur-ws.org/Vol-2510.

Kim, D. K. (2004). A Meta-Modeling Approach to Specifying Patterns. PhD Dissertation, Colorado State University.

Kim, D.-K. (2005). Evaluating conformance of UML models to design patterns. In *Proceedings of the 10th IEEE International Conference on Engineering of Complex Computer Systems* (pp. 30–31). Washington, DC: IEEE Computer Society.

Kim, D.-K., & Shen, W. (2008). Evaluating pattern conformance of UML models: A divide-and-conquer approach and case studies. *Software Quality Control*, 16, 329–359.

Lehman, M. M. (1980). Program life cycles and laws of software evolution. *Proceedings of the IEEE*, 68(9), 1060–1076.

Lehman, M. M. (1997). Metrics and laws of software evolution revisited. In *Proceedings of the 1996 European Workshop on Software Process Technology (EWSPT 1996)*, Nancy, France, Lecture Notes in Computer Science (vol. 1149, pp. 108–124).

Lehman, M. M., Perry, D. E., & Ramil, J. F. (1998). Implications of evolution metrics on software maintenance. In *Proc. of the 1998 International Conference on Software Maintenance (ICSM '98)*, Bathesda, MD.

Martin, R. (1994). OO design quality metrics an analysis of dependencies. In *Proc. Workshop Pragmatic and Theoretical Directions in Object-Oriented Software Metrics*. OOPSLA.

Meta Object Facility (MOF) (n.d.). Specification Version 1.4.1 formal/05-05-05. https://www.omg.org

Metsker, S., & Wake, W. (2006). *Design Patterns in Java*, Software Pattern Series. Boston, MA: Addison-Wesley.

Parnas, D. L. (1994). Software aging. Invited plenary talk. In *16th International Conference (ICSE* 1994*)* (pp. 279–287).

Schloss Dagstuhl (2016). Managing technical debt in software engineering. *Dagstuhl Reports*, 6(4). https://www.dagstuhl.de/16162

Schumacher, M., Fernandez-Buglioni, E., Hybertson, D., Buschmann, F., & Sommerlad, P. (2006). *Security Patterns: Integrating Security and Systems Engineering*. West Sussex: Wiley.

Shi, N. (2007). Reverse Engineering of Design Patterns from Java Source Code. Ph.D. Dissertation, University of California-Davis, Davis, CA, USA. https://www.cs.ucdavis.edu/~olsson/students/ShiNija.pdf.

Steel, C., Nagappan, R., & Lai, R. (2005). *Core Security Patterns: Best Practices and Strategies for J2EE Web Services, and Identity Management*. Upper Saddle River, NJ: Prentice Hall.

Strasser, S., Frederickson, C., Fenger, K., & Izurieta, C. (2011). An automated software tool for validating design patterns. In *ISCA 24th International Conference on Computer Applications in Industry and Engineering, CAINE '11*, Honolulu, HI.

The Object Management Group (n.d.). https://www.omg.org/spec/UML/2.5.1/About-UML/.

Tsantalis, N., & Chatzigeorgiou, A. (2006). Design pattern detection using similarity scoring. *IEEE Transactions on Software Engineering*, 32(11), 896–909.

Vuze (n.d.). https://www.vuze.com.

Wang, H., & Bieman, J. (2006). Design Pattern Coupling, Change Proneness, and Change Coupling: A Pilot Study, Computer Science Technical Report. Colorado State University.

Warmer, J. B., & Kleppe, A. G. (1998). *The Object Constraint Language: Precise Modeling with UML*, Object Technology Series. Boston, MA: Addison-Wesley.

Yau, S. S., Collofello, J. S., & MacGregor, T. (1978). Ripple effect analysis of software maintenance. In *Proc. COMPSAC 78* (pp. 60–65).

13 Test-Driven Development

Madhusudan Srinivasan and Upulee Kanewala

13.1 INTRODUCTION: WHAT IS TEST-DRIVEN DEVELOPMENT?

Test-Driven Development (TDD) is a software development methodology for writing unit tests even before implementation (Viktor and Garcia, 2018). Kent Beck developed it in the late 1990s as a part of extreme programming. *Extreme programming* is a software development methodology to improve the quality of software and the ability to adapt to changing customer requirements (Beck, 2000). TDD goes hand in hand with the agile software development methodology. An agile methodology is an iterative approach to project management and software development that involves building software incrementally from the start of the project instead of delivering the software all at once at the end. The agile projects should exhibit four main principles (Highsmith, 2009):.

1. Individuals and interactions over processes and tools
2. Working Software over comprehensive documentation
3. Customer collaboration over contract negotiation
4. Responding to change over following a plan

Among these, TDD follows the core agile value "working software over comprehensive documentation," thus focusing on producing functional code.

The traditional software development models revolve around writing the code first and then writing tests for testing that code. This practice of postponing the testing activities to the end of the software development process can decrease the software quality and increase the cost of testing and fixing defects. On the other hand, TDD focuses on implementing the tests first and code later.

13.2 HOW TDD WORKS

TDD consists of three phases, *Red, Green and Refractor,* which are repeated until the Software development is complete:

- **Red:** During the first (red) phase (as shown in Figure 13.1), the feature to be implemented is identified from the requirements specification of the software under development. This step aims to drive the implementation of the feature by implementing a test that would test the considered feature. Once the test is implemented and executed, the test must fail since the code for the feature is not yet implemented. Thus, the process is red. If the test passes, it signals a problem with the existing code or the test since the relevant code is not yet implemented. Writing the test first leads to understanding the feature better and removing any ambiguities associated with specifications written using natural language.
- **Green:** The purpose of the second (Green) phase (as shown in Figure 13.1) is to implement the code required to implement the requirement selected in the above red state and pass the test that was written. One should not go back to step 1 and introduce any new functionality until this test passes. It is also vital to run all the tests, including the last test, once the requirement is implemented. The execution of all the tests is done to ensure that any existing functionalities are not broken with the implementation of the new requirement. If all the tests pass (i.e., green), the green phase is completed since all the tests are passing, and the software is working as expected.

DOI: 10.1201/9781003188377-15

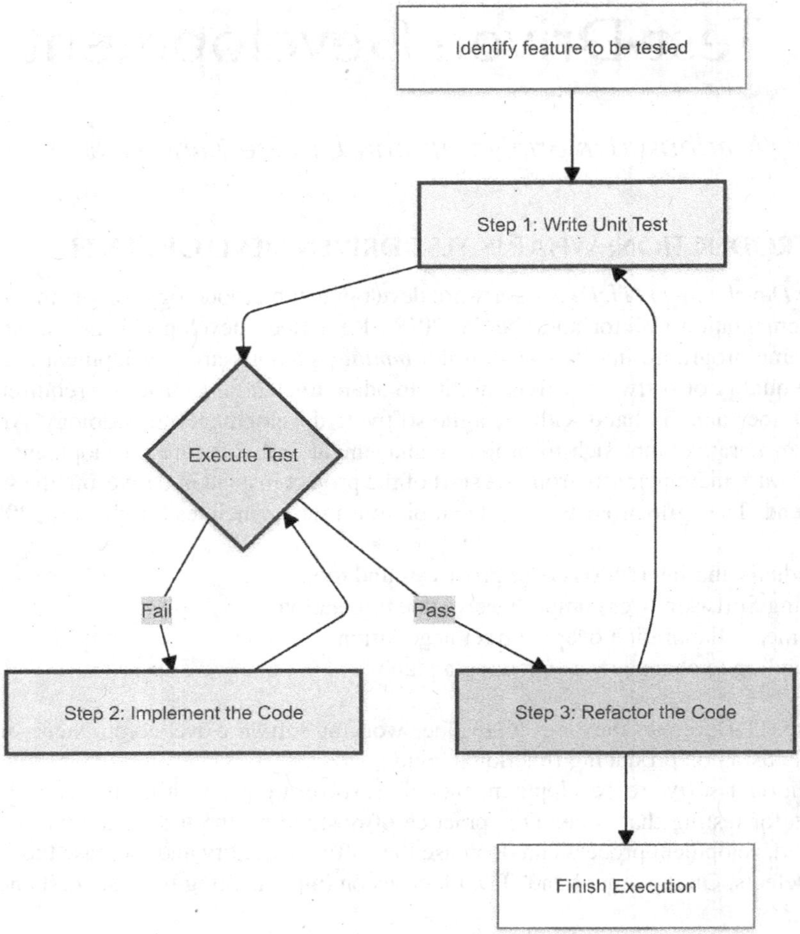

FIGURE 13.1 Steps of test-driven development.

- **Refractor:** The third (refactoring) step is optional and comes after the green phase if the developer thinks that the code quality can be improved. The refactoring phase can involve code modifications such as removing duplicate code, reducing the size of the code by breaking the code into smaller chunks, and also removing complicated logic, and control statements in a method. The refactoring process should not change the overall existing functionality.

13.3 BENEFITS OF TDD

- **TDD can Reduce the Cost of Fixing Bugs:** TDD helps developers resolve potential issues quickly and reduces the cost of fixing them. For example, assume that a developer introduced a new feature, feature B, and feature A that was already implemented broke due to the introduction of feature B. But once the developer executes all the tests in the green phase, tests associated with feature A would fail. Since the process is now in a red or failed state, the developer needs to fix the change that broke feature A before proceeding to the next test case or adding new features. As a result, TDD allows working on the fix right after modifying the code without waiting until further modifications, which would make fixing it much harder. Thus, TDD gives confidence to the developer to introduce new features without fear of breaking existing code.

- **TDD can Help Maintain Up-to-Date Tests:** With traditional software development practices, software documentation becomes easily outdated or incomplete as the code evolves. As a result, when the code is updated, any test cases written based on outdated documentation become irrelevant. With TDD, all the required functionalities are written as tests even before the code is written. Anytime when functionality changes, the TDD process would require the associated tests to be updated as well. Thus maintaining tests becomes a part of the development process and not an afterthought.
- **Improved Design:** TDD leads to less coupling and less cohesion and prevents adding long functions, leading to less testable and more buggy code. Cohesion represents the relationship between the modules, and coupling represents the relationship within the modules (Hitz and Montazeri, 1995). Good software must exhibit high cohesion and low coupling. With TDD, the developers get warning about the quality of the design early in the development cycle when a test cannot be written for any code changes due to tight coupling and low cohesion.

13.4 TOOLS FOR IMPLEMENTING TDD

Automation is important for making TDD seamless during software development and reducing the overhead of implementing and executing tests often. This automation can be achieved with the help of modern software development environments for many programming languages used today, as shown in Table 13.1.

13.4.1 Writing Junit Test Cases

Below we show an example of using writing test cases in Eclipse IDE using Junit.

1. Open Eclipse IDE.
2. Create a Project folder through the navigation flow: **File->New->Java Project**. A window opens and needs to enter the project name, as shown in Figure 13.2.
3. Click the Finish button at the bottom of the dialog box, as shown in Figure 13.2.

TABLE 13.1
Language, IDE, and Test Framework to Implement TDD

Language	IDE	Test Framework
Java	Eclipse[a]	Junit[b], TestNG[c]
Python	Spyder[d]	PyUnit[e]
C/C++	Microsoft Visual Studio[f]	Microsoft Unit Test[g]
Ruby	RubyMine[h]	Rspec[i]

[a] https://www.eclipse.org/ide/.
[b] https://junit.org/junit5/.
[c] https://testng.org/doc/.
[d] https://www.spyder-ide.org/.
[e] https://wiki.python.org/moin/PyUnit.
[f] https://visualstudio.microsoft.com/.
[g] https://docs.microsoft.com/en-us/visua-lstudio/test/unit-test-your-code?view=vs-2019.
[h] https://www.jetbrains.com/ruby/.
[i] https://rspec.info/.

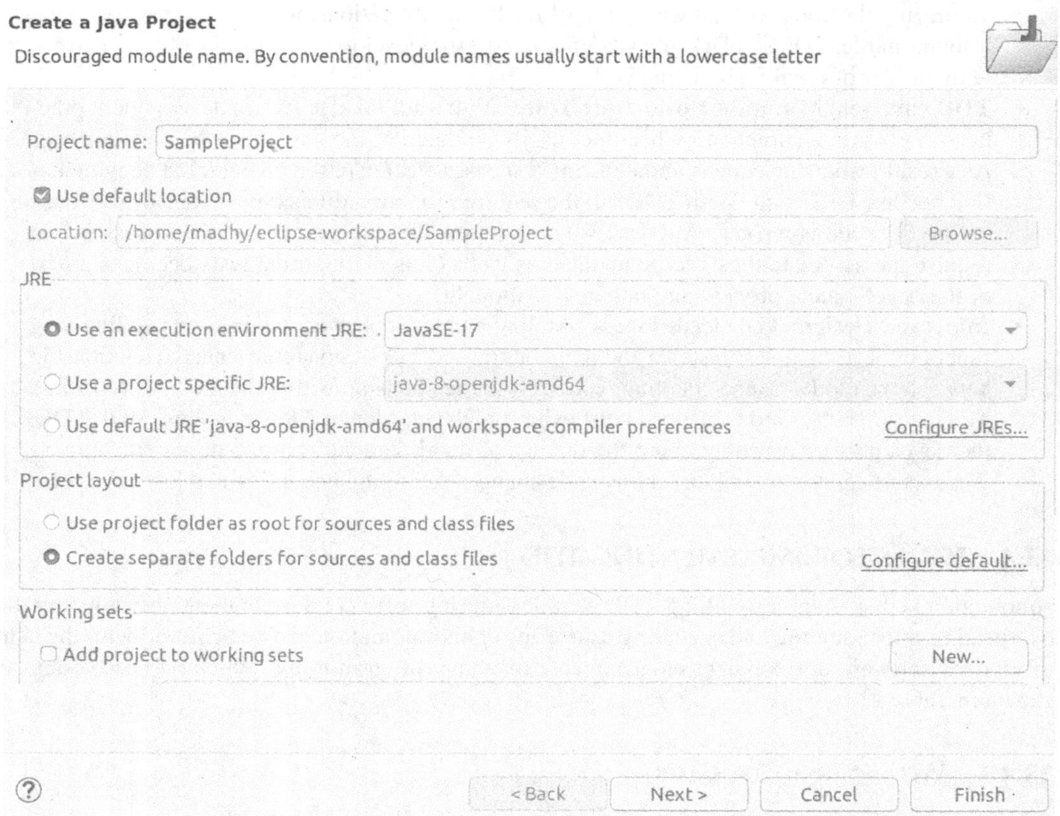

FIGURE 13.2 Create a new Java project window.

4. Click on the project and select **File->New->Junit Test Case.**
5. A window opens and enters the information, as shown in Figure 13.3.
6. The default template of the class file is generated, as shown in Figure 13.4.
7. We will use the class generated to write a Junit test case for an example program that checks if the expected and actual string match. In Junit, @Test annotation is added to define a test case in the Junit class file. Sample Junit test case is shown in Figure 13.5.
8. Right-click on the class file in the project and select run as Junit test to check if the test case passed or failed.

13.4.2 Writing Test Cases in Python

Unit test cases can be written in Python, as shown in Figure 13.6, and we use the *unittest* built into the Python standard library. Below are the steps for writing tests in Python.

1. Import unittest from the standard library.
2. Create a class called TestGreater that inherits from the TestCase class.
3. Convert the test functions into methods by adding self as the first argument.
4. Inside the method, use self.assertEqual() method on the TestCase class.
5. Provide the program entry point to call unittest.main().
6. Run the test to check for a pass or fail.

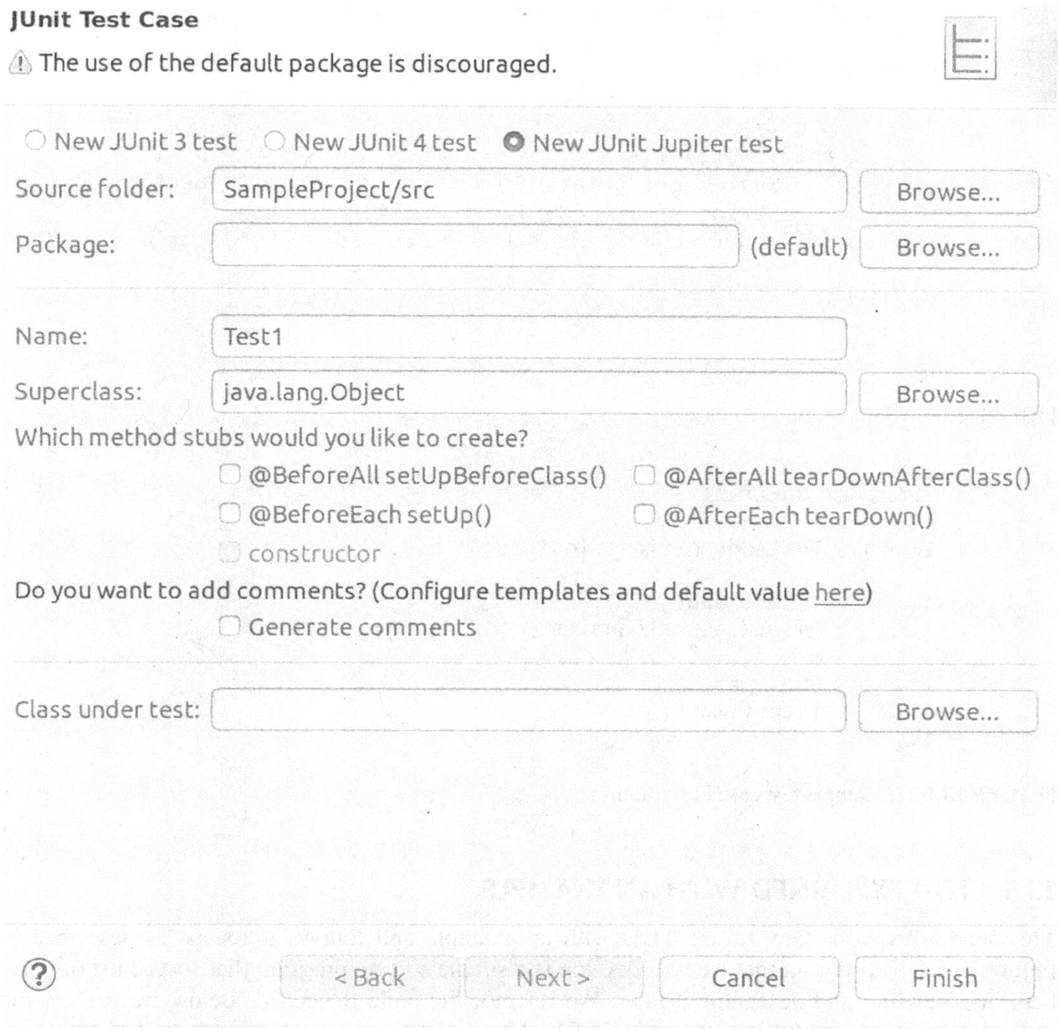

FIGURE 13.3 Create a new Junit test case window.

```
import static org.junit.jupiter.api.Assertions.*;

class Test1 {

    @Test
    void test() {
        fail("Not yet implemented");
    }

}

```

FIGURE 13.4 Default template of the class file.

```java
import static org.junit.jupiter.api.Assertions.*;

class Test1 {

    @Test
    void test() {
        System.out.println("My first Junit Test Case");
        String str1 = "My first Junit Test Case";
        assertEquals("My first Junit Test Case", str1);
    }

}

```

FIGURE 13.5 Sample Junit test case.

```python
import unittest

class TestAdd(unittest.TestCase):

    def test_add(self):
        self.assertEqual(sum([1, 2, 3]), 6, "Should be 6")

if __name__ == '__main__':
    unittest.main()

```

FIGURE 13.6 A sample test case in Python.

13.5 TDD EXPLAINED WITH AN EXAMPLE

This section explains how to use TDD with an example and follows the steps as described in Figure 13.1. The program that we will develop is a simple sorting program that sorts a list of numbers in ascending and descending order. We refactor the code if needed. Below we present the requirements and steps for implementing TDD.

Requirement 1:

The program must sort the numbers in a list in ascending order.

Step 1:

The requirement is written in the form of a test, as shown in Figure 13.7, and the test must fail since the implementation does not exist.

Run Test1. Test1 must fail and be in a red state, as shown in Figure 13.8.

Step 2:

Write the code to pass the last test, as shown in Figure 13.9.

Test1 passes and is in the green state, as shown in Figure 13.10.

Step 3:

Refactoring is an optional step, and we skip the step.

Requirement 2:

The sort method must return false and ask the user to provide input when an empty list is provided as input.

```
import static org.junit.jupiter.api.Assertions.*;
import java.util.ArrayList;
import java.util.List;
import org.junit.jupiter.api.Test;

class Requirement1 {

    @Test
    public void testPrintMessage() {
        List<Integer> list = new ArrayList<>();

        list.add(15);
        list.add(21);
        list.add(9);
        list.add(5);
        list.add(8);
        list.add(10);

        String result = implementation.sort_ascending(list);
        String sorted = "5, 8, 9, 10, 15, 21";

        assertEquals(result, sorted);
    }
}
```

FIGURE 13.7 A requirement written as a test.

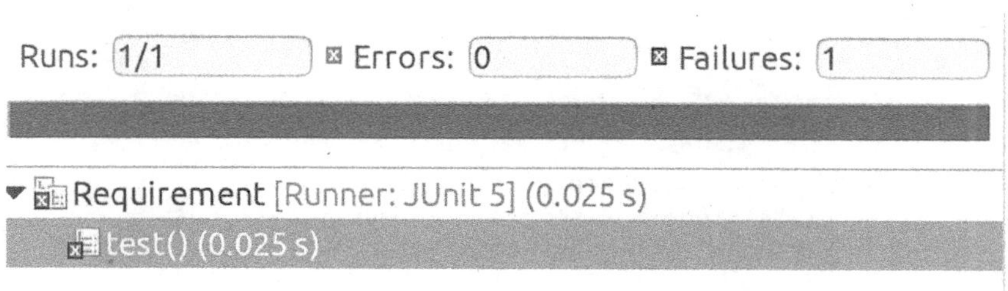

FIGURE 13.8 Test1 result.

Step 1:

Cycle 1 is completed, and cycle 2 starts with writing a test based on the requirement, as shown in Figure 13.11. The test is provided below.

Run Test1 and Test2 as shown in Figure 13.12. Test1 passes and Test2 fails.

Step 2:

Write the code to pass the last test, as shown in Figure 13.13.

Run Test1 and Test2 as shown in Figure 13.14. Test1 and Test2 pass.

Step 3:

Refactoring is skipped.

Requirement 3:

The program must sort the list in descending order, as shown in Figure 13.15. Below is the test for requirement three, and cycle three will continue.

```
import java.util.Collections;
import java.util.List;

public class implementation {

    public static String sort_ascending(List<Integer> list) {
        // Sort the list in ascending order
        Collections.sort(list);

        // Use StringBuilder to create a comma-separated string of the sorted numbe
        StringBuilder sortedList = new StringBuilder();
        for (int i = 0; i < list.size(); i++) {
            sortedList.append(list.get(i));
            if (i < list.size() - 1) {
                sortedList.append(", ");
            }
        }

        // Return the sorted list as a string
        return sortedList.toString();
    }
}
```

FIGURE 13.9 Implementation to pass the test.

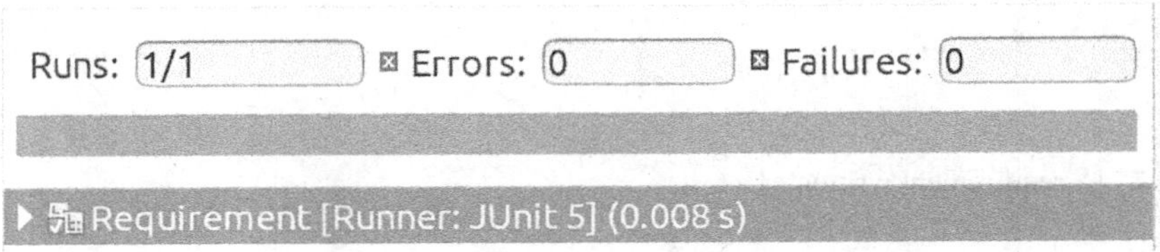

FIGURE 13.10 Test Result.

```
    @Test // Test2
    public void check_empty_input() {
        List<Integer> list = new ArrayList<>();
        String result = implementation.sort_ascending(list);

        assertEquals(result, "Please enter the input");
    }
```

FIGURE 13.11 Junit Test2.

Step 1:

Run Test1, Test2, and Test3 as shown in Figure 13.16. Test1 and Test2 pass. Test3 fails and is red.

Step 2:

Write the code to pass the last test, as shown in Figure 13.17.

Run Test1, Test2, and Test3 as shown in Figure 13.18. Test1, Test2, and Test3 pass.

Step 3:

Refactoring is carried out by combining the two methods sort_ascending and sort_descending to one method to remove code duplication and improve comprehension, as shown in Figure 13.19. The cycle continues until all the requirements are covered.

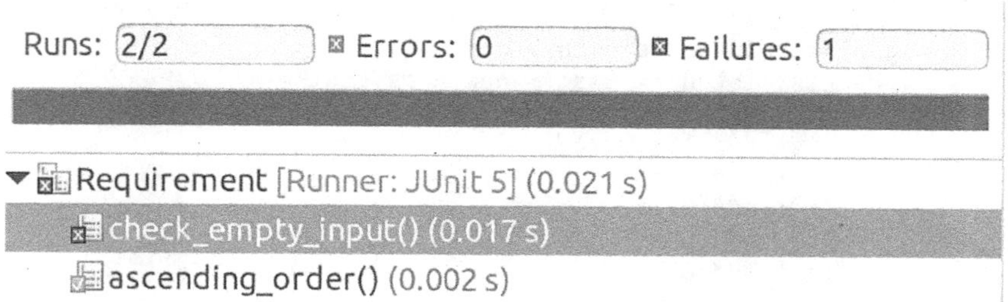

FIGURE 13.12 Test2 and Test1 result.

```java
import java.util.Collections;
import java.util.List;

public class implementation {

    public static String sort_ascending(List<Integer> list) {
        String concat = "";

        if (list.size() == 0) {
            String msg = "Please enter the input";
            return msg;
        }

        Collections.sort(list);  // Sort the list in ascending order

        for (int i = 0; i < list.size(); i++) {
            Integer val = list.get(i);
            concat = concat + " * " + Integer.toString(val);
        }

        // Remove leading " * " and any trailing spaces
        concat = concat.substring(3, concat.length());
        concat = concat.replaceAll("\\s+", "");

        return concat;
    }
}
```

FIGURE 13.13 Implementation to pass Test2.

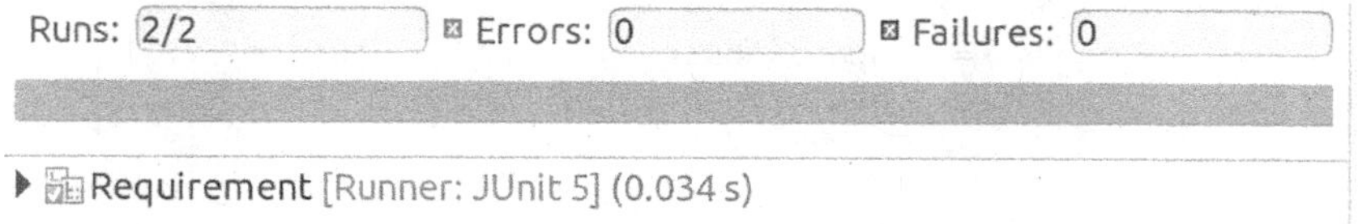

FIGURE 13.14 Test2 Result.

```java
@Test // Test3
public void descending_order() {
    List<Integer> list = new ArrayList<>();

    list.add(15);
    list.add(21);
    list.add(9);
    list.add(5);
    list.add(8);
    list.add(10);

    String result = implementation.sort_descending(list);
    String sorted = "21, 15, 10, 9, 8, 5";

    assertEquals(result, sorted);
}
```

FIGURE 13.15 Junit Test3.

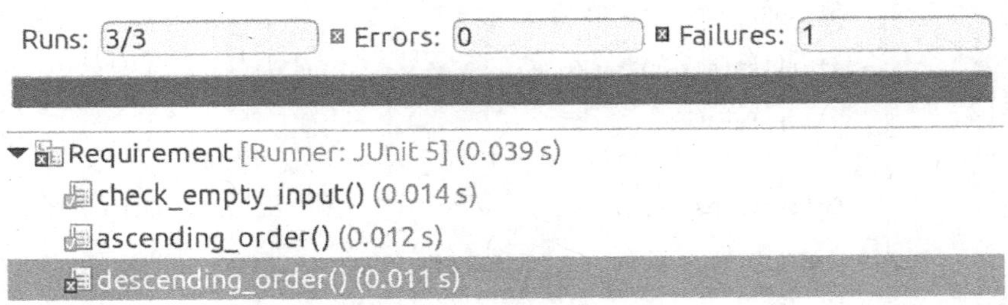

FIGURE 13.16 Test1, Test2, and Test3 result.

```java
public static String sort_descending(List<Integer> list)
    String concat = "";

    // Sort the list in descending order
    Collections.sort(list, Collections.reverseOrder());

    for (int i = 0; i < list.size(); i++) {
        Integer val = list.get(i);
        concat = concat + " * " + Integer.toString(val);
    }

    // Remove the leading " * " and any trailing spaces
    concat = concat.substring(3, concat.length());
    concat = concat.replaceAll("\\s+", "");

    return concat;
}
```

FIGURE 13.17 Implementation to pass Test3.

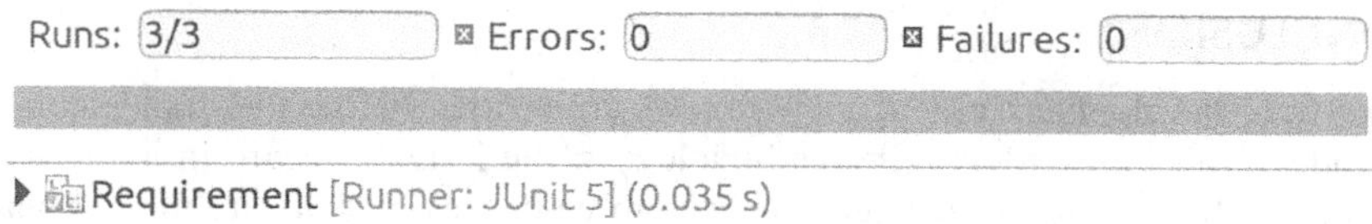

FIGURE 13.18 Test1, Test2, and Test3 result.

```
public static String sort(List<Integer> list) {
    String concat = "";
    Scanner sc = new Scanner(System.in);

    System.out.print("Enter AS for sort ascending and SD for descending: ");
    String val = sc.toString();

    if (list.size() == 0) {
        String msg = "Please enter the input";
        return msg;
    }

    if (val.equals("AS")) {
        Collections.sort(list);
    } else {
        Collections.sort(list, Collections.reverseOrder());
    }

    for (int i = 0; i < list.size(); i++) {
        Integer val1 = list.get(i);
        concat = concat + " * " + Integer.toString(val1);
    }

    return concat;
}
}
```

FIGURE 13.19 Refactoring code.

13.6 LIMITATIONS OF TDD

There are some limitations to TDD.

- Test and code need to be changed when requirement changes and involve fixing broken tests due to code change. As a result, additional time and resources are required to develop the software.
- TDD becomes time-consuming when the system design is constantly tweaked or unstable. Also, if the design is not straightforward at the beginning of the development, using TDD would lead to rewriting both test and code.
- Implementation takes time initially since it requires understanding the code, developing the test, running the test, and then implementing the code.
- When a requirement is not well understood, bugs can be introduced either in the test or the implementation code, which would be hard to detect.
- Often, novice programmers have a hard time with TDD since the tests need to be written first. Thus, some development teams would require additional training and resources before using TDD.

13.7 CONCLUSIONS

TDD is a software development practice where tests are written before the implementation of the functional code. TDD follows the agile methodology of building software in iterations and produces functional software after every iteration. The three phases of TDD are the red phase, where a test is written, the green phase, where the implementation is written to pass the written test, and an optional refactoring step for improving code design. The benefits of the essential side effect of TDD is that it results in 100% test coverage, I.e., every line of code is tested, and tests would serve as a functional specification for the code. TDD, when implemented correctly, can provide numerous benefits, particularly in reducing the software development cost and increasing the quality of software.

REFERENCES

Beck, K. (2000). *Extreme Programming Explained: Embrace Change*. Boston, MA: Addison–Wesley.

Highsmith, J. (2009). *Agile Project Management: Creating Innovative Products*. London, UK: Pearson Education.

Hitz, M., & Montazeri, B. (1995). Measuring Coupling and Cohesion in Object-Oriented Systems. In *Proceedings of the International Symposium on Applied Corporate Computing* (vol. 50, pp. 75–76).

Viktor, F., & Garcia, A. (2018). *Test-Driven Java Development*. Birmingham, UK: Packt.

Part 3

System Analysis

14 Reliability

Christian K. Hansen

14.1 INTRODUCTION

The reliability of a physical component or system is often defined in terms of its ability to function correctly for a specified operating period often measured as *time* (e.g. hours, days, and years) or *operating cycles* (e.g. miles). In its simplest form, a physical item can only be in one of two states, either *operating* or *failed*. An item that will never return to an operating state once it has failed is commonly referred to as a *component* or part and is *non-repairable* in nature. A component often fulfills a particular function as part of a larger *system* made up of many interconnected components. Similar to a component, a (repairable) system can also be classified as being either in an operating or in a failed state, but unlike a (non-repairable) component it can be restored to its operating state after failing, usually by means of replacing one or more components that have failed. Almost all systems can, at least in principle, be repaired although in some cases repair is more costly than replacing the entire system. For example, the circuit board of a modern home appliance is a system made up of integrated circuits and other electronic components. Upon failure, the circuit board could be repaired by identifying and replacing the particular component(s) that caused the failure. Often doing so is more expensive than replacing the board with a brand new one. Even if it is less expensive to repair a circuit board than to replace it, one must also consider the fact that a repaired (refurbished) circuit board will contain other aging components that make the board more likely to fail again in comparison with a brand new one. In this chapter, we define components as devices or smaller systems that either cannot (or typically will not) be repaired, whereas a system has the ability to be repaired. Often a repairable system will be repaired one or more times before it is ultimately retired from service.

In Sections 14.2 and 14.3, we discuss models and statistical methods related to the reliability of components, and in Sections 14.4 and 14.5, we discuss reliability at the system level. The study of reliability of components and systems differs substantially both in methodology and terminology. Attempts to use a common terminology to describe component and system reliability, such as the use of the term *failure rate,* have often led to serious misconceptions, especially when it comes to reliability estimation on the basis of statistical data (Ascher & Feingold, 1984).

14.2 COMPONENT RELIABILITY MODELS

As mentioned, the status of a component is represented by one of two states, either *operating* or *failed*. More precisely, we may describe the life cycle of a component by identifying one or more of the following events:

- The component is put into service at time T_O
- The component fails at time T_1
- The component is taken out of service at time T^* where $T_O < T^* \leq T_1$

When a component remains in service until it fails, we have $T^* = T_1$. In this case, we know the onset and failure times of the component, and the time elapsed between those two events is referred to as the *lifetime* $X = T_1 - T_O$. Likewise, if a component is taken out of service before it fails, the time

DOI: 10.1201/9781003188377-17

elapsed between T_O and T^* is referred to as the *censored lifetime* $X^* = T^* - T_O$. In this case, the actual lifetime X is unknown; however, we can deduct that $X > X^*$ (Figure 14.1).

When we observe a set of similar components, we often record n lifetimes:

$$X_1, X_2 \ldots X_n$$

or m censored lifetimes:

$$X_1^*, X_2^* \ldots X_m^*$$

Sometimes we have a combination of the two, when some of the components failed and some did not, e.g.

$$X_1, X_2^*, X_3 \ldots X_{n+m}^*$$

Another way to look at censoring is that we stop observing the component before it fails. This may occur when an aging component is replaced by a new one, or in a *life test* experiment in which we intend to observe components until they fail; however, some of the components are still operating when the experiment ends. Some literature, e.g. (Nelson, 1990b), refers to censored observations as *truncated* data, specifically as *right truncated* if a component is taken out of service before it fails ($X > X^*$) or *left truncated* if the component failed before observation started ($X < X^*$).

The standard model for a lifetime X of a component is to consider it as a continuous (non-negative) random variable with the *cumulative distribution function* (CDF) defined as the probability that the component fails at or before the time x:

$$F(x) = P(X \leq x) \; x \geq 0$$

F is a continuous function that is differentiable "almost everywhere", and the *probability density function* (PDF) f is defined as the derivative of the CDF (defined "almost everywhere"):

$$f(x) = F'(x) \quad x > 0$$

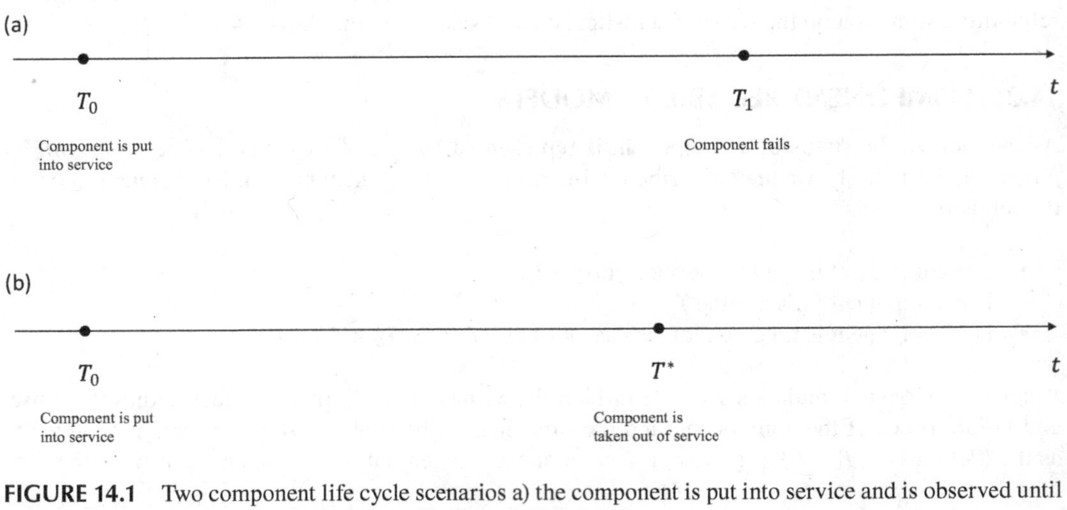

FIGURE 14.1 Two component life cycle scenarios a) the component is put into service and is observed until it fails, and b) the component is taken out of service before it fails.

A commonly used probability distribution used to model component lifetimes is the *Weibull* distribution with the CDF and PDF:

$$F(x) = 1 - e^{-\left(\frac{x}{\alpha}\right)^{\beta}} \quad x \geq 0$$

$$f(x) = F'(x) = \frac{\beta}{\alpha}\left(\frac{x}{\alpha}\right)^{\beta-1} e^{-\left(\frac{x}{\alpha}\right)^{\beta}} \quad x > 0$$

In addition to the CDF and PDF, two other functions are frequently used to define the probability distribution of X. The *reliability function* R is defined as the probability that the component is still operating at time x:

$$R(x) = P(X > x) = 1 - F(x) \quad x \geq 0$$

The *hazard rate* (HR) h is defined as

$$h(x) = \frac{f(x)}{R(x)} = \frac{f(x)}{1 - F(x)} \quad x > 0$$

In the case of the Weibull distribution, we have

$$h(x) = \frac{\beta}{\alpha}\left(\frac{x}{\alpha}\right)^{\beta-1} \quad x > 0$$

The parameter $\alpha > 0$ is known as the scale parameter, whereas the parameter $\beta > 0$ is known as the shape parameter. The distribution will have fundamentally different characteristics depending on whether β is greater than, less than or equal to 1.

HR is also sometimes referred to as a "force of mortality" (Ascher & Feingold, 1984) or a "failure rate." The latter is problematic since it is often confused with "rate of occurrence of failures" that we use to describe a repairable system. The HR is perhaps the most fundamental concept used to describe the reliability of a component. To understand the interpretations of the PDF and HR consider a small time interval$(x, x + \Delta x)$. The quantity $f(x)\Delta x$ is approximately the probability that the component will fail within the time interval $(x, x + \Delta x)$ whereas the quantity $h(x)\Delta x$ is approximately the probability that the component will fail within the time interval $(x, x + \Delta x)$ given that the component is still operating at time x.

Figure 14.2 shows the behavior of the HR for the Weibull distribution in three different cases. The HR is increasing when $\beta > 1$, decreasing when $\beta < 1$, and constant when $\beta = 1$. In the special case of $\beta = 1$, the Weibull distribution simplifies to an *exponential* distribution. The exponential distribution is the only continuous probability distribution with the property of a constant HR, often referred to as having a *memoryless* property. If the HR is increasing as function of time, it implies that the component becomes increasingly prone to failure as the age of the component increases, and consequently, the reliability of the component decreases with age. In the case of the exponential distribution, the constant HR and the memoryless property imply that the component has "no memory" of its age when it comes to failure, i.e. the component remains "as good as new" regardless of its age. Intuitively, one would expect the reliability of a component to naturally decrease with age, and therefore, the HR to be a non-decreasing function. However, there are scenarios that can explain a decreasing behavior of the HR such as heterogeneity among components (Follmann & Goldberg, 1988). Related to this is the phenomenon of *infant mortality* where a small

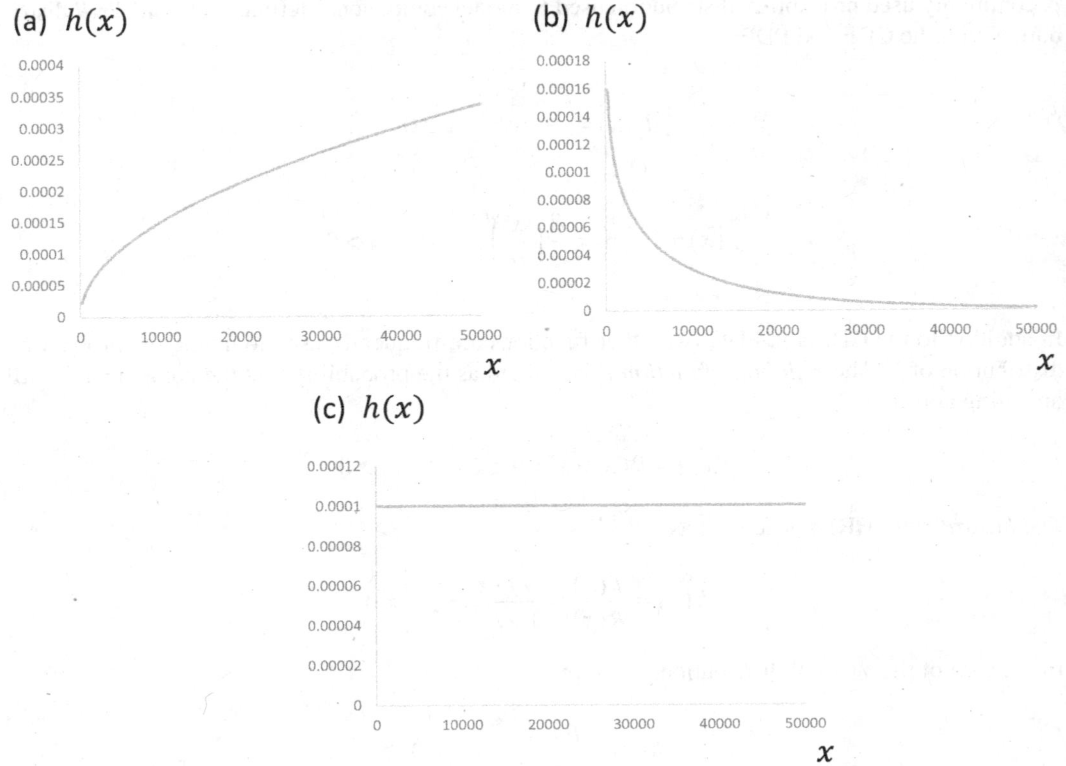

FIGURE 14.2 Behavior of the hazard rate (HR) for the Weibull distribution in three cases. (a) $\beta = 1.5$, (b) $\beta = 0.8$, (c) $\beta = 1$. The scale parameter is chosen as $\alpha = 10,000$ (hours).

percentage of components have manufacturing defects that make them prone to early failure (Jensen & Petersen, 1982). This small percentage will contribute to a higher than usual hazard rate initially, which then decreases after the period of likely early failures has been surpassed. The *bathtub* curve (Jensen, 1995) is a particular type of HR that takes into account both infant mortality and component wear-out as the component ages (Figure 14.3). The HR is decreasing during the initial operating period (often referred to as the burn-in period), followed by a period of constant HR (often referred to as the random failures period), and finally a period of increasing HR (often referred to as the wear-out period.)

The burn-in period of the bathtub curve serves as a basis for *life tests* or *accelerated life tests* where the intent is to "bypass" the burn-in period before the components are put into operation (Nelson, 1990a). During this exercise, weak components that are prone to early failure will likely fail and be eliminated from use, whereas those components that survive the life test are expected to be more reliable following the test with the left side of the bathtub curve being eliminated. In accelerated life tests, the burn-in period is shortened by means of subjecting the component to a higher than usual stress, e.g. increased temperature, accelerating the time to failure for the weak components.

Likewise, the wear-out period of the bathtub curve plays an important role in *preventive maintenance* strategies (Barlow & Hunter, 1960) that aim at replacing components before they fail and before the components start aging. With the combination of life tests and preventive maintenance strategies, the resulting HR becomes approximately constant after both the burn-in and wear-out periods have been eliminated. The assumed constant HR implies that component lifetimes become approximately exponentially distributed within that range.

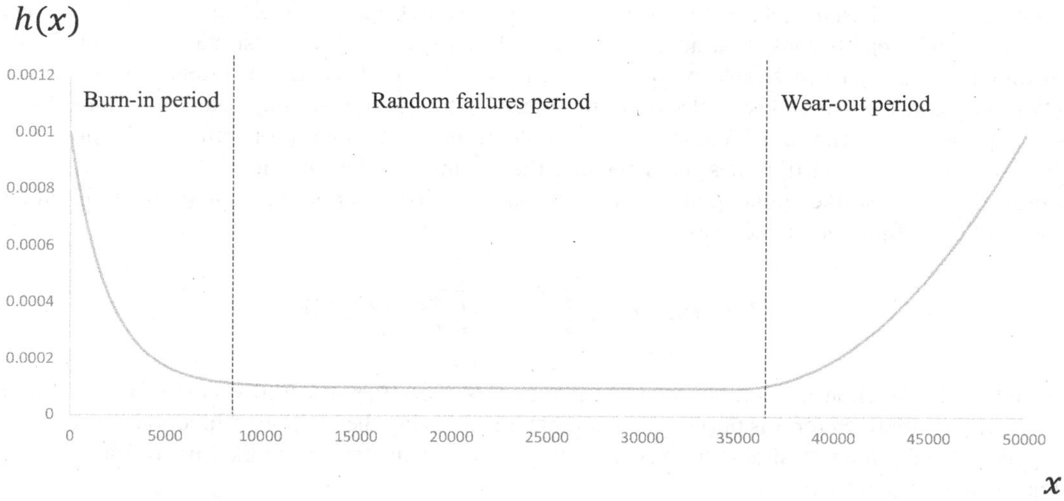

FIGURE 14.3 The bathtub curve as a model for a hazard rate with three different periods, burn-in, random failures, and wear-out.

Beyond the Exponential, Weibull and bathtub models for HR, various other models have been proposed in the literature, including the *roller-coaster model* (Wong, 1988) or *mixed distributions* and *competing risks* (Hansen & Thyregod, 1992).

14.3 STATISTICAL ANALYSIS OF COMPONENT FAILURE DATA

Component failure data is often used to estimate a model of the reliability of a component. Statistical methods related to the analysis of component lifetimes can be roughly divided into:

- Parametric model fitting methods
- Nonparametric model fitting methods

In parametric model fitting, a probability distribution (or family of distributions) described by the set of (unknown) parameters $\theta_1, \theta_2, \ldots \theta_k$ is chosen. The parameters are then estimated using statistical data. Often parameter estimates are supplemented by the construction of *confidence intervals* or specific values of the parameters are tested using *hypothesis testing.*

In nonparametric model fitting, no assumptions are made with respect to the type of distribution of the component lifetimes. Instead, graphical techniques are used to plot an estimate of one of the functions that identify the distribution, most commonly the CDF or the cumulative hazard f.unction (CHF), a function for which the derivative is the HR.

14.3.1 Parametric Methods: Maximum Likelihood Estimation

Component lifetime data will usually consist of a combination of observed lifetimes of the components that failed $x_1, x_2, \ldots, x_n$ and censored lifetimes for the components that did not fail $x_1^*, x_2^*, \ldots, x_m^*$. We shall assume that the $n+m$ components have lifetimes that follow the same probability distribution with a PDF and CDF determined by (unknown) parameters $\theta_1, \theta_2, \ldots \theta_k$ and that the lifetimes (censored or not) are all independent. The *maximum likelihood* method is most commonly used to estimate the unknown parameters. When estimating parameters, it is important to take into account both the actual failures and the censored observations. Methods based on *moment estimation* or *least squares,* if based exclusively on the actual failures and ignoring the censored observations, will inevitably lead to biased parameter estimates. Maximum likelihood estimators, although not always unbiased, have desirable properties in terms of being efficient, consistent, and sufficient (Wackerly et al., 2008) and typically have small bias if any. The underlying idea of maximum likelihood is to choose the set of parameters that will maximize the joint probability of the observed lifetimes and censored lifetimes, i.e. choosing the parameters that will make what was actually observed, the most likely among all sets of observations. The function that is maximized is called the likelihood function, in this case:

$$L(\theta_1,\theta_2,\ldots\theta_k)=\prod_{i=1}^{n} f(x_i)\prod_{j=1}^{m}\left(1-F\left(x_j^*\right)\right)$$

Note that the function L is a function of the parameters, since the observed values $x_1,x_2,\ldots,x_n$ and $x_1^*,x_2^*,\ldots,x_m^*$ are fixed for this purpose. Rather than maximizing the likelihood function, it is usually computationally more desirable to maximize the natural logarithm of the likelihood function (the log-likelihood function):

$$l(\theta_1,\theta_2,\ldots\theta_k)=\log L(\theta_1,\theta_2,\ldots\theta_k)=\sum_{i=1}^{n}\log f(x_i)+\sum_{j=1}^{m}\log\left(1-F\left(x_j^*\right)\right)$$

Example 14.1

Suppose 10 components were tested, and three of the components failed at times $x_1=3$, $x_2=7$ and $x_3=9$ (in thousands of hours) and the remaining seven components were all still operating when the test ended at 10,000 hours, i.e. $x_1^*=x_2^*=\ldots=x_7^*=10$ (in thousands of hours). We shall assume that all component lifetimes are exponentially distributed with a single unknown parameter α, also representing the mean time to failure (MTTF) $\alpha=E[X]$. The CDF and PDF take on the forms:

$$F(x)=1-e^{-\frac{x}{\alpha}} \quad x\geq 0$$

$$f(x)=F'(x)=\frac{1}{\alpha}e^{-\frac{x}{\alpha}} \quad x>0$$

The log-likelihood function can be simplified as follows:

$$l(\alpha)=\sum_{i=1}^{n}\log f(x_i)+\sum_{j=1}^{m}\log\left(1-F\left(x_j^*\right)\right)$$

$$=\sum_{i=1}^{n}\log\frac{1}{\alpha}e^{-\frac{x_i}{\alpha}}+\sum_{j=1}^{m}\log\left(1-\left(1-e^{-\frac{x_j^*}{\alpha}}\right)\right)$$

$$=-n\log\alpha-\frac{1}{\alpha}\sum_{i=1}^{n}x_i-\frac{1}{\alpha}\sum_{j=1}^{m}x_j^*=-n\log\alpha-\frac{1}{\alpha}\left(\sum_{i=1}^{n}x_i+\sum_{j=1}^{m}x_j^*\right)$$

In order to find the maximum of the log-likelihood function, we differentiate with respect to α and set the derivative equal to zero:

$$\frac{dl(\alpha)}{d\alpha} = -\frac{n}{\alpha} + \frac{1}{\alpha^2}\left(\sum_{i=1}^{n} x_i + \sum_{j=1}^{m} x_j^*\right) = 0$$

Solving for α yields:

$$\hat{\alpha}_{ML} = \frac{\sum_{i=1}^{n} x_i + \sum_{j=1}^{m} x_j^*}{n} = \frac{3+5+7+7\times 10}{3} = \frac{85}{3} \approx 28.33 \ (\text{in thousands of hours})$$

We note that the value of α for which $\frac{dl(\alpha)}{d\alpha} = 0$ is indeed a maximum since

$$\frac{dl(\alpha)}{d\alpha} > 0 \text{ for } \alpha < \hat{\alpha}_{ML} \text{ and } \frac{dl(\alpha)}{d\alpha} < 0 \text{ for } \alpha > \hat{\alpha}_{ML}$$

We can interpret this result as the estimated mean time to failure is equal to the collective operating time of all components combined divided by the total number of failures observed. This makes sense since the exponential distribution has a constant HR $h(x) = \frac{1}{\alpha}$ which is the reciprocal of the MTTF. The reciprocal of the maximum likelihood estimate of the MTTR therefore takes on the form of "total number of failures divided by the collective operating time of all components combined" which is most intuitively how one would estimate a constant HR.

When a more complicated probability distribution, e.g. one reflecting the bathtub curve is used, no closed form solution for $\hat{\theta}_{ML}$ can be derived, and instead the likelihood or log-likelihood function must be maximized using numerical optimization algorithms such as the ER algorithm (Dempster et al., 1977).

When the sample size is large, it can be shown that the maximum likelihood estimator is approximately normally distributed. The covariance of the estimator (in general a vector of dimension k) can be found using the observed Fisher information matrix

$$I_{\hat{\theta}} = -\begin{pmatrix} \frac{\partial^2}{\partial \theta_1^2} l(\hat{\theta}) & \cdots & \frac{\partial^2}{\partial \theta_1 \partial \theta_k} l(\hat{\theta}) \\ \vdots & \ddots & \vdots \\ \frac{\partial^2}{\partial \theta_k \partial \theta_1} l(\hat{\theta}) & \cdots & \frac{\partial^2}{\partial \theta_k^2} l(\hat{\theta}) \end{pmatrix}$$

Here $\hat{\theta} = \left(\widehat{\theta_1}, \widehat{\theta_2}, \ldots \widehat{\theta_k}\right)$ is a vector of the maximum likelihood estimates of the k parameters. The inverse of the observed Fisher information matrix provides large sample estimates of the variances and covariances of the parameter estimates, which can then be used to construct confidence intervals or hypothesis tests for the parameters (Arjas et al., 1991).

14.3.2 Nonparametric Methods: The Nelson-Aalen Method

When no assumptions are made with respect to the probability distribution of the component lifetimes, a nonparametric method may be used to estimate a distribution. Usually the attention is on estimating the HR $h(x)$ or the related CHF:

$$H(x) = \int_0^x h(t)\,dt \quad x \geq 0$$

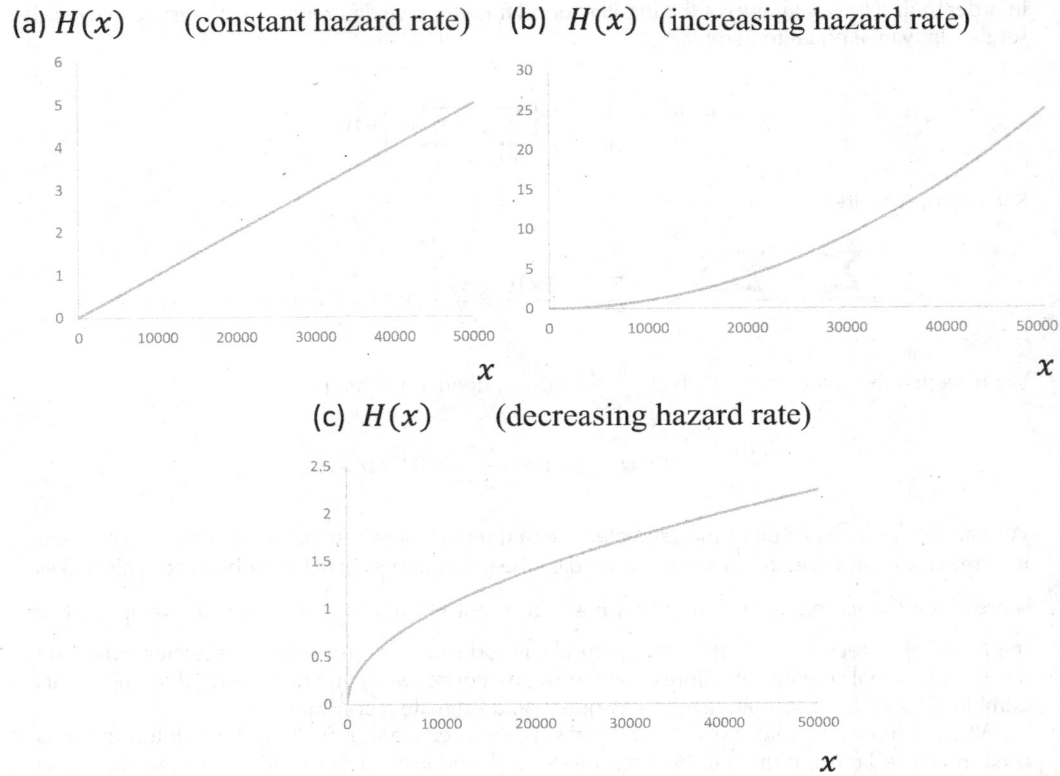

FIGURE 14.4 Examples of cumulative hazard functions (CHF) with (a) constant hazard rate, (b) increasing hazard rate, and (c) decreasing hazard rate.

Since the HR is simply the derivative of the CHF (whenever differentiable), important characteristics of the HR can be derived from the shape of the CHF. A linearly increasing $H(x)$ when plotted against x implies a constant hazard rate, whereas a concave up (down) shape of $H(x)$ when plotted against x implies an increasing (decreasing) hazard rate, see Figure 14.4.

The Nelson-Aalen estimator for estimating the CHF was introduced independently by American reliability practitioner Wayne Nelsen (Nelson, 1969) and Norwegian mathematical statistician Odd Aalen (Aalen, 1978) and can be expressed as:

$$\widehat{H(x)} = \sum_{x_i \le x} \frac{d_i}{n_i} \quad x \ge 0$$

We define $\widehat{H(0)} \equiv 0$, unless failures occur at time $x = 0$.

Here $x_1, x_2, \ldots, x_n$ represent the observed (non-censored) component lifetimes as previously defined, d_i represents the number of failures occurring at time x_i (usually $d_i = 1$ except when multiple failures occur at the exact same time) and n_i represent the number of components at risk of failing immediately prior to x_i. The resulting $\widehat{H(x)}$ is a stepwise constant function with jump discontinuities at each of the observed lifetimes $x_1, x_2, \ldots, x_n$. Censored lifetimes $x_1^*, x_2^*, \ldots, x_m^*$ (if any) do not contribute to a jump at these times, but must be taken into account when calculating n_i. Because $\widehat{H(x)}$ is a non-differentiable function at each of its jump discontinuities (and constant everywhere else), it is not meaningful to derive an estimate of the hazard rate directly from differentiating the

Nelson-Aalen estimator, however, important characteristics of the hazard rate can often be derived from visual examination of the Nelson-Aalen plot as demonstrated in the following example.

Example 14.2

Consider again 10 components that were subjected to a life test. In this case, five of the components failed at times $x_1 = 2$, $x_2 = 2$, $x_3 = 7$, $x_4 = 8$ and $x_5 = 10$ (in thousands of hours). The remaining five components were taken out of service before they failed resulting in censored lifetimes $x_1^* = 3$, $x_2^* = 7$, $x_3^* = 9$, $x_4^* = 12$ and $x_5^* = 14$ (in thousands of hours). When calculating $\widehat{H(x)}$, we only need to consider the values of x for which $\widehat{H(x)}$ will change, in this case, $x = 2,7,8$ and 10. The calculations can be summarized in the following table:

$x = x_i$	d_i	n_i	$\widehat{H(x)}$
0			0
2	2	10	0.2
7	1	7	0.342857
8	1	5	0.542857
10	1	3	0.87619

The corresponding Nelson-Aalen plot is shown in Figure 14.5. Although the differentiation of the actual Nelson-Aalen estimator $\widehat{H(x)}$, shown in blue in Figure 14.5, is not meaningful, we may construct a smooth (differentiable) function that approximates $\widehat{H(x)}$, as shown in Figure 14.5. This function appears to be initially concave down representing a decreasing hazard rate and then it turns concave up representing an increasing hazard rate. This type of behavior is consistent with the bathtub curve we previously discussed. While a smooth curve can often simply be drawn by hand, more sophisticated methods for approximating a piecewise constant function with a first-order differentiable $C^{(1)}$ function are discussed in numerical analysis textbooks, such (Chapra, 2018), and software tools the MATLAB curve fitting toolbox (MATLAB, n.d.). It should be noted that curve fitting using an nth order polynomial, especially with degrees of $n = 3$ or higher, often will result in an approximating function that exhibits more oscillation than a function drawn by hand and will often not maintain the monotone (increasing/decreasing) characteristics of the data points provided. Interpolated least-square cubic spline curves often provide good results in this case.

14.4 SYSTEM RELIABILITY MODELS

In the previous sections, we have focused on components considered to be non-repairable, i.e., once they fail they are discarded and cannot be restored to a working condition. Most systems are repairable in nature, i.e. they can be restored to service usually by means of replacing one or more components that have failed. For the purpose of this chapter, we assume that repair happens instantly, or repair times are negligible compared to the time elapsed between failures. Repair times play an essential role in the *availability* of the system. The study of availability will be the focus of Chapter 15 of this volume.

While the lifecycle of a component can be described in terms of a single lifetime X or a censored lifetime X^* and modeled by a probability distribution, modeling the lifecycle of a system is much more involved. The life cycle of a system typically is described in terms of the times the system failed $T_1, T_2, \ldots, T_n$ with each failure event being followed immediately by instant repair. In addition, the time scale is usually defined in such a way that the system is initially put into service at time $T_0 \equiv 0$. Sometimes the system is taken out of service at a time T_{n+1}^* when no failure occurs, and we will refer to T_{n+1}^* as a censored (right truncated) failure time. Assuming that the system is operational

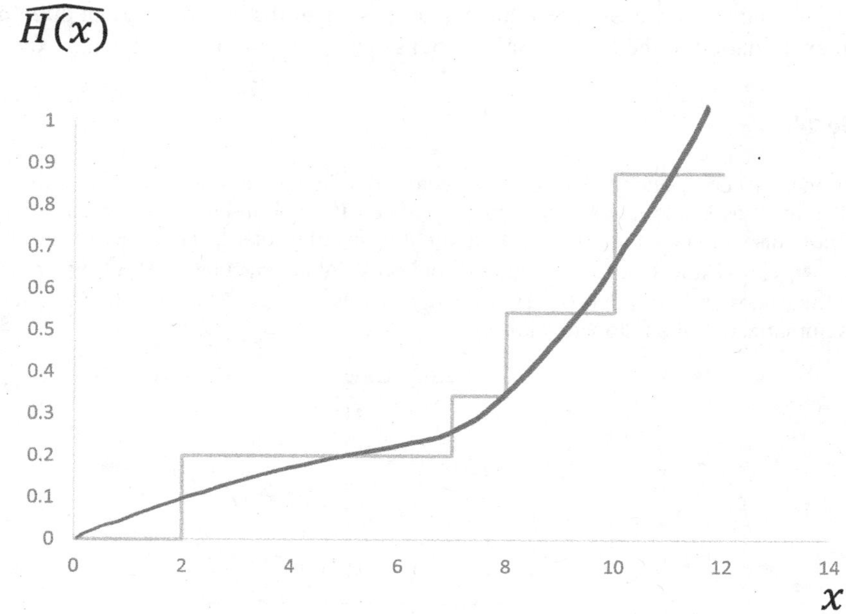

FIGURE 14.5 Nelson-Aalen plot (piecewise constant) approximated by a smooth curve.

at the time it is put into service, and that two failures can not occur at the exact same time, the n failures occur in sequence. As a result, we have $T_0 (\equiv 0) < T_1 < T_2 < \ldots < T_n < T_{n+1}^*$. If the system is taken out of service (or if the observation of the system stops) immediately following its last failure, there is no censoring taking place and no value of T_{n+1}^* is observed. The failure times are recorded on what we refer to as a *global time scale* $t$and is measuring the total *age* of the system, or time since it was first put into service. A *local time* scale x is often used to describe the times since the most recent failure, or the time between two consecutive failures. The interarrival times $X_1, X_2, \ldots, X_n$ measure the times between consecutive failures $X_i = T_i - T_{i-1}$, $i = 1..n$. If a censored failure time T_{n+1}^* is observed, likewise, a censored interarrival time can be defined as $X_{n+1}^* = T_{n+1}^* - T_n$.

It should be emphasized that, although we use the same symbols $X_1, X_2, \ldots, X_n$ to represent the *lifetimes* of n similar components, as well as n *interarrival* times of a single system, one should not assume in general that $X_1, X_2, \ldots, X_n$. is an i.i.d. (independent and identically distributed) sequence of random variables when the X's represent system interarrival times.

While a probability distribution is used to model (non-repairable) component lifetimes, a stochastic process is usually used to model failure times of a (repairable) system. A complete description of a stochastic process involves specifying the entire family of multivariate probability distributions of all combinations of failure times sequences $T_1, T_2, \ldots T_N$. This is generally not a practical approach, especially since the number of random variables N, i.e. the observed number of failures, in this sequence is also a random variable. Instead, a model is usually limited to describing the mean and variance of the number of failures as a function of t,

$$M(t) = E[N(t)]$$

$$V(t) = Var[N(t)]$$

In the following subsections, we discuss four commonly used models for repairable systems, as well as possible scenarios for the Mean Cumulative Number of Failures (MCNF) $M(t)$. The derivative $m(t)=M'(t)$ (when a such function exists) and is often referred to as the *rate of occurrence of failures* (ROCOF) or *failure intensity*. The ROCOF $m(t)$has some resemblance with the hazard rate $h(x)$, and in the reliability literature, both functions have been referred to as a *failure rate*. As mentioned previously, the quantity $h(x)\Delta x$ is approximately the probability that a component will fail within the time interval $(x,\ x+\Delta x)$ given that the component is operating at time x. Similarly, the quantity $m(t)\Delta t$ is approximately the probability that a system will fail within the time interval $(t,\ t+\Delta t)$. Since a system is assumed to be restored to service immediately following each failure, adding the condition "given that the system is operating at time t" does not change the interpretation for $m(t)$. With these similarities, it is not surprising that the two functions are being confused with one another. However, HR and ROCOF represent two fundamentally different concepts, and using the same name for both can lead to very misleading conclusions about the reliability of a component or system. Therefore, we strongly discourage the use of the term *failure rate* altogether.

14.4.1 Homogeneous Poisson Process

In the *homogeneous Poisson process* (HPP) model, it is assumed that the cumulative number of failures $N(t)$ that has occured at time t follows a Poisson distribution with mean λt.

$$P(N(t)=j)=\frac{(\lambda t)^{j}}{j!}e^{-\lambda t_{\circ}}\quad j=0,1,2,\ldots$$

$$M(t)=EN(t)=\lambda t$$

$$m(t)=M'(t)=\lambda$$

$$V(t)=Var[N(t)]=\lambda t$$

The HPP process has the property that the interarrival times X_i make up an i.i.d. sequence of random variables that follow an exponential distribution with mean $\frac{1}{\lambda}$. The HPP process as a model a repairable system shares some of the properties of the exponential distribution as a model for a non-repairable component. A component that has an exponential distributed lifetime has *no memory* of how long it has been in service in terms of its likelihood of failing in the next small time interval $(x,\ x+\Delta x)$. If the mean lifetime is $E[X]=\alpha$, then the corresponding hazard rate is a constant $h(x)=\frac{1}{\alpha}$. A system that fails in accordance with an HPP, likewise, has no memory of its age and/or when or how often it has failed in terms of its likelihood of failing in the next small time interval $(t,\ t+\Delta t)$. And the interarrival times $X_1, X_{2,}\ldots, X_N$ make up an i.i.d. sequence of exponential random variables. In this case, the hazard rate of the interarrival times and the ROCOF of the system failures take on the same constant function. It would therefore be tempting to think that in general there might be a simple relationship between the hazard rate of the interarrival times and the system ROCOF, but that is not the case. In fact, in general, the interarrival times of a system do not make up an i.i.d. sequence so it is meaningless to even talk about the "hazard rate of the interarrival times" (Ascher & Hansen, 1998).

14.4.2 Nonhomogeneous Poisson Process

In the HPP model, the ROCOF is a constant $m(t)=\lambda$. Therefore, the reliability of the system does not change with age. A natural generalization of the NPP is the *nonhomogeneous Poisson process* (NHPP) for which the ROCOF is assumed to vary with time $m(t)=\lambda(t)$,

$$M(t)=E[N(t)]=\Lambda(t)$$

$$m(t)=M'(t)=\lambda(t)$$

Another assumption made with respect to the NHPP process is the independent increments assumption, i.e. failures that occur in two or more non-overlapping intervals are independent. One consequence of this assumption is that the likelihood of a failure occurring in a small time interval $(t, t+\Delta t)$ does not depend on the history of failures that occurred prior to time t. This property is referred to as a *minimal repair* assumption. When a system is repaired it is restored to service by "minimal means" such that the system is neither more or less reliable after the repair than it was immediately prior to the failure/repair. If the system is subjected to a major overhaul, on the other hand, this assumption would not be reasonable and the NHPP would not be a suitable model.

It should be emphasized that the independent increments assumption does not imply that the interarrival times of the NHPP are i.i.d. In fact, with the exception of the special case where the NHPP is also an HPP, the interarrival times are neither independent nor identically distributed. A popular case of the NHPP is the *Power Law* Process (PLP) (Rigdon & Basu, 1989) in which

$$M(t)=E[N(t)]=\Lambda(t)=\left(\frac{t}{\alpha}\right)^{\beta}$$

$$m(t)=\lambda(t)=\frac{\beta}{\alpha}\left(\frac{t}{\alpha}\right)^{\beta-1}$$

In the special case $\beta=1$, this NPHH simplifies to an HPP where $\lambda=\frac{1}{\alpha}$. When $\beta<1$, the ROCOF is a decreasing function and when $\beta>1$, the ROCOF is an increasing function. Similar to the Weibull *distribution* which has a hazard rate of the same functional form as the ROCOF above, a decreasing ROCOF implies that the system reliability is improving with age (e.g. due to infant mortality) and an increasing ROCOF implies that the system reliability is decreasing with age (e.g. due to wear-out). In Ascher and Feingold (1984), the three types of systems, namely a) decreasing ROCOF, b) constant ROCOF, and c) increasing ROCOF are classified as *happy* system, *non-committal* system, and *sad* system, respectively. In a happy system, there is a tendency for interarrival times to become larger and larger (i.e. failures occur less and less frequently) and in a sad system, there is a tendency for interarrival times to become smaller and smaller (i.e. failures occur more and more frequently), see illustration in Figure 14.6.

In some literature, the PLP is also known as the *Weibull* Process, e.g. (Lee & Lee, 1978). This terminology is not recommended as it may lead to the misconception that the interarrival times for the PLP follow a Weibull distribution, which is not the case.

14.4.3 Renewal Process

In the HPP model, the interarrival times are i.i.d. and follow an exponential distribution. The *renewal process* (RP) is generalization of the HPP in which interarrival times are i.i.d. but can be

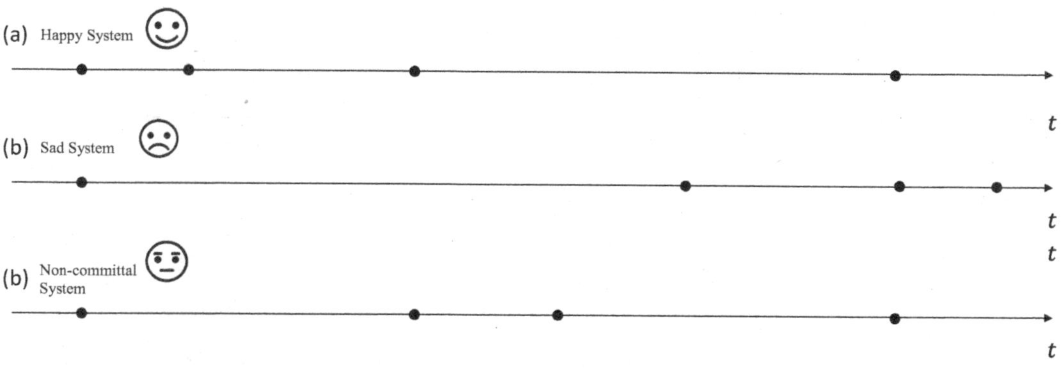

FIGURE 14.6 In the happy system (a) interarrival times have a tendency to become larger and larger. In the sad system (b) interarrival times have a tendency to become smaller and smaller. In the non-committal system (c) there is no such pattern in the interarrival times.

from any distribution with non-negative support. Except for the special case where the distribution of the interarrival times is exponential, an NHPP is not an RP and an RP is not an NHPP. In the terminology of happy, sad, and non-committal systems, systems that follow the RP model would fall into the non-committal category since, by definition, there is neither a tendency for interarrival times to become larger or smaller. This does not imply that the ROCOF will be constant, although it can be shown that the ROCOF for an RP is asymptotically constant,

$$\lim_{t\to\infty} m(t) = c > 0$$

As the name suggests, an implication of the RP is that the system is restored to an "as new" condition following each repair, i.e. it is restored to a condition that is exactly the same as if the system was brand new. This assumption is radically different from the minimal repair assumption related to the NHPP model. This type of repair is sometimes referred to as *perfect repair* (Rai et al., 2020). Other types of repair, such as *imperfect repair*, or *unsuccessful repair* (Hansen & Ascher, 2002) may be the result of intermittent failures for which the root cause is not accurately diagnosed.

The PDF of the i.i.d. interarrival times uniquely define an RP, and the ROCOF can be derived from the PDF using the following result (Cox, 1962)

$$m^*(s) = \frac{f^*(s)}{1 - f^*(s)}$$

where $m^*(s)$ and $f^*(s)$ denote the Laplace transforms or the ROCOF and PDF respectively.

14.4.4 Superimposed Renewal Process

The RP is particularly appealing as a model for a *socket*, which can be thought of as a particular location of a system in which a component is housed. If a failed component is always replaced by a similar one, then the failures in that socket will behave like an RP. A household lamp may be thought of as a socket if the light bulb is the only component that can fail. Sometimes a system can be decomposed into a set of sockets representing all of the different components that might fail. If we use an RP to model the failures occurring in each socket, and then superimpose all of those failures onto a timeline of failures for the system, as shown in Figure 14.7, we arrive at the *superimposed renewal process* (SRP) model (Cox, 1962). Note that while each RP is made up of

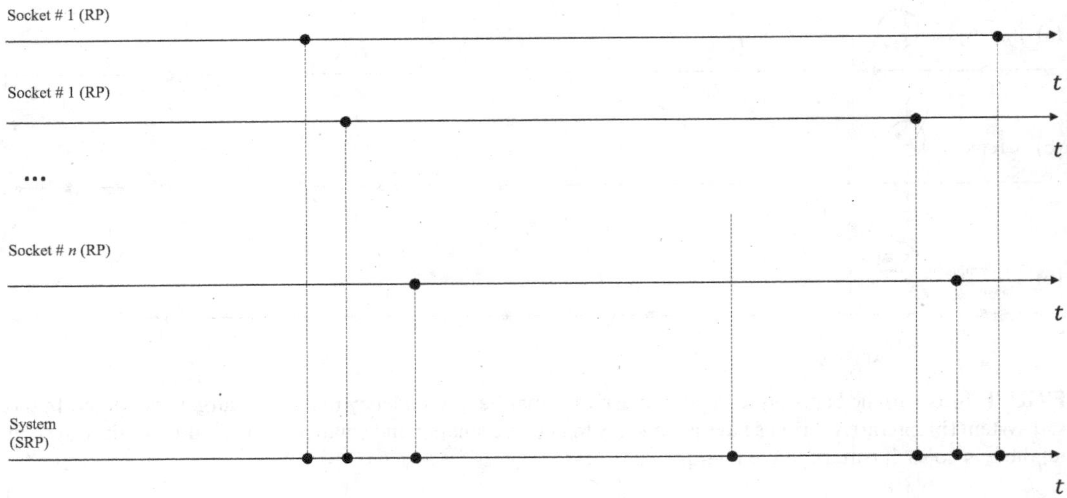

FIGURE 14.7 The superimposed renewal process made up of the failures observed from n renewal processes.

interarrival times that are i.i.d., the SRP does not have interarrival times that i.i.d., except in the special case where all component lifetimes are exponentially distributed in which case the SRP model simplifies to an HPP. The SRP is particularly appealing as a model for a system for which a physical layout of the system is well known. Hansen and Thyregod (1990) used the SRP to model field failures for printed circuit boards where records of the individual components replaced upon failure are recorded. When a system is made up of a large number of sockets, the NHPP model can be used as an approximation for the system failures, since replacing one single component may be considered a minimal repair. An example is an automobile, where small repairs such as replacing the brake pads or windshield wipers would reasonably be assumed to be minimal repairs. In contrast, a major restoration or overhaul involving rebuilding the engine and transmission and other major components might be considered an example of perfect repair, which restores the system to an "as new" condition.

14.5 STATISTICAL ANALYSIS OF SYSTEM FAILURE DATA

As in the case of component failure data, system failure data can be analyzed statistically using either a parametric or nonparametric method or some combination of the two.

14.5.1 Parametric Methods

The four models described in the previous section are all characterized by a set of parameters. In this section, we discuss methods by which these parameters may be estimated based on available failure data. Unless otherwise specified, we assume that failure data for a system is given as the set of times at which the system failed $t_1, t_2, \ldots, t_n, t_{n+1}^*$ where

$$t_0 (\equiv 0) < t_1 < t_2 < \ldots < t_n < t_{n+1}^*$$

We generally assume that no two failures occur at the exact same time, and that repair times are negligible in comparison with the times between any two failures. If the system is taken out of service immediately following the last failure, or if observation stops at that time, then no value of the censored failure time t^*_{n+1} is reported. Because the HPP model is a special case of an RP model, we shall consider those two cases together.

14.5.1.1 Estimating the Parameters in a Renewal Process

In a renewal process, the interarrival times (i.e. times between failures) are assumed to be i.i.d. (and specifically exponentially distributed in the special case of the HPP model). Interarrival times $x_1, x_2, \ldots, x_n$ are calculated as:

$$x_i = t_i - t_{i-1} \quad i = 1, 2, \ldots n$$

If a censored failure time t^*_{n+1} is reported, then we also obtain a censored interarrival time.

$$t^*_{n+1} = t^*_{n+1} - t_n$$

The maximum likelihood method discussed in Section 14.3.1 may now be used to estimate the model parameters associated with the distribution of the interarrival times. Once the parameters in the PDF $f(x)$ have been estimated, we can obtain the ROCOF using the previously mentioned formula:

$$m^*(s) = \frac{f^*(s)}{1 - f^*(s)}$$

where $m^*(s)$ and $f^*(s)$ denote the Laplace transforms or the ROCOF and PDF respectively.

In the special case of the HPP there is only one single parameter $\lambda = \frac{1}{\alpha}$ where λ represents the constant ROCOF $m(t) = \lambda$ and α is the Mean Time Between Failures (MTBF), $\alpha = E[X_i]$. It is easily shown that the maximum likelihood estimate for λ is

$$\widehat{\lambda_{ML}} = \frac{n}{\sum_{i=1}^{n} x_i + x^*_{n+1}} = \frac{n}{t^*_{n+1}}$$

if a censored failure time t^*_{n+1} is observed and otherwise

$$\widehat{\lambda_{ML}} = \frac{n}{\sum_{i=1}^{n} x_i} = \frac{n}{t_n}$$

14.5.1.2 Estimating the Parameters in a Nonhomogeneous Poison Process

The NHPP model is defined by either the MCNF

$$M(t) = \Lambda(t) = E[N(t)]$$

or the ROCOF

$$m(t) = \lambda(t) = M'(t)$$

Both functions are defined by a set of parameters $\theta_1, \theta_2, \ldots \theta_k$ that can be estimated by constructing a likelihood or log-likelihood function using the observed data $t_1, t_2, \ldots, t_n, t^*_{n+1}$. In this case, unlike for the RP, the interarrival times $x_1, x_2, \ldots, x_n, x^*_{n+1}$ are not observations of an i.i.d. sequence. Neither is the case for the failure times $t_1, t_2, \ldots, t_n, t^*_{n+1}$, however, the failure times are independent as result of the minimal repair property. This leads to the likelihood function

$$L(\theta_1,\theta_2,\ldots\theta_k) = \left(1 - F\left(t^*_{n+1}\right)\right)\prod_{i=1}^{n} f_i(t_i)$$

or the log-likelihood function:

$$l(\theta_1,\theta_2,\ldots\theta_k) = \log L(\theta_1,\theta_2,\ldots\theta_k) = \log\left(1 - F\left(t^*_{n+1}\right)\right) + \sum_{i=1}^{n} \log f_i(t_i)$$

Here $f_i(t_i)$ denotes the PDF of time to the ith failure evaluated at the observed value. The first term involving $F\left(t^*_{n+1}\right)$ can be ignored if no censored observation $t^*_{n+1}$is applicable.

In the case of the commonly used power law process where

$$m(t) = \lambda(t) = \frac{\beta}{\alpha}\left(\frac{t}{\alpha}\right)^{\beta-1}$$

The maximum likelihood estimates of the parameters are (Rigdon & Basu, 1989):

$$\widehat{\beta_{ML}} = \frac{n}{\sum_{i=1}^{n} \ln\left(t/t_i\right)}$$

$$\widehat{\alpha_{ML}} = \frac{t}{n^{1/\widehat{\beta_{ML}}}}$$

Where $t = t^*_{n+1}$ when a censored time is available and $t = t_n$ if not.

14.5.1.3 Estimating the Parameters in a Superimposed Renewal Process

In the case of the SRP model, we consider two situations

a. Failure times are observed at the socket level, i.e. a set of failure times for each individual socket is recorded
b. Failure times are observed at the system level, i.e. the socket at which the failure occurred is unknown.

In the first case (a), we may use the methods for the RP to estimate the parameters associated with the PDF of the interarrival times in each socket, which will establish the ROCOF for each socket $m_j(t) \quad j = 1,2,\ldots m$. The resulting ROCOF for the system is then obtained by adding the ROCOFs found for each socket.

$$m(t) = \sum_{j=1}^{m} m_j(t)$$

In the second case (b), we shall assume that the number of sockets is large, in which case the NHPP model may be used as an approximation.

14.5.2 Nonparametric Methods: The Nelson-Aalen Method

Although the Nelson-Aalen estimator discussed in Section 14.2.2 was originally developed as a tool for estimating the cumulative hazard function, a similar method exists for nonparametric estimation for the Mean Cumulative Number of Failures (MCNF) $M(t) = EN(t)$ of a repairable system (Nelson, 1988).

The nonparametric estimate of $M(t)$ is computed as follows:

$$\widehat{M(t)} = \sum_{t_i \le t} \frac{d_i}{n_i} \quad t \ge 0$$

where $t_1, t_2, \ldots, t_m$ denote the observed system failure times, $d_1, d_2, \ldots, d_m$ the number of failures observed at each of these times, and $n_1, n_2, \ldots, n_m$ the number of systems in service at these times. In the case of a single repairable system being repaired instantly after each failure where no more than one failure can occur at the exact same time, we have $d_1 = d_2 = \ldots = d_m = 1$ as well as $n_1 = n_2 = \ldots = n_m = 1$ this formula simplifies to:

$$\widehat{M(t)} = \sum_{t_i \le t} 1 \quad t \ge 0$$

i.e. the $\widehat{M(t)}$ is piecewise constant that jumps one unit at each of the observed failure times. We define $\widehat{M(0)} \equiv 0$, unless failure(s) occur at time $t = 0$.

Example 14.3

Consider a single repairable system with observed failure times $t_1 = 1$, $t_2 = 2$, $t_3 = 10$, $t_4 = 50$ and $t_5 = 120$ (in thousands of hours). Also assume that the system is taken out of service (or observation stops) at $t_6^* = 200$ (thousand hours). A casual examination of the corresponding interarrival times $x_1 = 1$, $x_2 = 8$, $x_3 = 40$, $x_4 = 50$, $x_5 = 70$ (and $x_6 > x_6^* = 80$) would suggest that times between failures are getting larger and larger and based on this, the system appears to behave like a "happy system" using the terminology of Section 14.3.2. Since we are using a nonparametric method, we are not making any assumptions about the system being modeled as an NHPP, although models such as the HPP and RP should be ruled out as appropriate models. Given the "happy system" behavior, it would be natural to expect the ROCOF to be decreasing and the MCNF to be concave down. The Nelson-Aalen plot shown in Figure 14.8 (and in particular the smooth curve approximating the $\widehat{M(t)}$ seems to confirm this expectation.

14.6 RELIABILITY PREDICTION

Reliability prediction is concerned with estimating the future reliability of a system through its various design phases. Reliability prediction may involve using data and/or models for similar components used in the design, and/or continuously improving the design of the system to achieve a desired level of reliability. Improving the reliability of a system may involve changing the choice of components used in the design (choosing components with proven high reliability records over, perhaps less expensive, components with historically lower reliability). Alternatively, a design may involve building in redundancies so that the system may continue to operate even after one or more of its components have failed. Adding redundancy to a design, increases the cost and complexity of the system, so wise trade-offs must be made with respect to cost, complexity, and reliability. We will discuss redundancy in more detail in the next chapter.

Since the 1960's, reliability predictions were frequently based on data from HIL-HDBK-217 which includes estimated hazard rates for electronic components under various user conditions.

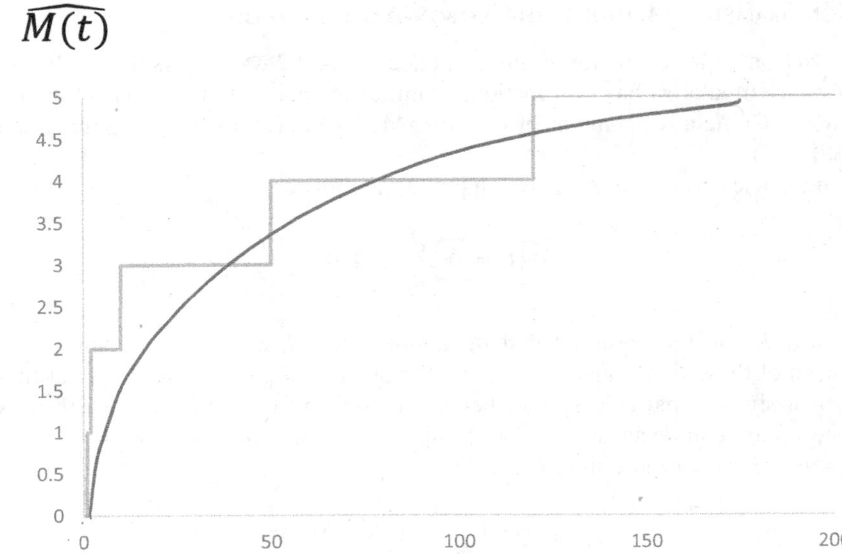

FIGURE 14.8 Nelson-Aalen plot (piecewise-constant) for a single repairable system approximated by a smooth curve.

Over the most recent decades, however, reliability predictions based on MIL-HDK-217 have been criticized for failing to accurately predict reliability under field conditions (O'Connor, 1983) and even characterized as "fraudulent and highly misleading" (O'Connor, 2000). More successfully, reliability predictions have been based on failure data observed in the field, such as failure data for in-orbit communication satellites (Hansen, 1995). Other models for reliability predictions have been based on the physics of failure taking into account such factors as how materials decay or break down over time, e.g. under the influence of high temperature (Lall et al., 1997).

REFERENCES

Aalen, O. (1978). Nonparametric inference for a family of counting processes. *Annals of Statistics*, 6(4), 701–726.

Arjas, E., Hansen, C. K., & Thyregod, P. (1991). Heterogeneous part quality as a source of reliability improvement in repairable systems. *Technometrics*, 33, 1–12.

Ascher, H. E., & Feingold, H. (1984). *Repairable Systems Reliability.* New York: Marcel Dekker.

Ascher, H. E., & Hansen, C. K. (1998). Spurious exponentiality observed when incorrectly fitting a distribution to nonstationary data. *IEEE Transactions on Reliability*, 47(4), 451–459.

Barlow, R., & Hunter, L. (1960). Optimum preventive maintenance policies. *Operations Research*, 8, 90–100.

Chapra, S. C. (2018). *Applied Numerical Methods with MATLAB for Engineers and Scientists.* Boston: McGraw Hill.

Cox, D. R. (1962). *Renewal Theory. London*: Methuen.

Dempster, A. P., Laird, N. M., & Rubin, D. B. (1977). Maximum likelihood estimation from incomplete data via the EM algorithm. *Journal of the Royal Statistical Society*, 39, 1–38.

Follmann, D. A., & Goldberg, M. S. (1988). Distinguishing heterogeneity from decreasing hazard rates. *Technometrics*, 30(4), 389–396.

Hansen, C. K. (1995). Reliability prediction and simulation for a communications-satellite fleet. In *Annual Reliability and Maintainability Symposium* (pp. 152–158). Washington, DC: IEEE.

Hansen, C. K., & Ascher, H. E. (2002). Modeling repair events under intermittent failures and failures subject to unsuccessful repair. *Quality and Reliability Engineering International*, 18(6), 453–465.

Hansen, C. K., & Thyregod, P. (1990). Estimation of the mean cumulative number of failures in a repairable system with mixed exponential component lifetimes. *Quality and Reliability Engineering International*, 6(5), 329–340.

Hansen, C. K., & Thyregod, P. (1992). Component lifetime models based on Weibull mixtures and competing risks. *Quality and Reliability Engineering International*, 8(4), 325–333.

Jensen, F. (1995). *Electronic Component Reliability*. Chichester: Wiley.

Jensen, F., & Petersen, N. E. (1982). *Burn-In*. Chichester: Wiley.

Lall, P., Pecht, M., & Hakim, E. B. (1997). *Influence of Temperature on Microelectronics and System Reliability*. Boca Raton: CRC Press.

Lee, L., & Lee, S. K. (1978). Some results on inference for the Weibull process. *Technometrics*, 20(1), 41–45.

MATLAB. (n.d.). *Curvefitting Toolbox*.

Nelson, W. (1969). Hazard plotting for incomplete failure data. *Journal of Quality Technology*, 1(1), 27–52.

Nelson, W. (1988). Graphical analysis of system repair data. *Journal of Quality Technology*, 20, 24–35.

Nelson, W. (1990a). *Accelerated Testing. Statistical Methods, Test Plans and Data Analysis*. Hoboken: Wiley.

Nelson, W. (1990b). Hazard plotting of left truncated life data. *Journal of Quality Technology*, 22, 230–238.

O'Connor, P. D. (1983). Microelectronic system reliability prediction. *IEEE Transactions on Reliability*, 32, 9–13.

O'Connor, P. D. (2000). Commentary: Reliability-past, present, and future. *IEEE Transactions on Reliability*, 49, 335–341.

Rai, R. N., Chaturvedi, S. K., & Bolia, N. (2020). *Repairable Systems Reliability Analysis: A Comprehensive Framework*. Hoboken: Wiley.

Rigdon, S. E., & Basu, A. P. (1989). The power law process: A model for the reliability of repairable systems. *Journal of Quality Technology*, 21(4), 251–260.

Wackerly, D., Mendenhall, R.L., & Scheaffer, R. (2008). *Mathematical Statistics with Applications*, 7th Edition. Belmont: Thomson, Brooks/Cole.

Wong, K. (1988). Off the bathtub onto the roller-coaster curve (electronic equipment failure). In *Annual Reliability and Maintainability Symposium* (pp. 356–363). IEEE.

15 Availability

Christian K. Hansen

15.1 INTRODUCTION

In Chapter 14, we discussed system reliability based on the assumption that a system is repaired immediately following each failure or that repair times are negligible. From that perspective, a repairable system is always available, and there are no losses incurred as result of system downtime. In reality, the repair of a system cannot be done instantaneously, and a system will experience downtime during the time the system failure is diagnosed, needed parts are ordered and installed, before the system is put back into service.

In this chapter, we define availability following the notation and terminology of (Barlow & Proschan, 1975). The foundation of the following definitions is the binary stochastic process:

$$X(t) = \begin{cases} 1 & \text{if the system is available (up) at time } t \\ 0 & \text{otherwise} \end{cases}$$

We define the *instant availability* as the ensemble mean of $X(t)$, or equivalently, the probability that $X(t)$ equals 1:

$$A(t) = E[(X(t)] = Pr(X(t) = 1)$$

The average availability over the period $0 \le t \le T$ is:

$$A_T = \frac{1}{T}\int_0^T A(t)\,dt$$

and the steady-state availability (if it exists) is:

$$A = \lim_{T\to\infty} A_T$$

If the limit $\lim_{t\to\infty} A(t)$ exists, the steady-state availability can also be found as

$$A = \lim_{t\to\infty} A(t)$$

Various other definitions of availability have been introduced in the literature such as

$$A = \frac{E[\text{Uptime}]}{E[\text{Uptime}] + E[\text{Downtime}]}$$

Or

$$A = \frac{MTBF}{MTBF + MTTR}$$

DOI: 10.1201/9781003188377-18

Both of these definitions lack clarity and require additional assumptions about the properties of the system failure and repair processes. Defining the quantities $E[\text{Uptime}]$ and $E[\text{Downtime}]$ rely on the existence of the expected values of two random variables representing uptime and downtime. This assumes that stationary probability distributions of such random variables can be constructed and that their expected values exist. Likewise, the definitions of MTBF (Meantime Between Failures) and MTTR (Meantime To Repair) are ambiguous. In addition to assuming that "times between failures" are identically distributed, it is also unclear exactly what "time between failures" means when repair is taken into account. If this is intended to represent an "uptime" cycle, then the name is misleading since technically the time elapsed between two consecutive failures includes one downtime and one uptime cycle. The term "time to repair" is equally ambiguous.

In this chapter, we consider "availability" only as it relates to a system and not a component. A component, by nature, is available until it fails, and subsequently, it is never available again. If we were to apply the previous definition to a component, the stochastic process $X(t)$ would be transient and the steady-state availability will be zero, since $\lim_{t\to\infty} A(t) = 0$.

Availability plays a particular important role in mission-critical systems where downtimes jeopardize the safety of individuals and/or cause severe financial loss as a result. Such systems include emergency response systems, power grids, communication networks, and systems performing financial transactions. Benchmarks for such systems may include availability in the order of "five nines" or higher, i.e. 99.999% corresponding to an average downtime of less than 6 minutes per calendar year. Usually, such systems must be designed using redundancy or with applicable backup systems that are almost instantaneously deployed once a failure occurs. We will discuss redundancy in Section 15.4.

In layman's terms, "reliability" and "availability" are often used interchangeably. For example, cellular network providers often advertise "reliability" of their network when they are actually referring to their network's "availability." System reliability and availability are related concepts, but are by no means the same. A system can have poor reliability (i.e. it fails often), but at the same time it has 100% availability if repairs are done instantly. Likewise, a system can have high reliability, i.e. failures are rare, but at the same time have poor availability if failures take excessively long to diagnose and correct.

15.2 MODELS FOR FAILURE AND REPAIR

Consistent with the notation in Chapter 14, we shall denote system failure times as $T_1, T_2, \ldots, T_n$ and a censored system failure time as T_{n+1}^* (if applicable). Similarly, we shall denote system repair times (i.e. times at which a repair is completed) as $S_1, S_2, \ldots, S_n$ where $T_0 (\equiv 0) < T_1 < S_1 < T_2 < S_2 < \ldots < S_n < T_{n+1}^*$. In the event that a system is taken out of service following the nth failure, there are no observed values of S_n or T_{n+1}^*. Likewise, in the event that the system is taken out of service before the nth repair is completed, we may replace S_n with a censored repair time S_n^*. In Chapter 14, we defined interarrival times as times between two successive failures. When repair time is considered, we may consider two types of interarrival times,

$$U_i = T_i - S_{i-1} \quad i = 2,3,\ldots n \quad (U_1 = T_1 - T_0)$$

$$D_i = S_i - T_i \quad i = 1,2,\ldots n$$

The interarrival times $U_1, U_2, \ldots U_n$ represent the system uptimes, whereas the interarrival times $D_1, D_2, \ldots D_n$ represent the system downtimes. In the event that a censored failure or repair time is observed, we may also consider a censored uptime U_{n+1}^* or a censored downtime D_n^*

$$U_{n+1}^{*} = T_{n+1}^{*} - S_n$$

$$D_n^{*} = S_n^{*} - T_n$$

Based on these definitions, we can express the instant availability using the failure and repair times as follows:

$$X(t) = \begin{cases} 0 & \text{if } \exists i : T_i \le t < S_i \\ 1 & \text{otherwise} \end{cases}$$

$$A(t) = E[(X(t)] = Pr(X(t) = 1)$$

Figure 15.1 illustrates an example of a timeline of events (failures and repairs) for a system. The corresponding stochastic process $X(t)$ is shown as alternating between 0 and 1 as the system alternates between downtime and uptime cycles.

It is often reasonable to assume that the uptimes times $U_1, U_2, \ldots U_n$ are independent of the downtimes $D_1, D_2, \ldots D_n$. In this case, we may consider failure and repair events separately and record them on separate time scales. Figure 15.2 shows the same events as in Figure 15.1, but with failures and repairs broken into two separate timelines. These separate time scales $\widetilde{T}_1, \widetilde{T}_2 \ldots, \widetilde{T}_n$ denote the failure times where

$$\widetilde{T}_k = \sum_{i=0}^{k} U_i \quad k \ge 1$$

$$\widetilde{T}_0 \equiv 0$$

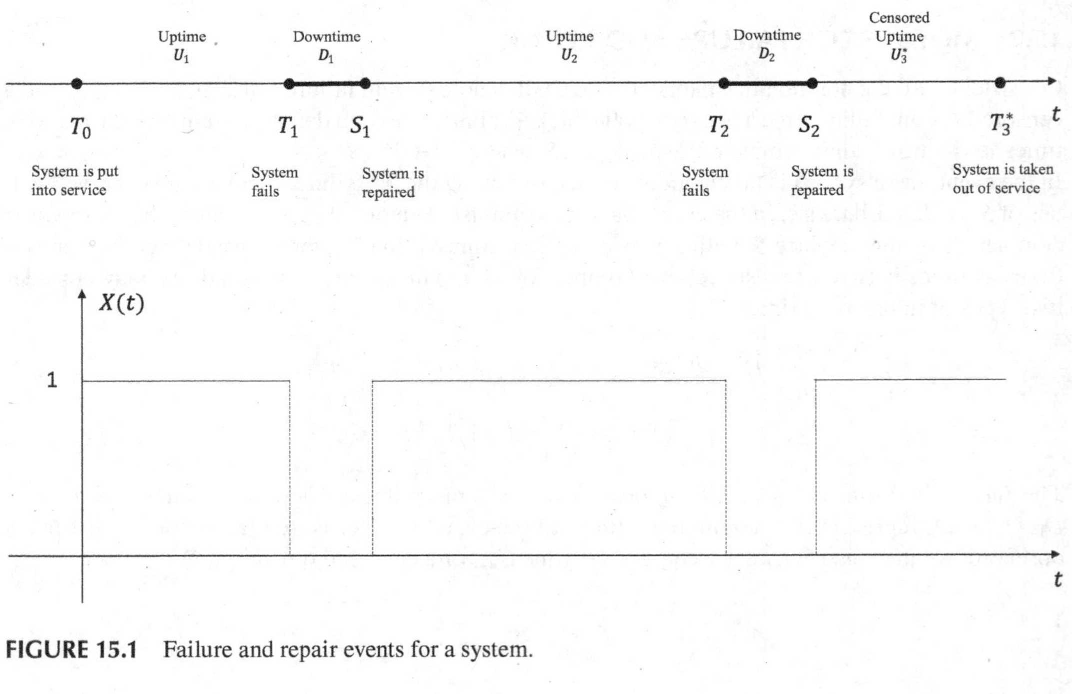

FIGURE 15.1 Failure and repair events for a system.

Likewise, $\widetilde{S_1}, \widetilde{S_2} \ldots, \widetilde{S_n}$ denote the repair times where

$$\widetilde{S_k} = \sum_{i=0}^{k} D_i \quad k \geq 1$$

$$\widetilde{S_0} \equiv 0$$

In Figure 15.2, we use t_1 to identify the time scale that tracks time during uptime cycles and time s_1 as the time scale that tracks time during downtime cycles. In this case, we assume that time t_1 is paused while the system is waiting to be repaired, and resumed once the repair is completed. Likewise, time s_1 is paused while the system is operational and is resumed only during the repair cycles. For the overall calendar time global time t, we have $t = t_1 + s_1$. In this scenario, we generally assume that a system does not fail while it is temporarily out of service. In some cases, one must take into account the possibility that a system could fail even while it is in the process of being repaired, for example, one component or subsystem fails while another component or subsystem is being replaced or repaired. To accurately model this situation, we must account for the possibility of a downtime immediately following another downtime. This happens as a result of a failure that occurred during another repair and not being discovered until the system was attempted to be put back into service.

One advantage of separating failures and repairs into separate point processes is that we can apply the models introduced in Chapter 14 to each of these separate processes. For example, we may use a nonhomogeneous Poisson process (NHPP) to model the failures and a renewal process (RP) to model the repairs. In the special case where both failures and repairs are modeled by an RP, it is easily shown that the steady-state availability can be calculated simply as:

$$A = \frac{E[U_i]}{E[U_i] + E[D_i]}$$

This formula is similar to the previously mentioned definitions.

$$A = \frac{E[\text{Uptime}]}{E[\text{Uptime}] + E[\text{Downtime}]}$$

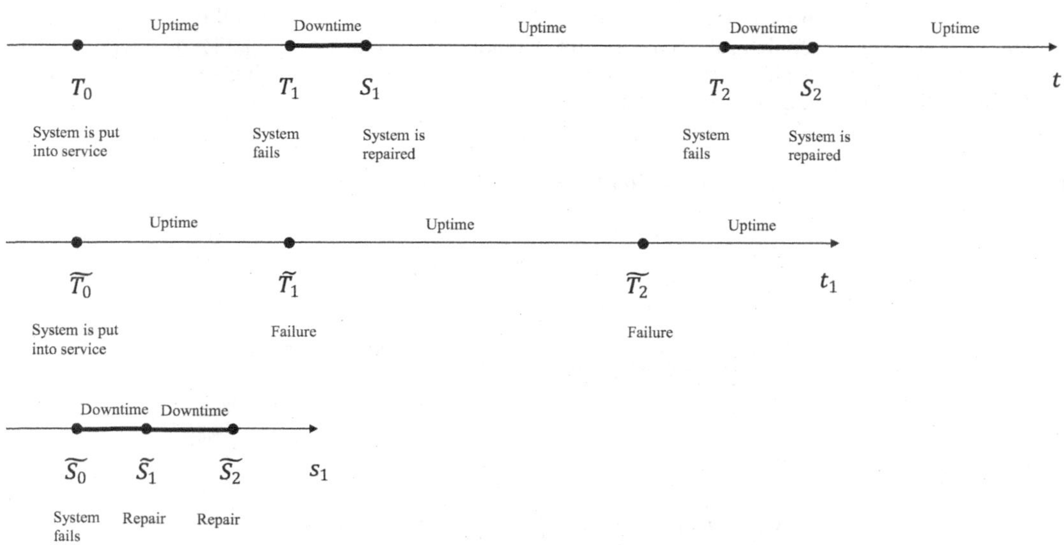

FIGURE 15.2 Failure and repair events recorded on separate time scales.

and

$$A = \frac{MTBF}{MTBF + MTTR}$$

but it offers more clarity about the conditions under which it is meaningful to talk about such terms as "expected uptime" or "mean time between failures" and what precisely these terms mean (Cox & Miller, 1965), refers to this model as an *alternating renewal process.*

Note that the existence of a steady-state availability does not imply that the instantaneous availability $A(t)$ is constant, only that $\lim_{T\to\infty} A_T = A$.

To obtain a general expression for $A(t)$ consider

$$A(t) = E[(X(t)] = Pr(X(t) = 1) = 1 - Pr(X(t) = 0) = 1 - \Pr(\exists i : T_i \le t < S_i) =$$

$$1 - (\Pr(T_1 \le t < S_1) + \Pr(T_2 \le t < S_2) + \ldots) = 1 - \sum_{i=1}^{\infty} Pr(T_i \le t < S_i)$$

In the case where $t \ll E[U_1]$, the initial portion of the instant availability function may be approximated as

$$A(t) = Pr(X(t) = 1) \approx Pr(T_1 > t) = 1 - F_{U_1}(t)$$

where $F_{U_1}(t)$ is the CDF of the first uptime, or equivalently the CDF of the time to first failure.

Example 15.1

Consider a system for which failures occur according to a homogeneous Poisson process (HPP) with constant ROCOF $m(t) = \lambda = 10$ [failures per hundred thousand hours]. If repair was instantaneous, thus the system would fail on average once every 417 days or 1.14 year. In this example, repair is not instantaneous. Rather, the time needed to repair follows a normal distribution with mean $\mu = 21$ [days] and standard deviation $\sigma = 2$ [days]. We shall assume that the system can not fail while it is under repair, and because of the memoryless property of the HPP, the system is in a same-as-new state each time the system is put back into operation following each repair. Using [hour] as the common unit of time, we have

$$E[U_i] = \frac{1}{\lambda} = 10{,}000 \text{ [hours]}$$

$$E[D_i] = \mu = 504 \text{ [hours]}$$

$$A = \frac{E[U_i]}{E[U_i] + E[D_i]} = \frac{10{,}000}{10{,}000 + 504} \approx 0.9520 = 95.20\%$$

For $t \ll \frac{1}{\lambda}$, we have

$$A(t) \approx Pr(T_1 > t) = 1 - Pr(T_1 \le t) = 1 - (1 - e^{-\lambda t}) = e^{-\lambda t}$$

Example 15.2

Consider a system for which the failures occur in accordance with a power law process (PLP) with ROCOF $m(t) = \lambda(t) = \frac{\beta}{\alpha}\left(\frac{t}{\alpha}\right)^{\beta-1}$. Assume that repairs follow an RP with finite mean $E[D_i] = \mu$ and finite variance $Var[D_i] = \sigma^2$. Unless $\beta = 1$, the uptimes $U_1, U_2, \ldots, U_n$ are not i.i.d. so the evaluation of the steady-state cannot be evaluated using the formula:

$$A = \frac{E[U_i]}{E[U_i] + E[D_i]}$$

However, if $\beta > 1$, we know that the ROCOF is increasing which implies that

$$E[U_1] > E[U_2] > \ldots > E[U_n] > 0$$

Consider the time $t = \widetilde{T}_n + \widetilde{S}_n = \sum_{i=0}^{n} U_i + \sum_{i=0}^{n} D_i$ at which time we have observed n failures and n repairs. Although the renewal process assumption does not hold, the following inequality holds:

$$0 < A < \frac{E[U_n]}{E[U_n] + E[D_n]}$$

As we let $n \to \infty$, also $t \to \infty$, we have $\lim_{n\to\infty} E[U_n] = 0$ and by the "squeeze theorem" (Stewart, 2015),

$$A = \lim_{t\to\infty} A(t) = \lim_{n\to\infty} \frac{E[U_n]}{E[U_n] + E[D_n]} = 0$$

By similar arguments, when $0 < \beta < 1$ we have $\lim_{n\to\infty} E[U_n] = \infty$

$$\frac{E[U_n]}{E[U_n] + E[D_n]} < A < 1$$

$$A = \lim_{t\to\infty} A(t) = \lim_{n\to\infty} \frac{E[U_n]}{E[U_n] + E[D_n]} = 1$$

The PLP (with parameters α, β) has the property that the time to first failure follows a Weibull distribution with the same parameters (α, β) as scale and shape parameters. Therefore, for $t \ll \alpha$, we have

$$A(t) \approx Pr(T_1 > t) = 1 - Pr(T_1 \le t) = 1 - \left(1 - e^{-\left(\frac{t}{\alpha}\right)^{\beta}}\right) = e^{-\left(\frac{t}{\alpha}\right)^{\beta}}$$

When failures and repairs are modeled by two independent homogeneous Poisson processes (HPPs), the binary stochastic process $X(t)$ also has the property of a (discrete space, continuous time) Markov process (Cox & Miller, 1965) with two states $X(t) = 0$ and $X(t) = 1$, see Figure 15.3. Here the conditional probability $\Pr(X(s) = i \mid X(t) = j)$ with $t \le s$ does not depend on any events that occurred prior to time s, i.e. the probability of any future state $X(s)$depends only on what state it is in at the current time t. The Markov process is described by two transition rates λ and γ, where $\lambda = \frac{1}{E[U_i]}$ and $\gamma = \frac{1}{E[D_i]}$. λ can also be described as the constant ROCOF for the failure HPP and γ similarly, the "rate of occurrence of repairs" in the HPP that models the repairs.

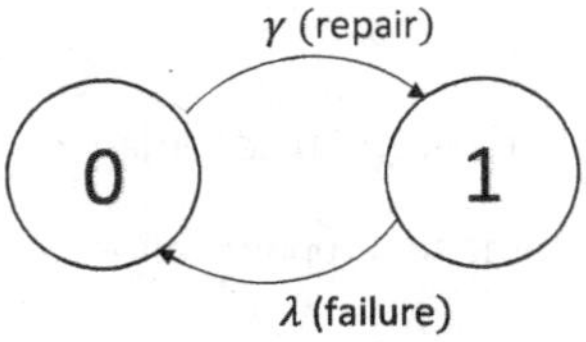

FIGURE 15.3 The two states $X(t)=1$ (system up) and $X(t)=0$ (system down) illustrated as a Markov process in the case where both failures and repairs occur in accordance with a homogeneous Poisson process (HPP). λ represents the rate (ROCOF) for system failures and γ the rate for repairs.

15.3 STRATEGIES FOR SYSTEM MAINTENANCE

So far, we have only considered downtime in the context of replacing a failed component immediately after a failure occurs. We may consider such downtime as *unscheduled* since the time of each repair is unknown. In some cases, however, it may be desirable to intentionally take the system down to replace or recondition components that may be prone to failure, *before* a failure actually occurs. We refer to such downtime as *scheduled* and the corrective action as *maintenance* rather than repair. The subject of maintainability was covered in detail in Chapter 16 of this volume. In this section, however, we shall consider simply the question of maintenance versus repair for the purpose of maximizing system availability and minimizing overall repair and maintenance costs.

A complex system, such as an automobile, is typically subjected to a combination of scheduled maintenance and unscheduled repairs during the course of its life cycle. Simple systems, such as kitchen appliance, may not require scheduled maintenance, but will be subject to unscheduled repair one or more times during its lifecycle. A question that arises is when it is optimal for a system to undergo scheduled maintenance, and another one, what the optimal balance between scheduled maintenance and unscheduled repairs is such that it maximizes availability or minimizes overall usage cost.

Consider the process of replacing a component within a system with a similar new one. The cost of this operation performed as scheduled maintenance is often less than the same operation performed as an unscheduled repair. Scheduled maintenance may utilize service personnel working off-peak hours, and parts can be ordered in advance so that they are in stock when needed. Maintenance may also allow taking advantage of mass discounts, and more the ability to "shop around" and purchase parts that are not needed right away. Similarly, the downtime resulting from scheduled maintenance is often less than the downtime as a result of the same operation performed as an unscheduled repair, as a repair may require additional time for diagnosis and time for ordering the parts needed.

Following the model described in the previous section, we assume that uptimes and downtimes are described as sequences of random variables, $U_1, U_2, \ldots, U_n$ and $D_1, D_2, \ldots, D_n$ that may or may not be i.i.d. When a system is subjected to scheduled maintenance (in addition to repairs), this result in additional censoring of uptimes as illustrated in Figure 15.4.

Strategies for system maintenance may be based on maximizing either the instant, average, or steady-state system availability as a function of the number or frequency of scheduled maintenance events. Alternatively, the strategies may be based on the total expected cost (over a period of time) of performing maintenance and repairs. Before we discuss an example, let's consider some situations when scheduled maintenance may not be optimal at all. If the type of maintenance performed meets the conditions for "minimal repair" as discussed earlier, then maintenance is never optimal, since the system is in the same state immediately after the maintenance as it was immediately before maintenance began. Intuitively, maintenance should only be performed if it increases the reliability of the system, that is, it makes it less prone to failure after (compared to before) maintenance. Similarly, if a system has a decreasing ROCOF, this implies that the system reliability gets better with age, and therefore maintenance is not optimal since it merely brings the system back to an "as

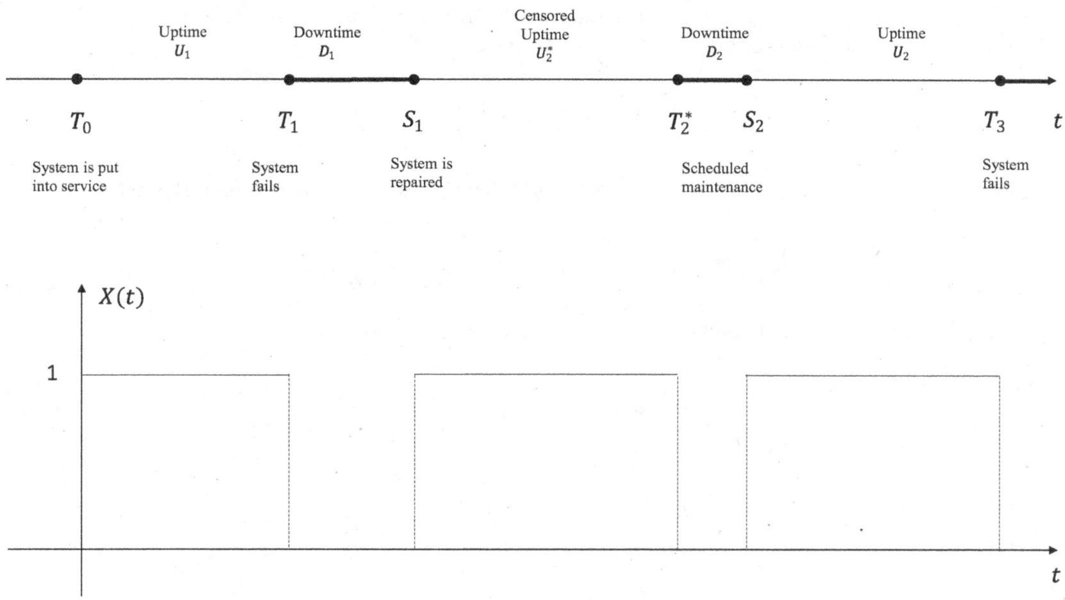

FIGURE 15.4 Uptimes and downtimes for a system subjected to both scheduled maintenance and unscheduled repairs.

new" condition, which is not desirable when "used is better than new." These factors considered, maintenance is most desirable when the system ROCOF is increasing (i.e. the system is wearing out with age) and when maintenance restores the system to an "as new" condition or something closer to "as new" than it was before the maintenance occurred. This should not be interpreted as saying that routine maintenance that resembles minimal repair, such as checking the oil or fluid level in a car or inspecting its tires for uneven wear, is not beneficial. From a system engineering perspective, such actions are more appropriately considered as "health monitoring" than maintenance. The topic of Prognostics and Health Management (PHM) is a very important topic in engineering management and will be covered in another volume of this book series.

Example 15.3

Consider a system that fails in accordance with a PLP with $\beta = 2$ and $\alpha = 200$ [days]. Unscheduled minimal repairs result in downtimes D_i which follow a normal distribution with $\mu = 2$ [days] and standard deviation $\sigma = 0.5$ [days]. Assume additionally that the system is subjected to unscheduled maintenance every x days. Each maintenance involves a complete overhaul that replaces any component that shows sign of wear and effectively restores the system to an "as-new" condition. The maintenance takes a fixed period of $\gamma = 7$ [days] to complete. We wish to determine the optimal choice of x that maximizes the average availability. In this case, the instant availability function $A(t)$ repeats itself every period of $x+\gamma$ [days]. For an optimized maintenance strategy, therefore, we wish to minimize the average availability over one period $0 \le t \le x+\gamma$,

$$A_{x+\gamma} = \frac{E\left[U_1 + U_2 + \ldots + U_{N(x)} + U^*_{N(x)+1}\right]}{x+\gamma} = 1 - \frac{E\left[D_1 + D_2 + \ldots + D_{N(x)}\right] + \gamma}{x+\gamma}$$

Here $N(x)$ denotes the random number failures that occur in the period $0 \le t \le x$. This breaks this period down into $N(x)$ pairs of uptimes and downtimes and one censored uptime that occurs after the last unscheduled repair and before the scheduled maintenance time x. The total downtime is the sum of the $N(x)$ observed downtimes plus the downtime γ observed as result of the scheduled maintenance. From the properties of the PLP, we have

$$E[N(x)] = \left(\frac{x}{\alpha}\right)^{\beta}$$

Since $N(x)$ is independent of each downtime D_i (which all have the same mean μ), we have

$$E\left[D_1 + D_2 + \ldots + D_{N(x)}\right] = E[N(x)]\mu = \left(\frac{x}{\alpha}\right)^{\beta}\mu$$

Using the given numerical values, we have for the average availability,

$$A_{x+\gamma} = 1 - \frac{\left(\frac{x}{\alpha}\right)^{\beta}\mu + \gamma}{x+\gamma} = 1 - \frac{\left(\frac{x}{200}\right)^{2} 2 + 7}{x+7} = \frac{x - 2\left(\frac{x}{200}\right)^{2}}{x+7}$$

Differentiating with respect to x yields,

$$\frac{d}{dx}A_{x+\gamma} = \frac{\left(1 - \frac{4x}{200^2}\right)(x+7) - \left(x - 2\left(\frac{x}{200}\right)^2\right)1}{(x+7)^2}$$

$$= \frac{\left(\frac{2}{200^2} - \frac{4}{200^2}\right)x^2 - \frac{28}{200^2}x + 7}{(x+7)^2}$$

$$= \frac{-\frac{1}{20000}x^2 - \frac{7}{10000}x + 7}{(x+7)^2}$$

Setting $\frac{d}{dx}A_{x+\gamma} = 0$ and finding the roots of the quadratic in the numerator yields only one positive root $x \approx 367.23$ [days] which is easily verified to yield a maximum of $A_{x+\gamma}$. Thus, the optimal maintenance strategy is to subject the system to a scheduled overhaul roughly once a year. On average, the system will experience $E[N(x)] = \left(\frac{x}{\alpha}\right)^{\beta} = 3.37$ unscheduled repairs as a result of random failures during each period between scheduled maintenance events and the average availability for the system is $A_{x+\gamma} = 0.9633 = 96.33\%$.

15.4 DESIGNING SYSTEMS WITH REDUNDANCY

In mission-critical systems, where any downtime either puts individuals at risk or is associated with a high financial loss, availability near 100% is required. For example, an emergency response system operating 24 hours a day, 99% availability means the system is down on average 14 minutes per day, which is unacceptable since every minute likely will make a difference in saving someone's life. While 100% availability may be unachievable, a "five nines" availability of 99.999% would imply an average downtime of less than 1 second per day which might be acceptable.

Designing mission-critical systems will typically require designs using the highest quality components available, and subjecting the system to frequent failure monitoring and maintenance, but in addition, some level of *redundancy* is necessary to avoid or limit downtime during scheduled maintenance as well as unscheduled repairs. Redundancy can be used both at the component or system level. For example, most systems require uninterrupted power to function, so typically a system that runs off a standard power grid must have at least one alternate power source, e.g. a battery or a gasoline-driven generator, available to keep the system running in the event of a power outage. Similarly, mission-critical software may reside on multiple computers or servers, so that service can be restored quickly in the event of a computer failure or required reboot. We may consider these as *system-level* redundancy designs. Other redundancy designs are made at the *component level*. This involves having multiple components perform the same function, so that if one fails, the system will continue to operate.

A *k-out-of-n-system* is a configuration in which n identical components perform one or more functions that require only k such components. Such configurations are often illustrated in a reliability block diagram. As an example, consider a redundant configuration of high power amplifiers (HPAs) on board an in-orbit communications satellite (Hansen, 1995). The satellite serves as a repeater of various communication lines each powered by an HPA. Without redundancy, the failure of an HPA will result in loss of service for that communication line. Since an in-orbit satellite cannot easily be repaired if a unit fails, redundant HPAs that can be activated to replace a failed unit are usually included in the design. Examples of such redundant designs are shown in Figure 15.5. In particular, a *1-out-of-n-system* (where $n \geq 2$) is known as a *parallel system*, i.e. a system that will function correctly if at least one of its n components function. Similarly, an *n-out-of-n-system* (where $n \geq 2$) is known as a *series system*, i.e. a system that functions correctly only if all of its n components function. Series system configurations extend beyond just n-out-of-n-systems (involving n identical components), in fact all systems that do not employ redundancy are considered series systems, since the system requires all of its various components to function in order for the system to function correctly.

Figure 15.6 shows an example, where major subsystems of a communications satellite make up a series system, since an (unrecoverable) failure of any subsystem will ultimately render the satellite unusable.

A k-out-of-n-system is fundamentally a non-repairable system unless repair in the form of replacement of the n components is possible. Suppose initially k of the n units are actively performing a function and as these units fail, the remaining n-k units are activated, one at a time, to keep the

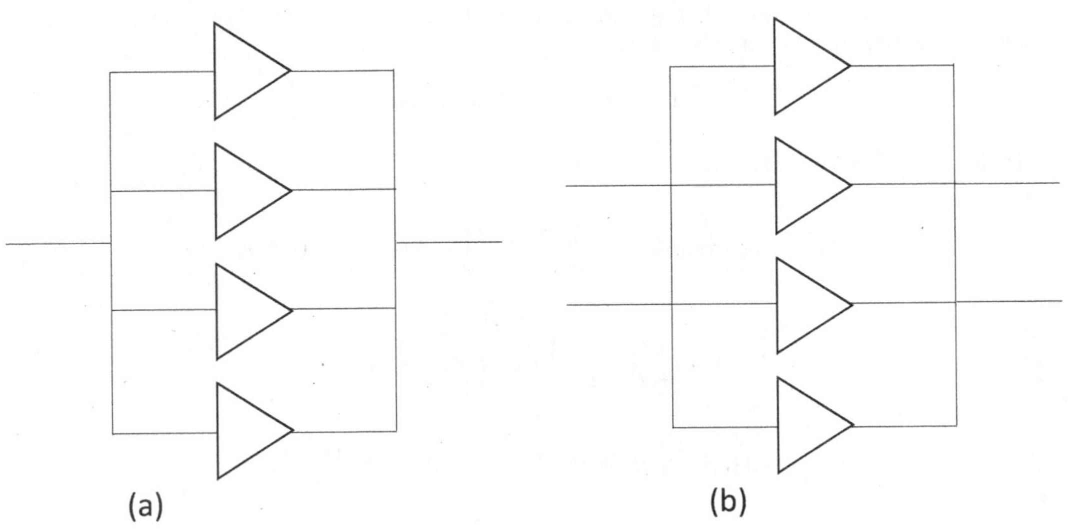

FIGURE 15.5 Examples of redundant configurations of high power amplifiers serving as part of the communications system of an in-orbit satellite, (a) 1-out-4 configuration (parallel system), (b) 2-out-4 system (two out of the four HPAs are needed to serve the two lines).

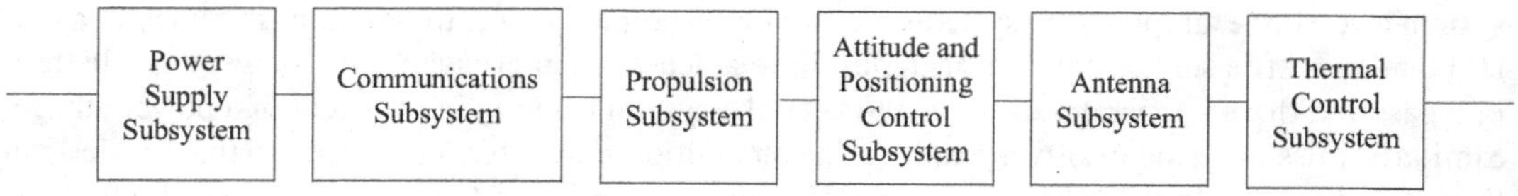

FIGURE 15.6 Reliability block diagram of the major subsystems of an in-orbit communications satellite. This represents a series system since all subsystems must function in order for the satellite to function.

system working. One may consider this type of system as a "partially" repairable system, considering the *n*-*k* units as "spare parts" that allow the system to restored to service *n*-*k* times. If we let $X^{(k)}$ denote the lifetime of the *k*-out-of-*n*-system (i.e. the time at which $k-1$ or fewer of the components working), and let $F(x)$ denote the CDF of the i.i.d. component lifetimes, $X_1, X_2 \ldots X_n$, then we have for the CDF of the system lifetime:

$$F^{(k)}(x) = Pr\left(X^{(k)} \le x\right) = \sum_{i=0}^{k-1} \binom{n}{i} (1 - F(x))^i F(x)^{n-i}$$

If no repair of the system is performed, the system instant availability is simply the probability that the system is working, a.k.a. the reliability function

$$A(t) = \Pr(X^{(k)} > t) = R^{(k)}(t) = 1 - F^{(k)}(t)$$

This function can be used to determine minimum redundancy requirements, in order to achieve a desired level of availability, e.g. "five nines."

Example 15.5

A system includes $k = 5$ identical components that are required for the system to function correctly. Suppose each of the components have 95% chance of operating without failure for 5 years. We wish to determine the level of redundancy needed to ensure "five nines" system reliability or, the total number of components n such that

$$\Pr(X^{(k)} > 5 \text{ years}) \ge 0.99999$$

To determine the value of n, we use

$$\Pr(X^{(k)} > 5 \text{ years}) = 1 - \sum_{i=0}^{4} \binom{n}{i} 0.95^i 0.05^{n-i} \ge 0.99999$$

$$\Leftrightarrow \sum_{i=0}^{4} \binom{n}{i} 0.95^i 0.05^{n-i} \le 10^{-5}$$

$$\Leftrightarrow 0.05^n + 0.95 \cdot 0.05^{n-1} n + 0.95 \cdot 0.05^{n-1} \frac{n(n-1)}{2}$$

$$+0.95^2 \cdot 0.05^{n-2} \frac{n(n-1)(n-2)}{6}$$

$$+0.95^3 \cdot 0.05^{n-3} \frac{n(n-1)(n-2)(n-3)}{24}$$

$$+0.95^4 \cdot 0.05^{n-4} \frac{n(n-1)(n-2)(n-4)}{96} \leq 10^{-5}$$

While this inequality is not possible to solve analytically, we know that n is an integer greater than or equal to 6, so we can simply evaluate $\sum_{i=0}^{4}\binom{n}{i}0.05^i 0.95^{n-i}$ for values $n = 6,7\ldots$ until we get a quantity that meets the "five nines" criterion.

n	$\sum_{i=0}^{4}\binom{n}{i}0.05^i 0.95^{n-i}$
6	0.017502
7	0.002866
8	0.000372
9	$4.12 \cdot 10^{-5}$
10	$4.09 \cdot 10^{-6}$

Thus a 5-for-10 configuration is the smallest level of redundancy needed to achieve the desired "five nines" reliability (0.999996).

15.5 FAILURE MODE AND EFFECTS ANALYSIS

One objective of designing systems with redundancy is to avoid or minimize the effect of *single-point failures*. Single-point failures refer to a component that by failing results in the entire system failing. By definition, any failure in a series system, i.e. a system with no redundancy, is a single-point failure. However, although all single-point failures result in some sort of system failure and system unavailability, not all system failures are of equal impact in terms of safety and financial loss incurred. *Failure Mode and Effects Analysis* (FMEA) is an engineering management tool, typically applied during the system design phase and is sometimes also referred to as Design Failure Mode and Effects Analysis (DFMEA). In this analysis, single-point failures and their effect on the system are identified and sometimes eliminated through design changes. *Failure modes* identify the different manners in which a component might fail, for example, a switch may fail to open or fail to close. The *effects* of the various failure modes identify the impact this failure has on other components and the system as a whole. For example, the failure of a thermostat may result in overheating of one or more other components that may then lead to a chain of other failures leading to a catastrophic system failure. Sometimes FMEA is supplemented by a *criticality* analysis and the related tool is referred to as *Failure Modes, Effects and Criticality Analysis* (FMECA). The acronyms FMEA, DFMEA, and FMECA are often used interchangeably, therefore in this section, we use the term FMEA whether or not it includes a criticality analysis. FMEA is an inductive "bottom-up" analysis that starts with a hypothetical failure of a single component, and then attempts to predict how this will impact the system as a whole. *Fault Tree Analysis* (FTA) is, in some sense, the opposite analysis. FTA is a deductive "top-down" analysis tool that starts with a system failure and then traces its root cause(s) down to the component level. FMEA may be *qualitative* in nature, i.e. it studies the possible chains of events that may occur, but may also be supplemented by a *quantitative* analysis in which probability and stochastic process models, such as those discussed in Chapter 14, are used to determine the probability of the failure and/or the expected time for it to occur. Other quantitative analyses that are part of an FMEA focus on quantifying the severity of the failure and the likelihood of it being detected. A *Risk Priority Number* (RPN) summarizes the potential risk as a function of the failure's severity, probability of occurrence, and ease of detection (Kiran, 2017).

FMEA is commonly used during the design of an automobile, where the various subsystems are studied and each component subject to possible failure is reported in an FMEA worksheet. An example of an FMEA worksheet is shown in Figure 15.7. The severity of a failure may be reported on a scale from 1 to 10, where "1" is unnoticeable and "10" presents extreme danger to the driver and passengers. A minor inconvenience such as the failure of the Bluetooth unit that connects the driver's cellphone to the car's audio system might be classified as a "2" or "3" on this scale, whereas an engine fire would be a "10". As an example, consider the braking system which typically includes the following components:

- Brake pedal and assembly
- Brake fluid reservoir
- Brake pipes
- Calibers
- Disk rotors or drums
- Brake pads

Without any redundancy employed in the design, an FMEA might identify that a broken brake line could lead to drainage of the brake fluid, and in effect prevent transfer of power from the brake pedal to the calibers and brake pads causing complete and catastrophic breakdown of the braking system, a resulting severity level "10". Government-approved standards, such as those published by the SAE International (formerly Society of Automotive Engineers), require various levels of redundancy in the design of an automotive braking system that essentially prevents catastrophic failure as the result of a single-point failure. For the reasons identified above, brake systems must include at least two independent brake fluid lines, so that if one line fails, the car will in the worst case only have reduced braking capability. A failure of a single brake line, thus might be reduced to a severity level of "4" or "5." Also, cars must be equipped with a mechanical emergency brake that functions independently of the hydraulic brake system. In addition to the brake system major components, there are additional components that monitor the correct function of the braking system such as the brake fluid control lamp, and sensors that monitor the remaining useful life of the brake pads. Modern cars may be equipped with additional subsystems that either enhance or monitor the function of the braking system, such as

- Power brake/servo system
- ABS control module
- Brake fluid control gauge

DFMEA
Design Failure Mode and Effects Analysis

Product Part			Team:											FMEA No.:						
Project:			Responsible:				Manager:							Original Date:		Revision Date:				
								Control Methods					Recommended Action				Result			
Item	Functions	Requirement	Failure Mode	Effects	Severity	Class	Causes	Prevention Control	Occurrence	Detection Control	Detection	RPN	Action	Responsibility	Target Finish Date	Actual Finish Date	Severity	Occurrence	Detection	RPN

FIGURE 15.7 Example of a (D)FMEA worksheet. (Retrieved from https://www.iqasystem.com/.)

In the era of autonomous vehicles, FMEA tools play an even more important role in automotive design, especially for the sensors that are designed to perform functions ordinarily performed by humans.

15.6 USING SIMULATION FOR RELIABILITY AND AVAILABILITY STUDIES AND SYSTEM DESIGN

Systems are often designed under simplified assumptions with respect to the models that describe random failures and the time to complete a repair. Simple models like the exponential distribution used for a component failure or the length of a repair, and the HPP used for system failures allow for exact calculation of probabilities related to reliability and availability studies, performed either using hand manual calculations, or using mathematical/statistical software. Often a design engineer may want to study the robustness of a design using more complex models such as the bathtub curve, an NHPP or SRP (discussed in Chapter 14). Simulation offers a simple method for studying reliability and availability when complex models do not allow for exact probability calculations.

Simulation is based on the generation of *pseudo-random numbers* (Ahrens, Dieter, & Grube, 1971) which models a continuous random variable U distributed uniformly over the interval (0,1). A uniform random variable can easily be used to generate a random variable drawn from virtually any probability distribution. Suppose we wish to generate observations from a probability distribution with CDF $F(x)$ and assume that the function $F$is monotonically increasing (i.e. it has an inverse function). The random variable $X = f^{-1}(U)$ will have a distribution with CDF $F(x)$ since

$$\Pr(X \le x) = \Pr\left(f^{-1}(U) \le x\right) = \Pr(U \le F(x)) = F(x)$$

Example 15.6

We wish to generate observations of a Weibull random variable X with parameters α and β. Here we have:

$$F(x) = 1 - e^{-x^{\beta}/\alpha} \quad x \ge 0$$

$$F^{-1}(u) = -\alpha\left(\ln(1-U)\right)^{\frac{1}{\beta}}$$

Therefore $X = -\alpha\left(\ln(1-U)\right)^{\frac{1}{\beta}}$ has a Weibull distribution with mean α. Since $1-U$ also has a uniform distribution over the interval (0,1), alternatively, the random variable $Y = -\alpha(\ln U)^{\frac{1}{\beta}}$ also has a Weibull distribution with parameters α and β.

Most distributions that do not have a monotonically increasing CDF can also be modeled using U, for example simulation of the discrete binary random variable

$$X = \begin{cases} 1 & \text{with probability } p \\ 0 & \text{with probability } 1-p \end{cases}$$

can be generated using pseudo-random numbers U as follows,

$$X = \begin{cases} 1 & \text{if } 0 \le U \le p \\ 0 & \text{otherwise} \end{cases}$$

Finally, many distributions that have a monotonically increasing CDF have an inverse function that is impossible or difficult to express analytically. Such distributions are sometimes more easily generated using multiple independent sets of pseudo-random numbers, as the following examples show.

Example 15.7

We wish to generate observations of a mixed exponential random variable X with the CDF:

$$F(x) = p\left(1 - e^{-\frac{x}{\alpha_1}}\right) + (1-p)\left(1 - e^{-\frac{x}{\alpha_2}}\right) \quad x \geq 0$$

This CDF is easily shown to be monotonically increasing, but the inverse function is not easily derived. Instead, we generate three independent pseudo-random numbers U_1, U_2 and U_3. As the exponential distribution is a special case of the Weibull distribution from the previous example we have that

$$X_1 = -\alpha_1 \ln U_1 \text{ and } X_2 = -\alpha_2 \ln U_2$$

are independent exponential random variables with means α_1 and α_2, respectively. Furthermore, the random variable X,

$$X = \begin{cases} X_1 & \text{with probability } p \\ X_2 & \text{with probability } 1-p \end{cases}$$

or

$$X = \begin{cases} X_1 & \text{if } 0 \leq U_3 \leq p \\ X_2 & \text{otherwise} \end{cases}$$

has the desired mixed exponential distribution. Figure 15.8 shows how observations of X are generated using an Excel spreadsheet.

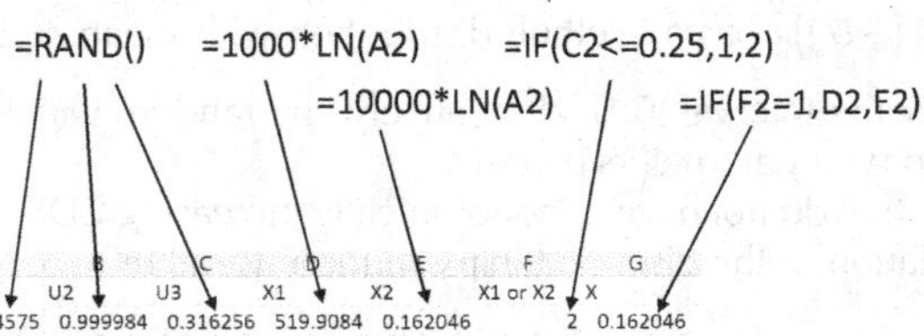

	A	B	C	D	E	F	G
1	U1	U2	U3	X1	X2	X1 or X2	X
2	0.594575	0.999984	0.316256	519.9084	0.162046	2	0.162046
3	0.86997	0.956215	0.047382	139.297	447.7208	1	139.297
4	0.973793	0.263842	0.272518	26.55658	13324.06	2	13324.06
5	0.356323	0.045916	0.100707	1031.918	30809.52	1	1031.918
6	0.480796	0.183026	0.203986	732.3116	16981.27	1	732.3116
7	0.332689	0.683011	0.869311	1100.546	3812.44	2	3812.44
8	0.959447	0.788064	0.06889	41.39806	2381.762	1	41.39806
9	0.50048	0.861628	0.323127	692.1875	1489.312	2	1489.312
10	0.413736	0.90267	0.480303	882.5282	1023.983	2	1023.983
11	0.877537	0.046267	0.666815	130.6356	30733.24	2	30733.24

FIGURE 15.8 Simulation of observations from mixed exponential distribution using an Excel spreadsheet. The parameters are $p = 0.25$, $\alpha_1 = 1{,}000$ and $\alpha_2 = 10{,}000$.

Example 15.8

We wish to generate observations of a log-normal random variable X. A log-normal distribution is particularly appealing to use as a model for failure and repair times since it has support on the positive real numbers, $X > 0$. Suppose Y is a normal random variable with mean μ and standard deviation σ then $X = e^Y$ has the desired distribution. In order to simulate observations from a normal distribution we note that the CDF for Y is:

$$F(y) = \int_{-\infty}^{y} \frac{1}{\sqrt{2\pi}\sigma} e^{-\frac{1}{2}\left(\frac{t-\mu}{\sigma}\right)^2} dt$$

This function does not exist in closed form, and neither does its inverse function. Therefore, a simulation using $Y = f^{-1}(U)$ will require numerical methods that make repeated calculations that are computationally intensive. A less computationally intensive approach is to use the Central Limit Theorem (Wackerly, Mendenhall, R.L., & Scheaffer, 2008) which implies that a random variable in the form,

$$Z = \frac{\bar{U} - \mu_U}{\sigma_U / \sqrt{n}}$$

where $\bar{U}$ is the average of n independent random variables $U_1, U_2, \ldots U_n$ with mean μ_U and standard deviation σ_U, is approximately standard normal for large n. In the case where $U_1, U_2, \ldots U_n$ are uniform random variables over the interval $(0,1)$, the approximation is quite accurate when $n = 10$ or higher, and in this case we have $\mu_U = \frac{1}{2}$ and $\sigma_U = \frac{1}{\sqrt{12}}$ (Figure 15.9). Thus we can generate a standard normal random variable using n pseudo-random numbers $U_1, U_2, \ldots U_n$ using

$$Z = \sqrt{\frac{12}{n}} \sum_{i=1}^{n} \left(U_i - \frac{1}{2}\right) = \sqrt{\frac{12}{n}} \left(\sum_{i=1}^{n} U_i - \frac{n}{2}\right)$$

and a log-normal random variable with parameters μ and σ using

$$X = \text{Exp}[\mu + \sigma Z]$$

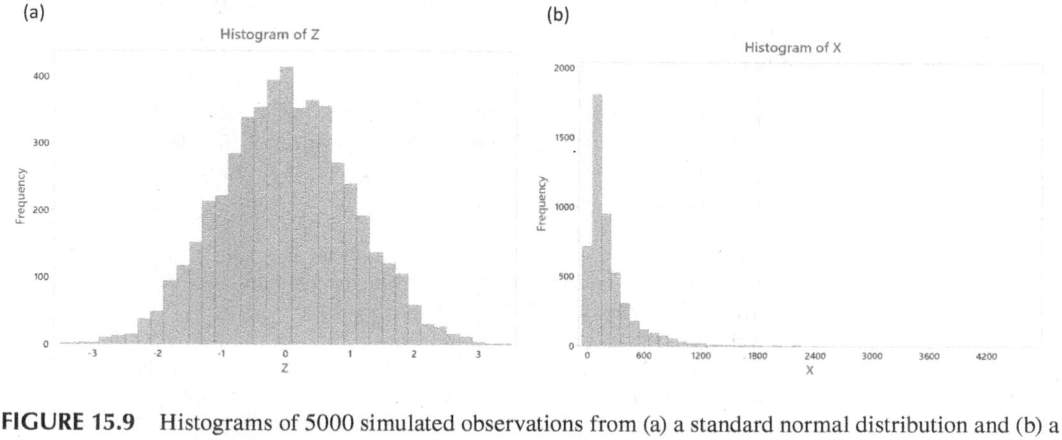

FIGURE 15.9 Histograms of 5000 simulated observations from (a) a standard normal distribution and (b) a log-normal distribution with $\mu = 5$ and $\sigma = 1$.

Example 15.9

An electric power facility requires continuous operation of five transformers. A redundant (spare) transformer can be momentarily engaged in the event of a single transformer failure. Each transformer in service will degrade over time and will eventually fail in accordance with a Weibull distribution with parameters $\alpha$and β. To reduce the chance of failure each transformer is subjected to a scheduled maintenance every m days at which time the transformer is restored to an "as-new" condition. This maintenance takes γ days to complete. Transformers are taken down for maintenance only one at a time, so if a transformer is due for maintenance at a time when another one is being serviced (for maintenance or repair), the maintenance is delayed until the first available time when this none of the transformers are being serviced. Once a transformer is repaired, it is in "as-new" condition and becomes the new "spare" that will replace the next transformer that is taken down for repair or maintenance. Repairing a transformer is much more time-consuming than a scheduled maintenance and can take weeks or months to complete. The number of days needed to complete a repair is assumed to follow a log-normal distribution with parameters μ and σ. In the event that a transformer fails while another one is under repair or maintenance, the entire electrical network will shut down causing power outages in the geographical area that it services. We wish to evaluate the average annual availability of this system. An exact calculation of the availability is difficult because it must take into account several scenarios, for example

- The schedule of the maintenance work varies depending on other transformers that fail or are under maintenance.
- The possibility of a transformer failing while another transformer is down due to maintenance or repair.

Using simulation, we can randomly generate unscheduled failure and repair times using the methods in examples 15.6 and 15.8. This will generate a series of events and between each of the events, we keep track of whether the system is available or not. For each simulation run, we calculate the average availability as

$$A = \frac{\text{Total Uptime}}{\text{Total Uptime+Total Downtime}}$$

And this average is then averaged over many repeated simulation runs.

There are a number of scenarios, in order from best to worst, that should be considered when analyzing the simulated data. Figure 15.10 shows examples of best and worst-case scenarios.

- Scenario 1 (best case):

  Transformer failures occur rarely and repair times are short enough so that they do not interfere with the scheduled maintenance. Initially, all transformers have scheduled maintenance around the same time of the year, but over time as failures and repairs are observed, scheduled maintenance events are spread out randomly over the year. Availability is at or near 100%.
- Scenario 2

  Failures occur regularly, but repair times are short enough that a new spare unit is available before another transformer fails. Occasionally a transformer may fail while a transformer is down for maintenance causing a short outage of no more than 24 hours. Availability is at 99% or higher.
- Scenario 3

  Failures occur regularly, and repair times are long so that occasionally a transformer fails while another one is under repair. This may cause outages lasting weeks or months, but in the long run, availability may be at 90% or higher.
- Scenario 4 (worst case):

  Failures occur frequently and repair times are long so that a failure may occur while two or more transformers are under repair. In this scenario, there is no option to perform scheduled maintenance and as a result, the remaining transformers in service will operate well past their useful life with an increased likelihood of failure as a result.

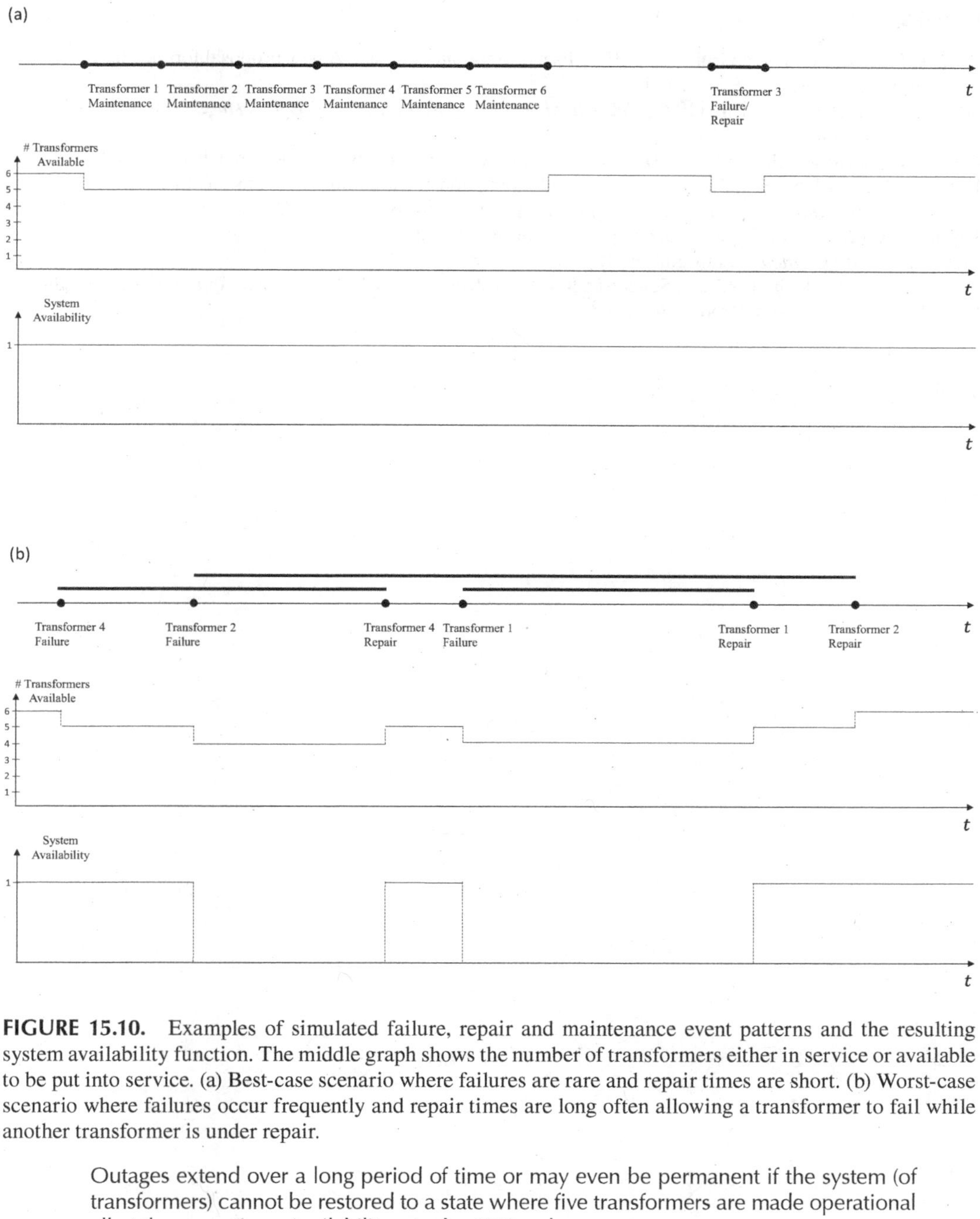

FIGURE 15.10. Examples of simulated failure, repair and maintenance event patterns and the resulting system availability function. The middle graph shows the number of transformers either in service or available to be put into service. (a) Best-case scenario where failures are rare and repair times are short. (b) Worst-case scenario where failures occur frequently and repair times are long often allowing a transformer to fail while another transformer is under repair.

> Outages extend over a long period of time or may even be permanent if the system (of transformers) cannot be restored to a state where five transformers are made operational all at the same time. Availability may be 10% or less.

Simulation allows us to study different strategies for the schedule and frequency of maintenance events as well as design changes, for example, adding a second redundant transformer. We may also vary parameters to study the effect of, for example, shortening the average repair times or changing the wear-out characteristics of the transformer failures.

REFERENCES

Ahrens, J. H., Dieter, U., & Grube, A. (1971). Pseudo-random numbers: A new proposal for the choice of multipliers. *Computing*, 6(1), 121–138.

Barlow, R. E., & Proschan, F. (1975). *Statistical Theory of Reliability and Life Testing.* New York: Rinehart and Winston.

Cox, D. R., & Miller, H. D. (1965). *The Theory of Stochastic Processes.* Boca Raton: Chapman Hall.

Hansen, C. K. (1995). Reliability prediction and simulation for a communications-satellite fleet. In *Annual Reliability and Maintainability Symposium* (pp. 152–158). Washington, DC: IEEE.

Kiran, D. R. (2017). *Total Quality Management.* Amsterdam: Elsevier.

Stewart, J. (2015). *Calculus*, 8th Edition. Boston: Cengage.

Wackerly, D., Mendenhall, R.L., & Scheaffer, R. (2008). *Mathematical Statistics with Applications*, 7th Edition. Belmont: Thomson, Brooks/Cole.

16 Maintainability

David R. Carey

16.1 INTRODUCTION

Maintainability is defined as the ability of an item to be retained in, or restored to, a specified condition when maintenance is performed by personnel having specified skill levels, using prescribed procedures and resources, at each prescribed level of maintenance and repair (Ebeling, 2019). There are typically two levels of maintenance and repair: Organizational level (O-Level), and Depot level (D-Level). O-level maintenance is performed at the end item level and involves removal and replacement of either the entire system or major assembly of the system. D-level maintenance is performed on modularized assemblies, which have been removed at the O-level, and involves removal and replacement of faulty piece part component(s).

Today, the single highest cost for the defense department is operations and maintenance (O&M) (Peter G. Peterson Foundation, 2022). In the 2022 fiscal year budget, the defense department spent twice as much to support the existing inventory as it did on new equipment (Office of the Under Secretary of Defense, 2021). The military recognizes this problem and reducing operations and support costs is a primary metric in the establishment of new weapon systems (Berenson, 2021). High O&M costs are due to many factors, that fall into the categories of testability and maintenance. The purpose of this chapter is to describe how system design can positively and adversely affect O&M and suggests designs that minimize these costs. The reader should come away with the importance of maintainability and the overall effect on test in the factory, field, and at the depot.

16.2 MAINTAINABILITY

Maintainability is the measure of the ability for a system or subsystem to be retained or restored to its desired operating condition (United States Department of Defense, 1988). Maintenance is any activity necessary for keeping or restoring a system to an operational readiness state. Maintenance will: detect, isolate, and repair defects; avoid unforeseen failures; maximize a systems life; and maximize efficiency, reliability, and safety. This all boils down to the two main reasons for maintainability: reduced system support costs and improved operational readiness.

16.2.1 Maintainability Influence on Reliability and Vice Versa

Reliability and maintainability combine to form metrics for system operational readiness or availability. Both reliability and maintainability are equally important. Reliability is the probability that a system will function correctly within defined mission and environmental requirements without failure (Blanchard et al., 1995). Reliability is a core requirement for any system. Maintainability can be expressed as the likelihood a system will operate within specification when maintenance actions are performed within a predefined schedule (Blanchard et al., 1995). The reliability definition is a fundamental part of maintainability.

Reliability can be designed into a system. The system design should address system failure modes and their causes. A system will fail for a variety of reasons, manufacturing defects, parts defects, insufficiency in design features, operating out of design tolerance (environmental and operational), improper maintenance, and operator error to list a few. The designer can affect reliability through parts selection, analyzing operational requirements, perform material stress analysis, derating of parts, employing new technology, adding system redundancy, and design simplification. The more

DOI: 10.1201/9781003188377-19

reliable a system the less likely it will fail, requiring less maintenance. System maintenance cannot be eliminated. Thus, maintainability merits special consideration.

Maintenance influences the system cost of ownership in labor and material. This in turn represents significant budgetary costs over the life of the system. Maintenance and repair costs can be greater than three times the cost of design and manufacturing (Peter G. Peterson Foundation, 2022). It is important that maintenance be discussed at the beginning of any system design activity to improve the life cycle of a system.

With advancing technology, system maintainability planning has had to respond to the increase in complexity, with maintenance strategies for fault detection, isolation, and repair that can support advanced electronic equipment functionality. The increased system complexity leads to the potential for decreased reliability, higher failure rates, and increased maintenance actions. As a result, repair activities can no longer be built on individual knowledge, decisions, and analysis.

For maintainability planning, the Mean Time to Repair (MTTR) of a system is no longer the sole requirement. Maintainability must include specific skill sets required to maintain, diagnose, and repair a system. Additionally, how to reduce or simplify maintenance activities has become equally significant.

Maintainability is a risk because of the reluctance of the design community to change from its emphasis on performance to include maintainability. If the maintainability engineer and the designer fail to accomplish this, then they will fail to achieve the objective of operational readiness and availability.

Implied from above, maintainability is a function of design or a design consideration. On the other hand, maintenance is the response to maintainability decisions made during design or a consequence. In other words, maintenance must accept with whatever maintainability level was established during design.

Another consideration involving maintenance is the difficult task of training new maintenance technicians when the experienced technician transfers out of the department, gets promoted, or otherwise leaves. With the increasing complexity and technological advances in electronics, maintenance techniques have become more complicated.

As previously stated, maintenance requires highly skilled personnel, using standard equipment to verify functionality and isolate faults within the system. The maintenance technicians may compensate for insufficiency and errors in specifications by creating ways to circumvent incompatibilities. Decisions on marginal performance may vary between maintenance technicians, leading to inconsistency. When the human element is introduced, the success of the maintenance operation is influenced by the intelligence and experience of the maintenance technician. Inadequate or insufficient maintenance activities can reduce reliability and operational readiness of a system.

16.2.2 Measuring Reliability and Maintainability

To measure reliability and maintainability, there is a need to quantify the availability or readiness of the system and the amount of work required maintaining the system. Reliability and maintainability affect availability, dependability, labor costs, and maintenance/support costs. For these categories reliability and maintainability measurement can be stated as shown in Table 16.1:

The most obvious metric would be the ease of maintenance and how this will influence the amount of work required to maintain a system. Ease of maintenance is a function of the speed with which maintenance operations can be performed to avoid malfunctions or correct them if they occur. Ease of maintenance is enhanced by any design consideration that will reduce the time and effort necessary to maintain equipment at highest operational readiness. This can be quantified by the time required to perform the maintenance activity.

Maintenance activities fall into two categories, corrective and preventive. Corrective actions are performed when equipment failure has been detected. Preventative actions are performed periodically, usually on a predefined schedule. An example of preventative maintenance would be periodic

TABLE 16.1
Reliability and Maintenance Measures

	Reliability	Maintainability
Availability (readiness)	Mean Time Between Failure	Mean Time to Repair
Dependability (mission success)	Mission Time Between Failure	Mission Time to Repair
Labor cost	Mean Time Between Maintenance	Hours/Action
Support cost	Mean Time Between Demand	Cost to Remove a Part

calibration. Corrective maintenance consists of three main activities, identification of the defective parts (fault isolation) using some form of test; perform the repair (replace the parts or correct the fault); and verify system operation (ensure the repair was successful).

Maintainability is primarily concerned with access and the mechanical removal and replacement of assemblies. What all these tasks have in common is that they increase the workload on the system maintainer. Workload equals cost. Operations and maintenance costs makes up more than 40% of the defense budget (Peter G. Peterson Foundation, 2022).

The designer affects two aspects of maintainability that are extremely important: the interchangeability of components and the elimination of manual alignments. Interchangeability is the ability to replace any Line Replaceable Unit (LRU) with a component of the same configuration without performing any manual alignments, calibrations, or adjustments. The difference between a calibration and an adjustment is that a calibration is expected to be performed on a periodic basis where an adjustment is only performed once after a specific event.

16.2.2.1 Alignment and Their Elimination

Alignments are primarily due to two circumstances:

1. The physical alignment of one assembly must have a specific relationship with the alignment of another assembly
2. The alignment of an assembly has a direct relationship to its function.

A first case alignment example could be between two optical assemblies that must align along a single path. An example for the second case could be a one-axis gyro that must be aligned on a specific axis for best performance.

There are two methods of eliminating the requirement for alignments:

1. Combine assemblies which must be aligned and eliminate the requirement.
2. Provide an automatic method of aligning the assemblies that do not require manual input.

Design is a factor in the execution of the second method. The specific designs can be much too different to provide a standardized example but there are common issues:

1. The designer must ensure that the required accuracies can be achieved and stored within the circuitry and stored in nonvolatile memory.
2. The designer must ensure that all the LRUs required to perform the alignment are identified so that their replacement results in a new auto-alignment.
3. The designer should recommend a sequence of alignment when multiple alignments are required.
4. Design must ensure that all voltages, currents, etc. required for alignment are accessible automatically through software.

The task is to eliminate the requirement for alignments or provide access to all controlling and monitoring circuitry needed to perform the alignment automatically.

The important message to the designer is:

1. Identify any required signal-related alignments (such as two signals that need to be in-phase).
2. Ensure that the circuitry necessary to perform signal-related alignments automatically exists and operates in the presence of expected noise, thermal, and environmental conditions.
3. Ensure any control circuitry designed can handle accuracies and resolutions required for inter-assembly's alignments.
4. Ensure software has been provided requirements for the performance of any necessary automated alignments.

The best practice is to eliminate the need for any alignments.

16.2.2.2 Calibration and its Automation

Calibrations are typically required when the capability of the devices in a circuit falls short of the accuracy or resolution required for circuit performance. Automatic calibration may provide a method of using less expensive devices in the assembly of a circuit and therefore reduce the unit cost of the assembly. It is only manual calibration that must be eliminated. Any required manual calibration will need to be scheduled for all systems within the inventory and represents a reduction in readiness over an extended period. Automatic calibration can be executed at power up with little adverse effect on the system (other than the seconds required for the automatic calibration). A common automatic calibration performed in systems today is the determination of offsets, quantification errors, etc. in a data acquisition circuit. Multiple references and a digital to analog converter (DAC) can be used to calibrate out noise and component aging-related tolerances. Devices with an inherent accuracy of a few percent can be calibrated to within 1% of actual.

Digital potentiometers, programmable clock delays, etc. may allow critical timing and phase relationships to be automatically calibrated.

The important message to the designer is:

1. Do not require any manual calibration of circuitry. This represents a large cost and potential degradation to system readiness.
2. If calibration is necessary, ensure that the circuitry required to automatically calibrate the signal exists and can provide critical accuracy and resolution.
3. Ensure the software knows of any requirement for automatic calibration and understand the thresholds indicating that automatic calibration has failed.

16.2.2.3 Automation and Elimination of Adjustments

Adjustments are basically the same as calibrations except that they do not represent a continuously required event. The best design will eliminate the need for adjustments but if they are required, they should be made automatically.

The message to the designer is:

1. The first choice for adjustments is through digital controls such as DAC output or Digital Potentiometers.
2. The next choice is an adjustment that can be made by the user while the system is in operational mode.
3. Manual adjustments in the field are unacceptable.

16.3 TESTABILITY

Maintainability cannot be discussed without an in-depth look into testability. Testability is defined as the property of a system, subsystem, assembly, or component to be effectively and efficiently tested to determine its operability (United States Department of Defense, 1988). For most military systems, this includes the development of a metric representing the probability of successfully detecting and reporting the presence of a fault (fault detection) and of determining the location of the fault within the system, subsystem, or assembly (fault isolation).

Testability is the property of a system that makes it possible to test (United States Department of Defense, 1988). The total cost of an item, over its lifetime, life-cycle-cost, is driven primarily by the cost of its maintenance. To reduce life-cycle costs, systems need to be developed with higher reliability and greater ease of maintenance. One of the greatest drivers toward ease of maintenance is testability. If a device cannot be tested it cannot be repaired. As previously stated, the first step in any maintenance action is to test the device to see if it is defective and if so to determine what component(s) need to be replaced.

As a device goes through its life cycle, tests are necessary to: ensure it meets its specifications, determine the operational state, detect the presence of faults, and isolate faults. Difficulty in testing, fault detection, and fault isolation will result in increased life cycle costs.

Built-In-Test (BIT), manual testing, and automatic testing capabilities must be considered. The user of the system must be able to test and repair the device easily. 100% of every failure must be detectable, isolatable, and repairable. This is a performance measure of the testability of the device.

If a system has been indicted for repair, it is unavailable for its intended mission. The system must be returned to an operational condition as quickly as possible. Fast, effective, and accurate maintenance test is a critical factor in performing this return to service.

When a system is in the field, BIT is one method used for the detection and isolation of faults. If the BIT performance metric is 95% fault detection and 95% fault isolation to one assembly then, this leaves approximately 10% of all failures as falsely detected and misdiagnosed.

If the depot test equipment performance metric is 100% accurate in the detection of failures, all the No Evidence of Failure (NOEF) hardware that is returned due to misdiagnosis can be tested, verified as operational, and shipped immediately back to the field. However, if the depot equipment detects only 95% of all faults, each "good" assembly must be tested with an alternative method to ensure that it is not in the 5% that cannot be automatically detected.

Non-performance of BIT and depot test equipment can compound each other. BIT and depot testability have become significantly more critical from a cost perspective.

Factory testing and depot testing are performed under very different circumstances with different goals and should not be considered as interchangeable. Factory test ensures that the equipment is suitable for delivery to the customer. It is not intended to show that the equipment meets all specified performance requirements. Qualification testing (performed once) is used to show conformance to specification requirements. Acceptance testing performed in the factory is used to identify component failures and manufacturing errors.

Because all component parameters degrade with exposure to the environments, the equipment must function correctly in the presence of parametric degradation and must not indicate failure of the component due to expected changes. For example, a 1% resistor should be tested as a 1% resistor by the factory test equipment, but BIT must allow for a 2%–3% change in resistor value over time due to environmental effects.

Factory test accuracy is not a large issue for the military customer since defects will become evident on use. It is a high-cost driver for the company since the complete system must be sent back, repaired, retested, and redelivered to the customer. It is an initial nuisance for the customer and bad press for the company but does not affect the long-term cost of the equipment. Therefore, factory test is not as critical as depot test from a customer perspective but is extremely critical to profitability.

TABLE 16.2
Testability Levels

Test Level	Environment	Tolerance Level	Failure Mode
Factory	Room Ambient	Initial	One or more failures; opens, shorts, wrong or defective components, components inserted incorrectly
Depot	Room Ambient	End of life	Single component failure
Field (BIT)	Full Specified Temperature Range	End of life	Single component failure

Recurring hardware testing is normally performed at three distinct levels and the environmental and failure mechanisms vary with each. The characteristics of each are defined in Table 16.2. The test philosophy of each of these levels is discussed in paragraphs 16.3.1–16.3.3.

16.3.1 Factory Test

Factory production testing is performed at room ambient and should not, if possible, require any special environmental conditioning, such as warm-up time, or temperature conditioning.

During factory testing, there may be multiple failures. Initial factory failures include wiring errors, opens, shorts, improperly installed components, improperly polarized components, wrong components, bad components, and solder splashes. After the system is placed into service only one component is expected to fail at a time.

Production testing does not qualify a design. It is used to verify that the system has been assembled correctly and that all the components are in proper working order. Production testing is used to detect manufacturing defects. The number of tests should be minimized and performed as quickly as possible to increase production rate. Statistical analyses should provide the basis for the determination of test limits to optimize the detection of faulty hardware.

The testing philosophy should be structured to detect failures at the lowest possible level since the cost to remedy a failure increases substantially at higher test levels. The factory test philosophy is usually driven by requirements and program direction.

In general, the test philosophy at each level of factory test assumes that the previous level of test has been performed successfully and duplicity of tests is discouraged since it decreases production rate and increases cost. Multiple failures will occur during production and the factory test philosophy should be structured to detect multiple failures. It is desirable to detect all failures, in each system, during one pass across the test set. However, this is usually not possible during functional testing and may not always be possible during in-circuit testing, depending on the type of test used. Multiple failures may have to be detected and isolated by sequential test runs.

Factory testing sometimes defers testing to higher levels for one reason or another. Usually, a test is elevated to a higher level only because it is difficult or cannot be performed at a lower level, an undesirable situation. It is preferred that tests be performed at the lowest possible level. If a test is omitted at the shop replaceable unit (SRU) level but performed at a higher level, critical SRU functions may not be verified at the lower level. This is not a problem if the SRU is installed into the next higher assembly and the missing tests are performed at the higher level. However, if the same SRU test is used for acceptance of a spare SRU, the critical test will not be performed.

The use of factory tests in the depot or field is often recommended but has shortcomings. Factory testing uses a hierarchical approach where if a function is proven at a lower level, it does not have to be checked again at a higher level. In the depot and field, there are no lower-level tests. When unmodified factory tests are used at these levels the danger exists that some important function, tested at a lower level in the factory, will not be tested in the depot or field.

Factory test is very similar to depot test. What succeeds at the depot will succeed at the factory. There are two significant differences:

1. Multiple failures may exist at the factory and the independence of circuitry is important. If a single failure makes multiple circuits fail to operate, all the components in those circuits will be removed and the cost of the assembly goes up accordingly.
2. The level of accuracy is higher at the factory since components should only reflect purchased tolerances. This may make noise levels much more critical during factory test.

When creating a Test Requirements Document (TRD), always consider the accuracy of measurements in the presence of noise. A 10-bit Analog to Digital Converter (ADC) is not considered extremely accurate. When working with +/–5 V references, the least significant bit represents 10 mV. If you expect 100 mV of noise in your circuit, requiring accuracies to third least significant bit (LSB) might be a mistake (ie. 80 mV). If you are going to use digital filtering, auto-calibration, or multiple references to eliminate noise and tolerance build-up, include the description of these processes in your TRD.

16.3.2 Depot Test

The government now operates a two-tier maintenance operation. The first level of maintenance is at the weapon system. This is known as the field, tactical or organizational level (O-Level) maintenance and is usually limited to running BIT and replacing black boxes. The black boxes are then taken to the second or depot level (D-Level). The D-Level is usually located a long distance from the O-Level and may employ both automatic test equipment (ATE) and manual test stations. The skill level of the test technicians increases as equipment becomes more complex.

Depot tests are normally structured to detect single component failures. Test routines isolate the failure to a group of components rather than a single component because of equipment limitations and system design. The group of components called out due to a failed test routine is called an ambiguity group. Repair action usually consists of replacing all components in the ambiguity group. Good testability design reduces the size of the ambiguity group.

The designer determines the ambiguity group size through the access to test points, the ability to control the initial conditions of test, and the ability to separately test individual functions.

Initial test conditions are usually more critical to digital design since automatic test generation software must start with a known vector. However, if the digital circuitry reacts to an analog input, the analog input must be brought to a known initial condition to ensure stable start-up by the digital hardware.

The two major reasons for inadequate resolution in depot testing are: lack of test points and inadequate circuit control. The lack of test points is easy to envision. If you use an operational amplifier to scale a DC input followed by an inverting unity gain amplifier, they form an ambiguity group. The tester must check the input to the buffer or failure will always result in the replacement of both components. An example of inadequate circuit control is the inability to take over an internal clock with an external clock. Without the external control of the internal clocking circuitry, the test equipment will not know the state of the equipment on power up and it is very probable that the state will appear pseudo random and dependent on the speed with which the test equipment can synchronize with the system. Lack of control of internal clocks is a significant problem for the test engineer.

Testing at depot must use test limits that include end of life tolerances. One problem with this is that differences between test systems can cause a system to pass either BIT or depot testing and fail at the other. Cases have occurred where systems tested and failed by BIT in the field pass in depot test. One method to prevent this from happening is to make the test limits slightly tighter at depot than in BIT. This will ensure that any system that fails BIT will also fail at the depot.

Commonality between the depot and factory is universally accepted as a good practice due to cost savings. Some programs have implemented vertical commonality between the factory and depot levels. This practice has some problems since the test philosophy in the factory is different from the philosophy used in the depot.

The use of depot or field tests in the factory sometimes are inadequate because field and depot testing assume that the system worked at one time and only one component has failed. No shorts or opens exist except in components. Fault isolation routines used at depot or field levels of maintenance rely on these facts to isolate defective components. Factory testing cannot make these assumptions. The factory system may never have worked. Solder splashes may cause shorts, conductor paths may be open, wires may be on the wrong pin, components may be inserted incorrectly, the wrong component may have been used, or defective components may have been inserted. Factory isolation routines must take all these possibilities into account.

16.3.3 Field Testing

Testing of hardware in the field relies extensively upon BIT. BIT is performed by hardware and/or software, which is included as an integral part of the system. Limits for this testing include initial, temperature, and end of life test limits. BIT functions may take several forms. BIT is the most critical test feature in military equipment. BIT capability determines the amount of unnecessary maintenance that will be performed at all levels of maintenance. BIT has many components - hardware (the circuitry and firmware), software (to access the information produced by hardware), and reporting (discrete analog and digital methods of indicating a detected failure and recommended isolation to the user).

In the field, a design is usually part of an entire system or subsystem. For example, entire systems may be composed of several assemblies, and each of these assemblies may be composed of several individual circuit cards. It is usually possible to structure a system's BIT, in addition to the overall system architecture, so that BIT can determine a fault's location within an assembly or individual circuit card. A preliminary testability analysis during the concept design stage is recommended in that it promotes functional partitioning. Functional partitioning, locating all circuitry and components on a single replaceable module, is preferred to circuit partitioning in that the location of faults within specific functions could, more easily, be unambiguously determined. There are usually significant cost and logistic benefits to BIT being able to determine a fault's location to the least expensive subassembly that can be replaced in the field. For example, for a system containing an assembly consisting of multiple circuit cards, it is preferable to have the system's BIT capable of determining which circuit card within the assembly a fault is located, so that only the faulty card needs to be replaced, instead of having to replace the entire assembly because BIT could not determine which circuit card was faulty.

A Level of Repair Analysis (LORA) is performed by Logistics Engineering to determine the optimal configuration of subassembly, circuit card, component, etc. that can be replaced in the field. A Partitioning Analysis is performed to establish how the various system functions will be allotted among the system's subassemblies, circuit cards, etc. The designer's participation is crucial to the success of the LORA and Partitioning Analysis.

These analyses typically address the following factors that have significant impact on a system's BIT design. These factors should be documented prior to the concept review phase of the design.

- Establish and define the boundaries between replaceable assemblies.
- The percentage limitations of circuit real estate to be allotted for BIT, depot, and factory test resources.
- Whether the measuring assets for performing BIT will be located within the design or external to the design.

- How BIT failures will be determined and reported (for example, will BIT measurements be processed by another assembly or will the assembly have the resources to determine test pass/fail itself).

If a signal or stimuli meets the following criteria (as determined by the LORA and Partitioning Analysis), then it should be monitored and evaluated by BIT resources.

- It is an input signal coming from another replaceable assembly.
- It is an output signal going to another replaceable assembly.
- It is necessary to know the state of this signal to determine if another signal is correct.

16.4 CHAPTER SUMMARY

In this chapter, the concept of maintainability and testability was introduced. The single highest expense for the defense department is systems O&M costs. The relationship between maintainability and reliability was considered. To reduce cost, systems need to be developed with higher reliability and greater ease of maintenance. Measurements to determine the level of reliability and maintainability were presented. It was shown that these measures are required to quantify the readiness of a system. The designer can significantly affect these measures by eliminating alignment, automating calibration, and automating or eliminating adjustment requirements. Finally, an in-depth look into testability and its effect on maintainability was presented. One of the greatest drivers toward ease of maintenance is system testability. The designer has a responsibility to perform design for testability analysis and look at all levels of test, factory, field, and depot.

REFERENCES

Berenson, D. (2021). *Getting the Department of Defense off the Hamster Wheel: Reducing Operation Costs to Invest in the Future of the Force.* Metamorphic Media LLC. https://warontherocks.com/2021/03/getting-the-defense-department-off-the-hamster-wheel-reducing-operating-costs-to-invest-in-the-future-of-the-force/. (Accessed June 10, 2022).

Blanchard, B. S., Verma, D. C., & Peterson, E. L. (1995). *Maintainability: A Key to Effective Serviceability and Maintenance Management*, 1st Edition. New York, NY: Wiley-Inter-Science.

Ebeling, C. E. (2019). *An Introduction to Reliability and Maintainability Engineering*, 3rd Edition. Long Grove, IL: Waveland Press Inc.

Office of the Under Secretary of Defense (2021). *Operation and Maintenance Overview US DoD FY22 Budget Request.* Washington, DC: United States Department of Defense.

Peter G. Peterson Foundation (2022). *Budget Basics: National Defense.* Peter G. Peterson Foundation. https://www.pgpf.org/budget-basics/budget-explainer-national-defense. (Accessed June 10, 2022).

United States Department of Defense (1988). *DOD-HDBK-791(AM) Maintainability Design Techniques.* Washington, DC: United States Department of Defense.

17 Performance Evaluation

Byeong Kil Lee

17.1 INTRODUCTION

Modern microprocessors are integrated with billions of transistors to implement state-of-the-art architectural features, including deep pipelines, parallel processing, speculation features, memory hierarchy, accelerators, etc. Microprocessors can be simply categorized into general-purpose processors (GPP) and application-specific processors (ASP). The ASP normally has a target workload, for which designers are focused on the functionality and performance of the specific workload such as multimedia, network processing, and machine learning. On the other hand, the GPP targets a group of workloads called a benchmark suite, which includes representative applications for a wide range of fields. The performance evaluation of general-purpose processors can be challenging and time-consuming work considering a variety of target applications.

In the early-design stage, most microprocessor design companies are doing performance modeling which is to implement the microprocessor in software for design exploration and performance estimation. This is because designing a microprocessor takes a long time to get an engineering chip from the fab. Architectural design decisions and performance evaluations are done in the stage of performance modeling before starting the actual microprocessor design. Therefore, performance modeling is an important step for achieving the target performance of the microprocessors. Once the design starts, both design analysis and performance evaluation will be done at each design stage. The performance evaluation results need to be compared to the ones from the performance model, and the required tuning of the performance model should be done to make the model more accurate.

This chapter presents various topics in the performance evaluation of microprocessors and computer systems. An overview of performance evaluation is presented in Section 17.1. Section 17.2 presents a comparison of performance prediction and performance measurement. Performance modeling and simulators are described in Section 17.3. Section 17.4 presents performance metrics that are used for evaluating microprocessors. Existing benchmarks for performance analysis and evaluation are described in Section 17.5. Section 17.6 presents SimPoint, which is an intelligent sampling approach that selects samples called simulation points, based on a program's phase behavior. The performance monitoring counters integrated into the chip for performance tuning are introduced in Section 17.7.

17.2 PERFORMANCE EVALUATION: PERFORMANCE PREDICTION VS. PERFORMANCE MEASUREMENT

Performance can be translated into the value of the product; therefore, performance evaluation is a very important step in microprocessors design or systems design. The performance should be evaluated and monitored at each development stage to accomplish the target performance. The result of the performance evaluation of a microprocessor or a system can be presented with a numerical metric that can be used for comparison to other microprocessors or systems.

Performance evaluation includes two different steps depending on the time to measure the performance – performance prediction and performance measurement. Performance prediction is to evaluate the performance before the actual product is available as an engineering sample or final product. Performance prediction can be done with a simulator that is implemented based on the performance model or a design module that is implemented with hardware description languages.

DOI: 10.1201/9781003188377-20

Performance measurement is to evaluate the performance of the actual product. Performance measurement provides an accurate metric value, while the accuracy of the prediction depends on how the performance model is accurately designed. Performance prediction should be included as one of the design procedures that needs to be done at each design stage and will impact the final performance. The difference between the predicted values and the measured values should be analyzed, and all the data needs to be accumulated for many years to make the performance model accurate and improve the prediction accuracy. Figure 17.1 shows the concept of performance prediction and performance measurement.

The required components for performance evaluation include workloads or benchmarks to be executed on the simulator or actual products, appropriate performance metrics and a performance model (e.g., simulator) as shown in Figure 17.2. Each component will be described in the following sections.

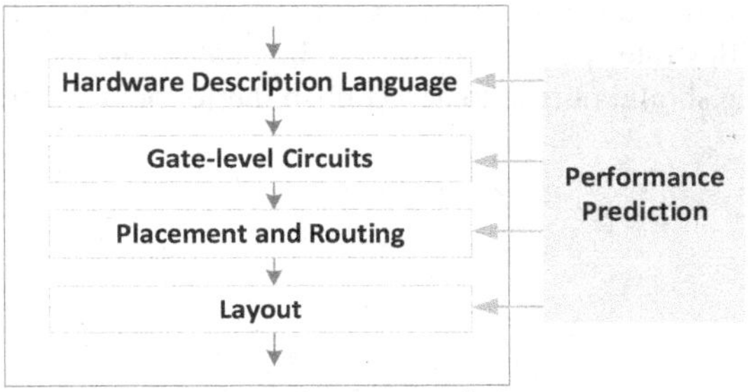

(a) Performance prediction

(b) Performance measurement

FIGURE 17.1 Performance evaluation: performance prediction in early-design stages vs. performance measurement on the actual chip. (a) Performance prediction. (b) Performance measurement.

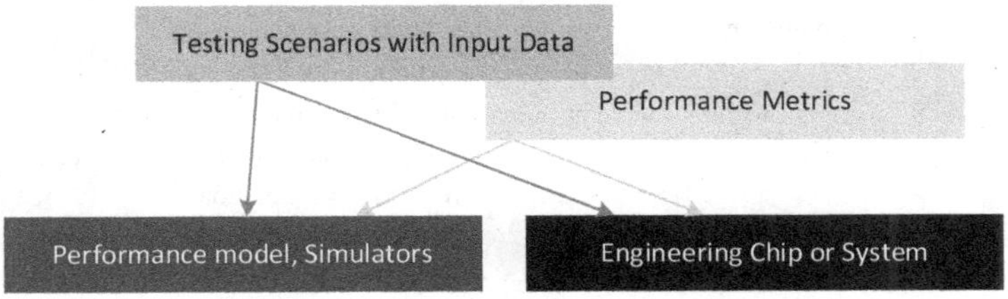

FIGURE 17.2 The required components for performance evaluation.

17.3 PERFORMANCE MODELING

Performance modeling, which is implemented as a simulator, is normally done in the architectural prototyping stage or early-design stage. The model will be used for performance prediction when actual systems are not available for performance measurement and design exploration for performance tuning to achieve target performance. There are two types of performance modeling such as simulation-based modeling and analytical modeling. Simulation models can further be classified into several categories: execution-driven simulation, trace-driven simulation, full-system simulation, etc. Analytical models are based on probabilistic models, queuing models, etc.

Performance models are used for both evaluating the performance and determining architectural configuration to maximize the performance, which is called design exploration or performance exploration. For the performance exploration, the simulation model should be configurable to change the scale of hardware resources to see the performance sensitivities. A large number of applications need to be executed on the simulator to evaluate the overall performance for a specific hardware configuration, which takes a long time to measure the performance values. The simulation time can be a bottleneck for the whole design period. To resolve the issue, there have been studied methodologies to effectively reduce the simulation time. We will introduce the methodologies in Section 17.6. Table 17.1 illustrates performance modeling categories.

Performance modeling should pursue to accomplish the following features:

- Accuracy
- Cost-efficiency
- Configurability
- Flexibility
- Scalability
- Wide applicability
- Fast simulation time
- Easy-to-use

17.3.1 Simulation and Simulators

Simulation is a software-based framework for performance exploration and performance evaluation of microprocessors and systems with configurable hardware resources. For better accuracy, cycle-accurate simulation methods are popularly used in microprocessor design. Normally, simulators are designed for new microprocessors to make sure that the architecture can achieve the target performance. As shown in Figure 17.3, a simulator as the target machine is executed on the host machine.

TABLE 17.1
Performance Modeling Categories

Performance Modeling	Modeling Type
Simulation	Cycle-accurate simulation
	Execution-driven simulation
	Trace-driven simulation
	Full-system simulation
	Event-driven simulation
	Statistical simulation
Analytical modeling	Queuing model
	Markov model

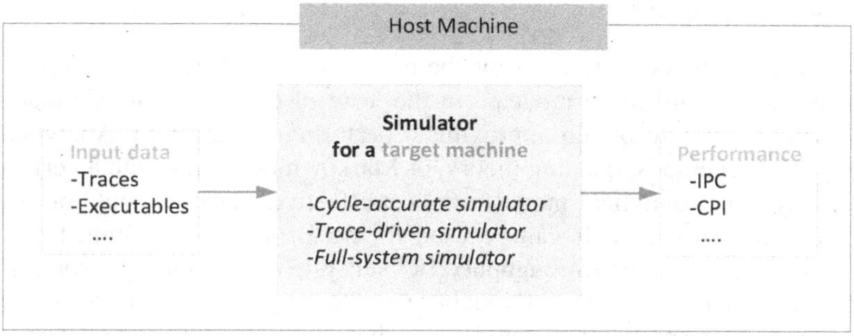

FIGURE 17.3 Simulator (for a target machine) executed on a host machine.

Depending on the objective of the simulators, they can be functional simulators or timing-based simulators. Depending on the input data type of the simulators, they can be trace-driven or execution-driven simulators. Also, simulators can be full-system simulators or component-level simulators based on the scope of simulations.

Functional simulators can be used to validate the functional correctness and can also be used for performance evaluation by adding additional timing information (e.g., clock cycle, etc.) for each component. The timing simulations have a longer simulation time than functional simulations.

In trace-driven simulations, input is modeled as a trace or sequence of information representing the instruction sequences or addresses that can be used in the target machine. For example, Cachesim5 is an example of a trace-driven cache simulator for evaluating cache performance only (Cmelik and Keppel, 1995). Trace-driven simulators are not timing simulators nor functional simulators and can be used for evaluating specific performance values of any components in the microprocessors, such as branch prediction, translation lookaside buffer (TLB), etc. Trace-driven simulators are straightforward, easy to design, and easy to debug. However, trace size can be big depending on the types of traces, and traces do not include all dynamic behaviors (e.g., mispredicted paths, etc.). There are many techniques to reduce the trace size such as skipping the warmup stage or sampling certain intervals.

Execution-driven simulators use the program executables (e.g., execution files) as input, not the trace. If the executables need input data and additional options, all information will be fed into the execution-driven simulator. Also, mispredicted paths can be accurately simulated which is different from trace-driven simulators.

SimpleScalar (Source: Simplescalar) is one of the execution-driven simulators. Execution-driven simulation is not easy to design and requires cycle-accurate design concepts, but it should be configurable and scalable (e.g., multi-core, etc.) like other simulators. Asim (Emer et al., 2002) and MicroLib (Perez et al., 2004)are examples of execution-driven simulators. The TRIMARAN is the execution-driven-simulator for VLIW (very long instruction word) architecture (Source: Trimaran).

Both trace-driven simulators and execution-driven simulators simulate the subsystem or the processor, which are not related to operating system (OS) activities and external devices. However, complete system simulators (or full-system simulators) include all the hardware components of the processor with the commercial operating system, memory subsystem, disks, network controllers, graphics controllers, etc. SimOS (Rosenblum et al., 1995) and Simics (Source: SIMICS) are examples of complete system simulators. SimOS models the MIPS instruction set architecture (ISA), while SIMICS models the SPARC ISA. Full-system simulators are difficult to design and take a long simulation time, but it is very accurate.

17.3.2 Analytical Models

Analytical performance models are used for the evaluation of relatively large computer systems. Since large systems are hard to be modeled in the form of cycle-accurate simulators, analytical modeling is normally used to obtain approximate performance numbers. Analytical models are based on probabilistic methods, queuing theory, or Markov models to create a performance model of the computer system. Analytical models are cost-effective, but they could not capture all the details. However, analytical models can be used for performance evaluation for larger systems focusing on overall performance (throughput), I/O subsystem, etc. There is some research work on analytical models such as Noonburg (Noonburg and Shen, 1997) with the Markov model, Sorin (Sorin et al., 1998) with the probabilistic modeling scheme, etc. Analytical models can also be used for performance modeling and evaluation at the early stage of microprocessor design.

17.4 PERFORMANCE METRICS

When evaluating the performance of microprocessors, choosing proper performance metrics is important. The metric can represent an overall performance or specific components' performance. Normally, a group of benchmark applications is used to evaluate the microprocessor or computer systems and obtain the performance metric values from simulations or measurements. The performance metric should provide one representative number, which can be aggregated from the individual benchmark, that can provide a relative performance among microprocessors. To find a central tendency of the metrics from individual benchmarks, mean values are used. The appropriate mean should be used for performance metrics among three kinds of means such as arithmetic mean, harmonic mean, and geometric mean. Arithmetic mean can be used for time-based metrics, harmonic mean for rate-based, and geometric mean is a valid measure of central tendency for ratios or dimensionless metrics.

Performance metrics that are popularly used in microprocessor design include CPI (cycles per instruction), IPC (instructions per cycle), MIPS (million instructions per second), MFLOPS (million floating-point operations per second), GOPS (giga [billion] operations per second), cache hit rate, branch prediction rate, TLB hit rate, MPKI (millions per kilo-instructions), etc. Depending on the number of cycles for specific instruction, a weighted mean should be used among different instruction types.

- **CPI**: CPI is the number of cycles per instruction of a benchmark suite. Several instruction types need a different number of cycles for execution.
- **IPC**: IPC is the number of instructions per cycle of a benchmark suite.
- **MIPS**: MIPS is the total instruction count in millions divided by the total execution time of a benchmark suite.
- **MFLOPS**: MFLOPS is the million floating-point operations per second that are used for the performance metric of the floating-point applications.
- **GOPS**: GOPS is the giga [billion] operations per second that are used for the performance metric of highly parallel computing applications such as deep learning applications. Even TOPS (tera operations per second) is also popularly used.
- **MPKI**: MPKI is the misses per kilo-instructions that are used for a metric of cache or branch prediction performance,

17.5 BENCHMARK FOR PERFORMANCE EVALUATION

A computer's performance should not be evaluated with a specific application because each application has a different workload and requires different hardware resources. A group of applications called a benchmark suite should be used for the performance evaluation of computers. The benchmarks should be representative of applications that are executed on computers in the real world, so it is hard to create benchmarks from tons of existing applications. An earlier version of benchmarks includes small kernels extracted from applications or synthetic programs that are based on existing applications. For example, Whetstone is a synthetic floating-point benchmark as the main workload extracted from several floating-point programs. Dhrystone is for integer applications. Dhrystone and Whetstone are simple programs and have been used for many years to measure the performance of computers and compare the performance among computers. However, both of them are not representing actual workloads considering real-world applications' complexity and compiler optimizations.

The Standard Performance Evaluation Cooperative (SPEC) (Source: SPEC) and the Transactions Processing Council (TPC) (Source: TPC) have made available several benchmark suites and guidelines for evaluating computer performance in 1988. Also, many other benchmarks have been released for embedded computing, server, mobile computing, network processors, etc.

17.5.1 CPU Benchmarks

SPEC CPU is an industry standard for CPU benchmarks. The SPEC CPU benchmark has a standardized suite of source code that comes from existing applications that are used in the real world. The SPEC CPU suite contains integer applications and floating-point applications. The SPEC benchmark suite contains several input data sets for each program - a large input set, test, and train.

Table 17.2 provides a list of the SPEC CPU 2017 benchmark suite (Source: SPEC CPU 2017). The SPEC CPU 2017 benchmark package contains SPEC's next-generation, industry-standardized, CPU-intensive suites for measuring and comparing compute-intensive performance, stressing a system's processor, memory subsystem, and compiler. SPEC designed these suites to provide a comparative measure of compute-intensive performance across the widest practical range of hardware using workloads developed from real user applications. The benchmarks are provided as source code (C, C++, or Fortran) and require the use of compiler commands as well as other commands via a shell or command prompt window. SPEC CPU 2017 also includes an optional metric for measuring energy consumption. The SPEC CPU 2017 benchmark package contains 43 benchmarks, organized into four suites:

- The SPECspeed 2017 Integer and SPECspeed 2017 Floating Point suites are used for comparing time for a computer to complete single tasks.
- The SPECrate® 2017 Integer and SPECrate® 2017 Floating Point suites measure the throughput or work per unit of time.

17.5.2 Embedded Computing Benchmark

The EDN Embedded Microprocessor Benchmark Consortium (EEMBC) released the performance benchmarks for processors in embedded applications in 1997. The EEMBC benchmarks (Source: EEMBC) include a suite of benchmarks designed to reflect each category of real-world embedded applications. The early version of EEMBC benchmarks has six categories including automotive/industrial, consumer, networking, office automation, and telecommunications markets. At the application level, the benchmarks include engine control, digital cameras, printers, cellular phones, modems, and similar devices with embedded microprocessors. Currently, EEMBC Benchmarks can be grouped into the following categories:

- Ultra-Low Power and Internet of Things
- Heterogeneous Compute
- Single-core Processor Performance
- Multi-core Processor Performance
- Phone and Tablet

TABLE 17.2
SPEC CPU Benchmark 2017

SPECrate 2017 Integer	SPECspeed 2017 Integer	Language	Application Area
500.perlbench_r	600.perlbench_s	C	Perlinterpreter
502.gcc_r	602.gcc_s	C	GNU C compiler
505.mcf_r	605.mcf_s	C	Route planning
520.omnetpp_r	620.omnetpp_s	C++	Discrete Event simulation
523.xalancbmk_r	623.xalancbmk_s	C++	XML to HTML conversion
525.x264_r	625.x264-s	C	Video compression
531.deepsjeng_r	631.deepsjeng_s	C++	Artifical Intelligence (Chess)
541.leela_r	641.leela_s	C++	Artifical Intelligence (Go)
548.exchange2_r	648.exchsnge2_s	Fortran	Artifical Intelligence (Sudoku)
557.xz_r	657.xz_s	C	General data compression
SPECrate 2017 Floating Point	**SPECspeed 2017 Floating Point**	**Language**	**Application Area**
503.bwaves_r	603.bwaves_s	Fortran	Explosion modeling
507.cactuBSSN_r	607.cactuBSSN_s	C++, C, Fortran	Physics: relativity
508.namd_r	-	C++	Molecular dynamics
510.parest_r	-	C++	Biomedical imaging
511.povray_r	-	C++,C	Ray tracing
519.lbm_r	619.lbm_s	C	Fluid dynamics
521.wrf_r	621.wrf_s	Fortan, C	Weather forecasting
526.blender_r	-	C++, C	3D reendering and animation
527.cam4_r	627.cam4_s	Fortran, C	Atmosphere modeling
-	628.pop2_s	Fortran, C	Wide-scale ocean modeling
538.imagick_r	638.imagick_s	C	Image manipulation
544.nab_r	644.nab_s	C	Molecular dynamics
549.fotonik3d_r	649.fotonik3d_s	Fortran	Computational Electromagnetics
554.roms_r	654.roms_s	Fortran	Regional ocean modeling

Source: SPEC CPU (2017).

Ultra-Low Power (ULP) and Internet of Things benchmark primarily focus on power and energy. The scores associated with the benchmarks are derived from measurements taken using the STMicroelectronics PowerShield (Source: PowerShield). The Heterogeneous Compute benchmark is EEMBC's original benchmark focused on single-core 8- and 16-bit embedded processors. These benchmarks aren't suitable for today's mixed-core heterogeneous platforms deployed for high-performance computation. The benchmarks in this category differ from single-core or symmetric multi-core performance benchmarks EEMBC has developed in the past, therefore they provide more sophisticated frameworks for asymmetric computing. Single-core Processor Performance is also the original EEMBC benchmark. These suites are still in popular use today for compiler testing, regressions in new embedded core RTL, and academic analysis. Symmetric Multi-core Processor Performance is built on the multi-instance test harness (MITH) which exploits the POSIX pthreads interface for testing both context-level and worker-level parallelism. Lastly, for Phone and Tablet, BrowsingBench™ measures browser performance with BrowsingBench, a collection of webpages loaded using a local Nginx server over a wired LAN connection.

Berkeley Design Technology, Inc. (BDTI) (Source: BDTI) has focused exclusively on digital signal processing. BDTI provides the industry standard BDTI Benchmarks™, which contain DSP routines such as FIR filter, IIR filter, FFT, dot product, and Viterbi decoder.

The MediaBench benchmark suite (Lee et al., 1997) consists of several applications belonging to image processing, communications, and DSP applications. Examples of applications that are included are JPEG, MPEG, GSM, G.721 Voice compression, Ghostscript, and ADPCM. JPEG is the compression program for images, MPEG involves encoding/decoding for video transmission, Ghostscript is an interpreter for the Postscript language, and ADPCM is adaptive differential pulse code modulation.

17.5.3 TPC Benchmarks

The Transaction Processing Council (TPC) released the benchmarks for transaction processing and database operations. The meaning of the transaction is a set of operations including disk accesses, operating system calls, database updates, or data transfer. The TPC produces benchmarks that measure transaction processing (TP) and database (DB) performance in terms of how many transactions a given system and database can perform per unit of time, which is the throughput, for example, transactions per second or transactions per minute. The TPC benchmarks can be classified into two categories: Online Transaction Processing (OLTP) and Decision Support Systems (DSSs). OLTP systems are used in day-to-day business operations (e.g., banks), while DSSs are primarily used for business analysis purposes, to understand business models and trends.

There are four benchmarks in TPC: TPC-C, TPC-H, TPC-R, and TPC-W. These benchmarks can be run with different data sizes or scale factors.

- TPC-C is an OLTP benchmark. It simulates a complete computing environment where a population of users executes transactions against a database.
- TPC-H is a DSS benchmark. A DSS database may consist of records of transactions from the previous several months of a company's operation, which can be analyzed to shape the future strategy. The TPC-H benchmark consists of a suite of business-oriented queries and concurrent data modifications.
- TPC-R is a DSS benchmark similar to TPC-H, but it allows additional optimizations based on advanced knowledge of the queries. It consists of a suite of business-oriented queries and concurrent data modifications.
- TPC-W is a transactional Web benchmark. The workload simulates the activities of a business-oriented transactional Web server in an electronic commerce environment.

17.5.4 Mobile Benchmarks

The market of mobile platforms is growing so quickly. The performance of mobile processors or mobile devices is getting attention from manufacturers and users as well. The Geekbench is a cross-platform utility for benchmarking the central processing unit of mobile computers. Geekbench (Source: Geekbench) began as a benchmark for Mac OS X and Windows and now Geekbench provides a benchmark score for Android devices as well. It uses a scoring system that separates single-core and multi-core performance and workloads that supposedly simulate real-world scenarios. The software benchmark is available for macOS, Windows, Linux, Android, and iOS.

The usefulness of Geekbench scores from earlier versions (up to Version 3) was heavily disputed because Geekbench combined disparate benchmarks into a single score. Later revisions starting with Geekbench 4 addressed these concerns by splitting integer, floating point, and crypto into sub-scores. Geekbench also provides a benchmark score for machine learning, deep learning, and artificial intelligence as Geekbench ML. There are several other mobile benchmarks, including AnTuTu benchmark (Source: Antutu), Android benchmark (Source: Android), etc.

17.5.6 OTHER BENCHMARKS

Among other benchmark suites, the network processor benchmark includes Commbench (Wolf and Franklin, 2000), NetBench (Memik et al., 2001), and NpBench (Lee and John, 2003). Scientific benchmark suites include SPLASH (Stanford Parallel Applications for Shared Memory), SPEC HPC 2002 and SPEC OMP, ASCI, and NAS. Java benchmarks include SPEC JVM98 and SPEC jbb2000, Java Grande, and VolanoMark. Other miscellaneous benchmark suites include 3DMark for 3D applications, BioBench for bioinformatics benchmarks, SYSmark for office productivity software, etc (Yi and Lilja, 2006).

17.6 SAMPLING SIMULATION POINTS

Performance evaluation and workload characterization are based on detailed simulations that take a long time (e.g., weeks or months) to complete the full execution of the benchmarks (e.g., SPEC CPU, etc.). It requires fast simulators and faster machines with parallel simulations, but the simulation times will be a bigger issue or critical bottleneck if your simulation needs sensitivity analysis or tradeoff analysis with different configurations and different input data.

Performance evaluation in the early simulation period needs to be excluded because all hardware resources are in cold states which means performance numbers (e.g., IPC, cache hit rates, etc.) will not be in a normal range of values. One of the simplest ways to reduce the simulation time is randomly picking a simulation period or multiple simulation periods, and performance should be evaluated with the picked simulation period(s) by skipping the cold states and unmeasured period(s). However, the accuracy will be varied depending on which simulation period(s) are sampled for performance evaluation.

As a more intelligent method to reduce the simulation time, SimPoint (Lau et al., 2005) is proposed and popularly used in industry and academia. Executing benchmarks show a small number of similar patterns, called phases, that are recurring in the full execution period. In SimPoint, representative phases are chosen and weighted appropriately to get overall performance. Simulating only a few simulation points can reduce the simulation time a lot.

The definition of phase in SimPoint is the recurring behavior of a program executing over time. As shown in Figure 17.4, the behavior of the CPI shows some repeated patterns. Each data point represents the average value for CPI over an interval (e.g., 10 million instructions). Based on the SimPoint paper, the behavior of all of the architecture metrics (branch prediction, cache misses, etc.) tends to show repeated patterns based on the characteristics of the application on the hardware resources.

The basic block vector (BBV), which is used in SimPoint, is a structure designed to capture information about how a program is changing behavior over time. A basic block is a section of code that is executed from start to end with one entry and one exit. The behavior of the program at a given time is directly related to the executing code, and basic block distributions provide this information as a signature. A BBV is a one-dimensional array, with one element in the array for each basic block in the program. By comparing the BBVs of two intervals, the similarity of the two intervals can be evaluated. If two intervals have similar BBVs, then we expect the performance of those two

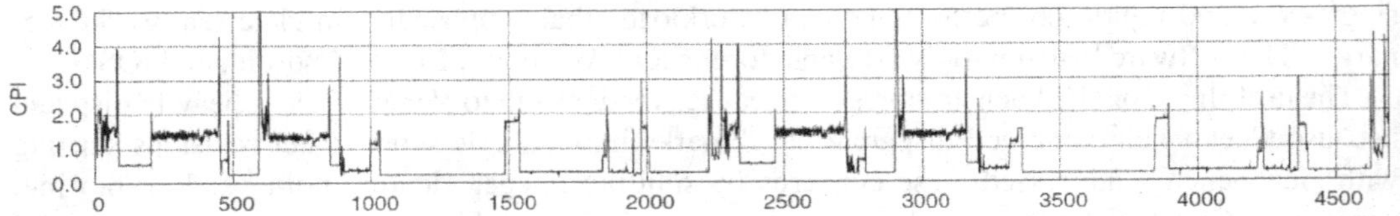

FIGURE 17.4 The behavior of the CPI from each interval of execution for gzip-166 at 10 million interval size. (Lau, 2005.)

intervals to be similar. There are several ways of comparing two vectors by finding the Euclidean distance or Manhattan distance. Also, the dimensionality of the frequency vector data needs to be reduced using random linear projection, which will speed up the clustering algorithms.

The goal of clustering is to divide a set of points into groups, or clusters, such that points within each cluster are similar to one another. The K-means algorithm (Jin et al., 2011) is a simple and efficient clustering algorithm, used in SimPoint to split program behavior into phases.

From each phase, we choose one representative interval that will be simulated to represent the behavior of the whole phase. To choose a representative, SimPoint picks the interval that is closest to the center of each cluster. SimPoint also gives a weight for each simulation point. With the weights and the detailed simulation results of each simulation point, a weighted average for the performance metric (e.g., CPI, miss rate, etc) will be computed. This weighted average of the simulation points gives an accurate representation of the complete execution of the program.

The final step in SimPoint is to combine the weighted simulation points to obtain an overall performance estimate for the program's execution. For example, if we have three simulation points and their weights are [35, 45, 20] and their CPIs are (2.2, 1.8, 1.1), then the weighted average of these points is $CPI=0.35*2.2+0.45*1.8+0.20*1.1=1.8$. The weighted average CPI is the estimate of the CPI for the full execution.

17.7 PERFORMANCE MONITORING COUNTERS

Performance monitoring counters (PMCs) in modern processors are built-in hardware counters to count certain events that take place at the CPU level, such as the number of cycles and instructions that a program executed, associated cache misses, access to off-chip memory, etc. The events that can be tracked with performance monitoring counters are usually simple behaviors. But if they are meaningfully combined, they can provide extremely useful insights into the behavior of a program. In the case of Intel Core CPUs (Source: Intel), processors have performance monitoring units (PMUs) that can be programmed to count many performance-related events. Generally, processors have two parts: core and uncore. The core part includes CPUs, registers, upper-level caches, TLBs, and instruction processing units, while the uncore part includes a last-level cache, memory controller, etc. Since there are hundreds of events in the core part, a PMU is supporting a logical core. Also, the uncore part has a PMU to monitor some events. For Intel CPUs, a core PMU has 2~4 programmable counters per logical core and 1~3 fixed-function counters per logical core. Some uncore and core programmable counters can be only programmable with certain types of events. The PMCs

TABLE 17.3
The Event from the PMCs

Type	Event
PERF_TYPE_HARDWARE	CPU Cycles, INSTRUCTIONS, BUS Cycles, CACHE References, CACHE Misses, BRANCH Instructions, BRANCH Misses
PERF_TYPE_HW_CACHE	L1D Prefetch Accesses, L1D Read Accesses, L1D Read Misses, L1D Write Accesses, L1D Write Misses, L1D Prefetch Misses, L1I Prefetch Accesses, L1I Read Accesses, L1I Read Misses, L1I Write Accesses, L1I Write Misses. L1I Prefetch Misses, LL Prefetch Accesses. LL Read Accesses, LL Read Misses, LL write Accesses, LL Write Misses, LL Prefetch Misses, DTLB Read Accesses, DTLB Read Misses, DTLB Prefetch Accesses, DTLB write Accesses, DTLB Write Misses, DTLB Prefetch Misses, ITLB Prefetch Accesses, ITLB Read Accesses, ITLB Read Misses, ITLB Write Accesses, ITLB Write Misses, BPU Read Accesses, BPU Read Misses, BPU Write Accesses, BPU Write Misses

Source: Pattee (2022).

can count numerous performance events when the microprocessor is executing some workloads. The event profiling data provides a good understanding of how applications are performing with all the hardware resources. The behavioral information can be used for tuning the performance.

Performance monitoring counters provide a low-overhead mechanism for collecting processor behaviors on the hardware resources in the microprocessor. The PMCs will count any of a set of performance events while the processor is running applications and the operating system. The performance events can be categorized into a few categories: program characterization, memory, pipeline stalls, branch prediction, and resource utilization. Table 17.3 shows some of the events that can be extracted from the PMCs.

The PMCs hardware is composed of two main components: event detectors and event counters. Event detectors can detect any architectural events (e.g., branch mispredictions). The performance event counters can count the number of events under certain conditions. By properly configuring the event detectors and event counters, various events under various conditions can be captured and used for performance tuning.

The PMC hardware can be used to create an event-based profile that locates a part of code that causes a performance bottleneck. The PMC hardware can be configured to interrupt the application after a specific number of performance events have occurred. The resulting event-based profile indicates the most frequently executed sections of code, called a hotspot, that caused the particular performance event.

Also, the PMCs and hardware can be used for tuning algorithms for different processor architectures. For example, algorithms developed for one processor do not perform well on different versions of processors. Since the execution behavior of the algorithms in the processor is kind of a black box, it is unclear how to tune the algorithm for a new processor architecture. However, the behaviors of the hardware resources collected from PMC hardware can provide clues to resolve the performance issues by changing the algorithm to reduce memory traffics, increase cache utilization, etc. Recently, PMCs have been widely used in various domains including system power estimation, firmware modification, and malware detection (Pattee et al., 2022). One of the primary drawbacks of using PMCs is the limited number of monitoring counters in a processor.

REFERENCES

Android. Benchmark. https://www.androidbenchmark.net/

AnTuTu. Benchmark. https://www.antutu.com/en/index.htm

BDTI. https://www.bdti.com/

Cmelik, B., & Keppel, D. (1995). Shade: A fast instruction-set simulator for execution profiling. *ACM SIGMETRICS Performance Evaluation Review*, 22(1), 128–137.

EEMBC. https://www.eembc.org

Emer, J., Ahuja, P., Borch, E., Klauser, A., Luk, C. K., Manne, S., …, & Juan, T. (2002). ASIM: A performance model framework. *IEEE Computer*, 35(2), 68–76.

Geekbench. https://www.geekbench.com/

Intel. Nehalem Core PMU. https://www.intel.com/content/dam/develop/external/us/en/documents/30320-nehalem-pmu-programming-guide-core.pdf

Jin, X., & Han, J. (2011). K-means clustering. In C. Sammut, & G. I. Webb (eds.), *Encyclopedia of Machine Learning*. Boston, MA: Springer.

Lau, J., Sampson, J., Perelman, E., Hamerly, G., & Calder, B. (2005). The strong correlation between code signatures and performance. In *IEEE International Symposium on Performance Analysis of Systems and Software*.

Lee, B., & John, L.K. (2003). NpBench: A benchmark suite for control plane and data plane applications for network processors. In *Proc. Int'l Conf. Computer Design*.

Lee, C., Potkonjak, M., & Smith, W. (1997). MediaBench: A tool for evaluating and synthesizing multimedia and communication systems. In *Proceedings of the 30th International Symposium on Microarchitecture.*

Memik, G., Mangione-Smith, W., & Hu, W. (2001). NetBench: A benchmarking suite for network processors. In *Proc. Int'l Conf. Computer-Aided Design (ICCAD).*

Noonburg, D. B., & Shen, J. P. (1997). A framework for statistical modeling of superscalar processor performance. In *Proc. 3rd Int. Symp. High Performance Computer Architecture (HPCA)*.

Pattee, J., Anik, S. M., & Lee, B. K. (2022). Performance monitoring counter based intelligent malware detection and design alternatives. *IEEE Access*, 10, 28685–28692.

Perez, D., Mouchard, G., & Temam, O. (2004) MicroLib: A case for quantitative comparison of microarchitecture mechanisms, In *Proc. MICRO*.

PowerShield. Firmware. https://www.st.com/resource/en/user_manual/um2269-getting-started-with-power-shield-firmware-stmicroelectronics.pdf

Rosenblum, M., Herrod, S. A., Witchel, E., & Gupta, A. (1995). Complete computer system simulation: The SimOS approach. *IEEE Parallel & Distributed Technology: Systems & Applications*, 3(4), 34–43.

SIMICS. Simulator, VIRTUTECH. https://www.virtutech.com.

SimpleScalar. Simulator Suite. https://www.cs.wisc.edu/~mscalar/simplescalar.html.

Sorin, D. J., Pai, V., Adve, S., Vernon, M., & Wood, D. (1998). Analytic evaluation of shared memory systems with ILP processors. In *Proc. Int. Symp. Computer Architecture*.

SPEC CPU 2017. Benchmarks. https://www.spec.org/cpu2017

SPEC. Benchmarks. https://www.spec.org

Transactions Processing Council. https://www.tpc.org

Trimaran. Home page. https://www.trimaran.org.

Wolf, T., & Franklin, M. (2000). Commbench—A telecommunications benchmark for network processors. In *Proc. Int'l Symp. Performance Analysis of Systems and Software*.

Yi, J. J., & Lilja, D. J. (2006). Simulation of computer architectures: Simulators, benchmarks, methodologies, and recommendations. *IEEE Transactions on Computers*, 55(3), 268–280.

18 Optimizing Complex Systems Operation

William Ross

18.1 INTRODUCTION TO METHODS FOR OPTIMIZING COMPLEX SYSTEMS

Complex systems come with a high initial investment cost so maximizing the useful life and reducing sustainment costs of a complex system are key objectives for the complex system's program managers. Many of the Department of Defense's major weapon systems are examples of high investment cost for highly technical systems that remain in the Department's inventory for 50 or more years. Granted, some sub-systems of a complex weapon system platform are upgraded or replaced driven by technology advances, obsolescence, and threat changes while other systems must be sustained through the life of the platform. Commercial airlines, power plants, and communication network systems are other examples of complex systems with high initial investment costs and long life cycles.

Complex systems begin with an analysis of alternative solutions where the cost and benefits of the alternative solutions are analyzed via the life cycle cost (LCC) model. The LCC model spans the development, manufacture, and sustainment of the system for its entire life cycle. Many of the methods and techniques for optimizing complex systems are logistical in nature, centered during the sustainment phase, and relate to some of the other late chapter topics such as "On-Board Health Management", "Automated Test", "Life Cycle Cost Modeling", "System Support", and "System Maintenance Planning". These topics with their associated actions are critical in the system development, fielding, and sustainment phases of systems to help ensure that proper logistics resources are put in place to efficiently sustain the system during its life cycle. A complex system is optimized when it provides or exceeds the expected system reliability, readiness, and performance at or below the expected cost.

This chapter is tutorial in nature providing an overview of system optimizing methods and techniques targeted for senior leadership and program managers but is also useful for electrical engineers, computer engineers, systems engineers, software engineers, as well as logisticians.

18.2 WHEN ARE THESE OPTIMIZING METHODS APPLIED?

Complex systems need to be designed upfront for testability, maintainability, and supportability for LCC reasons and to achieve the desired system readiness or availability. However, these critical attributes may not be optimized in the system's initial design phase, due to unknowns or sometimes they are traded off due to cost, schedule risks, or other reasons. It is important that these attributes be optimally addressed in the upfront design phase as it is much more costly to address the testability, maintainability, and supportability attributes after a system has been baselined and fielded.

In the system's development phase, designs are based on analysis –analysis of data from similar systems or analytical modeling techniques. During the system's sustainment phase, real performance, maintenance, and failure data become available for analysis. The real data and associated analysis will justify adjustments to the initial design and maintenance planning, leading to a better optimized system. At this phase of the system's life cycle, the testability, maintainability, and supportability attributes can be improved through engineering hardware, software, and firmware

DOI: 10.1201/9781003188377-21

changes and through logistics process adjustments. It is important that the historical maintenance and failure data are continually collected and analyzed throughout the complex system's life cycle, as sub-systems often degrade with age requiring the system's design and logistics processes to be adjusted. There are several other chapters that address processes for maintenance data collection and analysis to include automated decision techniques. This chapter summarizes some of the logistics methods that have been applied in recent years during the development and sustainment phases of a system's life cycle to help optimize the system's operation.

18.3 METHODS FOR OPTIMIZING COMPLEX SYSTEMS

18.3.1 System Development Phase

18.3.1.1 Modeling a Complex System

Ever increasing complexity, system-of-systems, compressed development times, system security, long life cycles, and other factors are pushing current methods of system engineering complex systems into obsolescence. Given today's greater computing power and modern digital tools, complex systems (and some less-complex systems) are beginning to see the use of Model Base System Engineering (MBSE). MBSE uses modeling techniques not only for the traditional requirements and design phases of the system but continually through the design validation and verification phases and the later life cycle phase. This digital model-based methodology focuses on information exchange within the set of system models rather than the traditional document-based information exchange. This approach eases some of the traditional diagnostics, cybersecurity, and other challenges upfront during the system design phase but also through the system life cycle phase. The information exchange greatly eases the engineering and logistics change process as well as the status information and analysis processes. This information exchange characteristic will exist throughout the system's life cycle, optimizing not only the engineering design processes but also the long-term sustainment processes. There will be greater use of MBSE in complex systems over the coming decades.

18.3.1.2 System Requirements and Performance Specification

While a complex system must meet its intended operational requirements and performance (as documented by its specification), the specification needs to include considerations for the long life cycle of the system. Design for testability, maintainability, and supportability to include Built-in Test (BIT) software/firmware, BIT Equipment (BITE) instrumentation, sensors, and data ports for capturing performance and failure data are also important to optimize a system. These are the tools to provide system health and prognostic monitoring and analysis. It is critical that these and other logistics-related design factors are used to set the design constraints and are captured in the design Requirements and Performance Specifications for inclusion in the initial design. Logisticians need to participate during the design phase and must be allowed to influence desired supportability characteristics during design reviews to achieve the operational requirements for a developmental system and be able to communicate the requirements to the design team.

These critical requirements are sometimes traded off by program manager given cost, schedule, or performance challenges. When this happens, it is often difficult to achieve and implement these requirements during the sustainment phase. However, there are situations where these optimizing techniques are traded off for good reason. For example, on defense aviation systems, weight is critical for performance so adding sensors or BITE that add weight could be a nonstarter. In these situations, the key to optimizing a system is to capture whatever data might be available (BIT, sensor, etc.) and transfer that information offline for analysis and use by the sustainment engineering and logistics teams as well as the technicians at the various levels of maintenance.

18.3.1.3 Maintenance Planning and Other Logistics for Long-Term Sustainment

You may have heard the expression "design then throw it over the fence to logistics". Unfortunately, this is a practice that still goes on sometimes today where the engineers design the system and then pass it along to the logisticians to figure out how to maintain and support it. Design interface is the process where the engineers and logisticians must work together from the initial concept to final design to optimize the maintenance planning and the long-term sustainment, driving design to optimize future sustainment.

Key to the success of a complex system with long life cycle is a robust upfront maintenance plan and other logistics plans. These logistics plans must be more than snapshot plans. The upfront logistics planning must allow for an evolving maintenance approach as the system ages. Structure needs to be put in place to continually capture maintenance and failure data, analyze the data, and to make design and logistics process changes. Applying the techniques addressed in this chapter along with proper team structure, program reviews, and milestone checkpoints can help assure the proper attention to long-term sustainment in the upfront design.

18.3.1.4 Design for Resilience

Innovation has been an effective formula for providing better and successful complex systems since the first industrial revolution in the 1700s. However, complex systems are often vulnerable because they could be brittle, often not well maintained, and are not resilient, meaning that they are not designed to recover from a failure or an outside attack. "Resilience" is the ability of a system to take a failure or an attack and continue operating. Resilient systems are well maintained (hardware and software), fixed as or before they fail with predictive maintenance, and are progressively strengthened at any weak points. Performance Specifications need to include built-in resilience as a first-order capability requirement. This will drive changes to engineering and logistics methods, unprecedented transparency, the use of predictive maintenance, and advanced techniques for system security. Vendors need to be incentivized for resilience, and resilient designs need continual active maintenance, especially for the software delivered by vendors.

Cyber-attacks to complex systems hardware, software, and data continue to evolve and are growing in volume, technique, and sophistication. Traditional reaction approaches are no longer satisfactory. Proactive and adaptive resilient approaches are necessary today. This is especially true for complex systems that are likely connected systems. Even if a complex system is not connected, cybersecurity strategies need to be addressed as systems can be compromised during maintenance and other situations.

18.3.2 System Sustainment Phase

18.3.2.1 System Design Changes

Optimizing a complex system after the system is fielded often requires engineering or logistics changes to the system. The engineering changes could be hardware, software, or firmware changes. A formal Configuration Management and Engineering Change process must be put in place. The changes are driven by the analysis of system performance and maintenance failure history while leveraging health management and prognostics tools. Changes can drive performance improvements, reliability improvements, and longevity of sub-systems but also could include BIT or BITE changes or adding sensors to improve diagnostics of the system's operation.

18.3.2.2 Reliability Centered Maintenance

Key to reliability improvements of complex systems is the establishment of a reliability centered maintenance (RCM) program, an engineering framework and strategy for optimizing reliability at the lowest cost. The RCM framework includes identifying up front, the operating context of the complex system, and preparing the Failure Mode Effects and Critical Analysis (FMECA). This analysis helps determine the appropriate maintenance task for each failure mode. Failure Reporting, Analysis, and Corrective Action System (FRACAS) is the process for continuous review and analysis of failures with corrective actions. A Failure Review Board (FRB), comprised of senior management members, reviews the FRACAS recommendations, sets priorities, establishes schedules, assigns responsibility, and authorizes funding to initiate reliability improvements to the system.

Given the "bathtub" effect of system failures – a high rate of failures early in a system's fielding and again late in the system's life cycle, as the system becomes tired or aged – it is critical that a formal Engineering Change process and the RCM framework be in place at the initial fielding of the system to address these early failures. Utilizing a robust process for failure analysis, root cause of failures, and implementing improved design changes as these early failures occur is an effective method for driving down future LCC and improving system readiness. Reliability improvements are often thought of as hardware improvements, but the failure and root cause analysis could lead to software/firmware, maintenance procedures, training, or other logistics element changes. Also, throughout the system's life cycle, there needs to be proactive monitoring and mitigation of sub-system and component obsolescence, especially for electronics systems. The primary metric in dealing with obsolescence is *time* – where in the systems lifecycle does the obsolescence occur? The obsolescence will drive program managers to decisions of "replacement", "life-of-type buy", or to "live with it" for the rest of the system's life cycle.

Driven by degrading capability, in 2019 Naval Aviation in the Department of Defense adopted the Reliability Control Board (RCB). The RCB identified, prioritized, executed, and measured the effectiveness of improvements to aviation weapon system readiness by ranking degraders that prevent weapon systems from being Mission Capable. This is a data-driven process that provides a process to place accountability with the responsible organization for action to address the root cause of the degraders and to improve their reliability.

18.3.2.3 System or Sub-System Redundancy

System redundancy is a traditional design technique to keep a system optimally operational. Redundancy can be applied at the system level or sub-system level. NASA is famous for applying this technique to space vehicles where maintenance cannot easily be performed. If not addressed in the system design phase, an engineering change to improve the failing component or to add redundancy in troublesome areas may be required. Redundancy can be a design characteristic to help achieve the resilient design noted prior.

For a complex system to perform optimally, the concept of redundancy needs to go beyond the system or sub-system itself. Complex systems will have complex support tools. Redundance also needs to be considered in the maintenance support area, eliminating single points of failure of maintenance tools, test equipment, and test programs. For example, complex electronics systems require sophisticated Automatic Test Equipment (ATE) that is used at the factory, at the depot, and at other maintenance support levels. ATE are complex systems themselves needing to be four times more accurate than the system supported and costing millions of dollars each. The complex system's test program that is run by the ATE can also be very complex with thousands of lines of code and will include the unique hardware needed to interface the system to the ATE. A system being tested with its test program on an ATE can take a few minutes for a simple system or sub-system or many hours for a more complex system or sub-system. The system's test program needs to be run at least twice – first to diagnose and find the failure(s), and again to verify that the system is fully operational. With one ATE and one complex system test program at a maintenance site, there is a single point of failure if either of these two fails. Redundance in the complex system and in its logistics material and processes drive higher initial costs but can provide the desired or required system readiness.

18.3.3 Other Opportunities to Optimize a System

Noted here are some other examples of optimizing techniques that can be applied during either the development or sustainment phases of a complex system.

18.3.3.1 "Directed Test" and "Most Likely Failure"

The example of automatic test equipment (ATE) was noted as a critical manufacturing and maintaining tool for complex electronics systems. Complex systems tested with ATE need to run the test program at least two times and complex systems could have system test programs that run for hours for each test program run time. Some complex defense aviation systems run for ten or more hours each run time. By optimizing the manufacturing and maintenance tools and processes will in turn contribute to optimizing the overall system.

The concept of "Directed Test" can save a significant portion of the initial diagnostics run time by being directed to where in the test program the failure most likely occurred. To do this, the system under test's Built-in Test (BIT) failure code is needed by the ATE. The test program is designed for convenient break in points (maybe each 100 lines of code) where the ATE directs the test program where to start diagnostics, saving diagnostics run time. After system failure is repaired, the system will be run again, end-to-end, to verify that the repair was successful. This "Directed Test" technique has proven to save maintenance hours, but the reduced run times could also save on the number of ATE and system test programs required at a maintenance site.

At the end of the diagnostics run, the ATE can report the target failed card(s) or component(s). However, with complex systems, the ATE is often not able to determine a specific card or component but only an ambiguity group of cards or components. With a reasoner analyzing the systems maintenance failure history data, the ATE can recommend to the operator, the "Most Likely Failure" card or component information with the historical percentage of times that the failure was that card or component. Maintenance failure and test results data need to be continually collected and provided to the reasoner for analysis. This reasoning technique has proven to help optimize maintenance by targeting the actual failed card or component with the first diagnostics test program run. This technique provides a secondary savings by eliminating the need to test and verify as good, the group of good cards or components that otherwise would have been removed and replaced as part of an ambiguity group.

18.3.3.2 Vertical Test Integration

Generally, complex systems today are electronics intensive and are tested multiple times for various reasons. Electronic systems require testing during and at the end of the manufacturing process for system acceptance. They are also tested at the operational system (self-test or with external test equipment) to verify operation and to diagnose failures, and again at the depot and other maintenance levels to further diagnose and repair failures. Historically, the factory and each of the levels of maintenance or support have developed their own unique test equipment that brings unique hardware and software development cost with associated long-term sustainment costs for each of the new test systems. There are a number of reasons for this including: different funding sources; different organizational responsibilities; timing differences; and technical challenges.

Vertical Test Integration is the concept of integrating the factory, field, and depot testing of a system. The optimum solution is to use the same test system (ATE or manual test equipment) starting with the factory and vertically using the same test equipment at the depot and the field level. Only one test system type needs to be acquired and sustained, and the initial factory test program can be re-used as is or adjusted for the testing needed at the specific maintenance level.

An alternative to re-using the same test equipment hardware could be to apply test program software standards where the same test program software is able to operate on different test equipment hardware sets. Different test equipment hardware sets may already exist at each maintenance site making this a viable approach. This approach allows vertical transfer of the test program but will require sustainment of multiple test equipment hardware sets.

Vertical Test Integration offers several other optimizing characteristics. Historically, there has been a problem with failures detected at one maintenance level that cannot be duplicated at the next higher level of maintenance. This is commonly referred to as "Can Not Duplicate" or CND. Often this is caused by maintenance levels using different hardware and software to perform similar tests. Integrating various maintenance levels with the same or very similar test equipment hardware and software can eliminate or reduce the CNDs.

Another characteristic of vertical test integration is the ability to move test data results from the lower maintenance level to the higher maintenance levels. The test results data could be used at the higher maintenance levels for directed test and most likely failure concepts with ATE, or simply for a technician's knowledge.

18.3.3.3 Digital Model and Test Programs

More and more systems are providing a Digital Model, Digital Twin or Model-Based System Engineering (MBSE). These digital models can provide a number of benefits that help optimize systems as noted previously. Another benefit is the ability to leverage the system's digital model to create the system's initial factory lower-level test programs as well as the system's acceptance test program. Much of the data needed for these test programs are in the digital model and tools are available that greatly simplify the test program development process. The factory test programs can then be modified as needed to be re-used at the depot and field levels of maintenance.

18.3.3.4 Serial Number Tracking

If a system or sub-system is marked with a specific item serial number or other specific reference, the opportunity exists to track the maintenance history of each serialized item. This analysis process can identify "Bad Actors" or components that exhibit repeated maintenance actions. Through serial number tracking and analysis, Bad Actors can be completely removed from the system and the logistics process, redesigned, or refurbished to improve their performance.

18.3.3.5 Other Logistics Considerations

A technique to help reduce maintenance down time of complex systems is to pre-position spare parts locally at the system along with the tools, labor, and procedures. Also, the defense department has increased the use of Performance Based Logistics (PBL) contracts with vendors for repairable components. The long-term (seven or more years) PBL contract optimizes parts availability at a minimum cost to the government. Improving the reliability of the repairable components is a natural incentive for the PBL contractor to reduce its future costs under the fixed price contract. While requiring strict performance, the vendor is permitted to tradeoff decisions concerning cost, useful life of a component, reliability improvements, quantity of spares retained, etc. The vendor can then collect and analyze the necessary data to make educated decisions for reliability improvements. The PBL contractor will often employ some of the reliability centered maintenance techniques noted previously.

3D Printing or additive manufacturing capabilities are ever increasing with systems providing material and procedures to leverage these technologies to optimize systems. These technologies provide the ability to fast prototype new designs or changes and to produce production parts. Increasingly, system programs are providing equipment and procedures locally at the system and authorizing the local creation of spare parts with Additive Manufacturing technologies, accelerating parts for maintenance.

18.3.4 Final Notes

Although not a comprehensive list of techniques or methods used to optimize complex systems, this chapter provided a summary of some of the more recent successful techniques. Complex systems start with consideration of alternative solutions, a LCC analysis, a performance specification, and logistics considerations. The system program tailors its requirements to meet cost, schedule and performance objectives while optimizing to provide or exceed the expected system reliability, readiness, and performance. Optimizing complex systems is a continuous process by analyzing historical operational and maintenance data that become available during the sustainment phase.

Part 4

System Testing

19 Types of Testing

Michael Seavey

19.1 WHAT IS A TEST?

A test is a process of measuring either a property of, or the performance of, a component, assembly, subassembly, subsystem, product, or system.

Take this simple example of a product, its specification, and test requirement for example (Figure 19.1):

The "Test" (which will be an implementation of the test requirement) for the receptacle may consist of hook-up instructions, performing the measurement, evaluating the measurement results, and if necessary, proceeding to a diagnostic path to determine the fault.

The "Test" is an example implementation of a test requirement that verifies the specifications of the receptacle (in this case, the product).

From the IEEE Standards definition perspective, a
Hardware Test (with or without Software/Firmware) is:

A. An action or group of actions that are performed on a unit under test (UUT) to evaluate its parameter(s) or characteristic(s). *(Derived from IEEE Std 771).*
B. An observed activity that may be caused to occur (e.g., stimulus-response) in order to obtain information about the behavior of a test subject. Derived from IEEE Std. 1671–2010 Standard for Automatic Test Markup Language (ATML) for Exchanging Automatic Test Equipment and Test Information via eXtensible Markup Language.
C. A set of stimuli, either applied or known, combined with a set of observed responses and criteria for comparing these responses to a known standard. *IEEE Standard for Artificial System Intelligence Diagnostic Exchange Data and Service Tie to All Test Environments (AI-ESTATE).*

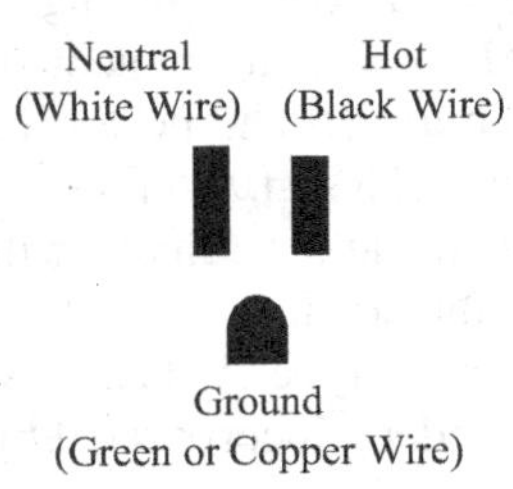

Product:
NEMA 5-15R
Receptacle
The standard 15-ampere electric receptacle (outlet) found in the United States

Specification:
1) Two-pole receptacle with a ground (meaning two sockets transmit electricity)
2) Current rating = up to 15 Amperes (A)
3) Voltage rating = up to 125 Volts Alternating Current (VAC)

Test Requirement:
1) Verify that the Voltage, as measured between the hot and neutral, is between 100 and 125 VAC.
1a) If the measured voltage is less than 100 VAC, first verify power is being applied from the breaker panel to the receptacle.

FIGURE 19.1 Example product, specification and test requirement.

DOI: 10.1201/9781003188377-23

19.2 WHAT ARE TYPES OF TESTS?

Throughout the manufacturing process, the various components, assemblies, subassemblies, subsystems, systems, and products are tested. Various types of testing are utilized to execute tests; from the building of the first assemblies through the assembly of the completed system or product.

The first time a system or product is usually tested is during the manufacturing process. The last time a system or product is usually tested is shortly before it will no longer be in service (due to obsolescence of components, damage, or advances in technology that will result in the replacement of the entire system/product).

Types of testing are:

a. **Alpha:** Alpha testing is a type of acceptance test carried out by the developer before a system/product is available for Beta testing.
b. **Beta:** Beta testing is a type of acceptance test carried out by a party unconnected with its development when the system/product is in the final stages of its development.
c. **Built-In:** Built-in test is a mechanism designed into the system/product that permits the system/product to test itself.
d. **Burn-In:** Burn-in is a testing practice utilized to detect early failures in a system/product by exercising the system/product under extreme operating conditions (such as elevated temperatures or voltages), over an extended time period (hours to days).
e. **Compatibility:** Compatibility testing of a subsystem or subassembly determines the degree that it can be used within a system/product, without modification.

   There are two types of compatibility tests:
   1. **Backward compatibility** is the testing of older versions of a subsystem or subassembly to verify it will operate in a newer system/product.
   2. **Forward compatibility** is the testing of a subsystem or subassembly to verify it will operate in upcoming or new versions of the system/product.
f. **Continuous Monitoring:** Sampling the state of a continuous stream of live data, either continuously or at a sample interval of one second or less, from which a real-time identification of problems can be determined.
g. **Destructive:** Destructive testing is a technique where the system/product is made to fail in an uncontrolled manner in order to find the point of failure.
h. **Diagnostic:** Diagnostics are programs written for the express purpose of examining the state, or locating problems with the hardware, or the software running on that hardware.
i. **EMI/EMC:** Electromagnetic Interference (EMI) is a measurement of the interference caused by an electromagnetic disturbance, which affects the performance of a product. Electromagnetic Compatibility (EMC) is a measure of a system/products ability to operate as intended in its shared operating environment, while not affecting the ability of other products in the shared environment to operate as intended.
j. **Functional:** Functional testing of a system/product is a type of black-box testing that bases its test cases on the specifications of the system/product; incorporated by supplying inputs and measuring outputs. There is no consideration of the internal structure of the system/product.

   Functional testing evaluates the compliance of a system/product with specified design document(s) or specification(s).
k. **Non-Destructive Inspection (NDI):** NDI is an evaluation of the properties of a material, component, or system without causing damage. For example, electronically testing for hairline fractures on an airplane wing without making physical contact with the wing.

l. **Nuclear:** Nuclear weapons tests are experiments carried out to determine a nuclear weapons' effectiveness and the amount of energy released when detonated.

m. **Self-Test (or Built-in-Self-Test (BIST)):** A self-test procedure within a system/product provides a means by which the system/product can execute the procedure to help detect subsystem degradation or identify other issues indicative of pending failure or unacceptable performance. Self-diagnostics allow the system/product to flag questionable data or communicate a need for an operator or field technician.

Embedded systems are good candidates for self-test implementations because they ordinarily have the processing power, memory, data conversion, and computational capabilities.

19.3 WHAT ARE TYPES OF TESTING?

There are three testing phases of a product or system's overall lifecycle: (a) system verification, (b) system testing, and (c) system maintenance.

a. **System Verification Testing (Also Known as Qualification Testing)**

System verification testing verifies the design and manufacturing process.

System verification measures include:

- Testing of the system hardware under conditions simulating expected real-life conditions.
- Testing of the systems software to ensure that the software performs as intended.
- Ensuring the system documentation is adequate and complete.
- Verifying that the system is capable of performing under expected normal conditions (abnormal conditions may be included).

b. **System Testing**

System testing is typically more detailed and thorough than system verification. System testing is required to ensure that every subsystem, subassembly, and assembly of the system is performing in accordance with the specified system requirements.

System testing measures include:

- Developing test criteria.
- Examining and reviewing all software and firmware.
- Applying functional tests to determine if the system meets the test criteria.
- Diagnosing faults in the event a functional test fails to meet the test criteria.
- Conducting tests in a variety of real-life conditions.
- Conducting tests over an extended period of time.
- Conducting load tests (exceeding conditions that are normally expected).

c. **System Maintenance**

After a system has been purchased and in use it should continue to be maintained. Maintenance routines will vary depending on the type and complexity of the technology.

System maintenance includes:

- Maintenance schedules (recommendations by the manufacturer may be available).
- Maintenance plans and procedures.
- Monitoring of the system (including feedback from users) to ensure that maintenance needs are being identified, addressed, and implemented.
- When modifications are made to a system, due to maintenance (or an upgrade), it may be necessary to initiate further system testing.

Each of these three phases utilizes one or more of the following testing types:

a. **Acceptance:** A test or set of tests to demonstrate the degree of compliance of a system or product with the purchaser's technical and/or contractual requirements.
b. **Black Box:** Black-box testing is the functional testing of a system or product (without knowledge of internal functionality). A tester provides an input and observes the output generated.
c. **Design Verification:** The process of evaluating whether or not the requirements of the design architecture are traceable to the verified functional architecture and satisfy the validated requirement baseline.
d. **Environmental:** The measurement and evaluation of the performance of a system or product subjected to environmental conditions such as temperature, humidity, and vibration.
e. **Environmental Stress Screening (ESS):** Testing of the usage of the system or product by simulating environmental conditions.
f. **Gray Box:** Gray Box refers to testing with limited knowledge of the internal workings. Gray box testing uses limited information to identify strengths and weaknesses.
g. **Hardware in-the-Loop:** A technique where real signals from a system or product (a radar for example) interface with a test system that is simulating reality (an aircraft for example). In this example, making the radar think it is in the aircraft.
h. **Hot Mock-Up:** A hot mock-up consists of all subsystems of a system or product (the subsystems are known as "gold" units) often constructed with the subsystems fully exposed in order to facilitate subassembly and component access and probing. The subassembly "being tested" replaces the equivalent "gold" unit in the test setup. The system or product is then tested.
i. **Integration:** Integration testing consists of the following: functional testing, bench testing, fault insertion, and design verification testing.
j. **Manufacturing Test:** Verification that the system or product passes key specifications before it leaves the factory. It is often the final step before the system or product is packed and shipped.
k. **Parallel:** Parallel testing is the process of executing multiple tests on different device combinations simultaneously. The goal of parallel testing is to resolve test time constraints.
l. **Product Validation:** Determines that the design of the system or product will meet its specifications and requirements. A ringing out of the design; by sweeping and cycling through various stimuli (input voltage limits, load limits) and environmental (temperature, humidity, and vibration) parameters to determine how the design will perform in the world. Product validation units are not sold as they have been exercised to the point of failure.
m. **Regression:** Regression testing is re-running functional tests on a system or product to ensure that the previously developed and tested system or product still performs after a change to the system or product. Changes that may require regression testing include bug fixes, software enhancements, configuration changes, and the substitution of electronic components.
n. **Reliability:** Reliability testing determines the probability that a system or product will perform its intended functions without failure, within design parameters, under specific operating conditions, and for a specific period of time. The probability time should be equal to or greater than the reliability specification of the system or product and/or the contractual requirement(s).
o. **Serial:** Serial testing is the process of performing time-sequential tests on a component, assembly, subassembly, subsystem, system, or product.
p. **Stress:** Stress testing is the process of testing beyond normal operational capacity, often to a breaking point, in order to observe the results, to determine the stability of the system or product. Stress testing determines modes of failure (how the device may fail).

q. **System:** System testing is the process of testing an integrated hardware and software system or product to verify and validate whether the system or product meets its design and/or contractual objectives.
r. **Unit:** Unit testing is the process of testing where individual subassemblies are isolated and tested to evaluate their compliance with specified requirements and that they behave as intended. Unit testing techniques include: Black Box Testing, White Box Testing, and Gray Box Testing.
s. **Verification:** The process of determining whether the requirements for a system, product, or subsystem are complete and correct. The products of each development phase fulfill the requirements or conditions imposed by the previous phase, and the final system or subsystem complies with specified requirements.
t. **Vibration and Shock:** Involves applying vibrations at various frequencies to a system or product, while shock tests involve subjecting a system or product to sudden, short and high levels of force.
u. **White Box**: White-box testing is a method of testing internal structures or workings of a system or product, as opposed to its functionality.

19.4 WHAT IS BEING TESTED?

Commonly tested systems or products include:

- Aircraft parts (landing gear, altimeters, etc.)
- Medical devices (ultrasound machines, X-ray machines, etc.)
- Military hardware (aircraft, tanks, missiles, etc.)
- Automotive parts (fuel pumps, electronic throttle controls, etc.)
- Train parts (diesel engines, AC alternators, brakes, etc.)
- Industrial equipment (HVAC systems, AC generators, turbines, etc.)
- Industrial machinery components (bearings, motors, actuators, etc.)

The following figure depicts the elements of each of these systems or products:

As depicted in Figure 19.2, what is tested can be:

a. **A System (or Product):** A collection of interacting components organized to accomplish a specific function or set of functions within a specific environment. A complete system or product includes all of the associated equipment, facilities, material, computer programs,

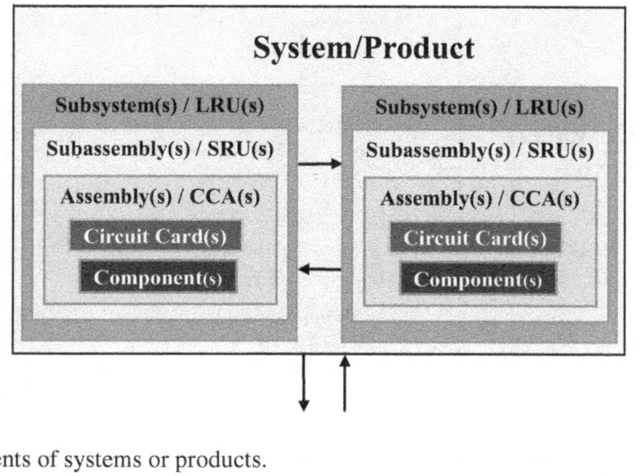

FIGURE 19.2 Elements of systems or products.

firmware, technical documentation, services, and personnel required for operations and support to the degree necessary for self-sufficient use in its intended environment.

b. **A Subsystem (or Line Replaceable Unit – LRU):** An element of the system or product's substructure that is a subordinate element of the system or product and is comprised of one or more subassemblies. The subsystem is replaceable as a whole; however, subassemblies are individually replaceable.
c. **A Subassembly (or Shop Replaceable Unit – SRU):** Two or more components that form a portion of a subsystem, system, or product. The subassembly is replaceable as a whole; however, components are individually replaceable.
d. **An Assembly (or Circuit Card Assembly – CCA):** A number of basic parts, components, or subassemblies, or any combination thereof, joined together to perform a specific function.
e. **A Circuit Card:** An insulating material, often multi-layered, constituted of epoxy-glass or phenolic resin and an etched copper foil(s) on which electrical components are mounted. The etched copper foil(s) pattern forms a circuit.
f. **A Component:** Any part or assembly of a system, subsystem, subassembly, assembly, or circuit card; such as a resistor, amplifier, power supply, or rack.

19.5 WHERE IS THIS BEING TESTED?

As depicted in Figure 19.3, where the testing takes place is:

a. **At the Factory:** Factory testing consists of a series of processes that evaluate the component/assembly/subassembly/subsystem/system/product during and after the assembly process. The processes verify that the component/assembly/subassembly/subsystem/system/

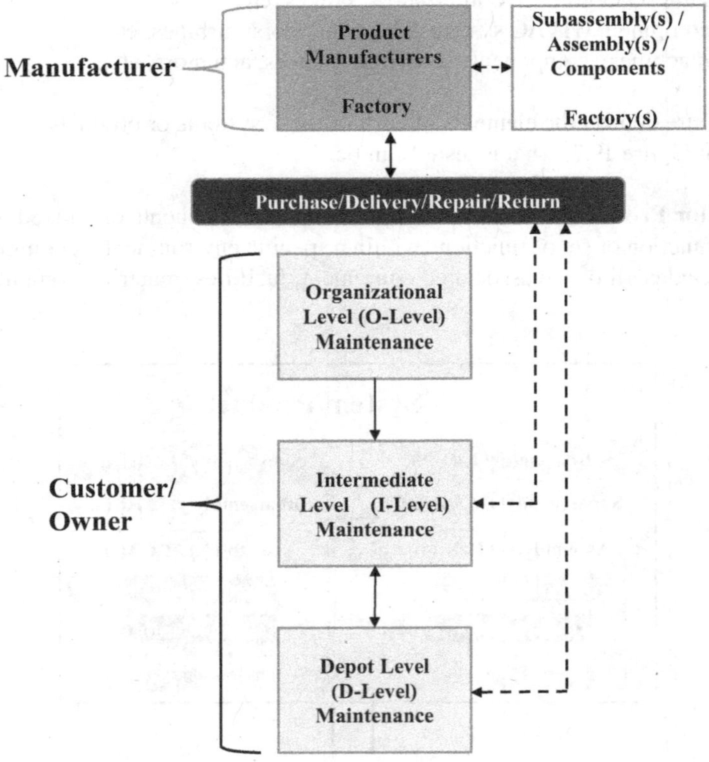

FIGURE 19.3 Where testing takes place.

product is built and operating in accordance with design specifications and/or contractual requirements before delivery.

b. **At the Organizational-Level (or O-Level):** Organizational-level maintenance (or basic-level maintenance) is the first level of maintenance that can be performed "on-equipment" (directly on the defense article or support equipment) without specialized training. It consists of repairing, inspecting, servicing, calibrating, lubricating, or adjusting equipment, as well as replacing minor parts, components, assemblies, and line-replaceable spares or units. This includes modifications, enhancements, or upgrades that would result in improving only the reliability or maintainability of the commodity (e.g., an increased mean time between failure (MTBF)) and does not enhance the basic performance or capability of the defense article.

   Definition taken from Reference (Electronic Code of Federal Regulations (e-CFR), n.d.)

c. **At the Intermediate-Level (or I-Level):** Intermediate-level Maintenance is second-level maintenance performed "off-equipment" (on removed parts, components, or equipment) at or by designated maintenance shops or centers, tenders, or field teams. It may consist of calibrating, repairing, testing, or replacing damaged or unserviceable parts, components, or assemblies. This includes modifications, enhancements, or upgrades that would result in improving only the reliability or maintainability of the commodity (e.g., an increased mean time between failure (MTBF)) and does not enhance the basic performance or capability of the defense article.

   Definition taken from Reference (Electronic Code of Federal Regulations (e-CFR), n.d.)

d. **At the Depot-Level (or D-Level):** Depot-level maintenance is third-level maintenance performed on- or off-equipment at or by a major repair facility, shipyard, or field team, each with necessary equipment and personnel of requisite technical skill. It consists of *providing evaluation or repair beyond unit or organization capability. This maintenance* consists of inspecting, testing, calibrating, repairing, overhauling, refurbishing, reconditioning, and one-to-one replacing of any defective parts, components, or assemblies. This includes modifications, enhancements, or upgrades that would result in improving only the reliability or maintainability of the commodity (e.g., an increased mean time between failure (MTBF)) and does not enhance the basic performance or capability of the defense article.

   Definition taken from Reference (Electronic Code of Federal Regulations (e-CFR), n.d.)

19.6 HOW WILL THIS BE TESTED?

Automatic (or automated) test equipment (ATE) is computer-controlled support equipment consisting of source, sensor, and load instrumentation that utilizes automation to quickly (and accurately) perform tests, diagnose faults, and evaluate test results.

ATE testing may be either semi-automated or fully automated. Semi-automated testing involves human interaction during some part of the testing process (to probe a specific point, to enter data from an instrument display, or to move a connector for example). In fully automated testing, the operator connects the device to the ATE and presses a "go" button, there is no human interaction until the testing is complete.

Manual Test Equipment is support equipment consisting of source, sensor, and load instrumentation that utilizes a human following a written test procedure to lead them through the testing process.

Performing tests consist of:

a. **Functional, Automatic:** Functional testing (also referred to as black-box testing) is where test scripts, test adaptors, holding fixtures, cable sets, and automatic test equipment are utilized to test per test specifications of the product or system (autopilot, radar, automobile, etc.) to verify the operation of the product or system per operational specifications. The testing ignores the internal mechanisms of the product or system and focuses on the outputs generated in response to selected inputs and execution conditions. In the event a test fails, the performance of diagnostic tests isolate and identify the failure(s).

b. **In-Circuit, Automatic:** In-circuit testing is where test scripts and electrical probes (also referred to as a bed-of-nails) make physical contact with components on a populated printed circuit board to test for opens, shorts, resistance, capacitance, polarity, and other quantities in order to verify the assembly was correctly fabricated before conformal coating is applied to the circuit card.
c. **Manually:** Manual testing is a situation where a human is solely responsible for performing the test(s) without the assistance of automated instrumentation, software, or test scripts. Manual testing is more practical and cost-effective than automated testing where testing is seldom performed.
d. **Semi-Automatic:** Semi-automated testing is a combination of automatic and manual testing, where a human operator is involved during some part of the automatic testing process such as to probe a specific point, move a connector or cable, or evaluate/interpret a screen display and manually enter the content.
e. **In Simulation:** The intent of the simulation is to present a situation(s) to a product or system (Autopilot, Radar, Automobile, etc.) that the product or system may (or may not) have been exposed to in a controlled environment. The purpose of the simulation is to see/evaluate how that product or system reacts to the situation.

19.7 WHY IS THIS BEING TESTED?

The testing of a component/subassembly/subsystem/system or product occurs because they are generally complex, often consist of many parts, can be expensive, and may cause issues when (and if) they do not work.

Hardware generally falls into one or more of the following categories:

1. A safety-critical component/subassembly/subsystem/system or product.
   a. "Safety-critical" implies that a failure or malfunction may result in
      i. Death or serious injury to people.
      ii. Loss or severe damage to equipment/property.
      iii. Cause environmental harm.
2. A product or system that customers expect to work flawlessly out of the box.
3. A mission-critical single-point-of-failure.
   a. A "single-point-of-failure" is part of a subassembly/subsystem/system or product that, should it fail, stops the entire subassembly/subsystem/system or product from working.
4. Regulated by a governing body.

REFERENCE

Electronic Code of Federal Regulations (e-CFR) (n.d.). Title 22 - Foreign Relations / CHAPTER I - DEPARTMENT OF STATE / SUBCHAPTER M – INTERNATIONAL TRAFFIC IN ARMS REGULATIONS / PART 120 - PURPOSE AND DEFINITIONS / 120.38 MAINTENANCE LEVELS / (c).

20 Software Test

Upulee Kanewala

20.1 INTRODUCTION: WHAT IS SOFTWARE TESTING?

Software testing is the process of executing the software using a set of inputs to check whether the produced outputs are correct according to the expected behavior of the program. Testing is the most widely used approach for quality assurance of software since many steps can be automated and integrated into the software development process. Figure 20.1 shows the main components involved in testing.

Typically, the tested program is the program under test (PUT) or system under test (SUT). t_1, t_2, …, t_n are *test inputs,* and o_1, o_2, …, o_n are *test outputs.* A *test oracle* determines whether a particular test has passed or failed by looking at the provided test input and the produced test output. A test oracle can be a human, assertions, or another program.

The effectiveness of testing depends on the test inputs and the test oracle used for testing. Using a well-designed set of test inputs that exercise the functionality of the PUT with a test oracle that can determine whether those tests have passed or failed will help increase confidence in the program. The terms *testing* and *debugging* are often used interchanging; however, it is essential to note that *debugging* is finding the root cause after observing a test failure.

A *test case* consists of the test input and its expected output. For example, consider a program *sort* that takes a set of integers as the input and sorts them in ascending order. A test case for *sort* looks as follows:

TEST CASE FOR *SORT:*

- Test input: {12, –45, 5}
- Expected output: {–45, 5, 12}

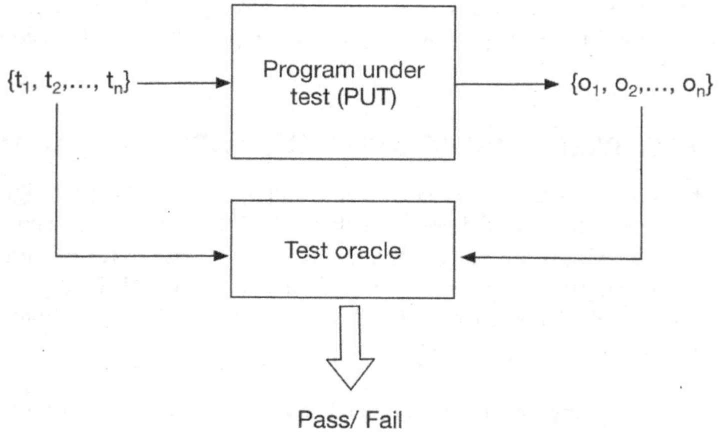

FIGURE 20.1 Main component of testing.

DOI: 10.1201/9781003188377-24

A *test suite* is a set of test inputs developed for a PUT and will consist of many test cases designed based on different goals.

With the increasing complexity of software and its use in safety-critical applications such as in medicine and autonomous vehicles, the emphasis on testing as a quality assurance method has also increased. Further, with the increased use of agile methods for developing software, programmers must write their tests during day-to-day development which have become an essential part of development.

20.2 LIMITATIONS OF TESTING

Software testing cannot be used to prove that a program is correct since perfect testing is impossible. Consider a program *p(i)* where the input to *p* is *i*. Assume that the domain of *i* is *D*. First, *D* could be infinite; therefore, it is not practically possible to test *p(i)* with all the possible inputs. Second, for input *i*, we cannot determine whether *p(i)* will finish running or continue to run indefinitely. This is known as the *halting problem,* a well-known undecidable problem in computability theory. Therefore, testing aims to increase the confidence about the program that it would produce correct outputs by systematically conducting the testing process.

20.3 FAULT/FAILURE MODEL

A *fault* in a program is a static defect in the source code. Depending on the location and the nature of the fault, not all the inputs trigger that fault which would result in an incorrect test output known as a *test failure*. During testing, three conditions must be satisfied for a test failure to occur.

1. **Reachability:** the location of the fault in the source code must be executed (i.e., reached) by the test input.
2. **Infection:** the execution of the fault must create an incorrect *program state*. The program state consists of the variables in the program and the program counter. If any of these holds an incorrect value at a given time, the program is said to be in an incorrect state.
3. **Propagation:** the infected state should propagate to an observable output causing a test failure.

Therefore, the three conditions, reachability, infection, and propagation, should be satisfied for a test failure to occur. This is also known as the RIP model in software testing.

According to the RIP model, the first condition that should be satisfied to detect a failure is reachability. Thus, there is a lot of emphasis on software testing on designing test cases to satisfy the reachability condition. In this chapter, we discuss two approaches widely used for creating test inputs.

20.4 INPUT SPACE PARTITIONING AND BOUNDARY VALUE ANALYSIS

In this approach, the program is treated as a black box, and the test inputs are designed based on the input specification. The first task is to identify the domain *D* of each input *i*. Then, D is partitioned into blocks where similar inputs are placed in a single block. These partitions are also known as *equivalence classes*. Then, representative values are chosen from each block to conduct testing. Ideally, a partition *p* with blocks b_i where $i=1, \ldots, n$ created on a domain *D* should satisfy the following two conditions:

1. There should not be any common element between any two pairs of blocks. That is $b_i \cap b_j = \varnothing$ where $i \neq j$.
2. When all the blocks are combined, they should make up D. That is $\bigcup_{i=1,\ldots,n} b_i = D$.

Once blocks are created, representative values from each block are selected to create test inputs. When selecting representative values, in addition to choosing a value inside the block, values on and near the boundaries are also selected. The institution behind selecting values on and near the boundary is that programmers usually make coding mistakes near boundaries. For example, it is a common mistake to use $<$ instead of $<=$. Therefore, selecting input values on and near the boundaries would help in uncovering these types of faults. When a program has multiple input parameters, those parameters can be partitioned independently of each other or considered together when creating the blocks. This decision depends on the program's semantics, which would lead to creating better test inputs.

When applying this approach in practice, it is common to have sub-partitions. For example, D is partitioned into two blocks as valid and invalid, and then those are partitioned into sub-partitions based on how behave those valid/invalid inputs functionally. This scenario is depicted in Figure 20.2. The domain is first partitioned into two higher-level partitions called *Valid* and *Invalid*. Then the *valid* partition is partitioned into four sub-partitions b_w, b_x, b_y, and b_z. Similarly, the *Invalid* partition is sub-partitioned into b_p, b_q, b_r, and b_s. Then, we pick representative test inputs from each of these partitions, as shown with blue dots in Figure 20.2. Finally, we choose the test inputs at boundary values and near the boundary values (i.e., just inside and outside of the considered boundary) as shown with red dots in Figure 20.2.

Consider a program that accepts the month as an integer and output the number of days of the month. Based on this description of the program, we can create the partitions as follows:

$$Valid: 1 \le i \le 12 \ where \ i \ is \ an \ integer$$

$$Invalid: i \langle 1 \ or \ i \rangle 12$$

Then, these two partitions can be sub-partitioned based on the characteristics of the program. For example, we know that certain months have 30 days, 31 days, and 28/29 days. This characteristic of the program can be used to create three sub-partitions of the valid partition as follows:

$$p_1^v : integer \ values \ of \ months \ with \ 30 \ days$$

$$p_2^v : integer \ values \ of \ months \ with \ 31 \ days$$

$$p_3^v : integer \ values \ of \ months \ with \ 28/29 \ days$$

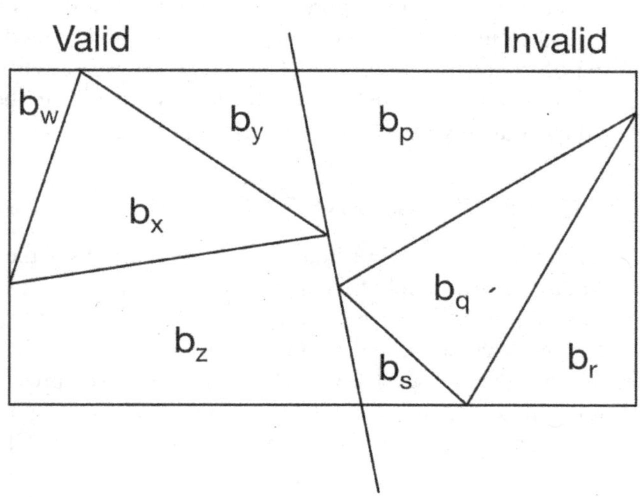

FIGURE 20.2 Input space partitioning.

Similarly, we can create sub-partitions for the invalid partition as follows:

$$p_1^{in} : i < 1$$

$$p_2^{in} : i > 12$$

Once the partitions are created, the next task is to select the representative test inputs from the partitions and their boundary values. The following table shows the test inputs selected from each partition:

Partition	Test Input
p_1^v	9
p_2^v	5
p_3^v	2
p_1^{in}	−15
p_2^{in}	55

The final step of this process is to identify the test inputs representing the boundary values. The following table shows the test inputs selected to represent the boundary values:

Boundary	Test Inputs
Between Valid and Invalid	1,0,2,11,12,13
Between p_1^v and p_2^v	3,4
Between p_2^v and p_3^v	1,2

Consider the boundary between *Valid* and *Invalid* partitions. Two integers lie on this boundary, 1 and 12. We also pick inputs that are just inside and outside of the boundary. Therefore, we choose 0 and 2 with 1 and 11 and 13 with 12 as test inputs.

There are several advantages of using this approach for creating test inputs. Mainly, the person who is testing does not need to understand how the program is implemented and only needs to understand the inputs and outputs of the program. Therefore, there is less learning curve to apply this method. Further, depending on the resources available for testing, the number of test inputs can be increased or decreased by adjusting the number of partitions created.

20.5 CREATING TEST INPUTS TARGETING COVERAGE

As opposed to using only the input/output information, this approach uses the source code of the program to create test inputs. Thus, to apply this approach, the source code of the PUT should be available. The goal of this approach is to create test inputs that execute various elements of the source code. For example, these elements could be statements, branches, or paths in the source code.

The common approach for creating coverage-based test inputs is to represent the program as a *directed graph* and then create test inputs that *visit* the elements in the graph, such as nodes and edges. The most widely used graph representation is the *control flow graph* (CFG).

20.5.1 Control Flow Graph

The CFG of a given program represents all the possible executions of the program by describing its control structures. The nodes in the graph represent the *basic blocks* in the program,[1] and the edges represent how the control transfers between basic blocks. A *basic block* is a sequence of statements in the program where the control to that sequence of statements only enters at the beginning of the sequence and leaves only at the end of the sequence. Thus, there cannot be any branching in or out from the middle of the basic block. The *start node* where the execution starts for the program is denoted using an incoming arrow. The *exit node* where the execution ends is denoted using a node with a thicker border (see Figure 20.6).

Figures 20.3–20.5 show the CFG representations of commonly used control structures *if, if-else,* and *while,* respectively. The first step in creating the CFG is to identify the basic blocks in the program. For example, consider the code snippet in Figure 20.3 corresponding to *if.* There are three basic blocks in this code snippet represented with nodes *a, b,* and *c* (statements belonging to each of those basic blocks are listed next to the nodes for clarity). Once the basic blocks are identified, the next step is to add edges between those basic blocks to show how the control or execution flows from one basic block to the other. For example, from basic block *a,* execution can proceed to basic block *b* or basic block *c* depending on whether the condition in basic block *a* is evaluated as *true* or *false.*

Following the above process, we can derive the CFG for the sample program *computeAverage* listed in Figure 20.6. There are six basic blocks in this program. Node *a* is the start node, and nodes *b* and *g* are the exit nodes for this program.

Any execution on the program corresponds to traveling a *path* in the CFG that starts with a start node and ends with an exit node. When executed on the program, a path taken by a test input is called a *test path.* A test path should start from a start node and end from an exit node. For example, for the program in Figure 20.6, the test path corresponding to the test input [2] is *[a, c, d, e, d, g].* Thus, this test path visits the nodes *{a, c, d, e, g}* which corresponds to executing statements 1, 2, 3,

```
if (x < y)
{
    y = 0;
    x = x + 1;
}
z = x
```

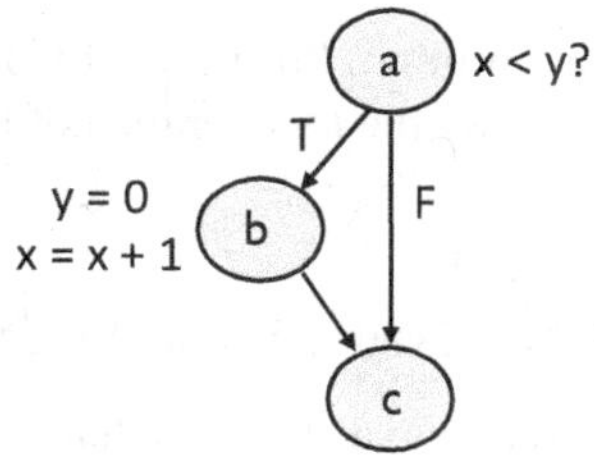

FIGURE 20.3 CFG representation of If.

```
if (x < y)
{
    y = 0;
    x = x + 1;
}
else
{
    x = y;
}
z = x
```

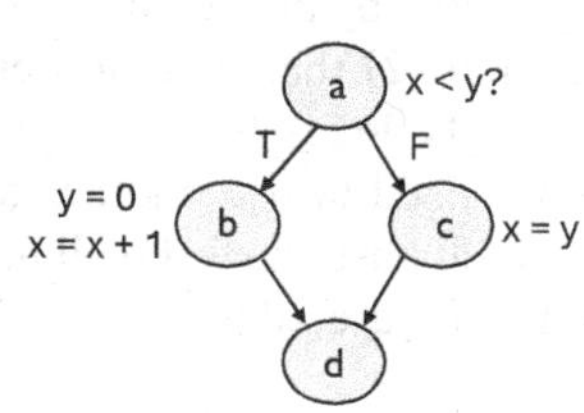

FIGURE 20.4 CFG representation of If-Else.

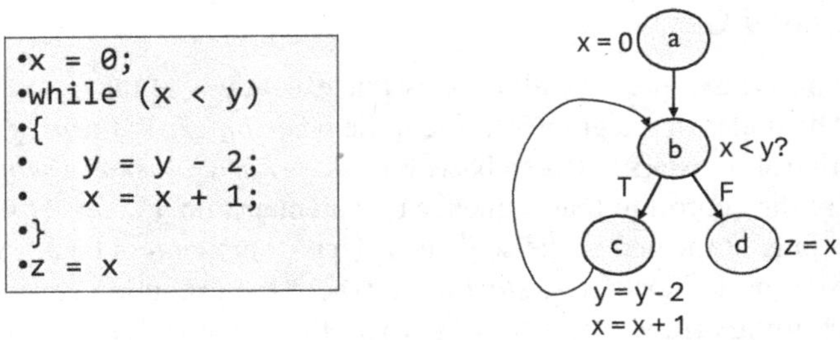

FIGURE 20.5 CFG representation of While.

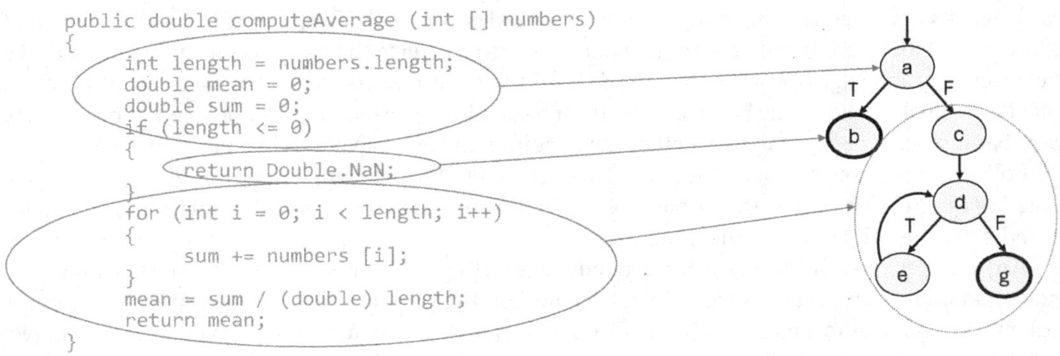

FIGURE 20.6 CFG representation of computeAverage Program.

4, 6, 7, 8, and 9 in the program. Similarly, it visits the edges *{(a, c), (c, d), (d, e), (e, d), (d, g)}*. Thus, this test path executes only the *false* branch of the *if* statement in line 4 and executes the *for loop* in line 6 one time.

20.5.2 Creating Test Inputs Targeting Statement Coverage

This approach aims to create test inputs that *execute all the reachable statements in the PUT.* This requirement can be mapped to visiting all the *reachable nodes* in the CFG by at least one test input. Note that programs may have statements that could not be executed with any input that we typically refer to as *dead code.* When trying to achieve statement coverage, it is not required to create test inputs to execute such statements. It is important to note that 100% statement coverage does not guarantee that the program does not have any faults. It only ensures that all the reachable statements in the program are executed by at least one test input.

Consider the test input {2} that we considered for the program in Figure 20.6. As we discussed above, this test input visits nodes *{a, c, d, e, g}* and does not visit node *b.* Therefore, to satisfy the statement coverage, we need to use another test input that visits node *b.* For this, we can use an empty array since it would have the test path *[a, b].* Thus, the test inputs {2, {}} execute all the statements in the program in Figure 20.6.

20.5.3 Creating Test Inputs Targeting Branch Coverage

This approach aims to create test inputs that *execute all the reachable branches in the PUT.* This requirement can be mapped to visiting all the *reachable edges* in the CFG by at least one test input.

Like statement coverage, there can be branches that could not be executed with any input. Further, 100% branch coverage does not guarantee that the program does not have any faults. It only ensures that all the reachable branches in the program are executed by at least one test input. The test paths of the two test inputs {2, {}} visit all the edges in the CFG in Figure 20.6. Thus, these two test inputs execute all the branches in the program in Figure 20.6.

20.5.4 Other Coverage Types

Specific problems with statement and branch coverage have led to developing more complex coverage requirements. For example, particular faults are revealed when executed with certain combinations of branches (i.e., paths). Some faults exist in logical expressions in programs. Thus, the following test requirements are developed focusing on those complex requirements:

- **Path Coverage:** requires test inputs to visit reachable paths in the CFG.
- **Definition-Use (Def-Use) Coverage:** this is a family of test requirements focused on exercising the program's data flow. Thus, def-use coverage requires the test inputs to exercise how values are created and used within the program.
- **Logic Coverage:** this family of test requirements focuses on exercising the logical expressions in a program in various ways.

Due to the complexity of these requirements, and the increased number of test inputs required to satisfy these test requirements, and the lack of tool support, these are not widely used in practice as statement and branch coverage discussed above. However, different aspects of logic coverage have been incorporated into standards accepted by the United States Federal Aviation Administration (FAA) for safety-critical aviation software.

20.6 TESTING PHASES/LEVELS

Testing is conducted at different phases of the software development and may use various artifacts produced at these phases to assist testing.

- *Acceptance Testing* is conducted to evaluate the software with respect to the end-user requirements to ensure that they are met at the delivery. Thus, acceptance testing is performed on the completed software and is often done before the software is handed over to the stakeholder. Typically, acceptance testing involves users who understand the software and its domain very well.
- *Unit testing* is testing of a unit such as a single function, procedure, or class. Unit tests are usually written by the developers who develop the corresponding unit of code. Since unit tests are often executed and are tightly integrated with the code, they are written using testing tools such as JUnit[2] for Java. Unit testing can be conducted in parallel by testing many units at the same time. The most significant advantage of unit testing is the easiness of debugging once a failure occurs since the testing focuses on one unit.
- *Integration testing* is conducted to ensure that units tested in isolation work as expected when working together. Thus, it focuses on testing the interfaces between different units that communicate correctly, bypassing the required information between them. Often it is common to have components that are not yet implemented when conducting integration testing. In those situations, those missing components are simulated using *drivers* and *stubs*.
- *System testing's* goal is to ensure that the whole software system works as expected. During system testing, it is assumed that the components that make up the software system work

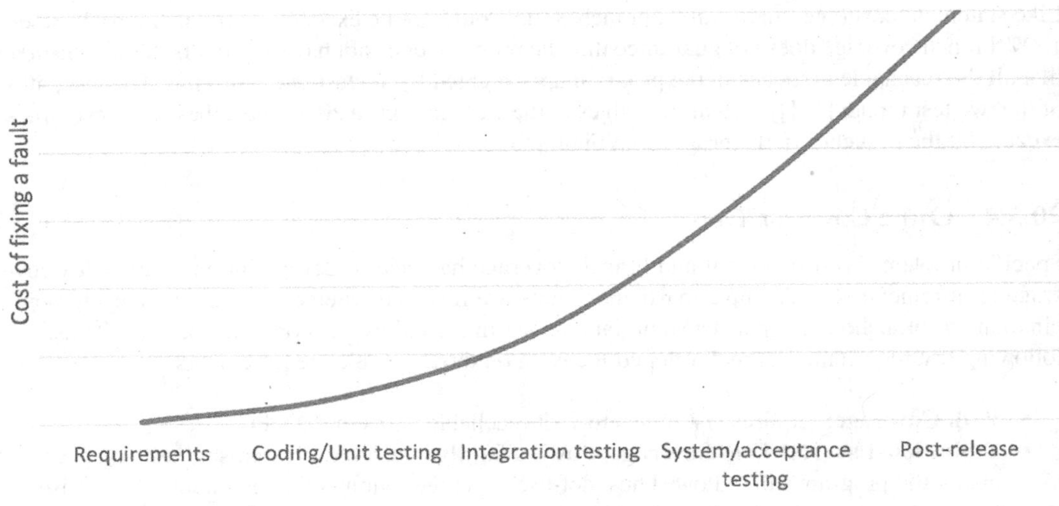

FIGURE 20.7 Cost of fixing a bug.

correctly. Usually, system testing is not done by the programmers but by a separate testing team.

- *Regression Testing* is conducted when the software system changes due to adding new functionality or fixing a fault. Testing is performed to ensure that the software still behaves correctly after the modifications.

As shown in Figure 20.7, the cost of fixing fault increases exponentially with time during software development, as shown. The faults detected during later phases of development, such as at post-release, will cost significantly higher to fix than faults detected in early phases such as during unit testing. Thus, there are many emphases placed on writing effective unit tests and integration tests so that potential faults can be detected earlier.

20.7 TEST AUTOMATION

Typically, 50% of a software development budget is spent on testing-related activities. Manual testing is still widely used since it is cheaper and easy to implement. Also, some testing, such as user interface testing, is tough to automate.

Recently, there is significant interest in developing methods for automating some testing activities such as test input generation. The most naive approach for test input generation is the random generation, where the input such as a number or a string is generated randomly. Random test input generation is quite attractive since it is easy to implement with any programming language and effectively detects faults. However, random test input generation can lot of redundant and invalid inputs. As a solution to this, *feedback-directed random test generation* is developed where the randomized creation of new test inputs is guided by the feedback of the execution of previously generated test inputs (Pacheco et al., 2007). RANDOOP[3] is a Java-based testing framework that is developed following this approach.

Compared to using a random approach for test input generation, *concolic testing* focuses on generating inputs based on coverage. (Sen, 2007). This approach uses symbolic execution guided by concrete inputs to simplify complex and unmanageable constraints generated by symbolic execution. In symbolic execution, symbolic values are used as inputs instead of concrete values, and the program is executed. Then the variable values in the program are represented using symbolic expressions over symbolic input values. The program output would also be expressed as a function of the symbolic values. As the program is executed with these symbolic values, path constraints are developed as formulas over symbolic expressions. Depending on the PUT, these generated constraints can be too complex to be solved in a reasonable time using constraint solvers. In such situations, concrete inputs are used for simplifying these constraints. DART (Godefroid et al., 2005), SAGE (Godefroid et al., 2008), CUTE (Sen and Agha, 2006), jCUTE (Sen and Agha, 2006), and CREST (Burnim and Sen, 2008) are several tools that implement concolic testing.

Search-based software testing (SBST) is another approach that uses metaheuristic search techniques for generating test inputs (Ali et al., 2009). Here, generating inputs is modeled as an optimization problem; from many potential test inputs for a PUT, we need to select test inputs that satisfy particular objectives such as meeting coverage criteria or fault reveling test inputs. EvoSuite[4] is a Java-based automatic test generation tool that uses SBST and can generate test cases for a given program that would satisfy statement and branch coverage.

20.8 TEST ORACLES AND ORACLE PROBLEM

As we discussed at the beginning of this chapter, a test oracle is a mechanism that determines whether a test case has passed or failed and is an essential part of software testing. However, there are programs for which test oracles are unavailable due to the complexity of the program and its outputs. Also, it might be practical difficulties to develop an oracle. Such programs are referred to as programs facing the *oracle problem* (Barr et al., 2014). Since there is no automated method for determining the correctness of test outputs, the main challenge faced when testing those programs is test automation. Thus, often a limited number of tests are used for testing these programs where the correctness of the test outputs is checked manually.

One of the widely used techniques to overcome the oracle problem is *metamorphic testing* (MT) (Chen et al., 2018). Instead of checking the correctness of individual test outputs, this technique uses *metamorphic relations (MRs),* a necessary property of the program under test, and relates multiple inputs and outputs. For example, let's consider testing a program that computes the sine of a given angle using randomly generated inputs. Assume that the randomly generated input is 27° While we might not know the value of sin (27°), we know that sine function follows the property $\sin(-\theta) = -\sin(\theta)$. Thus, we can use this property to test the program using MT, where we execute the program with inputs 27° and –27° check the produced outputs satisfy the above property. Thus, without know the correct value of sin (27°), we can conduct testing on this program. MT is used to successfully detect faults in programs across various domains such as Compilers (Tao et al., 2010), navigation applications such as Google Map (Zhou et al., 2018), citation databases, web search engines (Zhou et al., 2015), machine learning applications (Xie et al., 2009) and autonomous vehicles (Zhou and Sun, 2019).

20.9 CONCLUSIONS

Testing is the most widely used approach for assuring the quality of software. It entails executing the software with test inputs and checking the produced outputs to ensure that they are correct according to the expected behavior of the program. Fault detection effectiveness of testing highly depends on the test inputs and the test oracle used in testing. Input space partitioning and coverage base test inputs can be used to create effective inputs while reducing the number of test cases that need to be executed. There exist several approaches to automatically generating test inputs, such as feedback-directed random test case generation and concolic testing. A test oracle is another essential component that is required to conduct testing. Testing techniques such as metamorphic testing can be used when test oracles are not available.

NOTES

1 The control flow graph can also be created with each node representing a statement in the program.
2 https://junit.org/junit5/
3 https://randoop.github.io/randoop/
4 https://www.evosuite.org/

REFERENCES

Ali, S., Briand, L. C., Hemmati, H., & Panesar-Walawege, R. K. (2009). A systematic review of the application and empirical investigation of search-based test case generation. *IEEE Transactions on Software Engineering*, 36(6), 742–762.

Barr, E. T., Harman, M., McMinn, P., Shahbaz, M., & Yoo, S. (2014). The oracle problem in software testing: A survey. *IEEE Transactions on Software Engineering*, 41(5), 507–525.

Burnim, J., & Sen, K. (2008). *Heuristics for Scalable Dynamic Test Generation.* Technical Report, EECS Department, University of California, Berkeley.

Chen, T. Y., Kuo, F. C., Liu, H., Poon, P. L., Towey, D., Tse, T. H., & Zhou, Z. Q. (2018) Metamorphic testing: A review of challenges and opportunities. *ACM Computing Surveys (CSUR)*, 51(1), 1–27.

Godefroid, P., Klarlund, N., & Sen, K. (2005). Dart: Directed automated random testing. In *Proceedings of the 2005 ACM SIGPLAN Conference on Programming Language Design and Implementation* (pp. 213–223).

Godefroid, P., Levin, M. Y., & Molnar, D. A. (2008). Automated whitebox fuzz testing. In *Proceedings of Network and Distributed Systems Security.*

Pacheco, C., Lahiri, S. K., Ernst, M. D., & Ball, T. (2007). Feedback-directed random test generation. In *Proceedings of the 29th International Conference on Software Engineering (ICSE'07)* (pp. 75–84).

Sen, K. (2007). Concolic testing. In *Proceedings of the Twenty-Second IEEE/ACM International Conference on Automated Software Engineering* (pp. 571–572).

Sen, K., & Agha, G. (2006). CUTE and jCUTE: Concolic unit testing and explicit path model-checking tools. In *Computer Aided Verification* (pp. 419–423).

Tao, Q., Wu, W., Zhao, C., & Shen, W. (2010). An automatic testing approach for compiler based on metamorphic testing technique. In *2010 Asia Pacific Software Engineering Conference* (pp. 270–279).

Xie, X., Ho, J., Murphy, C., Kaiser, G., Xu, B., & Chen, T. Y. (2009). Application of metamorphic testing to supervised classifiers. In *2009 Ninth International Conference on Quality Software* (pp. 135–144).

Zhou, Z. Q., & Sun, L. (2019). Metamorphic testing of driverless cars. *Communications of the ACM*, 62(3), 61–67.

Zhou, Z. Q., Sun, L., Chen, T. Y., & Towey, D. (2018). Data from: Metamorphic relations for enhancing system understanding and use. *IEEE Transactions on Software Engineering*, 46(10), 1120–1154.

Zhou, Z. Q., Xiang, S., & Chen, T. Y. (2015). Metamorphic testing for software quality assessment: A study of search engines. *IEEE Transactions on Software Engineerin*, 42(3), 264–284.

21 Error Handling

Stephyn G. W. Butcher

21.1 INTRODUCTION

Computer programming languages generally provide some mechanisms for handling error conditions such as division by zero or attempting to read a file that doesn't exist. The three main mechanisms are error codes, exception handling, and error types. The enterprise languages (Java/C#/C++) all use exception handling as do many popular interpreted languages such as Python, Ruby, and JavaScript. The C programming language relies on error codes. Lisp, of course, can have anything and sometimes has everything. In this chapter, we will be concentrating on exception handling because it is the most widespread.

Being widespread, one would think that exception handling is straightforward. And this turns out not to be the case. First, there's what the language designers think we should do. Next, there's the advice of experts in the field, who try to balance the idealism of the language designers with the pragmatism of work-a-day programmers. Lastly, there's what programmers on the front lines actually do. We will spend most of the last part of this chapter trying to make some sense of it all.

The aim of this chapter is as follows. We will look at the basics of exception handling, as described by the documentation for several languages. This gives us a baseline of syntax for discussion and exploration variations. It also allows us to see what the "party line" is for these languages as far as exception handling is concerned.

For a topic as large as exception handling, there is no shortage of advice. In this next section, we will look at some of the advice provided by professionals for professionals on exception handling.

We will then look at some empirical research into exception handling in actual code. This is what programmers actually do. Some of the results are not surprising: work-a-day programmers do not follow the advice of professionals. Other results are surprising: work-a-day programmers are not using exceptions for their intended use.

21.2 LANGUAGE DESIGN

The Java Language Tutorial (Oracle, 2022a) lists several advantages to using exceptions. The first advantage echoes the most commonly cited rationale for exception handling: separation of program logic from error handling. In that explanation, they describe the nearly canonical example of opening a file (see Listing N.1):

Listing N.1

```
readFile {
    open the file;
    determine its size;
    allocate that much memory;
    read the file into memory;
    close the file;
}
```

DOI: 10.1201/9781003188377-25

and enumerate all the possible ways this code can fail:

1. What happens if the file can't be opened?
2. What happens if the length of the file can't be determined?
3. What happens if enough memory can't be allocated?
4. What happens if the read fails?
5. What happens if the file can't be closed?

We might only note that there may be several, independent reasons that a file can't be opened: it might not exist or the program might not have proper permissions.

The example goes on to show the mess of code that would be required to handle all of those errors using error codes in a C-like language (Listing N.2).

Listing N.2

```
errorCodeType readFile {
    initialize errorCode = 0;

    open the file;
    if (theFileIsOpen) {
        determine the length of the file;
        if (gotTheFileLength) {
            allocate that much memory;
            if (gotEnoughMemory) {
                read the file into memory;
                if (readFailed) {
                    errorCode = -1;
                }
            } else {
                errorCode = -2;
            }
        } else {
            errorCode = -3;
        }
        close the file;
        if (theFileDidntClose && errorCode == 0) {
            errorCode = -4;
        } else {
            errorCode = errorCode and -4;
        }
    } else {
        errorCode = -5;
    }
    return errorCode;
}
```

And it is difficult to understand what the code is trying to do versus identifying what went wrong. The idea of separating the "happy path" from error conditions lies at the heart of most design principles surrounding exception handling. The same code with exception handling looks much cleaner (Listing N.3):

Listing N.3

```
readFile {
    try {
        open the file;
```

```
        determine its size;
        allocate that much memory;
        read the file into memory;
        close the file;
    } catch (fileOpenFailed) {
        doSomething;
    } catch (sizeDeterminationFailed) {
        doSomething;
    } catch (memoryAllocationFailed) {
        doSomething;
    } catch (readFailed) {
        doSomething;
    } catch (fileCloseFailed) {
        doSomething;
    }
}
```

So exception handling introduces a new flow control mechanism into the programming language. Although the exact keywords can differ from language to language the basic idea is that we can try to execute some "happy path" code, catch any exceptions that occur, handle them or, if we cannot, pass the exception to code earlier in the call stack. If there is any code that needs to execute whether or not there is an error (releasing a resource, for example), we can take care of it in a finally block. If we need to throw our own exceptions or re-throw the one at hand, we can do that as well. We can see some of this above.

21.2.1 Differentiating Exceptions

The Java Tutorial identifies another advantage to using exception handling: differentiating exceptions. This essentially boils down to the ability to use a class hierarchy of exceptions to either filter all exceptions at a high level or handle the specific problems, exception by exception.

For example, we could create an exception hierarchy that starts with Exception, creates several subclasses including IndexOutOfBoundsException and further subclass this to IndexExceedsBoundsException and NegativeIndexException. Another hierarchy might start with the subclass IOException and create subclasses for FileNotFoundException and BadFilePermissionException. The consumer of code that can throw these exceptions can decide whether to handle IndexOutOfBoundsException or the two more specific exceptions. This is especially useful if one specific exception could theoretically be handled by the caller and the other needs to be propagated up the call stack.

Our running example catches every very specific exception. With the existence of a class hierarchy of exceptions, we could simplify the code above to (Listing N.4):

Listing N.4

```
readFile {
    try {
        open the file;
        determine its size;
        allocate that much memory;
        read the file into memory;
        close the file;
    } catch (fileFailed) {
        doSomething;
    }
}
```

It's worth noting that we never see examples of "doSomething" beyond logging of some kind (Oracle (2022b), Listing N.5):

Listing N.5

```
} catch (IndexOutOfBoundsException e) {
    System.err.println("Caught IndexOutOfBoundsException: " +
e.getMessage());
} catch (IOException e) {
    System.err.println("Caught IOException: " +  e.getMessage());
}
```

We will have cause to examine this in detail later.

21.2.2 Checked Exceptions

With its particular implementation of exception handling, Java introduced the controversial idea of checked exceptions (Oracle, 2022c). If a method can throw a checked exception, this fact must be declared in the method signature. Additionally, if the caller does not catch a checked exception, it must declare the checked exception in its method signature. The basic idea is that the exceptions a method can throw is part and parcel of the method's contract, just as much as the parameter types and definitions.

For our running example, if all the exceptions we might encounter were checked exceptions and the code handled none of them, we would need to define readFile as follows (Listing N.6):

Listing N.6

```
readFile throws fileOpenFailed, sizeDeterminationFailed,
memoryAllocationFailed,
                readFailed, fileCloseFailed {
    open the file;
    determine its size;
    allocate that much memory;
    read the file into memory;
    close the file;
}
```

Checked exceptions became controversial because programmers don't like declaring all of those exceptions in the method declaration. They rightly note that all code can throw NullPointerExceptions or ArrayIndexOutOfBoundExceptions, and yet programmers are not required to declare them. The language designers respond, yes…but nobody expects the caller to recover from a NullPointerException. They sum up the bottom line (literally), "If a client can reasonably be expected to recover from an exception, make it a checked exception. If a client cannot do anything to recover from the exception, make it an unchecked exception." But the controversy uncovers an uncomfortable truth: most, if not all, exceptions are unrecoverable.

21.2.3 Exceptions in the Key of C#

Benefiting from the experience of Java, Microsoft added a few features to exception handling in C# that Java doesn't have (Wagner, 2023). For example, they decided not to include checked exceptions (or removed them, depending on the perspective). They introduced exception filtering. Exception filtering allows the programmer to add fine-grained control over exceptions by allowing catch blocks to reference program state related to an exception. As an example, imagine we would like to differentiate two different cases for an ArrayIndexOutOfBoundException. Without exception filtering, we would need to do the following (Listing N.7):

Listing N.7

```
try {
    a[index] = 2;
} catch (ArrayIndexOutOfRangeException e) {
    if (index < 0) {
        log("Indices must be positive")
    } else {
        log("Index {index} must be less than {length(a)}")
    }
}
```

However, with exception filtering, we have (Listing N.8):

Listing N.8

```
try {
    a[index] = 2;
} catch (ArrayIndexOutOfRangeException e) when index < 0 {
        log("Indices must be positive")
} catch (ArrayIndexOutOfRangeException e) {
        log("Index {index} must be less than {length(a)}")
}
```

As far as advice is concerned, Microsoft echoes many of the same themes as Oracle (and originally Sun). It is worth noting, however, that just as with the Java documentation, C# examples do not ever show us how we might recover from an exception in a catch block. The examples show:

1. Throwing a new Exception (Listing N.9)

    Listing N.9

```
catch (IndexOutOfRangeException e) {
    throw new ArgumentOutOfRangeException(
        "Parameter index is out of range.", e);
}
```

2. Logging the Exception and re-throwing it (Listing N.10)

    Listing N.10

```
catch (UnauthorizedAccessException e) {
    // Call a custom error logging procedure.
    LogError(e);
    // Re-throw the error.
    throw;
}
```

    and, finally,

3. Do something truly horrific with an Exception filter (Listing N.11)

    Listing N.11

```
{
...
    catch (Exception e) when (LogException(e)) { }
    Console.WriteLine("Exception must have been handled");
}
```

```
private static bool LogException(Exception e) {
    Console.WriteLine($"\tIn the log routine. Caught {e.GetType()}");
    Console.WriteLine($"\tMessage: {e.Message}");
    return false;
}
```

This last example is horrifying because the logging occurs as a side effect of a boolean test and the empty braces swallow the exception. This seems like a Bad Idea.

21.2.4 Why Aren't Retries Built-In?

When an exception occurs, we have three possible options: handle it, let it propagate up the call stack, or re-throw a new exception. Let's say we've decided to handle the exception where it is thrown.

It seems strange then that so few languages with exception handling actually have a retry mechanism. In fact, we are only aware of one language that does: Ruby (Community, n.d.).

Consider the case of a missing file. In the pseudocode above, this is covered by "doSomething". This "doSomething" might ask the user for the location of the file. Once we have asked for a new filename, however, how do we get back on the happy path? There doesn't seem to be a built-in way to accomplish this in either Java or C#. Using Java's syntax, Ruby essentially adds a retry keyword that restarts the try block (Listing N.12):

Listing N.12

```
readFile {
    try {
        open the file;
        determine its size;
        allocate that much memory;
        read the file into memory;
        close the file;
    } catch (fileNotFound) {
        ask for new file name;
        retry;
    }...
}
```

Still, the solution isn't perfect. What if there are no files that will work? In that case, we basically have to roll our own (Listing N.13):

Listing N.13

```
attempts = 1;
while (attempts < 6) {
    readFile {
        try {
            open the file;
            determine its size;
            allocate that much memory;
            read the file into memory;
            close the file;
        } catch (fileNotFound) {
            ask for new file name;
            attempts += 1;
        }...
    }
}
```

Unfortunately, this approach adds another level of indentation. It might have been better if this were built-in, given the goals of exception handling. To some, it might seem that the language designers don't really want us to handle exceptions. We will discuss this point later.

21.3 ADVICE ON EXCEPTION HANDLING

One commonly cited resource for Java programmers is Effective Java (Bloch, 2018). Despite the title, the advice is broadly applicable to all exception handling, except for the advice on checked exceptions. The advice in Effective Java can be neatly summarized as:

1. Use exceptions only for exceptional conditions
2. Use checked exceptions for recoverable conditions and runtime exceptions for programming errors
3. Avoid unnecessary use of checked exceptions
4. Favor the use of standard exceptions
5. Throw exceptions appropriate to the abstraction
6. Document all exceptions thrown by each method
7. Include failure-capture information in detail messages
8. Strive for failure atomicity
9. Don't ignore exceptions

Bloch #2 and #3 are only applicable to Java programmers. Bloch #6 is interesting because he suggests documenting all exceptions, checked and unchecked (runtime). Given the nature of checked exceptions, they are almost completely self-documenting, since they must be declared as part of the method signature. However, given the Java documentation's rationale for checked versus runtime exceptions, why should runtime exceptions be documented as well?

The rationale behind this advice is both interesting and telling. Bloch contends that runtime errors in a method must originate from either the parameters of that method or the object state, absent a complete programming failure. For example, if a method can throw an ArrayIndexOutOfBoundsException, this must be because some precondition for either a method parameter or object field was violated. Otherwise, the code would have been properly executed. As a result, the caller should be informed about those preconditions in the documentation.

Failure atomicity (Bloch #8) means that a method call that fails should leave the object in the same state it was in when the method was called.

In "Towards Exception-Handling Best Practices and Patterns", Wirfs-Brock (2006) provides her own advice for exception handling:

1. Don't try to handle coding errors.
2. Avoid declaring lots of exception classes.
3. Recast lower-level exceptions to higher-level ones whenever we raise an abstraction level.
4. Provide context along with an exception.
5. Assign exception handling responsibilities to objects that can make decisions.
6. Use exceptions only to signal emergencies.
7. Don't repeatedly throw the same exception.

It's interesting to examine the similarities and differences of the two lists as well as the prioritization. Bloch #1 and WB #6 are the same. They both agree that exceptions should be exceptional, but it has a lower priority for Wirfs-Brock.

Bloch #5 and WB #3 are the same ("recast lower-level exceptions in higher-level code"). Why could this be important? The very specific exceptions in the standard libraries often reveal implementation details and "computerese". If a FileNotFoundException is thrown and caught in lower-level code, that's fine. However, if a FileNotFoundException is thrown in lower-level code and caught in code looking for a configuration file, we have moved "up" a level of abstraction. We've gone from looking for a file to looking for a configuration file. At that higher level of abstraction, the error should indicate that the configuration file wasn't found. The most egregious example of this is an error dialog that simply displays the raw exception: java.lang.FileNotFoundException.

Bloch #4 and WB #2 are also similar. Programmers are more familiar with the standard or common exceptions defined by the language than any custom exceptions we might design. Therefore, we should use those standard exceptions whenever possible and customize the message to our application.

Another thing on which both Bloch (#7) and Wirfs-Brock (#4) agree is that we need to provide as many details or as much context, respectively, as possible when an exception occurs.

Finally, Bloch #1 and WB #6 appear to be quite similar: exceptions should be unlikely events, not the run-of-the-mill. Additionally, Wirfs-Brock warns that we shouldn't use exceptions to handle bugs in our code.

So given the desires of language designers and the advice of professional programmers, how do work-a-day programmers such as ourselves fare?

21.4 EXCEPTIONS IN THE WILD

In order to determine what programmers actually do, we look at two empirical studies of large codebases. The first is "Analysis of Exception Handling in Java Projects: An Empirical Study" (Nakshatri, Hegde, and Thandra, 2016) and the other is "How Developers Use Exception Handling in Java" (Asaduzzaman et al., 2016).

In "Analysis of Exception Handling", Nakshatri et al. looked at 605,556 Java projects on Github and SourceForge. These projects contained a combined total of 26,129,222 catch blocks of which 16,172,462 were analyzable. Their main finding was that programmers have generally started to use exception handling as a reactive debugging aid instead of a means of handling rare errors proactively.

The key findings were:

1. general exceptions were caught more often than specific exceptions with Exception, IOException, and SQLException being the most common.
2. a full 20% of catch blocks were empty.
3. the vast majority of non-empty catch blocks contained some kind of logging (print, print-stacktrace, or explicit logging) and no other action but nothing else most of the time.
4. checked exceptions were often converted to unchecked exceptions and re-thrown.

Based on these findings, we can conclude that a majority of programmers are:

1. Not catching the most specific exception to the circumstances and trying to recover.
2. Ignoring exceptions (contra Bloch #9) and, in some cases, providing no context for them (Bloch #7 and WB #4).
3. In those cases where exceptions are caught, they do at least log exception context (Bloch #7 and WB #4).
4. Ignoring the distinction between recoverable (checked) and runtime (unchecked) exceptions and converting the former to the latter and re-throwing them.
5. Using exceptions to handle coding errors (against Bloch #1 and WB #1 and #6).

In short, programmers are using exceptions to find bugs. To a certain extent, this makes sense because runtime errors are not generally recoverable. As we noted above, even if an exception is recoverable, getting back on the happy path is not trivial. On the other hand, one has to be a little shocked, have we just given up on understanding how our code works?

In "How Developers Use Exception Handling in Java", Asaduzzaman et al. sought to examine the prevalence of specific bad exception handling practices. Unlike the previous study, they had very specific good practices (and thus bad practices) in mind. For example, as with Bloch #1 and WB #1, Asaduzzaman et al. believed that runtime exceptions were unrecoverable and therefore should not be caught.

The study broke the developers out into five skill levels, novice to expert. The following list specifies each bad practice; the range indicates the low and high percentages of "total exception handling examples" out of each of those five skill levels.

1. Ignoring exceptions (11%–23%)
2. Catching unchecked exceptions (5%–12%)
3. Not preserving original exceptions (8%–13%)
4. Using generic exception handlers (25%–37%)
5. Catching errors (2%–5%)

They concluded, "Our study reveals that improper exception handling coding practice is not uncommon in Java applications and regardless of experience all developers use them."

Of course, this depends on what we think the purpose of exception handling is. Nakshatri et al. were a little more neutral in their analysis and thus concluded that developers had shifted their views on exception handling.

One interesting result of the Asaduzzman research is that developers typically subclass exception when creating new exception classes – thus making them checked–even though they often don't catch checked exceptions in their own code. This seems to be a case of "do as we say not as we do".

21.5 THE TAKEAWAY

One of the most frustrating things in programming is errors. While there are several mechanisms available in programming languages to handle errors, the mainstream programming languages use exception handling. As a mechanism of error handling, exception handling promotes the idea of code proceeding as if everything is just going to work ("the happy path") and then "catching" errors as exceptions. Although only Java makes a compile time distinction between recoverable errors (checked exceptions) and unrecoverable errors (unchecked exceptions or runtime errors), it is possibly an important distinction. Returning to a previous observation, we note that nearly every single example of exception handling in these basic language references leaves the actual exception handling to the reader's imagination. They either do nothing, suggest doing something, or simply log the exception. It seems that if exceptions were generally recoverable, there would be more examples of recovering from them.

We can acknowledge that this is hard and ask the experts. We looked at the advice of two software engineering consultants and book authors. The advice seems generally sound:

1. don't use exception handling as a kludge for bad code
2. document exceptions
3. don't ignore exceptions
4. capture the application context
5. don't let exceptions leak implementation details
6. recover from exceptions we can and let unrecoverable exceptions halt the app.

This all seems like fairly good advice. But what do programmers actually do? We looked at two papers and they basically show that programmers do none of this most of the time. Nakshatri,

Hegde, and Thandra (2016) found that most programmers, most of the time, do not use exceptions for error recovery but for debugging. While the papers conclude that this is "bad", we have to ask ourselves, under what circumstances might this be reasonable? Under what circumstances might this be cringeworthy?

21.5.1 Defensive Programming

The hopes and dreams of programming language designers aside, the happy path is probably not as naive as we've been led to believe. This suggests a strategic use of defensive programming. By "defensive programming", we mean assuming that things are not necessarily right rather than the more modern meaning of programming to prevent security issues, although that's important. If we revisit Listing N.3, our original example of textbook exception handling, this means checking for things that might reasonably go wrong (Listing N.14).

Listing N.14

```
readFile {
    try {
        if not file exists?
            doSomething;
        if not file readable?
            doSomething;
        open the file;
        determine its size;
        allocate that much memory;
        read the file into memory;
        close the file;
    } catch (readFileFailed) {
        doSomething;
    }
}
```

Here we check for the existence of the file and for read permissions. This simple example just pushes the hurt around. What is "doSomething" in this case? We could prompt the user for a new file name…if there's a user available. Most probably, we should check for the existence of the file as early in the program operation as possible and not just before we read it. If the file was written by another section of the code, that code probably should have verified that the file was written, that it would be readable, and that we didn't write a file so large we couldn't read it back into memory. This goes for all resources. Make sure that peripherals (testing equipment) are connected as soon as we can rather than right before it's needed.

Consider a program that must communicate with a remote service. It uses HTTP POST and retrieves JSON formatted data. Our control over this service may be complete (we wrote it), partial (another team wrote it), or non-existent (it's maintained by a different company). The format of that data could change accidentally or on purpose, without our knowledge. For example, a key might have added or deleted. A naming convention may leak from the database so that some keys are all UPPERCASE. Should we check the data when we first read it or when we go to use it? Should we check it at all or just let the program fail? And the answer it is, it depends.

If, for our application, it makes sense to check for the existence of required keys and to downcase all keys because that's what our code expects, we should do that. But for another application, it makes more sense to let it fail. In that case, we would catch the exception, log the context, and exit. The big change over the last 10 years (but possibly longer) is the increase in the number of programs that run "in the cloud" and spot instances. Recovering from an exception assumes a backup plan and that backup plan often involves a user. In the case of these cloud based server side applications, there's no user. The best we can do is log an exception, quit, and try again tomorrow (or according to the cron scheduling).

21.5.2 Testing

Exceptions should not be a substitute for understanding how our code works. It's a bit disheartening to learn that so much exception handling is used to debug code. We should know how our code works. Bloch suggests that methods should document their runtime errors as possible violations of parameter preconditions. But how do those preconditions fail to begin with? We should endeavor to make sure our programs can't enter bad states. This is undoubtedly a challenge.

We can start by making sure we have full unit test coverage and integration tests. There are generally only a handful of possibilities. For collections, it is usually zero, one, few, lots, boom. For numbers, it is usually values at either end of some range, for example, an unexpected negative number.

We should also test our exception handling code. There is nothing quite worse than seeing the log message, "the following exception occurred while handling the above exception". Sometimes this can mean that we have no logging about an exception at all. One way to test to our exception handling code is to delegate to a function that we can test (Listing N.15).

Listing N.15

```
readFile {
        open the file;
        determine its size;
        allocate that much memory;
        read the file into memory;
        close the file;
    } catch (readFileFailed) {
        log_to_aws(readFileFailed)
    }
}
```

We can test log_to_aws() separately and make sure that it does what we expect. We don't want any nasty surprises with respect to AWS credentials…

21.5.3 Network Recovery

As we've alluded to above, it seems that recovery is unlikely in many cases. We can check for some potential error conditions and we can test but any time there are inputs that are out of our control, there is the possibility of an error. The stark reality of the situation has led some wonder, are exceptions even exceptional? Is this not the normal course of events? It does seem that way.

There does seem to be one exception: network activity. Of all the code we write, the main time (and perhaps the only time) that retries make sense is when reading/writing over the network. A database may be under high load and thus timeouts may be larger than normal. For ourselves, this seems to be the only recovery code we write. It is best to use an exponential backoff so we don't cause our own denial-of-service attack.

21.6 CONCLUSION

To summarize, error handling is difficult. Despite the best intentions of language designers, it's rarely the case that we can actually recover from exceptions and most exception handling is becoming much more reactive. However, we should avoid using exceptions as a substitute for understanding how our code works. We should test our code, and we should make sure we test our exception handling code as well. If there is an unrecoverable error, we should log as much contextual information as possible. The logging code should be tested, too.

In those cases where it makes sense to be defensive, we should determine that some action can't be done sooner rather than later. Don't wait to check for a configuration file when we go to read it. Don't wait to check for a diagnostic peripheral when we're going to use it. If we do find a recoverable error, it should be conveyed at the right level of abstraction: "the configuration file is missing" not "file cannot be read".

Engineering is about knowing and understanding the trade-offs. We should know the benefits and costs of each.

REFERENCES

Asaduzzaman, M., Ahasanuzzaman, M., Roy, C. K., & Schneider, K. A. (2016). How Developers Use Exception Handling in Java? (pp. 516–519).

Bloch, J. (2018). *Effective Java*. Addison-Wesley.

Community, R. (n.d.) "Exception Handling." Exceptions - Documentation for Ruby 2.4.0. https://docs.ruby-lang.org/en/2.4.0/syntax/exceptions_rdoc.html. (Accessed January 31, 2023)

Nakshatri, S., Hegde, M., & Thandra, S. (2016). Analysis of Exception Handling Patterns in Java Projects: An Empirical Study (pp. 500–503).

Oracle (2022a). "Advantages of Exceptions." Advantages of Exceptions (The Java™ Tutorials > Essential Java Classes > Exceptions). https://docs.oracle.com/javase/tutorial/essential/exceptions/advantages.html. (Accessed January 31, 2023).

Oracle (2022b). "Putting It All Together." Putting It All Together (The Java™ Tutorials > Essential Java Classes > Exceptions). https://docs.oracle.com/javase/tutorial/essential/exceptions/putItTogether.html. (Accessed January 31, 2023).

Oracle (2022c). "Unchecked Exceptions - the Controversy." Putting It All Together (The Java™ Tutorials > Essential Java Classes > Exceptions). https://docs.oracle.com/javase/tutorial/essential/exceptions/putItTogether.html. (Accessed January 31, 2023).

Wagner, B. (2023). "Exception Handling - C# Programming Guide." Exception Handling - C# Programming Guide | Microsoft Learn. Microsoft. https://learn.microsoft.com/en-us/dotnet/csharp/fundamentals/exceptions/exception-handling. (Accessed January 31, 2023).

Wirfs-Brock, R. J. (2006). Toward exception-handling best practices and patterns. *IEEE Software*, 23(5), 11–13.

22 Automated Test System Design

David R. Carey

22.1 INTRODUCTION

An automated test system (ATS) uses computer-controlled stimulus and measurement instruments to perform tests and evaluate the results associated to the functionality and performance of electronic devices and systems [1]. The device tested on the ATS is referred to as a unit under test (UUT). A UUT can be defined as an electronic module, circuit card assembly, electronic subassembly, electronic assembly, or system. Electronic weapon systems, subsystems, and components are tested by ATS to ensure acceptable performance, functionality, safety, and mission readiness (MIL-STD-1309 D, 1992).

The significant difference between ATS and manual testing is that, with ATS, the test control and decision logic coded in the computer program frees up the test technician to perform other tasks. However, the introduction of automated test does not eliminate the need for an intelligent test technician or analysis of the UUT. In many ways, it complicates the matter because the analysis and decision logic for test must be determined at the time the ATS is being developed. An automated test process cannot rely on the intelligent reaction to circumstances that the technician has always provided in the manual test process. The advantages to the ATS are the speed, precision, and repeatability possible only with a computer.

Automated testing requires less skilled technicians than is required for manual testing. This does not mean that the tests are simpler. In fact, the tests are usually more complex than those available during manual testing. In automated testing, knowledge and experience are embedded into the test program (TP) software by the test engineer/programmer at the time the test program was developed. Automated testing is used in the factory environment to obtain a high production rate using low skill level test personnel. The customer may also find it advantageous to test units automatically that the factory may decide to test manually. Even items that only need to be tested on rare occasions may be needed in a hurry for mission readiness. Automated testing must also test all units that cannot be manually tested due to their high component density or high complexity.

In this chapter, a high-level process by which engineering develops, reviews, integrates, validates, and accepts the ATS for factory, depot, or field testing will be presented. The design process for ATS is presented using standard systems engineering principles. No prior knowledge of ATS development is assumed on the part of the reader.

22.1.1 Automated Test System

The ATS consists of the following: a control device or computer, stimulus source devices, response measurement devices, input/output switching devices, input/output control devices, and the computer program.

The test operator loads the test program onto the ATS. This can be accomplished through several methods. The test program can be loaded from a variety of memory devices such as compact disk (CD), digital video disc (DVD), or flash memory stick. It could also be loaded from a computer network or server.

The ATS computer directly controls the test process through the test program. It will select the stimuli: power supplies, signal generators, and digital signals from the various stimulus devices and set them to the required levels. The stimulus devices are routed to the UUT via an automatically controlled switching device. Simultaneously, the responses from the UUT are conditioned in the

DOI: 10.1201/9781003188377-26

ID and routed through the automatic switching device to the appropriate measurement devices. Measurement scales and calibration factors are applied where appropriate. For example, signal amplitudes may need to be scaled and path loss through ID and ATS cabling may require compensation. The measurement comparisons are made against preprogrammed limits. The operator monitors the test sequence. Results, errors, or operator intervention messages (adjustments, connections, or manual observations) are displayed on the computer monitor. Supplemental instructions may be contained in a Test Program Instruction (TPI) manual (MIL-STD-1309 D, 1992).

22.2 REQUIREMENTS PHASE

Before any hardware and software development can begin, the requirements for the ATS must be clearly established. The early identification of requirements that influence the ATS performance parameters and the ATS configuration from a support standpoint will insure a complete and effective test strategy with minimal cost of ownership.

Best practices to observe during requirements specification include:

1. Maintain all requirements in a single requirements document rather than the statement of work (SOW), communications, and design review notes. If this is not possible (for example, mechanical requirements are embedded in a drawing file), reference this file from the requirements document.
2. Requirements allocation to lower-level assemblies is maintained in this document.
3. Verification and validation requirements are specified in this document.

Requirements are recorded in the document and clarified as needed to fully capture all expectations. Any internal requirements (use of custom hardware, particular commercial off the shelf (COTS) equipment, and software language) are also recorded in the same document.

As requirements evolve over time, new requirements are added and others are deleted (crossed out, maintaining the history of when changed, by whom, and why).

Requirements gathering is the most important part with respect to producing the most cost-effective ATS design. The design team has an opportunity to positively influence the biggest cost and quality drivers for the ATS. The design team must be very aggressive in identifying and acting on opportunities (test at lowest level, test time reductions, and tolerance funneling) that will reduce program recurring costs.

ATS requirements should be of a general purpose, cover reconfigurable testing where applicable, and define the level of repair capability that facilitates the diagnosis and repair of systems, subassemblies, and components. The ATS designer needs to consider additional requirements: ATS throughput, maintenance, calibration, reconfigurability, intended use (factory, field, depot, or engineering development), customer, contractual, regulatory, test engineering functional, and documentation.

22.2.1 ATS Throughput

ATS throughput is defined as the test rate. The faster the speed of the ATS, the fewer the number of ATSs required to meet the production volume requirements. Test yield also forms a part of this equation. A reduction in the number of ATSs invariably drives down costs in other areas such as personnel and space required.

22.2.2 Maintenance and Calibration

Time for routine maintenance and calibration must be planned. Adequate training and backup for the test equipment is vital. A major problem in test can be catastrophic, as it will invariably halt production. Additionally, self-test requirements should be described in enough detail to convey the level of test necessary for the system. This does not need to be a step-by-step definition of testing. As with self-test, the calibration requirements should be defined. This would include any required algorithms. The specific details of pin numbers and paths would be deferred to the design effort.

22.2.3 Reconfigurability

Manufacturing and maintenance lines must be capable of supporting multiple products with as short a changeover time as possible. In a typical scenario, these changeovers are often out of the control of the test engineer but there must be methods for updates and the provision to reconfigure the ATS as different products are pressed into service.

22.2.4 Intended Use Requirements

The intended use of the ATS needs to be defined. Will the ATS be used in the factory, field, maintenance depot, or for engineering development? Each location has its own unique set of requirements.

22.2.5 Customer Requirements

Customer requirements define the features of an ATS that are needed by the customer. Customer requirements differ by product type and product need. These requirements will address price, quality, functionality, reliability, performance, maintainability, sustainability, ergonomics, efficiency, safety, compatibility, options, and control.

22.2.6 Contractual Requirements

Contractual requirements are specified in the statement of work and purchase order agreement. They will define the terms, conditions, and restrictions along with operating agreements, equipment leases, software licensing agreements, or other contractual commitments.

22.2.7 Regulatory Requirements

Regulatory requirements are defined by the government. Regulations direct how a business manages its business. The primary regulations for the DoD are the Federal Acquisition Regulation (FAR) and the Defense Federal Acquisition Regulation (DFAR). Government contractors are required to roll FAR and DFAR clauses down to their subcontractors.

22.2.8 Test Engineering Functional Requirements

Test engineering functional constraints must be evaluated in conjunction with the UUT test requirements. Functional requirements of the ATS may be factors in the determination of the effectiveness of the solution. For each alternative considered, the cost benefits should summarize as to how each satisfies or does not satisfy the functional requirements that must be met.

22.2.9 ATS Documentation Requirements

An ATS documentation package is required by the ATS test engineer/programmer to assist in test software development. ATS documentation includes:

- ATS operating characteristics
- ATS programming manual
- ATS operation and maintenance manual

Before test programming is initiated, the test engineer/programmer should be familiar with the first two items, particularly the ATS operating characteristics and software coding, and should be exposed to ATS operation and maintenance concepts. This provides insight into the overall automatic test function and often provides a better understanding of how stimuli are generated and response measurements are taken, stored, and evaluated. For convenience, ATS operating characteristics and other pertinent programming information can be summarized on a separate capabilities chart or integrated into a separate document, such as the ATS programming manual.

As ATS hardware is developed, new and better test techniques may be implemented or even required in the event certain critical performance parameters are to be achieved. If improvements or modifications to the ATS system are implemented during test software development, complete updates to the ATS documentation must be prepared and distributed to the test engineer/programmer.

If the ATS is still under development, UUT test program development should not be initiated until system parameters are completely defined and have been dynamically verified. If this is not possible, draft versions of each ATS document must be prepared and distributed to the test engineer/programmer. As the ATS system is developed, and design objectives are changed or waived, the original ATS draft document must be updated accordingly to keep the test engineer/programmer informed. Because some changes in ATS operating characteristics can be expected, UUT test software development conducted during ATS development requires special consideration. These early programs should be written to reflect the ATS interim specifications, with the limits on tests requiring high accuracies being left open until system tolerances have been verified. Integration of ATS self-test programs at early stages of development will aid in establishing the true system performance and confidence.

22.2.9.1 ATS Operating Characteristics

The ATS operating characteristics should contain information about the computer operating characteristics, and measurement and stimulus device information. This information is necessary to allow the test engineer/programmer to develop the best strategy to effectively use the ATS to test the UUT and determine if a fault condition is present.

It is critical that the parametric characteristics of the ATS be made known to the test engineer/programmer. The characteristics should be at a point of reference close to the measurement plain of the UUT. This means that the impedance, accuracy, signal amplitude, and timing of the ATS resources must be specified as they appear at the ATS interface to the UUT and not at the internal connections of the ATS devices. This will allow the degradation of ATS signals imposed by system characteristics to be incorporated into calculations made while developing the UUT tests.

22.2.9.2 ATS Programming Manual

The programming manual should serve as the basic communication link between ATS hardware and the test engineer/programmer. In addition, it must serve as an integrated systems manual describing the coding for various programmable stimuli and measurement devices. It must also describe for the test engineer/programmer the capabilities of the source language; and how the ATS is used to achieve desired results. It is the single most important document for consolidating ATS operation and programming information and, should be organized for quick reference.

The programming manual should include all signal routing and switching data, all stimulus and measurement characteristics, and must identify all ATS input/output connector terminals. It should also describe each ATS programmable stimulus, measurement, and switching function in terms of the operational ATS and options for the system configuration. If the details of how to program functional tests in terms of stimuli parameters and measurement capabilities are not clearly specified with statements defining limitations, i.e., frequency, voltage, granularity, and accuracy the test engineer/programmer may have difficulty programming complex tests.

The programming manual should be the responsibility of the ATS systems engineer with both hardware, software, and test design experience.

22.2.9.3 ATS Operation and Maintenance Manual

This manual should contain all the instructions and procedures for powering the ATS, loading or operating and control software, and operating the ATS in its different modes. It should also include details of any ATS operating procedures or special hardware and software aids which are of assistance during UUT test program integration. These integration aids are useful in single stepping tests, locating and reading test stimuli or response data from system memory, and configuring the instruments and connection paths used by the ATS for a particular test. The manual should also contain descriptions of procedures or software available for ATS troubleshooting and corrective maintenance. The ATS operating manual must also include recovery procedures from unanticipated events such as power interruptions.

22.3 ATS DESIGN PHASE

The success of any engineering project is influenced by the accuracy and clarity with which tasks are defined. The more definitive the task description and the more integrated the overall project structure, the higher the probability of achieving the desired results.

In ATS design, the need for specific tasks and responsibilities is equally important. These tasks and responsibilities are designed to support reaching the test objectives and requirements. To achieve these objectives the test engineer must understand the UUT, any software tools available to assist in design and development, and the design policy that states the requirements and standards to be applied to the design. To accomplish this, the test engineer must obtain complete and concise documentation on these various areas. Incomplete or inaccurate documentation can increase development time and costs.

In the ATS design phase, the UUT performance requirements are reviewed. Critical issues are identified, long lead commercial test instruments, tight tolerance requirements, or custom vs. COTS design.

ATS requirements would map into tasks that should be performed during the design process. The first task is to define the ATS environment factory, depot, field, or engineering development. Determine the scope of the ATS, test a single UUT type or multiple UUT types.

The designer will identify all the stimulus and measurement instruments from a review of the signal and measurement requirements and specify the instrumentation communication busses, AXIe, GPIB, USB, PCI/PCI Express, PXI/PXI Express, Ethernet/LAN, and MXI.

Once the instruments and communication types are determined then the following can be identified: ATS calibration instruments, software issues (legal, licenses, operating system), ATS long lead items (COTS, special items), establish ATS development, integration, and certification needs, preliminary parts list, and establish verification and validation (V&V) concept and certification criteria.

At this point, the conceptual design will be ready for review. The concept design review would be the first several required design reviews. How the ATS will satisfy the requirements is presented. Following a successful concept design review, the concept will get handed off to detail ATS hardware engineers. Furthermore, software development can begin by generating the test program software flow diagram. Maximum efficiency in budget is realized because the detail hardware designer and test engineer/programmer have clear direction to proceed.

The project manager will release an updated ATS budget, schedule, and any special consideration. The design team will create the ATS detailed bill of materials (BOM) and technical documentation package (TDP). They will also generate the preliminary V&V Plan, calibration procedure, functional test procedure, and update certification requirements. The complete detail design (hardware) should be ready for a detailed hardware design review or preliminary design review (PDR).

At the same time, the test engineer/programmer could hold a test program concept review. The successful PDR will trigger the project manager to authorize procurement, fabrication, and fabrication verification for long lead items.

Following the software concept design review, test program development can begin. Development of the test program requires a combination of talents from a variety of personnel. The ATS test engineer/programmer must be aware of any design and development software available that would be applicable to this task. This will enable the ATS test engineer/programmer to readily make changes or modifications, if necessary, to adapt existing software to a particular application.

22.4 INTEGRATION PHASE

In this phase, the process of debugging the operation and performance of the software and hardware is accomplished. Care should be taken when making changes to any element of the test program software because any one change may affect numerous tests.

Integration is the process of debugging the operation and performance of the ATS and test program; that is, test program execution on the ATS with the UUT, interface hardware, and documentation. Exercising a fault-free UUT on the ATS with the associated test program is the only means for conclusively verifying the performance and effectiveness.

The individual integrating the ATS is usually the test engineer/programmer, who is familiar with the UUT, the ATS, and the test program. Whoever the integrator is, they must be able to run the program on the ATS, to evaluate the test results, and to update or supervise the updating of the software, as required. It is important to maintain a complete log of changes or modifications made to the test program to prevent the need for duplicate effort during integration.

During check-out of the performance routines, the integration of operating instructions described in the TPI must be performed concurrently with test program integration. All printed or displayed material must be checked for clarity and accuracy. Since it is best to simulate the intended use case of the ATS as closely as possible, the ATS operator skill level should be close to the intended operator skill level and training. In this way, many otherwise undiscovered ambiguities or unclear operator instruction statements displayed or referenced can be clarified before program acceptance testing.

22.5 VALIDATION AND ACCEPTANCE PHASE

The purpose of validation is to ensure the accuracy, repeatability, reliability, and compliance of the ATS and associated test program, and operating procedures prior to certification. A validation plan should be developed.

The intent of ATS validation is to demonstrate that the ATS adequately performs all testing required by the requirements document and that appropriate procedures and operating instructions are in place to allow the use of the ATS for hardware testing.

During ATS development the designer establishes general guidelines in accordance with ATS requirements, outlining the means and procedures by which they plan to acceptance test the ATS. These general guidelines are approved in advance as an Acceptance Test Plan (ATP).

Acceptance is the final phase of the ATS development process. In this phase, the finished product is demonstrated to the customer and must meet all specifications that were set forth in the requirements document. Any information, text, logic diagrams, and schematics necessary for support of the test software are included in the final package as supplementary data.

22.5.1 ACCEPTANCE TEST PLAN

The ATP is developed to specify the demonstration of conformance to the ATS requirements and identify the level of confidence for the demonstration for acceptance tests. The quality assurance provision of the requirements documents should be an acceptance test matrix (ATM). Test procedures will be developed that will completely exercise the ATS to all provisions in the ATM. The test procedures will be designed to identify errors such as run time errors, procedural errors, and errors that inhibit performance. The procedures will be developed to exercise each branch of the software including fault isolation procedures (both manual and automatic). The procedures will be developed to validate self-test, calibration, program signature, and performance tests with a fault-free UUT.

22.5.2 Final Acceptance Procedures

The final acceptance procedures (FAP) include the successful completion of the validation testing, the approval of the design, and the verification of all corrective actions. Final acceptance is complete when the design review records indicate approval of the design, the ATS operation has been demonstrated, and the required corrective actions for all identified problems have been implemented.

22.6 SUMMARY

With the advent of more powerful computers, not only are traditional instruments becoming smarter; but entirely new types of instruments are appearing with far more capabilities, versatility, and accuracy. The ATS reduces the chance of human error by minimizing the need for manual intervention. Under automatic control, both test performance and fault isolation can be achieved with considerable speed and precision. In addition, the ATS provides consistency in the test operation, alleviates the impact of variation made during the test of the UUT, and reduces the amount of training required. The reader should have gained a better understanding of the systems engineering design process for an ATS. Additionally, they should be able to recognize the top requirements needed for an ATS and test program development effort.

Finally, not addressed in this chapter, a risk analysis should be performed to identify any potential areas of risk after the ATS requirements have been established. For an ATS, this may be new measurement techniques, design risk with the use of new or unproven equipment, supply chain risk, or customer communication risk leading to scope change. The ATS designer is responsible for defining technical risk areas, and the program manager is responsible for business risk. It is left to the reader to research more into risk assessment and management techniques.

WORKS CITED

MIL-STD-1309 D (1992). *Definition of Terms for Testing, Measurement and Diagnostics, Military Standard.* Washington, DC: Department of Defense.

23 Interoperability and Design/Test Standards

Michael Seavey

23.1 WHAT IS INTEROPERABILITY?

In a minimal sense, the capability to transfer data from one computer system to another without transformation loss is interoperability. In a broader sense, interoperability describes the extent to which one computer/software system purposefully connects and interchanges interpretable information (e.g., interact) with another in a timely manner. This interaction is between two different systems, subsystems, subassemblies, or software applications from a range of sources including manufacturing factories and repair facilities. This applies even when different manufacturers are involved.

Specifically within the aviation industry, interoperability is the capability of two or more networks, systems, components, or applications working together through exchanges of information between them, without any restriction, and with the ability to use the exchanged information for technical or operational purposes without restriction. It relies on uniform principles and global standards; achieved through procedures, hardware and software interfaces. Interoperability makes aviation systems, procedures, equipment, and training compatible around the world, facilitating seamless services to airspace users.

Being able to exchange information between applications, databases, and other computer systems is crucial for the modern economy.

Interoperability is not to be confused with integration, compatibility, or portability.

23.2 WHY IS INTEROPERABILITY NEEDED?

Within the aviation industry, global interoperability is a necessity. Aircraft are flying all over the world; they cannot afford to have non-standardized certifications.

Crews trained in non-standard ways, or facing many different procedures is inefficient, time-consuming, and prone to error. Standardization allows interoperability and prevents aviation from becoming more expensive and less safe.

While different regions may need different solutions, solutions should be adaptable and globally compatible.

23.3 BENEFITS OF INTEROPERABILITY

Lowered costs are a natural result of improved interoperability, given the improved efficiency in data sharing.

Interoperability makes testing more efficient by avoiding redundant tests and improving the sharing of the results, in both time and effort.

Interoperability results in easier access to items such as details in electronic records, but it will only work to the extent that various software vendors will agree to share what often can be proprietary system information. What's more, stakeholders must be willing to share details over networks for access to different databases.

DOI: 10.1201/9781003188377-27

23.4 IMPORTANCE OF INTEROPERABILITY

Recognizing the importance of interoperability is crucial. Many systems, subsystems, and pieces of equipment are computer-based and connected with one another over wired and wireless networks.

Interoperability is more a philosophy instead of just a standards-based interaction between computer systems. The intent of interoperability is to boost efficiency. It permits the presentation of data to be consistent.

More people connect with one another over networked computerized systems, and enhanced connections speed up reaction times as people can communicate more easily.

To reflect the growing importance of interoperability as a requirement for the product within the design phase of technical systems, the introduction of a design phase called "Design for Interoperability" has occurred; motivated by the ever-growing importance of information technology transfer, for both current and future products.

23.5 LEVELS OF INTEROPERABILITY

23.5.1 Foundational Interoperability

Foundational interoperability enables one information system to exchange data with another information system. The system receiving the information does not need to interpret the data; it will be instantly available for use.

23.5.2 Structural Interoperability

Structural interoperability defines the format of the data exchange. Standards govern the format of the messages sent from one system to another so that the purpose of the information is evident and passes through without alteration.

23.5.3 Semantic Interoperability

Semantic interoperability is the highest level of connection. Two or more different systems or parts of systems exchange data with an unambiguous, shared meaning and use the information readily. Semantic interoperability is therefore concerned not just with the packaging of data (syntax), but the simultaneous transmission of the meaning with the data (semantics).

23.5.4 Syntactic Interoperability

Syntactic interoperability is a prerequisite for semantic interoperability. Syntactic interoperability deals with the packaging and transmission mechanisms for data, as defined either by the XML or by SQL Standards.

23.6 INTEROPERABILITY STANDARDS

Interoperability Standards help foster communication between dissimilar systems. Interoperability Standards have been developed to identify how computer-based systems communicate. The Standards focus on interoperability between elements and common item data interfaces among elements. The Standards establish the specific terms that an implementation will be disseminating, establish protocols for networking, and establish security and encryption schemes, in order to describe the system and their connections.

Standards enable a variety of providers. Increasing commonality among providers while decreasing unique configurations has the potential to reduce the scope of the development effort.

The standards focus on interoperability between elements and common item data interfaces among elements.

The ability of components, systems, or vehicles delivered from multiple sources to work together as an effective system is important to the success of any mission. Good collaboration can make technology development and system maturation more efficient, by sharing the lessons learned and failures that drive requirements. Using standard assumptions can also make development more efficient by making tests conducted by one partner relevant and valid to multiple partners.

23.7 EXAMPLE INTEROPERABILITY SCENARIOS

A French Operational Repair Facility attempts to send failure history data to a German Intermediate Repair Facility. Without a standardized transmission protocol, the attempted transfer would be similar to two people speaking different languages trying to communicate. Similarly, should a Depot Repair Facility requiring 12 packets of data, only receive four, there will be no interoperability between facilities (Figure 23.1).

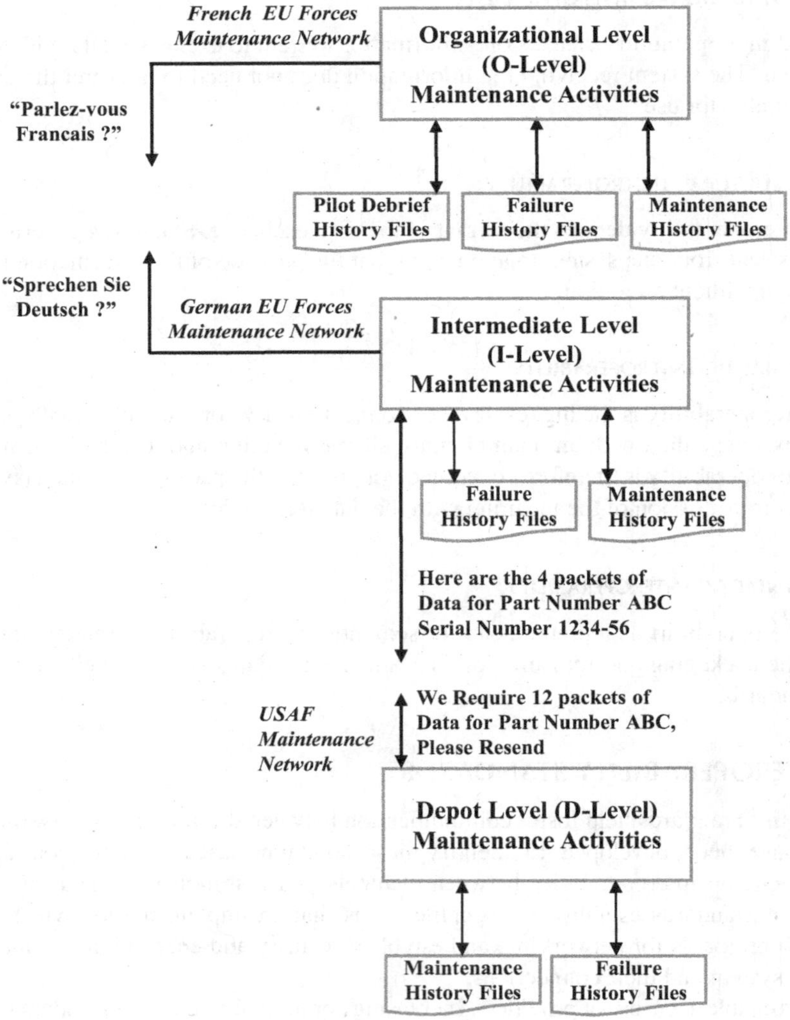

FIGURE 23.1 Maintenance activity interaction.

Figure 23.2 represents a high-level block diagram of a radar. Interoperability allows timing and control to transmit to the transmitter, receiver, or signal processor, even if they are using different software systems internally or if they are from different manufacturers.

A family doctor sends a patient to a specialist to address a specific medical issue. The family doctor can transmit the patient's records to the specialist, and the specialist's team can later quickly send over the results to the family doctor's office, even if the two practices are using different software systems internally (Figure 23.3).

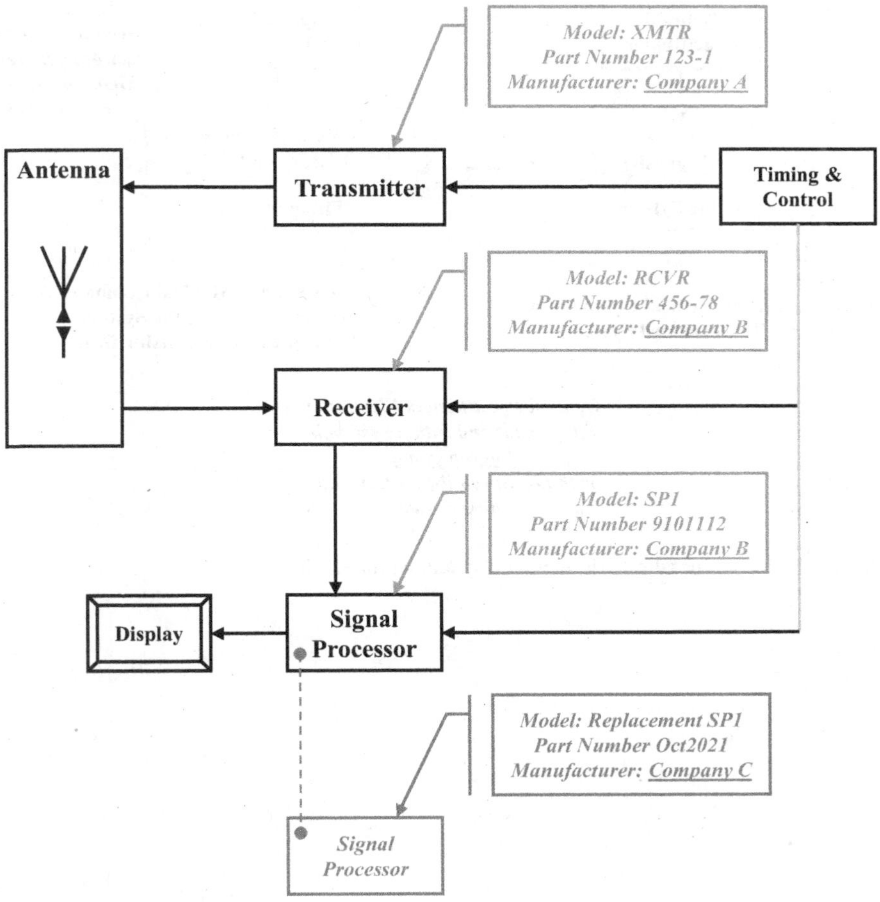

FIGURE 23.2 High level radar block diagram.

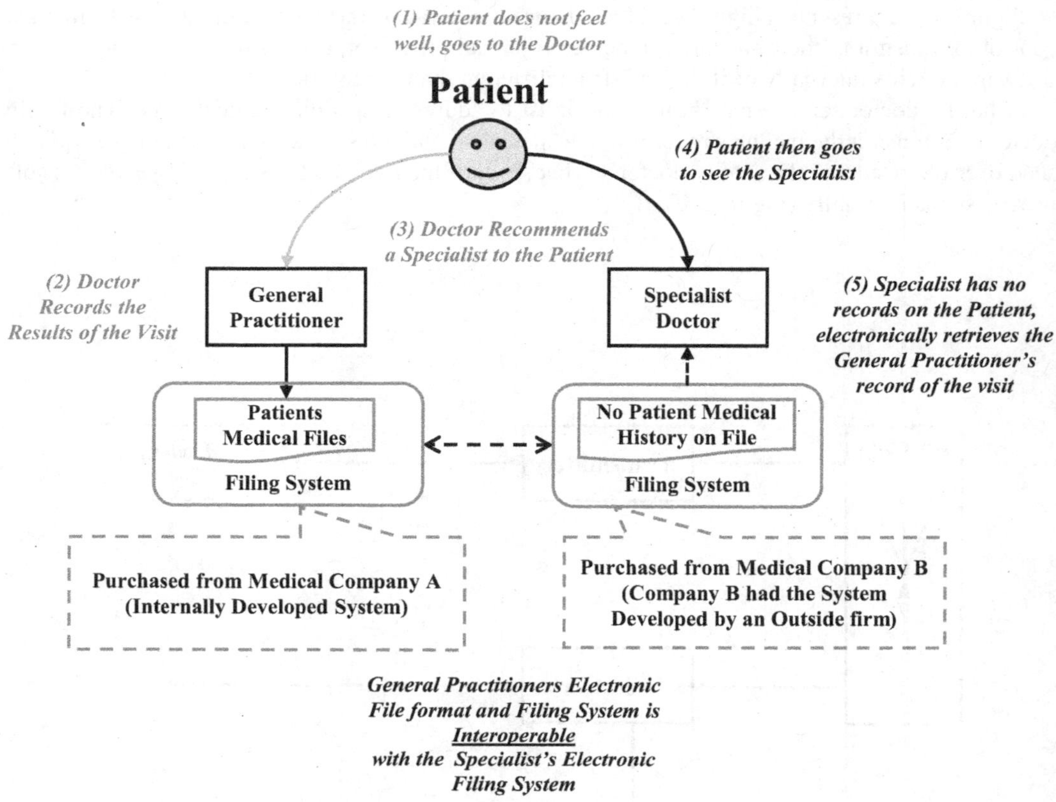

FIGURE 23.3 Patient, general practitioner, and specialist interaction.

Part 5

System Health

24 Uncertainty and Uncertainty Propagation

Alessandro Ferrero

24.1 INTRODUCTION

It is often believed that a good measurement result can be obtained by simply reading one or more instruments. Actually, this is only one step of the whole measurement process, and obtaining a correct measurement result requires more complex activities than the mere reading of instruments.

To fully understand this concept, one of the fundamental concepts in *metrology*, that is the science of measurement, two important documents should be referred to: the *International Vocabulary of Metrology* (*VIM*) (BIPM JCGM 200:2012, 2012) and the *Guide to the expression of Uncertainty in Measurement* (*GUM*) (BIPM JCGM 100:2008, 2008), both published by the BIPM (Bureau International des Poids et Mesures).

The VIM, in its clause 2.1, defines *measurement* as the "*process of experimentally obtaining one or more quantity values* that can reasonably be attributed to a quantity", where the *quantity value* is defined in clause 1.19 as a "*number and reference* together expressing magnitude of a quantity" (BIPM JCGM 200:2012, 2012).

It is worth noting that the VIM refers to *one or more* quantity values that can *reasonably* be attributed to the quantity to be measured (the *measurand*), thus implicitly stating that a single quantity value cannot fully represent the measurand.

This is confirmed by the definition of *measurement result*, given in clause 2.9: "*set of quantity values* being attributed to a measurand together with any other available relevant information". This is clearly stating that a *set* of values is needed and that *other available relevant information* must be associated to this set of values.

The VIM adds notes 1 and 2 to this clause, to refine the concept of *other* information. Note 1 states: "*A measurement result* generally contains "relevant information" about the set of quantity values, such that some may be more representative of the measurand than others. This may be expressed in the form of a probability density function (pdf)".

Note 2 states: "*A measurement result* is generally expressed as a single measured quantity value and a measurement uncertainty", where *measurement uncertainty* is defined in clause 2.26 as: "*non-negative parameter characterizing the dispersion of the quantity values* being attributed to a measurand, based on the information used".

All above-mentioned definitions start drawing a clear picture of the amount of information a measurement result may provide about the measurand. Since a single quantity value is not enough to fully represent a measurand, and additional information is needed to assess whether a set of quantity values may be more representative of the measurand than others, a measurement result cannot provide complete information about the measurand, but only a finite amount of information. This concept reflects the more philosophical assumption about the impossibility to obtain the *true value* of the measurand, which remains unknown and, quite often, unknowable (Ferrero and Petri, 2015).

One of the most important tasks of metrology is, therefore, that of evaluating the degree of completeness of the information provided by a measurement result by characterizing the dispersion of the quantity values being attributed to a measurand. Modern metrology has defined the measurement uncertainty concept as the main tool for obtaining a quantitative evaluation of such a dispersion.

DOI: 10.1201/9781003188377-29

24.2 A MODEL FOR THE MEASUREMENT ACTIVITY

The uncertainty concept can be fully understood if a model for the measurement activity is considered, capable of showing where and how uncertainty in measurement is originated so that each relevant contribution can be identified and evaluated.

To derive such a model, it is worth considering that measurement is a goal-oriented activity, since measurement results are generally employed as relevant input elements in a *decision-making process* aimed at identifying the best *actions* needed to achieve established *goals* while satisfying given *conditions* (Ferrero and Petri, 2015).

The main processes involved in these activities are represented by the blocks in Figure 24.1. It can be readily perceived that there is only one experimental process, and all other ones are descriptive processes, since they do not involve any experimental activity, but are aimed at better describing the measurement context. It is worth analyzing them briefly, to understand how they contribute to measurement uncertainty.

24.2.1 Identification and Modeling Process

It is well known that any measurement activity depends not only on the measurand but is also influenced by a number of other quantities, typically environmental ones, that must be correctly identified.

The identification process is therefore aimed at identifying the best suited measurand and measurement system to achieve the goals and the *environmental properties* that may affect the measurement result.

The effects of such environmental properties (such as, for instance, temperature, humidity, pressure, and gravity) and *time* appear as variations in the measurement result caused by their own variations.

It is then necessary to identify which properties, called *influence quantities*, affect the measurement result in a significant way, so that they can be controlled during the measurement process, or, in the case they cannot, *corrections* are applied. This is the case of time, that cannot be controlled and may cause drifts in the instrument: hence, drift entity since last *calibration* should be estimated and proper corrections applied.

The identification of measurand, measurement system and influence quantities is, however, useless unless a *mathematical model* is also defined to describe the interactions between these items. In such a model, each identified property is described by a *mathematical variable*, and the interactions by *mathematical relationships* among such variables.

The obtained model is generally deterministic, although probabilistic, fuzzy, neural, and neuro-fuzzy models are also used when the measurement context is too complex to be described by a deterministic model.

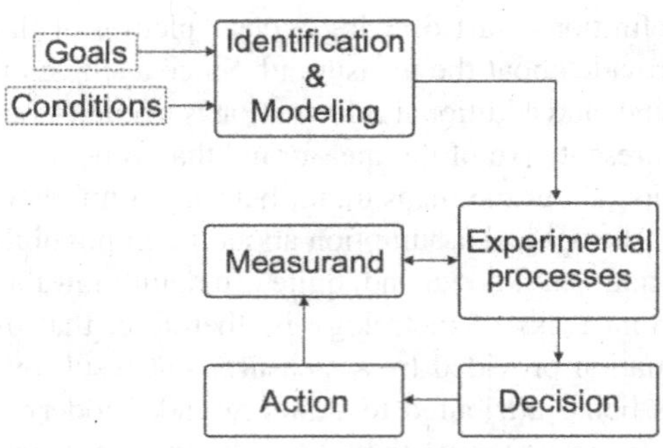

FIGURE 24.1 A measurement model showing the processes involved in the measurement activity.

The identification and modeling process provides a symbolical description of the whole measurement process and is, therefore, a critical element being all further steps based on this model.

Its criticality derives, mainly, from the fact that it can only provide an *approximated description* of the measurement context, due to the incomplete knowledge of all involved quantities and interactions and the intrinsic approximation of the mathematical representation of variables and relationships. This is the first reason for uncertainty in measurement, since, due to the approximations in the model, only a limited amount of information can be retrieved by the measurement process, regardless of the accuracy of the employed measurement system.

24.2.2 Experimental Processes

The experimental processes are, basically, the practical implementation of the measurement model obtained by the previous step. These processes are generally regarded as the main measurement processes, probably because they are the most visible ones, from the operative point of view.

They too represent a reason for uncertainty in measurement, due to the non-ideality of the employed equipment.

24.2.3 Decision Processes

As aforementioned, measurement results are mainly used as input elements in decision-making processes. Generally, the decision process is the result of the comparison of the measurement result with one or more threshold values (for instance given by a tolerance range, or limits set by regulations or laws) so that a decision can be taken according to where the measurement result is located with respect to the threshold values.

This is an apparently trivial process, if a single quantity value is incorrectly taken as the measurement result. However, a measurement result is correctly expressed by a set of quantity values and, consequently, the whole set of values has to be compared with the threshold values. As it will be shown in the next Section 24.7, the comparison is not immediate and implies the risk of making a wrong decision, risk that depends on measurement uncertainty and can be estimated starting from the evaluated uncertainty value.

24.3 UNCERTAINTY SOURCES

The model presented in the previous section allows us to identify the processes that give origin to the uncertainty contributions: the identification and modeling process and the experimental processes. These processes originate the two main contributions to uncertainty, defined by the VIM (BIPM JCGM 200:2012, 2012) as *definitional uncertainty* and *instrumental measurement uncertainty*, respectively.

24.3.1 Definitional Uncertainty

As already discussed in previous Section 24.2.1, the mathematical description of the measurand, environment, measuring system and their interactions provided by the identification and model process is imperfect.

Let us suppose, for instance, that the measurand is the diameter of a shaft and that the identified model, for the shaft, is a geometrical cylinder. This implies the assumption that the diameter is constant along the shaft, and a single diameter measurement taken on a randomly chosen point of the shaft is needed to measure the shaft's diameter.

However, due to manufacturing imperfections, the diameter may not be the same along the whole length of the shaft, and therefore, the single measured value, alone, does not represent the possible dispersion of the measurand values along the shaft. This may lead to an incorrect decision, for instance, in setting the tolerances for the internal diameter of a bearing with which the shaft shall be coupled.

This contribution to uncertainty is called *definitional uncertainty* by the VIM and defined, in its clause 2.27 (BIPM JCGM 200:2012, 2012), as the "*component of measurement uncertainty resulting from the finite amount of detail in the definition of a measurand*". Since only a *finite amount of details* is known about the measurand, and not the *infinite* amount of details that a complete knowledge requires, any model derived from these details is inevitably uncertain, thus contributing to measurement uncertainty.

Moreover, since the identification and modeling process represents the preliminary step of every measurement process, it affects all subsequent steps. This implies, as stated in Note 1 to VIM's clause 2.27 (BIPM JCGM 200:2012, 2012), that "*definitional uncertainty* is the practical minimum measurement uncertainty achievable in any measurement of a given measurand". Its correct evaluation is, hence, of critical importance, also from a mere economical perspective, because even the most accurate and expensive measurement equipment fails to reduce measurement uncertainty below the definitional uncertainty.

24.3.2 Instrumental Measurement Uncertainty

The experimental processes considered in Section 24.2.2 imply the use of measurement systems – that is the whole set of instruments and devices needed to practically perform a measurement activity – and their components are not, of course, ideal. This originates a contribution to measurement uncertainty that the VIM, in its clause 4.24 (BIPM JCGM 200:2012, 2012), calls *instrumental measurement uncertainty* and defines as the "*component of measurement uncertainty arising from a measuring instrument or measuring system in use*".

As explained in Note 1 to this clause, "*instrumental measurement uncertainty is obtained through calibration* of a measuring instrument or measuring system". While this is not the place to discuss *calibration* in depth, and we address the reader to (Ferrero, 2015) for more details, it is worth recalling that calibration implies the comparison of the measurement result provided by the measuring system in use with that provided by a *reference* system and the result of such a comparison is employed to correct the measurement result provided by the system in use.

Since the reference system, no matter how accurate it can be, cannot be ideal, the result of a calibration is affected by uncertainty too. Hence, this *calibration uncertainty* represents the minimum measurement uncertainty achievable by the measurement system in use, after calibration, if used in the same operating conditions as those at calibration time[1].

Proper calibration as well as a careful selection of the calibration laboratory are therefore extremely important in assessing instrumental measurement uncertainty. Since calibration uncertainty represents the lower bound for this uncertainty component, it has to be suitably lower than the expected uncertainty contribution provided by the measuring system in use; otherwise, the expected uncertainty can be never achieved.

24.4 THE EFFECT OF UNCERTAINTY

The discussion of the presented model for the measurement activity and the analysis of the uncertainty sources leads to a first important conclusion: if a measurement procedure is repeated several times on a measurand that is assumed to be constant throughout the entire process, different measured values will be obtained. None of them can be considered as representative of the measurand, but the whole set of obtained values shall be considered to represent the measurand.

This conclusion can be graphically represented, as shown in Figure 24.2, by a target, whose center is the measurand value, and each shot on the target represents a single measured value. It can be readily perceived that the gray dots representing each measured value are dispersed about their center of gravity (the mean value of the measured value) and that this point does not coincide with the target's center.

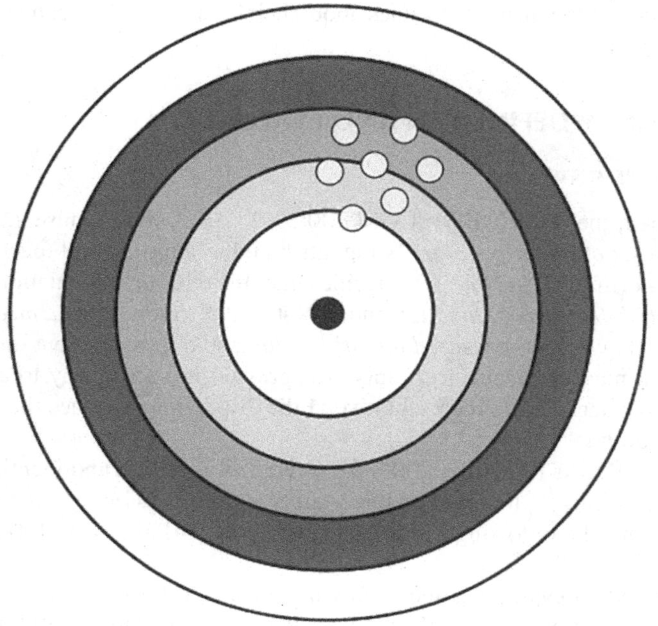

FIGURE 24.2 Graphical representation of repeated measurements on the same measurand. The measurand value is represented by the black target's center. Each gray dot represents the value returned by each measurement.

The displacement of the center of gravity with respect to the target's center, or, abandoning the target example and using stricter terms, the deviation of the mean value of the measured values with respect to the measurand value, represents a *systematic effect*. Such an effect does not change during repeated measurements and affects the measured value in a constant way.

Systematic effects are generally due to imperfections in the employed instruments, such as, for instance, the different length of the two beams of a weighbridge, or to influence quantities, such as, for instance, the temperature on a caliper.

Since they affect the measured values in a constant way, they cannot be easily detected and repeating the measurement procedure is ineffective. Generally, they can be detected by changing the measurement procedure, repeating the measurement procedure after having modified the influence quantities, or by comparing the measurement results obtained from two independent measuring systems, one of which is known to be not affected by systematic effects.

Once systematic effects have been recognized, suitable corrections can be applied, since they remain constant until the measurement conditions do not change. Indeed, the best practice in measurement requires that systematic effects are recognized and corrected for, as stated by the GUM in its clause 3.2.4 (BIPM JCGM 100:2008, 2008): "*it is assumed that the result of a measurement has been corrected for all recognized significant systematic effects and that every effort has been made to identify such effects*".

Going back to Figure 24.2, the dispersion of each measured value about their mean value represents the result of *random effects*. These effects are due to unpredictable, stochastic variations of the influence quantities that reflect themselves into variations of the values returned by repeated measurements.

Due to their stochastic nature, random effects cannot be compensated. On the other hand, their statistical *expectation* is zero, which means that their effect on the mean value of the results of a statistically significant number of repeated measurements tends also to be zero. Moreover, the statistical *standard deviation* of the measured values provides a measure, in a strict mathematical sense, of the dispersion of the measured values about their mean value and can be usefully exploited to define and evaluate uncertainty.

24.5 UNCERTAINTY DEFINITION AND EVALUATION

24.5.1 The Requirements

As already mentioned, the VIM (BIPM JCGM 200:2012, 2012), in its clause 2.9, defines the measurement result as a *set of quantity values* being attributed to a measurand together with any other available relevant information. In Note 1, it specifies that this relevant information *may be expressed in the form of a probability density function* and in Note 2 it also states that *a measurement result is generally expressed as a single measured quantity value* and a measurement uncertainty. We may therefore expect that measurement uncertainty is expressed in such a way to allow a probability density function to be identified as representative of the dispersion of values that can reasonably be attributed to a measurand.

The GUM (BIPM JCGM 100:2008, 2008) followed this approach and identified the conditions to grant the practical utility of the defined uncertainty in any possible measurement applications. According to the GUM, the following conditions must apply (BIPM JCGM 100:2008, 2008):

- The ideal method for evaluating and expressing the uncertainty of the result of a measurement should be *universal*. Hence, the method should be applicable to all kinds of measurements and to all types of input data used in measurements.
- The actual quantity used to express uncertainty should be:
    - **Internally Consistent:** it should be directly derivable from the components that contribute to it, as well as independent of how these components are grouped and of the decomposition of the components into subcomponents;
    - **Transferable**: it should be possible to use directly the uncertainty evaluated for one result as a component in evaluating the uncertainty of another measurement in which the first result is used.

Moreover, the GUM (BIPM JCGM 100:2008, 2008) states that in many industrial and commercial applications, as well as in the areas of health and safety, it is often necessary to provide an interval about the measurement result that may be expected to encompass a large fraction of the distribution of values that could reasonably be attributed to the quantity subject to measurement. Consequently, it considers an additional requirement: the ideal method for evaluating and expressing uncertainty in measurement should be capable of readily providing such an interval, in particular, one with a coverage probability or level of confidence that corresponds in a realistic way with that required.

24.5.2 The GUM Approach

According to the previous considerations, it is apparent that the GUM refers to probability as the mathematical tool to represent and evaluate measurement uncertainty. This is correct, from a mathematical and physical point of view, because, as already mentioned, the GUM assumes that *the result of a measurement has been corrected for all recognized significant systematic effects* and that every effort has been made to identify such effects. Therefore, if the only remaining effects that can influence the measurement result are the random ones, due to their stochastic behavior, a measurement result can be represented by a random variable, that, in turn, is described by a probability density function (pdf).

According to this model, the measured values can be considered as realizations of this random variable. Consequently, the standard deviation of this pdf represents a quantitative measure of the dispersion of these values about their mean value. Similarly, the mean value of the pdf can be assumed to be the best estimate of the measurand value.

24.5.3 Standard Uncertainty

According to the previous considerations, The GUM defines, in its clause 2.3.1 (BIPM JCGM 100:2008, 2008), *standard uncertainty* as: "*uncertainty of the result of a measurement* expressed as a standard deviation". Therefore, it satisfies the VIM definition of being a non-negative parameter characterizing the dispersion of the quantity values being attributed to a measurand.

The suggested notation for the standard uncertainty associated to a measured value x is $u(x)$.

Since the pdf of the distribution of measured value is assumed to be known, it is possible to define the coverage interval $\left[\overline{x} - u(x); \overline{x} + u(x)\right]$ about its mean value $\overline{x}$, taken as the best estimate of measurand x, whose coverage probability depends on the pdf. If the pdf is normal, as shown in the example of Figure 24.3, the coverage probability is 68.27%.

However, the GUM (BIPM JCGM 100:2008, 2008) requires that such an interval encompasses a *large* fraction of the distribution of values that could reasonably be attributed to the measurand, and that its coverage probability corresponds in a *realistic way* to that required.

It can be readily perceived, also by observing Figure 24.3, that, at least in the case of a normal probability distribution, the interval provided by the standard uncertainty does not encompass a *large* fraction of the distribution of values that could reasonably be attributed to the measurand, and that the related coverage probability is not *realistic*, since the probability that the measurand value lies outside this interval is greater than 30%.

24.5.4 Expanded Uncertainty

To obtain a more effective coverage interval about the measured value, the GUM, in its clause 2.3.5 (BIPM JCGM 100:2008, 2008), defines the *expanded uncertainty* as a "*quantity defining an interval about the result of a measurement* that may be expected to encompass a large fraction of the distribution of values that could reasonably be attributed to the measurand". In note I to this definition, the GUM also states that "*the fraction may be viewed as the coverage probability* or level of confidence of the interval".

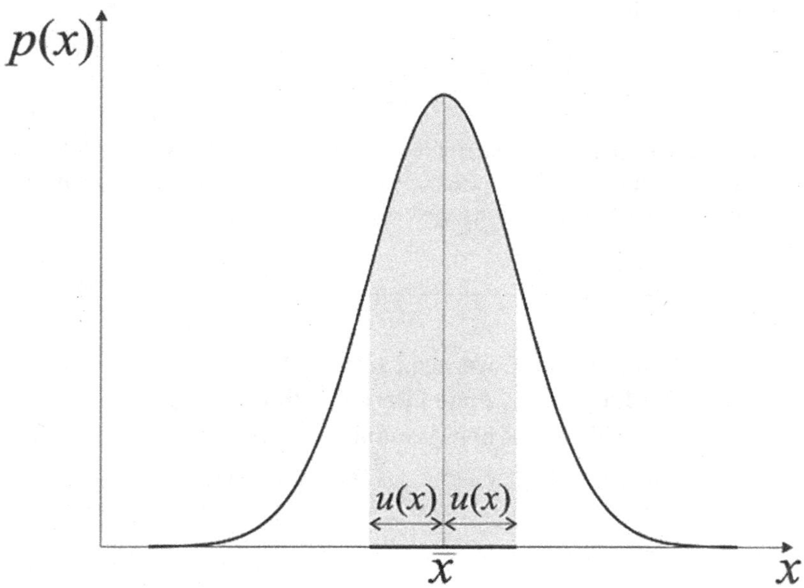

FIGURE 24.3 Example of distribution of the measured values under the assumption of normal pdf centered on $\overline{x}$, with standard uncertainty $u(x)$. The coverage interval $\left[\overline{x} - u(x); \overline{x} + u(x)\right]$ is also shown, whose coverage probability is given by the gray shaded area.

The suggested notation for the expanded uncertainty associate to a measured value x is $U(x)$. In general, the expanded uncertainty is obtained from the standard uncertainty as:

$$U(x) = k \cdot u(x) \tag{24.1}$$

where k is called the *coverage factor*.

Regardless of how the expanded uncertainty is computed, a confidence interval can be correctly associated to interval $\left[\bar{x} - U(x); \bar{x} + U(x)\right]$ if and only if the pdf representing the distribution of the measured values is known, or if *implicit* or *explicit* assumptions are made about the pdf itself, as clearly stated by the GUM (BIPM JCGM 100:2008, 2008).

Consequently, a coverage probability can be associated to the coverage factor k, as the probability that can be assigned to coverage interval $\left[\bar{x} - k \cdot u(x); \bar{x} + k \cdot u(x)\right]$ if and only if a pdf is known or assumed. For instance, Table 24.1 shows the coverage probabilities associated to some typical coverage factors when a normal pdf is considered.

24.5.5 Evaluation Methods

The GUM suggests two methods for evaluating standard uncertainty, defined as follows (BIPM JCGM 100:2008, 2008).

- *Type A evaluation* (GUM clause 2.3.2) *method of evaluation of uncertainty* by the statistical analysis of series of observations.
- *Type B evaluation* (GUM clause 2.3.3) *method of evaluation of uncertainty* by means other than the statistical analysis of series of observations.

24.5.5.1 Type A Evaluation Method of Standard Uncertainty

It is well known that the best estimate of the *expectation* (or *expected value*) μ_X of a random variable X, when N independent observations x_k, $k=1, \ldots, N$, are available, is given by the mean value of such observations, given by:

$$\bar{x} = \frac{1}{N}\sum_{k=1}^{N} x_k. \tag{24.2}$$

The dispersion of the values that can be attributed to X is given by the standard deviation σ, which is the positive square root of the random variable variance σ^2. An estimate of σ^2 is provided by the experimental variance of the N observations, given by:

TABLE 24.1
Values of the Coverage Factor k_p that Produces a Coverage Interval with Coverage Probability p When Assuming a Normal pdf

Coverage Probability p [%]	Coverage Factor k_p
68.27	1
90	1.645
95	1.960
95.45	2
99	2.576
99.73	3

$$s^2(x_k) = \frac{1}{N-1} \sum_{k=1}^{N} (x_k - \bar{x})^2. \tag{24.3}$$

It is known from statistics that the variance of the mean value $\bar{x}$ is given by:

$$s^2(\bar{x}) = \frac{s^2(x_k)}{N}. \tag{24.4}$$

This value represents the best estimate of the variance of $\bar{x}$ and therefore the best estimate of standard uncertainty $u(\bar{x})$ associated with measurement result $\bar{x}$ is given by:

$$u(\bar{x}) = \left|\sqrt{s^2(\bar{x})}\right|, \tag{24.5}$$

and represents standard uncertainty evaluated according to a type A evaluation.

The advantage of type A evaluation is that it requires only repeated measurements, under the same measurement conditions, to evaluate measurement uncertainty under those conditions. Its major drawback, however, is that it requires a number of independent repeated measurements, large enough to be statistically significant. This condition cannot be always met, mainly due to time constraints or difficulties in keeping the measurement conditions constant throughout the repeated measurements.

24.5.5.2 Type B Evaluation Method of Standard Uncertainty

There are many situations in which repeated measurements are not possible, and only one measured value is available. Nevertheless, uncertainty shall be evaluated and therefore the standard deviation of the distribution of values that could reasonably be attributed to the measurand shall still be evaluated.

The GUM (BIPM JCGM 100:2008, 2008) states that this can be done by *scientific judgment* based on all of the available information on the possible variability of the measured values (that are represented by a random variable *X*). The pool of information may include (BIPM JCGM 100:2008, 2008):

- previous measurement data;
- experience with or general knowledge of the behavior and properties of relevant materials and instruments;
- manufacturer's specifications;
- data provided in calibration and other certificates;
- uncertainties assigned to reference data taken from handbooks.

According to the available information, the operator shall derive, through his or her scientific judgment, the probability density function that best describes the variability of the measured values. It is then possible to evaluate the standard uncertainty as the standard deviation of the obtained probability density function, and the expanded uncertainty, so that a coverage interval about the measured value can be built with the desired coverage probability.

Of course, the correctness of the obtained uncertainty value and coverage probability are strongly dependent on the reliability of the available information and the way it is used. To this purpose, the GUM (BIPM JCGM 100:2008, 2008) states that the proper use of the pool of available information for a type B evaluation of standard uncertainty calls for insight based on experience and general knowledge and is a skill that can be learned with practice. Therefore, when performed by a skilled operator, a type B evaluation of standard uncertainty can be as reliable as a type A evaluation, especially in a measurement situation where a type A evaluation is based on a comparatively small number of statistically independent observations (BIPM JCGM 100:2008, 2008).

24.5.5.2.1 Numerical Example

Type B evaluation of standard uncertainty can be fully perceived with a numerical example. Let us suppose to measure a DC voltage V using a five-digit digital voltmeter in the 10 V range. Let us suppose that the instrument reading is V_r=7.2648 V. Let us also suppose that repeated measurements cannot be performed and that the only available data about the instrument are the manufacturer's specifications.

For the considered range, the manufacturer provides an accuracy interval whose half-amplitude a is given by 0.02% of the reading plus 6 digits. In the selected range, the weight assigned to the least significant digit is 100 µV and, therefore, we have:

$$a = 7.2648 \cdot 2 \cdot 10^{-4} + 6 \cdot 10^{-4} = 2.1 \cdot 10^{-3}\ \text{V}$$

According to the manufacturer's specifications, interval $V_r \pm a$ encompasses the whole distribution of values that could reasonably be attributed to the measurand, and no other information is available. To estimate the probability density functions of these values, it is possible to refer to the *maximum entropy principle* that, when no other information is available on the way values distribute over a closed interval, leads to consider a uniform probability distribution over that interval, whose standard deviation is given by $a/\sqrt{3}$.

Under this assumption, the standard uncertainty associated with V_r is given by:

$$u(V_r) = \frac{a}{\sqrt{3}} = 1.2 \cdot 10^{-3}\ \text{V}.$$

24.6 UNCERTAINTY PROPAGATION

In many practical applications, a measurand Y is not measured directly, but is derived from N other quantities $X_1, X_2, \ldots, X_N$, according to a functional relationship:

$$Y = f(X_1, X_2, \ldots, X_N), \tag{24.6}$$

where $X_1, X_2, \ldots, X_N$ are measurand themselves.

One of the GUM requirements for uncertainty is that it should be possible to use directly the uncertainty evaluated for one result as a component in evaluating the uncertainty of another measurement in which the first result is used (BIPM JCGM 100:2008, 2008). Therefore, if standard uncertainty has been evaluated for each measurement result of $X_1, X_2, \ldots, X_N$, it should be possible to derive the standard uncertainty of the measurement result of Y from those values.

24.6.1 The Combined Standard Uncertainty

The GUM (BIPM JCGM 100:2008, 2008) defines the uncertainty associated to the measurement result of Y as *combined standard uncertainty* and, to evaluate it, reconsiders (24.6). It can be immediately recognized that an estimate y of measurand Y can be obtained by applying (24.6) to the estimates $x_1, x_2, \ldots, x_N$ of the input quantities $X_1, X_2, \ldots, X_N$. Hence, the measured value for Y is given by:

$$y = f(x_1, x_2, \ldots, x_N) \tag{24.7}$$

Since $x_1, x_2, \ldots, x_N$ are measured values, each of them has an associated standard uncertainty $u(x_i)$, $i = 1, \ldots, n$. To evaluate the combined standard uncertainty $u_c(y)$ associated to the measured value y, the GUM assumes that function $f(\cdot)$ in (24.7) is linear about the measured value y, at least for small variations of each input quantity X_i about their estimates x_i.

Under this assumption, function (24.7) can be approximated by a first-order Taylor series expansion about the expectations $E(X_i) = \mu_i$ of the input quantities. The deviation of the measured value y with respect to its expectation $E(Y) = \mu_y$ is given by:

$$y - \mu_y = \sum_{i=1}^{N} \frac{\partial f}{\partial x_i}(x_i - \mu_i) \tag{24.8}$$

where $\mu_y = f(\mu_1, \mu_2, \dots, \mu_N)$.

The derivatives in (24.8) should be evaluated in the unknown expected values of the input quantities. However, in practice, they are approximated by the values computed in their estimates $x_1, x_2, \dots, x_N$, since the difference is considered negligible.

Moreover, higher order terms should be considered in the Taylor series expansion if all derivatives in (24.8) are nil, or if function $f(\cdot)$ cannot be considered linear about the measured value y. Since these cases are quite rare, the GUM considers the linear approximation as valid (BIPM JCGM 100:2008, 2008).

Equation (24.8) can be squared, thus obtaining:

$$(y - \mu_y)^2 = \left[\sum_{i=1}^{N} \frac{\partial f}{\partial x_i}(x_i - \mu_i)\right]^2 \tag{24.9}$$

that can be expanded as:

$$(y - \mu_y)^2 = \sum_{i=1}^{N} \left(\frac{\partial f}{\partial x_i}\right)^2 (x_i - \mu_i)^2 + 2\sum_{i=1}^{N-1} \sum_{j=i+1}^{N} \frac{\partial f}{\partial x_i}\frac{\partial f}{\partial x_j}(x_i - \mu_i)(x_j - \mu_j) \tag{24.10}$$

Let us now consider the following mathematical expectations:

- $E\left[(y - \mu_y)^2\right] = \sigma_y^2$ is the variance of y;
- $E\left[(x_i - \mu_i)^2\right] = \sigma_i^2$ is the variance of x_i;
- $E\left[(x_i - \mu_i)\cdot(x_j - \mu_j)\right] = \sigma_{i,j} = \sigma_{j,i}$ is the covariance of x_i and x_j.

It is then possible to write (24.10) in terms of such expectations and, taking into account that standard uncertainties are standard deviations by definition, the following equation is obtained for the combined standard uncertainty:

$$u_c^2(y) = \sum_{i=1}^{N} \left(\frac{\partial f}{\partial x_i}\right)^2 u^2(x_i) + 2\sum_{i=1}^{N-1} \sum_{j=i+1}^{N} \frac{\partial f}{\partial x_i}\frac{\partial f}{\partial x_j} u(x_i, x_j), \tag{24.11}$$

where $u(x_i, x_j) = u(x_j, x_i)$ is the estimated covariance associated with x_i and x_j. This covariance can be also expressed in terms of the *correlation coefficient* r defined as:

$$r(x_i, x_j) = \frac{u(x_i, x_j)}{u(x_i)u(x_j)}, \tag{24.12}$$

where $r(x_i, x_j) = r(x_j, x_i)$.

It can be immediately recognized that $-1 \le r(x_i, x_j) \le +1$, where $r(x_i, x_j) = 0$ means that x_i and x_j are independent of one another, $r(x_i, x_j) = 1$ means that they are fully correlated in a direct way, and $r(x_i, x_j) = -1$ means that they are fully and inversely correlated.

Taking into account (24.12), (24.11) can be re-written as:

$$u_c^2(y) = \sum_{i=1}^{N} \left(\frac{\partial f}{\partial x_i} \right)^2 u^2(x_i) + 2 \sum_{i=1}^{N-1} \sum_{j=i+1}^{N} \frac{\partial f}{\partial x_i} \frac{\partial f}{\partial x_j} u(x_i) u(x_j) r(x_i, x_j). \quad (24.13)$$

This last equation is also known as the *uncertainty propagation law* (BIPM JCGM 100:2008, 2008).

An important particular case of (24.13) occurs when all estimates x_i and x_j are independent of each other. In such a case, $r(x_i, x_j) = 0 \ \forall i, j$ and (24.13) simplifies as:

$$u_c^2(y) = \sum_{i=1}^{N} \left(\frac{\partial f}{\partial x_i} \right)^2 u^2(x_i) \quad (24.14)$$

24.6.2 The Combined Expanded Uncertainty

The mathematical derivation that yields (24.13) and (24.14) provides, in a strict mathematical way, the combined standard deviation of a random variable Y obtained by combining N random variables $X_i, i = 1, \cdots, N$. However, these same equations cannot provide any information about the pdf associated with Y, even if the pdfs associated with the input variables X_i are known.

Since the pdf associated to Y is not known, the pdf representing the possible distribution of the values that can be attributed to the estimates y of variable Y is not known too. Therefore, a coverage probability cannot be assigned to any interval about the estimate y.

To associate a pdf to the distribution of values that could reasonably be attributed to y, the GUM (BIPM JCGM 100:2008, 2008) suggests referring to the *Central Limit Theorem* (CLT). This well-known theorem of probability proves that, if function $f(\cdot)$ in (24.6) is linear, none of the considered σ_i^2 variances dominates over the others, and $N \to \infty$, then the pdf of Y is normal, regardless of the pdf associated with X_i. Of course, N cannot assume an infinite value. In this case, the pdf of Y will only approximate the normal pdf, and the larger is N, the better is the approximation.

It can therefore be concluded that, if and only if these assumptions are valid, the expanded uncertainty $U_c(y)$ can be obtained from the standard uncertainty provided by (24.13) or (24.14) multiplying it by a suitable coverage factor k, and the coverage probability listed in Table 24.1 can be associated to the interval $y \pm U_c(y)$.

24.6.3 The Monte Carlo Method and GUM Supplement 1

The validity conditions of the CLT cannot be always satisfied. Several practical cases exist in which function (24.6) is not linear and the number of X_i variables is limited as, for instance, in the case of power measurements, where only two variables are multiplied.

In such cases, the pdf associated to the resulting variable Y is not normal and, therefore, the coverage probability of any interval $y \pm k \cdot u_c(y)$ does not correspond to the coverage probability of the same interval subtended by a normal pdf. The consequence is that the expanded uncertainty is wrongly estimated.

To overcome this problem, Supplement 1 to the GUM (BIPM JCGM 101:2008, 2008) suggests referring to a Monte Carlo simulation to evaluate the expanded uncertainty.

The method is, in principle, quite simple. Having evaluated or assumed a pdf for every input variable X_i in (24.6), a single realization x_{ij} is simulated, taking also into account the possible correlation between the input variables. According to (24.7), a single realization:

$$y_j = f(x_{1j}, x_{2j}, \ldots, x_{Nj}) \quad (24.15)$$

is obtained for Y. By repeating this procedure for $j = 1, \ldots, M$, with M large enough to be statistically significant, an estimate of the distribution of values that could reasonably be attributed to Y is obtained.

The average value:

$$\bar{y} = \frac{1}{M} \sum_{j=1}^{M} y_j \tag{24.16}$$

and standard deviation:

$$u(\bar{y}) = \sqrt{\frac{1}{M-1} \sum_{j=1}^{M} \left(y_j - \bar{y}\right)^2} \tag{24.17}$$

are taken, respectively, as an estimate y of Y and the standard uncertainty $u(y)$ associated with y (BIPM JCGM 101:2008, 2008).

In order to evaluate the coverage probability of any interval $\bar{y} \pm k \cdot u(\bar{y})$, the number Q of values y_j that satisfy condition $\bar{y} - k \cdot u(\bar{y}) \le y_j < \bar{y} + k \cdot u(\bar{y})$ shall be computed. The desired coverage probability is then estimated as $p = Q / M$.

The major drawback of this method is that there is no guarantee that the chosen number of trials M is large enough to provide an accurate estimate of the desired distribution of Y. In general, values $M \ge 10^6$ are supposed to provide reliable estimates (BIPM JCGM 101:2008, 2008), but better results can be obtained by implementing a procedure that selects M adaptively, as the trials progress, following the guidance in Berthouex and Brown (1994).

Of course, the larger is M and the more complex is function (24.6), the higher is the computation time required to obtain the desired result. This makes the Monte Carlo method quite impractical for real-time uncertainty estimation.

24.7 UNCERTAINTY AND DECISIONS

As stated in Section 24.2.3, measurement results are used as input elements in decision-making processes that involve the comparison of the measurement result with one or more threshold values. Since a measurement result is not represented by a single quantity value, but by a pdf, the comparison is not a trivial task and implies the risk of taking the wrong decision. It is intuitive that such a risk is related to measurement uncertainty, and it is important to understand if and how it can be quantified, starting from a correct evaluation of measurement uncertainty.

This problem can be framed as follows. Let us suppose that a value x_m is measured for quantity x, that standard uncertainty $u(x_m)$ is evaluated and that a pdf $p(x)$ is either estimated or assumed for the distribution of values that could reasonably be attributed to the measurand.

Let us also suppose that the measured value shall be compared with a given threshold t and let us suppose that $x_m > t$. Is this enough to conclude that the measurand is surely above the threshold? Actually, a probability exists that the measurand falls below the threshold, given by:

$$Pr(x \le t) = \int_{-\infty}^{t} p(x)\,dx \tag{24.18}$$

This equation can be also geometrically interpreted as the area subtended by function $p(x)$ from $-\infty$ to t, as shown in the example of Figure 24.4, where $p(x)$ is assumed to be normal, and the area is shaded in gray.

The same probability can be more easily obtained as:

$$Pr(x \le t) = F_X(t), \tag{24.19}$$

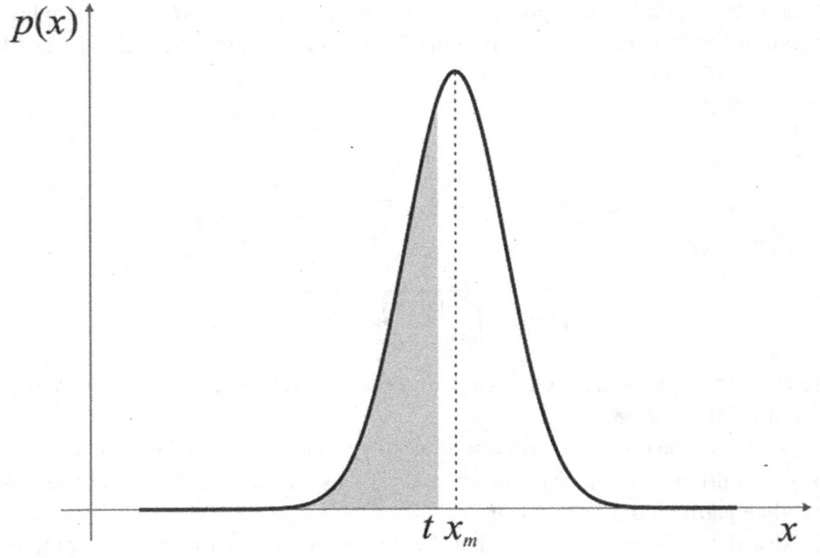

FIGURE 24.4 Geometrical representation of the probability that the measurand value x is lower than threshold t, given a measured value x_m and a normal distribution for $p(x)$. The probability of $x \leq t$ is given by the gray area.

where F_X is the cumulative probability function of variable X and is related to the probability density function by:

$$F_X(x) = \int_{-\infty}^{x} p(\xi)\,d\xi \tag{24.20}$$

Of course, the probability that the measurand is greater than the threshold is given by:

$$Pr(x > t) = 1 - F_X(t) \tag{24.21}$$

It is worth noting, once again, that the obtained probability value is strongly dependent on the assumed probability distribution $p(x)$. Therefore, the assumptions made to identify this probability distribution are of critical importance in the correct representation of the measurement result and the correct evaluation of the risk (probability) of wrong decision. This reaffirms the need for a *scientific judgment based on all of the available information on the possible variability of* the measurand, and hence for experienced operators.

24.8 CONCLUSION

This chapter has briefly recalled the reasons for excluding that a measurement result may provide the true value of the measurand, and has proposed a model for the measurement activity useful in identifying the main sources for uncertainty in measurement.

The uncertainty concept has been then discussed, together with the definition of standard uncertainty and the way different uncertainty contributions can be combined to evaluate uncertainty on a measurement result obtained as a combination of other results.

It can be concluded that the uncertainty concept has general validity and provides a scientifically sound, quantitative evaluation of "how good" a measurement result is. On the other hand, its practical implementation, as recommended by the GUM (BIPM JCGM 100:2008, 2008) and its

Supplement 1 (BIPM JCGM 101:2008, 2008), may have some limitations, mainly due to the considered mathematical theory: probability.

It is worth closing this chapter by mentioning that different solutions have been proposed – as summarized in Salicone and Prioli (2018) – to overcome these limitations. Although the discussion of new proposals for uncertainty expression and evaluation is beyond the scope of this chapter, I wish to conclude it by reminding that metrology, as every other science, is continuously evolving and nowadays limitations shall be seen as a motivation for investigating new theories, methods, and solutions that may become tomorrow's best practice.

NOTE

1 Any deviation from the calibration conditions introduces an additional contribution to uncertainty that shall be evaluated.

REFERENCES

Berthouex, P. M., & Brown, L. C. (1994). *Statistics for Environmental Engineers*. Boca Raton, FL: CRC Press.

BIPM JCGM 100:2008 (2008). *Evaluation of Measurement Data: Guide to the Expression of Uncertainty in Measurement,* 1st Edition. https://www.bipm.org/documents/20126/2071204/JCGM_100_2008_E.pdf/cb0ef43f-baa5-11cf-3f85-4dcd86f77bd6?version=1.7&download=true.

BIPM JCGM 101:2008 (2008). *Evaluation of Measurement Data — Supplement 1 to the "Guide to the Expression of Uncertainty in Measurement" — Propagation of Distributions Using a Monte Carlo Method*, 1st Edition. https://www.bipm.org/documents/20126/2071204/JCGM_101_2008_E.pdf/325dcaad-c15a-407c-1105-8b7f322d651c?version=1.5&download=true.

BIPM JCGM 200:2012 (2012). *International Vocabulary of Metrology: Basic and General Concepts and Associated Terms (VIM),* 3rd Edition. https://www.bipm.org/documents/20126/2071204/JCGM_200_2012.pdf/f0e1ad45-d337-bbeb-53a6-15fe649d0ff1?version=1.11&download=true.

Ferrero, A. (2015). The pillars of metrology. *IEEE Instrumentation & Measurement Magazine*, 18(6), 7–11.

Ferrero, A., & Petri, D. (2015). Measurement models and uncertainty. In A. Ferrero, D. Petri, P. Carbone, & M. Catelani (eds.), *Modern Measurements: Fundamentals and Applications* (pp. 3–45). Hoboken, NJ: John Wiley & Sons.

Salicone, S., & Prioli, M. (2018). *Measuring Uncertainty within the Theory of Evidence*. New York: Springer.

25 Fault Detection, Localization, and Isolation

Jason Rupe

25.1 INTRODUCTION: BACKGROUND AND DRIVING FORCES

Fault detection, localization, and isolation are necessary for any replaceable component or repairable system. Manual inspection is the simplest approach. But in complex systems, sensors and components provide data (telemetry) about the performance of components and systems while periodic tests and logs capture additional information, all of which can be valuable for determining whether a fault has emerged or is growing, and how severe. Faults are localized by identifying the fault type, the location in the system or network where the data were measured or the test performed, etc. Isolation of a fault removes it from the functional system or network, but that may have further impact on system performance, reliability, security, etc.

Fault management in different applications and contexts has similarities and differences. Faults are addressed in testing and deployment, software and hardware, systems and networks, and more. The larger the scale of deployment, or the more important the function, the more important it is to have the telemetry to manage faults to assure reliability, availability, security, and more concerns are met.

There are many foundational tools and techniques to fault management which can be used. Fundamental knowledge of the system or network is important for creating fault management solutions. Model-driven and data-driven approaches (including statistical methods) work in most application areas, and there are emerging techniques that use forms of artificial intelligence including machine learning tools to further improve fault detection and localization, often through identification.

As our networks and systems become more complex, automated, and relied upon, we must continue to develop automation, robustness, reliability, and security in our fault management solutions.

25.2 FAULT MANAGEMENT CONCEPTS AND DEFINITIONS

In this section, we define some foundation for fault management, forming the base for later sections on applications, techniques, and tools.

25.2.1 Faults

A fault is generally an undesirable feature. For our purposes, a fault is a deviation between an expected result or outcome and the experience or actual result or outcome, which is often defined by context such as a use or test case, and is often referred to as the residual. For a fault to be measured, some quantifications are required. For such a deviation or residual to be considered to be a true fault, the deviation or residual should, in some way, be significant. The significance is related to the effect of the deviation such as the failure mode, effect, or its criticality and is often subject to the use case, mission, or desired reliability, availability, or like measure of performance of the system or network.

The use case and context impact the expected results. In complex systems and networks, the appearance of a fault, and even its measurement, may be difficult to define and predict. Some faults generate failure, whereas others may increase the risk of failure or make other potential faults lead

DOI: 10.1201/9781003188377-30

to failure. So, setting benchmarks can be complex in actual use cases. Because the problem is complex, much research has been conducted on the subject.

Faults can appear slowly over time or nearly instantaneously, be transient or permanent, or even be causally related to conditions outside the control of the system such as weather, moisture, temperature, vibration, use cases, and more.

For the sake of this chapter, consider the fault as just defined, and recognize that a failure is a fault in the extreme, a failure being a complete cease of function.

By classifying types of faults, we can provide some examples as well.

- **Software Faults**: a fault in software is an error in coding or design that is manifested through the execution of the code, so it may remain latent.
- **Hardware Faults**: some physical features of the hardware are different than the intent or need and can be an indication of a quality problem, manufacturing failure, damage, degradation, design failure, etc.
- **Input or Output Faults**: the output of one component or subsystem may be the input to another, and an input fault may be incorrect raw material, a flaw, a value outside the anticipated range, etc.
- **Process Faults**: an error in a process or poor quality in a process can lead to faults in a process, such as work in the wrong process step, incorrect work done, etc.
- **Signal Faults**: a fault in an analog or digital signal may propagate through a system in the form of cascading faults and even failure.
- **Data Faults**: an error in data can cause many undesirable consequences including poor decisions, system failure, etc.
- **Intentional**: various sabotage and intentional faults are possible, such as putting a hole in a pressure line, cutting a transmission line, or flooding a network with traffic to block function (denial of service attack).

It is important to note that resiliency is what allows for faults to not become failures. Degradation can be revealed as a fault that can continue to lead to failure once any resiliency mechanisms are no longer sufficient. Degradation is described in other chapters of this book, but consider the familiar example of milk spoilage.

Because a component failure can lead to a subsystem or system fault, and a component or subsystem or system fault can lead to service failure, the possible relationships get complicated. We'll focus on the general concept of faults in this chapter and recognize failure as an extreme fault condition. Note that a single fault can cause multiple effects in the system or network.

Beauty is in the eye of the beholder; what constitutes a fault is determined by the use case and its impact on performance. One person's failure may be another person's fault. Further, faults can be defined statically, or at times be contextual as in dynamical or otherwise complex systems.

25.2.2 Detection

A fault is detected once its existence is identified in some way, be that by a human who can intervene, a sensor that can notify, a test that can be run, or a system that can act in some way on the information it can monitor. Referring to our previous example, milk will degrade through spoilage, and the fault is indicated through smell, texture, and other properties which are determined at use, assuming the user is paying attention.

A fault is often detected by the significance of the residual, the deviation of the actual outcome from an expected outcome, which is the definition of a fault. The milk stinks and is clumpy. Or the milk just smells a bit off. The significance can be defined statistically, by magnitude, or by its relation to the mission or services being supported. Spoiled milk may be dangerous to consume. In short, it is the impact of the fault that makes it important enough to address.

Generally, there are a few types of fault detection methods available, but often you will see others categorize them into two types: processing-based and model-based.

Processing-based methods of detection involve some mathematics or statistics which are applied to a signal or input, and the output is evaluated based on the expected outcome.

Model-based methods of detection involve some representation of the system, such as a truth table, fault tree, mathematical model, or benchmark, used to compare the actual outcome.

Considered another way, a fault can be detected by a statistical or mathematical deviation or a deviation from a specification.

There are several important features of fault detection to consider.

- **Stability**: fault detection should remain stable and consistent when undisturbed.
- **Robustness**: fault detection should be consistent against noise and disturbances as well as incomplete models for comparison.
- **Reliability**: fault detection should be reliable, with few false positives and false negatives; the probabilities of false positive and false negative are the most common concerns.
- **Availability**: fault detection should be highly available so as to not miss the occurrence of faults.
- **Speed**: fault detection should be sufficiently rapid to reduce any potential for propagation, or failure.
- **Security**: fault detection must be made secure against sabotage, spoofing, and other attacks.
- **Trustworthiness**: ultimately, fault detection must instill trust.

Fault detection rate is a critical measure of the effectiveness of a fault detection solution and is generally the performance measure of concern. But measurements of the above are key factors; for example, false positive and false negative rates impact the fault detection rate. Likewise, fault test coverage rate during production is the pre-deployment equivalent of the fault detection rate from the deployed system.

25.2.3 Localization

A fault is localized when it is located. Ideally, this degree of location is to the replaceable, rebootable, or repairable component so that no further localization is needed. In other words, the goal of localization is to place the fault to a component or subset that can be acted on quickly to isolate and restore quickly.

Turning again to spoiled milk, the smell is coming from the milk container, which helps isolate the fault to the container of milk. However, a further look may indicate that the cause is a fault in the refrigeration system leading to the premature degradation of the milk, which may require further fault detection and localization.

Localization can be determined by scope of the sensor, failure mode, type of fault, information (telemetry) obtained from the sensor, information obtained from the system or network response (say through passive or automatic isolation), a triggered test, distance to fault means, and other methods often specific to the system or network.

25.2.4 Isolation

Once a fault is localized, it can be isolated and removed from the system. A fault is isolated once it cannot propagate further, and its impact is limited. In the literature, isolation may be referred to as reconfiguration because the system or network has to respond to the fault to isolate it and possibly to continue to function in a resilient way. After a fault is isolated, additional actions can take place to recover the system or network or protect the use case or mission.

The spoiled milk from our example is removed from the refrigerator and disposed so that it can't lead to other food spoiling, once it is confirmed that the cause is not the refrigeration but that the milk has simply expired and the owner did not notice the expiration date on the carton.

Safety can be a key concern here, as some faults can be dangerous to the system, network, or life. Failure to notice the smell of the spoiled milk may result in illness. Fast isolation of faults in the power grid protects humans from being harmed when cutting power cables, for example. Near instantaneous fault detection and isolation protects life, but often at the cost of an outage. In less safety critical cases, more precise but complex localization may take place.

Once a fault is detected, localized, and isolated, it can be mitigated through a system or network reconfiguration (say protection switching), and then repaired. If the reconfiguration fully protects the service, then the repair that can follow is considered to be proactive; otherwise, it is reactive and often urgent.

25.2.5 Reconfiguration, Protection, Restoration

After a fault is isolated from the system or network, resiliency mechanisms may take over to assure the network or system can continue to deliver its function, fully if possible. Systems may reconfigure, software may reboot, signals may reroute, and other changes to the system or network take place to assure continued function. Failure protection and service restoration are critical for assured reliability and availability at the service and mission level. Automating the reconfiguration, protection, and restoration is what makes a system or network resilient.

At higher levels of systems and networks, isolation begins to involve reconfiguration and the related concepts of protection and restoration. A detailed treatment of these topics will be left to another chapter, though we will touch on it later in this chapter as it relates to fault management and fault tolerance.

25.2.6 Control and Management

To discuss fault management (including detection, localization, isolation, and possibly more), we have to discuss the mechanisms that allow fault management. In software and systems, there are often parts of the system that have the role of controlling the system. Some parts of that system will have the role of fault management, either integrated with other control roles or possibly separated.

Take the example of a fuse. Its purpose is to disconnect a circuit from the system to prevent a more critical failure from happening. In fulfilling its purpose, the system or network enters a degraded state, and may not fully function. When the fuse fulfills its purpose, due to its nature, the fault is immediately detected, partially localized, and isolated all at once. However, it is not the fuse that failed; there remains the cause of the circuit fault, which may still exist. Replacing the fuse may result in the immediate failure of the new fuse if the true cause of the fault is not isolated and repaired. So, while the fuse "failed" and needs replacement, the reality is the fuse has done its job and succeeded instead of failed. If the fault which caused the detection, localization, and isolation by the fuse is not found and removed, the fault condition may remain. The fuse therefore is a part of the fault control or management system.

Depending on secondary actions taken, the system or network can recover to partial or full functionality. In many systems, fuses have been replaced by circuit breakers. These breakers serve the same purpose, but repair does not require fuse replacement, only the repositing of the breaker to the functioning state. This results in less downtime overall, increasing availability. However, if the cause of the breaker trigger is not first fixed, the breaker can cycle to off and on states repeatedly until the cause is addressed or the breaker wears out.

In networks, those roles are delegated to the management plane. However, hardware and software in the network and its systems will contain components that are not always cleanly separate. Component layer detection, localization, and isolation will be integrated quite often, whereas the network management systems will handle higher level fault management functions. Networks are evolving, however. Network function virtualization and software defined networking are transitioning to more general hardware running software functions, allowing RESTful architectures for stronger management functions and better telemetry management.

We do need to discuss the failure modes of fault detection, localization, and isolation.

- Fault detection can have the following types of errors associated: false (positive) detection, missed (false negative) detection, and incorrect (the wrong fault) detection. Fault detection itself can be faulty.
- Localization can fail to sufficiently localize, incorrectly localize, or fail to localize at all.
- Isolation however can fail to isolate, isolate the incorrect component or part of the system, or even initiate bad effects on the overall system, network, or the services they support. Isolation can have undesirable effects but still be necessary for safety or other reasons.

Management systems need to consider these fault management failure modes and operate to continuously improve against them.

25.2.7 Fault Tolerance

Fault-tolerant systems and networks respond favorably to faults. Resiliency mechanisms built into systems and networks are what prevent a fault or failure from becoming a network or system wide failure, and to allow a service to be without fault or failure. Note that the system or network retain the fault or failure, but the service recovers to normal, without fault or failure.

For example, Data-Over-Cable Service Interface Specifications (DOCSIS®) defines several required resiliency mechanisms including echo cancellation, cyclic prefix, profile management, and forward error correction. Each of these mechanisms allows control and automation of how the network responds to faults, making the network fault-tolerant and the service more reliable. Echoes in the transmission line, which are RF energy reflected from impedance mismatches, can disrupt transmission by reducing the main signal energy and reflecting some of it back, causing interference to the desired signal. Cyclic prefix and echo cancellation allow the burst receiver to adjust to the impairment so that the analog signal can be received, cleaned up, and properly interpreted. Profile management further allows adjustment of the resolution needed in the analog to digital conversion process so that a noisy system can reduce the transmission bit rate to allow reliable transmission to continue to be resilient to the noise, while forward error correction (FEC) does its best to recover lost bits. The net result is a set of complimentary resiliency mechanisms built into DOCSIS® networks, allowing for reliable service over a transmission system subject to impairments.

It is worth mentioning the work of Geraci et al. (1991), which identifies that fault tolerance, security, and survivability overlap each other in part, with reliability at the center of all three concerns. Also, it is worth mentioning that resiliency is the manifestation of the automation of reliable fault detection, localization, and isolation as a part of the overall system or network. The result of a more resilient system or network is it can support higher reliability and availability. A relationship between faults, failures, and resiliency is depicted in Figure 25.1.

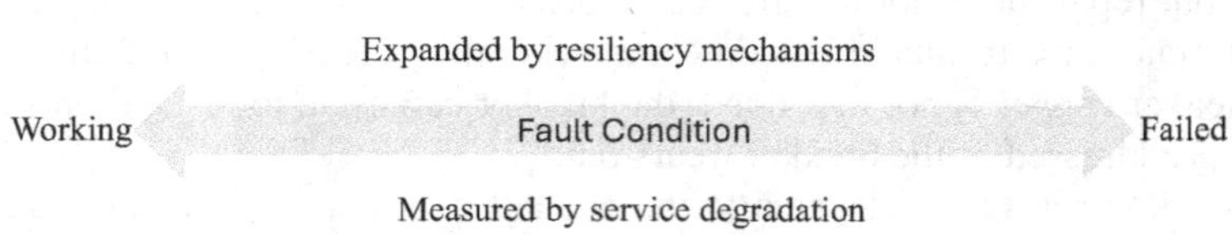

FIGURE 25.1 A depiction of the relationship between fault, failure, and resiliency.

25.3 SYSTEMS AND NETWORKS

Complex systems and networks require control as part of the main function. The control functions handle many different tasks and therefore can contain faults themselves. But among their functions are fault detection, localization, and isolation, not to mention protection and restoration. Sometimes these functions are embedded into the systems and at other times they are separate, or yet still they can be mixed.

In Ju et al. (2021), the authors present a survey of fault detection and more in networked dynamical systems, focusing on model-based methods mostly. The authors classify fault detection methods for dynamical systems into full state observers, interval observers, set membership filters, and iterative learning observers. Because networks and systems are often dynamical, several approaches have been studied and applied for optimal results.

In the rest of this section, we will cover several application areas, and explore the similarities and differences in how fault management is conducted.

25.3.1 SOFTWARE

Software fault management happens during test and development, but also can be built within the software. In this section, we mostly focus on software fault detection, localization, and isolation in the test and development process. A full treatment of software testing is covered in another chapter of this book.

Software Product Line is a concept where software modules intended for reuse in different software products and configurations are tested across a number of different use cases to detect, localize, and isolate faults (Jayamaha et al., 2019). This use complicates the definition of a fault.

In software testing, detecting faults requires software execution. Use case considerations inform what aspect the software must do best, and the design informs what the desired outcome is against the result. Some software faults are obvious as they cause the software to stop running, or deliver obviously poor results. But many software faults can require comparison to a model to determine the expected outcome. That model can be mathematical, a prototype, or an earlier proven version.

But for software, the challenge is to locate and isolate to fix the fault. In Abreu et al. (2009), the authors used model-based fault detection and localization in a software context they called spectrum-based fault localization. This method abstracts the program traces as the model for fault detection. The result is faster and more accurate localization of faults.

From the point of view of deployed systems, detection, localization, and isolation are very much related. Through software testing, a lot can be learned which can inform fault detection and localization in the final system (Arrieta et al. 2018). Testing can also reveal information to inform estimation of the reliability of the software which can inform the decision to deploy.

Once the software is sufficiently tested and found faults are isolated and removed, the software is deployed where it may be used in ways not yet imagined.

Local recovery can allow the software to recover quickly without disruption when faults are detected and properly localized. By the nature of the design, the isolation is simplified if not enforced by the design (Sözer et al., 2010). A software system experiencing a failure can have control elements that detect faulty behavior, restart suspected components, and confirm correction. If correction is not confirmed, that means either an outside component may be the cause, or the control software may be at fault. Some software systems will then cause the restart of a larger set of components to try again to address fault correction, potentially repeating until the entire system is restarted from a secure image; however, this action could result in an outage if taken to this extreme. As such, local recovery managed in this way is a recursive localization and isolation process.

In software, there are many fault localization methods. Sözer et al. (2010) suggested these can be classified as statistical methods which rely on context information about what software is running to determine cause likelihood to inform localization, and static methods which rely on a static model of the software to compare with the collected telemetry.

25.3.2 Hardware

This section treats fault detection, localization, and isolation in hardware in a general sense, and then addresses some of the more common complex systems in more detail.

25.3.2.1 Detection

Any added sensors add complexity. It is common to optimize design by using the fewest sensors necessary that can detect and localize efficiently. But sensor failure is an issue to consider, and false positive and negative rates and outcomes must be considered too, leading to a need for redundant sensors in some designs.

In electronics, examining changes in current or looking at the current differential (application of Kirchoff's law) are methods to consider. Methods that estimate and look for changes in impedance are common too.

Methods that search for changes in the frequency of an electrical wave in a cable are of common use too.

In RF transmission over coaxial cable, impairments in the RF signal can indicate faults in the network hardware. We will discuss this in more detail later in the chapter.

25.3.2.2 Localization

Physical localization is important for two reasons: minimizing the impact from the fault, and reducing isolation and repair time.

Manual localization in a system is conducted by isolating parts of the system and measuring properties against the expected results. Deviations indicate the location of a fault. A general strategy is a divide-and-measure approach where the subsystem that has an unexpected result is the subsystem holding the fault, then iteratively divides the indicated subsystem further and repeats until the fault is isolated to a replaceable section. Automating localization is very similar in that sensors are ready to provide the information that would otherwise be obtained through testing. By checking the sensor information, the localization can be done rapidly in many cases. Automobiles and complex household appliances are a convenient example today, as they can provide error codes that indicate a system fault, and help localize it. The heating system in a clothes drier has multiple temperature sensors which are located at different parts of the appliance. A high temperature at one location may indicate an air flow problem, for example. The location of the sensor localizes the fault by its position in the system or network and the fact that it detected the fault.

Time domain reflectometry (TDR) and traveling wave fault location methods use frequencies sent through a circuit, optical link, transmission line, or electrical connection to time the reflected energy which can indicate a fault. A pulse or similar waveform is transmitted through the device under test; when energy from the pulse is reflected, an impedance mismatch is indicated. By timing the reflected energy, and using a velocity of propagation calculation for the material, one can estimate the distance to an impedance change, which is a potential fault. Cable faults can be found using impedance methods, for example. The fault signature may at times indicate the cause, which may help localize the fault too. For a survey of localization methods for ship electronics, see (Babaei et al., 2018).

25.3.2.3 Isolation

Isolation in hardware is often made automatic through the use of fuses, breakers, latching relays, and other hardware components that can reliably (and often passively) respond in desirable ways to a fault condition.

Fault isolation in hardware electronics can be built into the component or made external to it. Such solutions in either case involve the use of fuses, breakers, or relays. Some microprocessors with internal protection circuits are an example of isolation built into the component. But there may be components designed to behave natively to isolate faults, such as in a voltage source converter.

In RF systems, it is common to use methods that avoid the RF fault or impairment by avoiding the use of the impeded frequencies or using them in a degraded state. Exclusion bands at times are a way to isolate impaired frequencies, and installing filters to attenuate ingress interference was once very common too. But these methods reduce the function of the network, so repair is still important. See Figure 25.2 for an example spectrum capture from a cable modem which indicates unused frequencies (labeled OFDM Exclusion Band) that are avoided perhaps due to common ingress issues in those frequencies.

25.3.3 Power Systems

Document (Babaei et al., 2018) provides a good survey of fault management in power systems on ships. A lot of the material covered in that paper applies to terrestrial power systems as well. In this work, the authors classify fault detection and localization of electrical systems into the following categories.

- **Distance Protection**: using impedance measurements to detect and localize faults.
- **Overcurrent Protection**: the application of fuses and breakers to respond to overcurrent conditions.
- **Current Differential**: Kirchhoff's current law states that the current into a node equals the current out of a node, so if the sum of the current connecting to a component deviates from zero significantly, a fault has occurred because some current is leaking that is not being monitored.
- **Active Impedance Estimation**: inject short duration current over a defined spectrum of frequencies to measure the impedance to find abnormal conditions.
- **Wavelet Analysis**: signals are examined in the time domain for changes indicating a fault has occurred (Jayamaha et al., 2019).

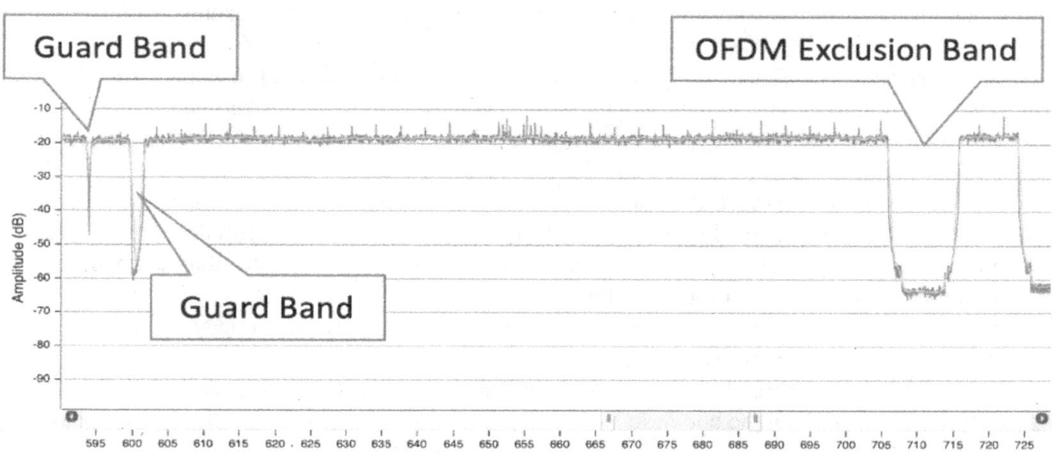

FIGURE 25.2 Downstream spectrum capture from a cable modem indicating unused frequencies to avoid interference issues, obtained from source, as published in (© SCTE, 2021b.)

- **Traveling wave**: combining time synchronization and global positioning systems with a voltage drop measurement can localize a fault.
- **Artificial Neural Network (ANN)**: use voltage and current waveforms over time to feed a neural network (Jayamaha et al., 2019).

Of course, this is a survey of known methods at the time of the survey's publication. Artificial intelligence in general, including machine learning, has been applied as well, in power systems and beyond.

This same work separates isolation into breaker and breaker-less methods which rely on power converters. Further, there are different ways of organizing fault detection, localization, and isolation methods, as we discuss later in this chapter.

25.3.4 Communication Networks and Systems

Communication networks transmit signals on analog connections of various types, and these days usually convert the analog signals to digital data.

25.3.4.1 Analog

An analog signal can experience multiple types of faults with multiple causes.

- **Clipping**: when the energy peak of the signal is beyond the capabilities of the line system, the peak is truncated and information is lost.
- **Fading**: variation in the signal strength, which may be caused by signal energy absorption.
- **Echoes and Multipath Propagation**: energy in the signal is reflected, creating an echo of the signal and less energy in the main signal. The signal can also follow multiple paths, resulting in dissipated signal energy, resulting in ghosting and jitter.
- **Power Loss**: the signal's energy is too low to be received with sufficient resolution and fidelity.
- **Noise**: other signals and energy combine with the intended signal, making it difficult to interpret the intended information.

25.3.4.2 Digital

Digital data from an analog signal can suffer faults as well. Data bits (1 or 0) are interpreted from analog signals, and formed into sets of data bits referred to as symbols or packets.

- **Bit Errors**: analog faults can lead to an incorrect interpretation and thus a bit error.
- **Packet Loss**: likely due to insufficient buffer sizes in the network, packets can be dropped; but also, lots of noise in the system can cause packets to not be recoverable by FEC or other means, and thus be lost.
- **Latency and Buffering Delays and Out of Order Packets**: during network congestion, which is when there are large amounts of traffic, incorrectly configured or insufficient buffer sizes in the network can cause packets to delay and arrive later than needed, and potentially in a different order than they were sent.
- **Bandwidth**: insufficient data rates (not intentional) in the network can lead to packet loss, retransmission, out of order, or latency.
- **Unreliability**: network elements that fail can lead to packet loss, bit errors, or latency.
- **Attacks**: an attack on the network, such as a denial of service (DoS) attack can lead to lost or late packets.

Interestingly, attacks can both impact and be addressed through fault detection, localization, and isolation.

25.3.4.3 Fault Management Approaches in Communications

Several approaches to fault detection, localization, and isolation exist in communication networks.

Fault detection can be essentially continuous to create event triggered approaches. Near constant monitoring makes this possible. But near constant monitoring is not always possible. Therefore, periodic polling or tests are often relied upon.

When fault management is not nearly continuous, intelligent scheduling of the fault procedures is important. There can be round-robin approaches where network nodes and elements are polled on a circular list. More common are approaches that poll data or request a test to be performed on a schedule or periodically. Due to limitations in networking, various protocols are employed to enable fair scheduling and treatment of fault management.

- **Logs**: many network elements will capture logs of certain events, which can be later retrieved or reviewed.
- **Simple Network Management Protocol (SNMP)**: still in heavy use, SNMP is a very old protocol for network telemetry based on the request-response model, where a request from a network management entity to a network element should get a response of the requested information which may be continuously collected data or the result of an executed test by the network element.
- **Remote Access**: some network elements allow remote access for a person or network management system to obtain helpful telemetry.
- **User Services Platform (USP)**: through this protocol, end network elements can be instructed and telemetry can be obtained.
- **Streaming Telemetry**: a newer approach to network telemetry, the network elements are enabled to provide telemetry through a modeling language, and network management systems or applications need only to subscribe to the data stream, which can be delivered continuously, at change, on a periodic basis, etc. YANG modeling is a popular approach for providing model-driven streaming telemetry.

Of course, there must be consideration for the amount of telemetry data extracted from the network, and the resources required to provide the telemetry. The network's primary job is not to provide telemetry; but a reliable network must provide the important telemetry to support network management. Edge compute and intelligence versus collection into the core is an important development and consideration for fault management. If fault detection, localization, and isolation can be conducted closer to the source, the network resources are not as taxed, and corrective action may take place sooner.

Closed-loop fault management is an important part of communication networks. These closed-loop systems often follow an approach like Observe, Orient, Decide, and Act (OODA) (Rupe and Zhu, 2019), in very rapid succession. Take for example the proactive operations (ProOps) platform, which is a framework for turning data into action in support of operations. Deployed as a software system, it is fully flexible. But the architecture guides construction in layers: data collection (tests and queries) and translation (including fault identification), correlating across time and system or network elements or sensors, analysis of the severity of the fault, and packaging the information to facilitate action.

25.3.4.4 Fault Management Examples in Communication Technologies

Here we provide a few details about fault management in optical, DOCSIS®, and Wi-Fi® technologies for example.

In optical links, several analog data elements can be captured to help monitor the quality of the optical link for faults. For a readily available reference of fault management in optical systems, see (Cable Television Laboratories, Inc., 2020) which specifies coherent optics for cable systems. For a coherent optics termination device, events are logged, loss of signal is reported, and asynchronous fault notifications are generated. Fault relevant data available include signal degrade threshold (bit error ratio, BER), state of polarization, transmitted power, received power (with maximum and minimum thresholds), link state, transceiver state and statistics, differential group delay, chromatic dispersion, symbol rate, pre-FEC BER, optical signal to noise ratio, polarization dispersion loss, fault event logs, and more. Event logs are prioritized as emergency, alert, critical, error, warning, notice, informational, and debug.

In addition, external test devices are capable of capturing the coherent symbols, so that the constellations can be displayed, and measurements of the difference between the target and the actual value can be measured and monitored as additional fault data. See Figure 25.3 for a constellation plot of simulated optical constellation data.

Many optical systems have included TDR functionality in the optoelectronics of the line system to allow rapid and automatic fault location.

In DOCSIS® RF links, several analog data elements can be collected and tests conducted periodically to indicate faults (Cable Television Laboratories, Inc., 2022).

The architecture for DOCSIS® is a tree structure of coax connected by active and passive electronics, where the trunk end is a cable modem termination system and the branches hold cable modems and service edge equipment. The upstream measurements are different from the downstream measurements because the devices on the ends have different roles and capabilities. In the

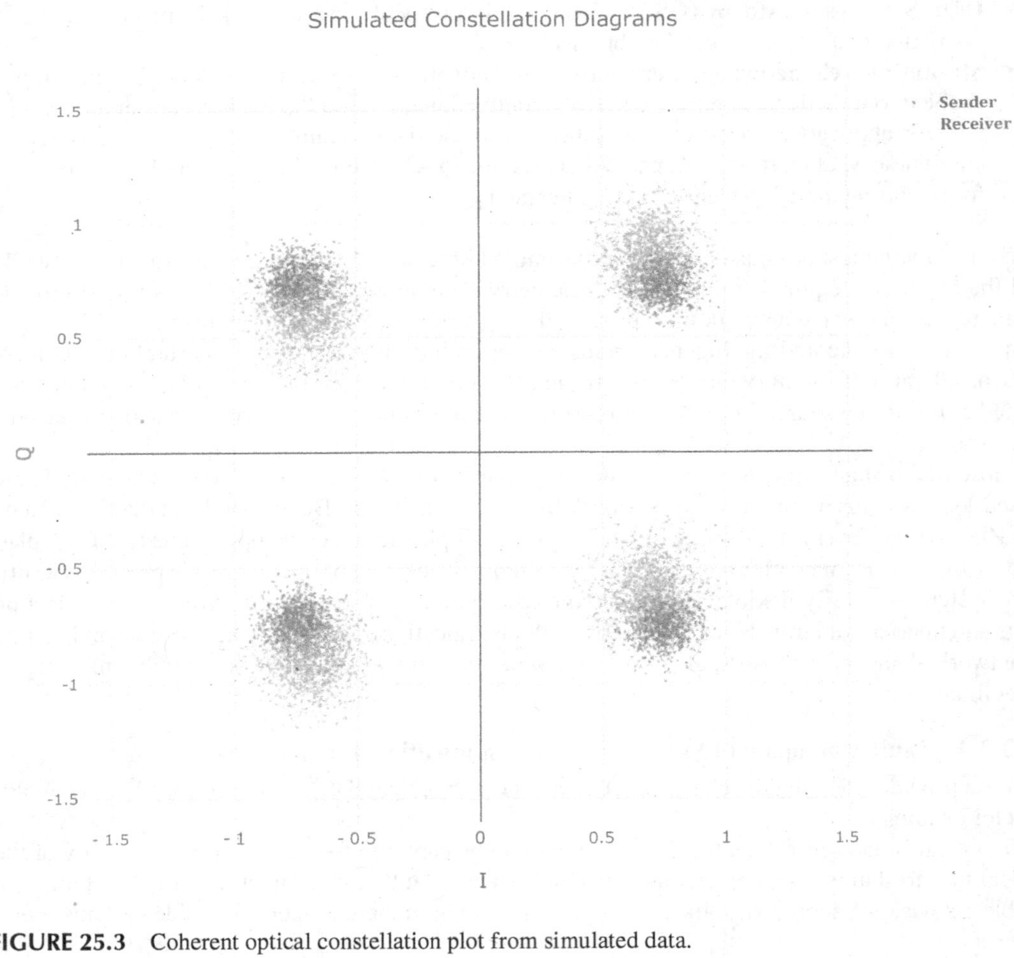

FIGURE 25.3 Coherent optical constellation plot from simulated data.

downstream, several measurements and tests help identify faults: transmit power, received power, wideband spectrum capture, receive modulation error ratio (RxMER) per subcarrier, FEC statistics, channel estimation coefficients, noise power ratio, constellation display, symbol capture, histogram (power level captured over time and delivered in histogram form), signal to noise margin, and more. Most useful for detecting cable plant faults are wideband spectrum capture and RxMER per subcarrier. FEC stats can indicate whether faults are impacting service or not by the number of uncorrectable codewords or bits. Several of these measurements and tests can indicate specific faults that are more likely to happen in certain parts of the network, which helps to localize.

Figure 25.4 shows wideband spectrum data that indicates water is present in the network (top left plot). The variability of the signal suggests the severity of the problem, and if a slope showing more attenuation in higher frequencies is apparent, then the water is likely in a cable. In this example, the data are filtered and autocorrelation is calculated (top right). The data are ordered and fit to a CDF, and those parameters are used to differentiate between amplitude ripple caused by a pair of impedance mismatches versus water in the network (Fox et al., 2021).

Localization can be done by identifying the likely location of the fault indicated based on knowledge (like failure modes, effects, and criticality analysis, FMECA), but also multiple cable modems can be used to match impairments, which indicates a shared problem in a shared part of the network, or isolated to a single customer which suggests the water is in the drop cable serving that customer.

In Wi-Fi®, several new data elements have recently been defined (Wi-Fi Alliance, 2021). Several useful data elements are provided for network fault management and troubleshooting including several configuration details, transmit and receive statistics, data rates, error counts, retransmit counts, RF noise, signal strength, bandwidth, and several events including failed connections. With these data elements, fault localization can happen more efficiently. It is most important to differentiate a Wi-Fi®-related problem from an Ethernet-related problem from a DOCSIS® network-related problem, etc.

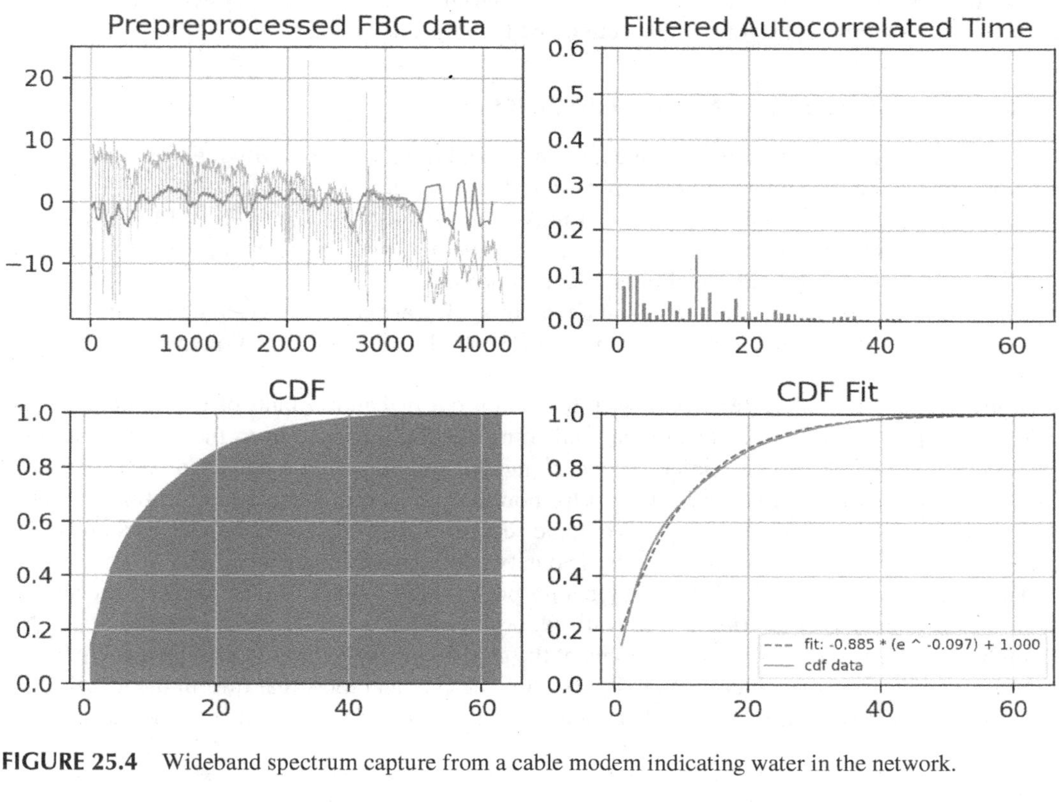

FIGURE 25.4 Wideband spectrum capture from a cable modem indicating water in the network.

25.3.5 Mechanical Components and Systems

The mechanical components of systems will falter in ways that can be identified through physical telemetry and means, and through physical tests and inspections.

Vibration analysis has the advantage of identifying faults during regular use of the mechanical part or system. The system can also be stressed or tested in certain modes that may better reveal faults in a safe way. Various sensors are used to capture vibration in machines at different locations, across multiple axes. The vibrations are analyzed for patterns, statistics extracted, and multiple approaches for detecting changes in the vibration that indicate a fault have been developed for various situations. Most popular are methods for detecting bearing faults before failure.

Fast Fourier transforms (FFT) and time-frequency analysis are used to analyze the vibration data. Vibration data collected over time are in the time domain. An FFT translates the time domain data into frequency domain data. By looking at the strength of consistent energy in a full spectrum of frequencies, changes in the vibration can be quantified. By looking at both the time and frequency domain data, it can be easier to identify patterns that indicate faults.

Imaging and non-destructive evaluation can be used to find cracks and other faults in materials and mechanical parts. By using x-ray, infrared, and other frequencies to image the material in greater detail or below its surfaces, the condition of materials can be determined and that information used to detect and locate faults. Likewise, sending certain vibration frequencies into the material can also reveal otherwise latent faults.

Mechanical systems often require fluids for lubrication and cooling. You may be familiar with maintenance indicators in automobiles, where sensors monitor fluids for their level and condition, and indicate a need for maintenance. Condition-based maintenance is more optimal than time-based maintenance in mechanical systems generally, but monitoring fluids and coolants for leakage can indicate wear out and damage faults that require attention.

The research and application field of prognostics and health management (PHM) has developed many other physical means for finding faults in materials and mechanical systems. That chapter contains more details useful for fault detection and localization.

25.4 GENERAL TOOLS AND TECHNIQUES

The first goal of fault management is to detect faults and mitigate them. Mitigation requires some level of localization, and potentially isolation, in order to mitigate.

Local and system wide fault management must work together efficiently. Local fault detection and mitigation handles fast response to keep systems and networks functioning, while system level fault management handles additional work to the same ends.

Closed-loop systems will automate much of this, if warranted in the system. But even in closed-loop systems, additional manual confirmation, localization, and isolation may be necessary to effect repairs.

There is merit to having sensors relay information out of band; a fault or even failure in the system may prevent the sensor from communicating needed information to the decision system, thus preventing localization or isolation, and certainly mitigation and repair. See Gridmetrics as an example of an out of band sensor system for power systems utilizing communication networks (Cruickshank, 2021). Another example is a home security system that uses cellular backup, so the system may be available even when the physical networks supporting it is not.

The sensors will need to be distributed in a purpose-driven design. But the decision process for fault management can be distributed, centralized, or a combination.

In the rest of this section, we discuss some of the general methods for fault management, particularly for fault detection and localization. We cover some classical tools that inform the localization and isolation, then discuss classes of detection methods. Data-driven fault detection is sometimes separated into multivariate statistical methods and machine learning methods, both of which are discussed. Then there are some model-based methods we will discuss as well. We will cover a few related methods that are evolving too.

25.4.1 FMECA and Truth Tables

FMECA is a method for capturing the needed information that will help align the fault to the condition, which will help system designers create the detection, localization, and isolation solution needed. Sometimes you may see this referenced as simply failure modes and effects analysis (FMEA) or in like terms. Based on the effects and criticality of each failure mode or fault mode, the fault tolerance solution can be designed to properly address the most critical faults as necessary for the mission or services supported by the system or network. Not every fault can be covered, but the important ones must be.

See Figure 25.5 for an example of an FMECA, using our spoiled milk example from earlier. This example is incomplete, but illustrative of the approach, and shows how an open door sensor may be helpful, but not helpful enough if the sensor is not monitored and heeded.

See Figure 25.6 for an example of a truth table. In this example, a fictional dryer with three heat sensors tied to the control system so that a overheat sensed ("bad") at any one sensor results in a "power off" outcome, whereas all three sensors indicate safe heat ("good") results in power to the system. Note that the truth table can come in multiple forms including tables or flow charts most commonly. But the truth table holds the information explaining how sensors are translated into fault conditions. Knowing that sensor A is next to the heating element, sensor B is next to the vent filter, and sensor C is inside the dryer, a condition where all sensors are good except for B may indicate an airflow problem.

Example Failure Modes, Effects, and Criticality Analysis Worksheet (partial)

System: Refrigerated Gallon of Milk
Subsystem: Gallon of Milk
Component: Milk
Function: Nourish

Fault-Failur	Cause	Local Effect	Higher Effe	Top Effect	Use Case Ef	Probability	Severity	Duration	Detection	Protection	Repair	Criticality
Clumpy	Age	Dangerous	Smell, Mess	Loss of Milk	No milk	0.0001	2	2 hr	0.9	none	buy milk	1
	Warm-Open Door	Dangerous	Smell, Mess	Loss of some groceries	No milk, groceries	0.00005	3	2 hr	0.9	door alarm	close door, dispose bad, buy new	3
	Warm-No Power	Dangerous	Smell, Mess	Loss of some groceries	No milk, groceries	0.00005	3	2 hr	0.9	lights off	close door, dispose bad, buy new	3
	Warm-Refrigerat or Failed	Dangerous	Smell, Mess	Loss of most groceries	No milk, groceries	0.000001	3	48 hr	0.9	spare refrigerat or	close door, dispose bad, buy new	9
Solid	Frozen - thermosta t set incorrect	Frozen Milk	Carton damaged	Loss of some groceries	Some loss of food short term	0.00001	1	1 hr	1	none	Adjust temperat ure, thaw milk, replace carton	1

FIGURE 25.5 An example of partial FMECA for our milk spoilage example.

Sensor Inputs			
A	B	C	Logic Output
good	good	good	power on
good	good	bad	power off
good	bad	good	power off
good	bad	bad	power off
bad	good	good	power off
bad	good	bad	power off
bad	bad	good	power off
bad	bad	bad	power off

FIGURE 25.6 An example truth table for a fictional set of sensors in a heat blower system for say a fictional clothes dryer.

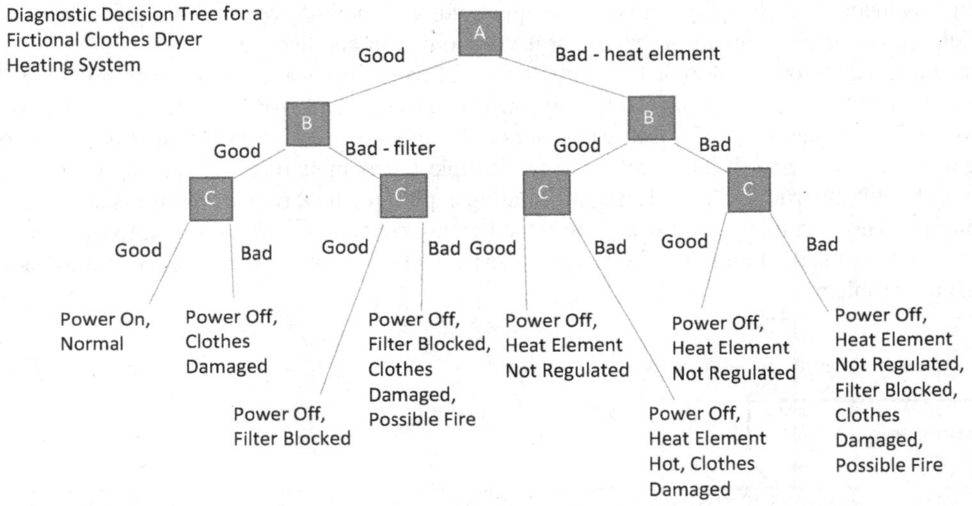

FIGURE 25.7 A diagnostic decision tree for our fictional clothes dryer example.

We can break down faults in terms of hardware or software or interface, but we can also break down faults in terms of physical or measurement or logic. In any case, the use case, environment, human error, poor installation or calibration, and other failure mode categories can apply in each category.

25.4.2 Diagnostic Decision and Fault Trees

Fault trees are explained in another chapter of this book. But those fault trees can be used for diagnosis, so have an important role in fault management. See Figure 25.7 for an example of a diagnostic decision tree. Note that this example does not use fault tree notation and symbols, but could be built that way too. Note that the diagnostic tree here is consistent with the truth table of the previous figure.

25.4.3 Statistical Methods

Statistical methods can be applied to fault detection and localization to enable the application of rules to the statistics calculated from telemetry, leading to fault detection and localization. Several statistical methods are common in fault management, and many are related to linear models and regression.

- Principal component analysis can help identify the best indicators for a fault by examining the data that best explain most of the variance. The method involves calculating Eigenvalues for each potential indicator data, then sorting those to find the largest Eigenvalues. The data associated with the largest Eigenvalues explain the most variance and are good candidates for fault indication.
- Partial Least Squares regression instead looks at the covariance of two data elements, the fault indicator and the potential predictors.
- Independent component analysis assumes that the observational data can be explained by multiple independent components, with the goal to find the components and their mix that explain the observed data.
- Linear discriminant analysis is an analysis of variance with continuous independent variables (telemetry) and a discrete categorical dependent variable (fault indication).

Statistical process control is a field that uses statistical methods to look for outliers in production systems or other series data. Aside from the many quality-oriented tools used, there are control charting techniques for variables and attributes. Specification limits and rules for determining outliers will help capture faults in systems as they worsen or sharply appear. In basic form for variables, relying on the central limit theorem, sample means and variances or ranges are calculated from samples over time. These statistics are plotted on control charts (with statistically based rules and limits) and specification charts (with specified rules and limits). Rather than looking for a simple threshold crossing, this approach allows for more sophisticated rules such as run rules to determine whether a fault in the system has occurred. Attribute control charts count nonconforming units or defects on the produced units to determine whether a fault in the system exists which is impacting the output. Like variables charts, attributes charts can allow for complex rules to determine the risk of a fault occurrence, and to use specifications and statistical rules and limits to determine outliers. Once a fault is indicated, the first step is to rule out any special causes to confirm that the indicator is due to a cause to address, to determine if a false positive has occurred first. The rules are set to control false positive and false negative rates to optimally manage risk.

25.4.4 Processing-Based Methods

The discrete Fourier transform (DFT), usually estimated using the FFT algorithm, is an example of a processing-based method. By processing complex (i,q) data sampled over time using a Fourier transform method, one can determine the frequencies represented in that signal. Letting, x_i for $i=0$ to $N-1$ be complex numbers, the component frequencies X_k can be determined by

$$X_k = \sum_{n=0}^{N-1} x_n e^{-i2\pi kn/N}$$

The result is a translation of the complex values to magnitudes in each frequency.

RxMER per subcarrier is another good example (SCTE, 2021a). RxMER is the average signal power to the average error power for constellation data. Power is calculated by the sum of the squares of the (i,q) values from the constellation data.

$$\text{MER} = 10\log_{10}\left[\sum_1^N \left(i^2 + q^2\right) \Big/ \sum_1^N \left(i_e^2 + q_e^2\right)\right]$$

As a ratio, when the error overpowers the signal in a given frequency subcarrier, the MER value drops, indicating a fault impacting the frequency band. Because this measurement is a ratio of signal to error (related to noise), it is a statistic indicating the deviation of the constellation point to the intended value; see Figure 25.3 for an example of constellation plots in the optical realm, which are very similar and equivalent.

25.4.5 Machine Learning Methods

Machine learning is a field encompassing data methods that exhibit self-learning in some form. While there are several methods developed, extended, and applied to fault management, a few of the core methods are briefly discussed here.

Support vector machines are a way of classifying variables into groups; a set of N-dimensional data may be separated by an $N-1$-dimensional plane, whose maximum thickness before contacting any points defines the support vectors. Data under normal operation and under faulty operation are used, and the goal is to find the plane that separates the data effectively between normal and fault. For an example application to wind turbines, see (Laouti et al., 2011).

K nearest neighbor is an algorithm that determines the most likely classification of each datum into any of K clusters. See (Zhou et al., 2019) for an example application of this approach to process monitoring.

Artificial neural networks (NN) are computational networks that simulate neural processing. Convolutional neural networks are a subset of NNs which are well suited to finding visual anomalies. Therefore, they are well suited for finding impairments in data plots and related frameworks. For an example of the application of this approach to the problem of finding and classifying impairments in RxMER data from cable modems for fault detection and isolation, see (Zhu et al., 2020).

25.4.6 Model-Based Fault Detection

A treatment of model-based fault detection and diagnosis is given in (Isermann, 2005). Model-based fault detection is a concept where a residual signal is created by taking the difference between the expected model signal and the actual, called a residual. Comparing that residual to a threshold then indicates a fault requiring an alarm to sound. Note that thresholds can be dynamic or adaptive, but are often fixed.

Due to noise in the system, the residual will also contain noise, which means false positive and false negative results are possible. To minimize false outcomes, it is important that the residual be made robust against noise sources as much as possible. In many cases, the fault indicator can be designed to be more robust.

For some in-depth discussion of these subjects, consider references such as (Wang et al., 2007) and (Li and Zhou, 2009). As such, it is important to mention here that fault detection, localization, and isolation overlaps control theory, though a detailed treatment is not necessary for our purposes.

TDR, discussed earlier in this chapter, is a test method for finding anomalies in transmission lines (power, coax, optical fiber, etc.). Recall that this test sends a pulse down the transmission line, measures the reflected energy over time, and uses this information to locate impedance changes. Given an impedance change reflects energy, which is undesired, the TDR is finding faults in the line. These faults, if minor and stable, may just be the result of the implementation of the system. But if severe, or if changing over time, they may indicate a fault that is undesirable and must be removed.

The field of statistical process control, discussed earlier in this chapter, is an example of model-based fault detection. By taking the average of data samples such that a process in control would result in averages that are normally distributed, a deviation from that model indicates a fault.

It is important to note that by looking at the differences between several expected and resulting outputs across many inputs, creating several residuals, much can be developed to determine the fault and diagnose its cause. Understanding the failure and fault modes of the system or network can point to the cause and therefore allow mitigation.

25.4.7 Digital Twins and Simulation

The concept of a digital twin, which is emerging in popularity, is an option worth consideration for addressing faults in complex systems and networks. A digital twin is a model copy (virtual representation) of a system or network or element useful as an experimental environment. It is often updated in real time with information and data from the real system it is a twin to.

A simulation of the system, or a digital twin model of the system, can be used to provide the expected result to compare the actual result against. But also, the digital twin or a simulation can be used to replicate some complex fault conditions and lead to knowledge useful for localization and isolation. It can potentially be used to train machine learning solutions too.

REFERENCES

Abreu, R., Zoeteweij, P., Golsteijn, R., & van Gemund, A. J. C. (2009). A practical evaluation of spectrum-based fault localization. *Journal of Systems and Software*, 82(11), 1780–1792.

Arrieta, A., Segura, S., Markiegi, U., Sagardui, G., & Etxeberria, L. (2018). Spectrum-based fault localization in software product lines. *Information and Software Technology*, 100, 18–31. doi:10.1016/j.infsof.2018.03.008.

Babaei, M., Shi, J., & Abdelwahed, S. (2018). A survey on fault detection, isolation, and reconfiguration methods in electric ship power systems. *IEEE Access*, 6, 9430–9441. doi:10.1109/ACCESS.2018.2798505.

Cable Television Laboratories, Inc. (2020). Point-to-Point Coherent Optics, Coherent Optics Termination Device OSSI Specification, P2PCO-SP-CTD-OSSI-I02-200501, May 1.

Cable Television Laboratories, Inc. (2022). Data over Cable Service Interface Specification Proactive Network Maintenance Primer for PNM Best Practices in HFC Networks (DOCSIS® 3.1), CM-GL-PNM-3.1-V03-220114, January 14.

Cruickshank, R. (2021). What's Smart about Smart Power? Modernizing the Power Grid and HFC Networks: Power Outage Notifications and Advanced Sensing. SCTE CableTec Expo.

Fox, K., Rupe, J., Zhu, J., Hranac, R., Zedan, N., Kolcun, J., & Wolcott, L. (2021). Water Can Run, But It Can't Hide: PNM Finds Soaked Cables. SCTE•ISBE Fall Technical Forum.

Geraci, A., et al. (1991). *IEEE Standard Computer Dictionary: A Compilation of IEEE Standard Computer Glossaries*. Piscataway, NJ: IEEE Press.

Isermann, R. (2005). Model-based fault-detection and diagnosis: Status and applications. *Annual Reviews in Control*, 29(1), 71–85. doi:10.1016/j.arcontrol.2004.12.002.

Jayamaha, D. K. J. S., Lidula, N. W. A., & Rajapakse, A. D. (2019). Wavelet-multi resolution analysis based ANN architecture for fault detection and localization in DC microgrids. *IEEE Access*, 7, 145371–145384. doi:10.1109/ACCESS.2019.2945397.

Ju, Y., Tian, X., Liu, H., & Ma, L. (2021). Fault detection of networked dynamical systems: A survey of trends and techniques. *International Journal of Systems Science*, 52(16), 3390–3409. doi:10.1080/00207721.2021.1998722.

Laouti, N., Sheibat-Othman, N., & Othman, S. (2011). Support vector machines for fault detection in wind turbines. In *Proceedings of the 18th World Congress, the International Federation of Automatic Control*, Milano (Italy), August 28–September 2.

Li, X., & Zhou, K. (2009). A time domain approach to robust fault detection of linear time-varying systems. *Automatica*, 45(1), 94–102.

Rupe, J., & Zhu, J. (2019). Kickstarting Proactive Network Maintenance with the Proactive Operations Platform and Example Application, SCTE Cable-Tec Expo.

SCTE. (2021a). Mathematics of Cable, SCTE Operational Practice, SCTE 270 2021.

SCTE. (2021b). Understanding Cable Network RF Spectrum Displays, SCTE Operational Practice, EC-2020-028r1, November 21, 2021.

Sözer, H., Abreu, R., Akşit, M., & van Gemund, A. J. C. (2010). Increasing system availability with local recovery based on fault localization. In *10th International Conference on Quality Software* (pp. 276–281). doi:10.1109/QSIC.2010.29.

Wang, J. L., Yang, G., & Liu, J. (2007). An LMI approach to H-index and mixed H-/H∞ fault detection observer design. *Automatica*, 43(9), 1656–1665.

Wi-Fi Alliance. (2021). Wi-Fi Data Elements Specification, Version 2.0, September 9.

Zhou, Z., Li, Z., Cai, Z., & Wang, P. (2019). Fault identification using fast k-nearest neighbor reconstruction. *Processes*, 7, 340. doi:10.3390/pr7060340.

Zhu, J., Sundaresan, K., & Rupe, J. (2020). Proactive network maintenance using fast, accurate anomaly localization and classification on 1-D data series. In *2020 IEEE International Conference on Prognostics and Health Management* (ICPHM) (pp. 1–11). doi:10.1109/ICPHM49022.2020.9187045.

26 Risk and Risk Analysis

Robert Taylor

26.1 INTRODUCTION

Managing risk in any endeavor is fundamentally accepting the fact that something will go wrong, that it will probably seem insurmountable at first glance, and will cost time and resources to resolve. This chapter will begin with the premise that for every automated test system (ATS) or test program set (TPS) development project something will go wrong. In most cases, the thing that went wrong was unplanned, not even thought to be relevant to the project, and of course, should not have happened – but it did. Given the premise above, the astute ATS project manager and management team will accept the fact that something will go wrong, and some inadvertent events will happen. The manager and the team should spend significant quality time at the project definition stage to look for things that could go wrong. This chapter describes techniques to help define and manage risk by considering and analyzing causal events and manage those causal events by using four key strategies: accept the risk, mitigate, or control the risk, transfer ownership of the risk and avoid the causal events. Finally, if the risk must remain in the project, some techniques are offered to manage the risk by reporting the risk as a sub-project, assigning a risk manager, and allocating appropriate resources to minimize (or accept) the results of the risk when it becomes a problem.

This chapter focuses on the ATS/TPS development project environment, and the reader may think that the risk issues discussed in this chapter do not apply to test. However, if one considers ATS/TPS design and execution as a project, this chapter is very relevant.

26.2 RISK MANAGEMENT

Risk management is a management discipline that attempts to increase the probability of any design project's success by identifying threats to the desired project, assessing the nature of those threats, and defining and implementing actions that eliminate or limit the impacts of these threats to the project goals (Association for Project Management, 2004). Risk management is a subset of project management functions, and it uses the same three fundamental metrics that project management uses to ensure success: cost, schedule, and requirements. This author considers those three metrics as the "magic triangle" of project management. They must be considered at the definition stage of a project, and they exhibit the characteristics of triangles: a side or angle cannot be changed without causing a reshape of the triangle. If one element of a test design changes – cost, scope, or schedule – the resulting two elements will have to change. Risk management tries to foresee risk issues in the future to avoid modifying the triangle.

The desired end of risk management is to minimize adverse events and maximize positive events. Therefore, positive events will be discussed as opportunities. The author believes that opportunities are more than just positive risks, and they need to be considered as taking as much, if not more, effort to define and execute as does the effort to reduce or eliminate risk.

Risk/opportunity management must be considered in earnest at the beginning of the project. It is at the beginning of the project that the number of solutions for risk and opportunities are greatest, and the cost to manage them is lowest than at any time in the project.

Finally, the designing of ATSs and TPSs should be considered as the project, and the remaining part of this chapter will continue with that thought. A partial list of risks that might be considered of the ATS/TPS discipline are as follows:

DOI: 10.1201/9781003188377-31

- Selection of an incorrect or erroneous test of a given parameter.
- Scaling a test measurement incorrectly resulting in an acceptance of a failed parameter or rejecting an acceptable parameter.
- Over designing a test such that the measurement is too accurate for the needed parameter resulting in a test that is overly expensive.
- Designing a test system that measures a parameter multiple times when only one time is necessary.
- Designing a test that is not necessary.
- Specifying a test that uses equipment that is over specified for the test being required and erroneous test results occur.
- New unproven measurement techniques.
- Design risk with the use of new or unproven equipment.
- Test and Measurement Equipment supply chain risk.
- Changing test requirements, not communicated by the customer, leading to scope change.

26.3 RISK AND OPPORTUNITY STRATEGIES

While risks and opportunities are closely related, each requires a different mindset for both to be successful. Risk strategies tend to eliminate events while opportunity events are often added to the project design and execution. In addition, there are situations in which a risk can ultimately lead to an opportunity. As mentioned above, there are four generic strategies for risk and opportunity management, and they are summarized in Table 26.1.

Each of the strategies will have actions that the risk and opportunity managers can use to eliminate or manage them both. Eliminating uncertainty is easy to understand. Since both risk and opportunity have a probability of occurring and have a resulting project impact if they occur, it is easy to understand that to eliminate uncertainty means to take actions that make the resulting probability zero. That means for a risk, the action will be to avoid the risk event, by taking an alternate path or eliminating a causal event leading to the impact. For an opportunity, removing the probability means to take the opportunity, in other words, make it happen.

A second strategy is to move the risk from the existing project and transfer it to another party. For a risk, the only real option is to transfer the risk to the customer which is the only party that can accept the risk by removing event conditions or take responsibility for the event outcome. Subcontractors or vendors still hold the responsibility of the project, and therefore, the subcontractor or vendor cannot assume the risk for the contracting party. For an opportunity, usually, the terms of the contract will have to be modified and accepted by the project customer because in most cases, the requirements will be changed leading to a shortened schedule or reduced cost. There is no motivation to change the contract unless there is an advantage to the customer. Hence, an ownership action for an opportunity is to share the resulting impact.

TABLE 26.1
Comparing Risk & Opportunity Management (Hillson, 2001)

Generic Strategy	Risk Strategy	Opportunity Strategy
Eliminate Uncertainty	Eliminate root cause or probability	Capture the opportunity
Allocate Ownership	Transfer the risk to the Customer	Share with Customer or Supply Chain as appropriate
Modify Exposure/Outcome	Manage/control the risk event and/or the outcome(s)	Enhance the probability or benefit value
Include it in the Baseline	Accept the risk outcome in the plan	Opportunity is no more; benefit has become part of plan.

A third strategy is to modify the exposure to the risk or opportunity. For a risk, this means controlling the risk probability and or impact, and usually for a risk, the action is to reduce the probability of the event or by changing the event which can reduce the probability of occurrence. This is the action that most project manager tends to follow. For an opportunity, the project team will want to increase the exposure to the opportunity by taking actions to improve the chances that the opportunity event will occur.

The final strategy that a project team can take is to accept a risk event and place the impact cost in the plan. This action is normally done when the probability of the risk event is very high, and the risk event is considered likely to happen. Consequently, the project team must accept an increase in the project cost, schedule or change in requirements, and usually, the choice is increased cost and/or extended schedule. For an opportunity, if the team captures the event with the first strategy action, it is already in the plan and the opportunity no longer exists and is "ignored."

For the test community, the term project is the test design and execution function, and the risk and opportunity event relate directly to issues to test design and execution.

There are four fundamental issues that influence automatic test equipment design that relate to correct or incorrect test results: test results recording a failure when there is in fact a failure; test results recording a failure when in fact there is not a failure; test results recording a passed test when in fact there is a failure; test results recording a passed test when in fact there is a good result. The real risks involve the cases above when a test result records an incorrect result. The test issues do not occur once but will occur with every test executed on a piece of equipment. Therefore, there are many risk and opportunity events that need to be evaluated for efficient test design. The next section discusses a process that can be successfully used to support the identification and management of risk and opportunity events.

26.4 RISK/OPPORTUNITY PROCESS

This section recommends a process that considers an entire test system and related testing execution. A key to successfully managing risks is to have a well-communicated process that will identify risks and assign a qualified team to manage the risks. Below are the steps recommended to identify and manage key risks for a project.

1. Brainstorm with a knowledgeable team to identify as many things that could go wrong. Create a list of "risks."
2. Evaluate all risk items on the list and eliminate those risks that either are duplicates or that really do not make sense to keep/manage.
3. List the remaining events in order of approximate impact on the program from highest to lowest value.
4. Evaluate the probability of occurrence for each of the events.
5. Determine the Projected Impact Value for each of the impacts and list them from highest value to lowest.
6. Categorize the risks as technical, cost, or schedule risks.
7. Using the four strategies of risk management, identify how each risk should be categorized.
8. Assign key project staff members to manage the risks.
9. Meet regularly to evaluate risk status and ultimately to eliminate the risks or minimize the impact to the program.
10. Consider the same steps for opportunities.

Finally, when the Projected Risk/Opportunity Values are determined for each risk and opportunity, sum the values to create an estimated or projected management reserve to account for what might happen.

26.5 THE ANALYTICS OF RISK

One of the most used systems for communicating risks and opportunities is the "IF, THEN" concept: an event leads to an impact 2. That means that something might happen which leads to an impact on the project. "IF" suggests that there is a probability of an event occurring and if it does, there is an outcome (or impact). That is, if we know a bad event that can cause a problem, we will want to identify what the event really is and what causes the event and estimate the "cost' of the event (Office of the Deputy Assistant Secretary of Defense for Systems Engineering, 2017).

Since risk is the description of something that could go wrong, we can establish some definitions that will relate the events and probability of happening. For example, consider an event that causes a failure in a test. We can define the variables as follows:

Risk Event: the event that occurs and causes an outcome with an impact on the project.
Outcome or Impact: the result of the project when the event occurs; this is usually described in dollars.
Probability that the event will occur, expressed in percent.
Probability that the risk event can be accurately measured. This is especially important for test processes since some testing techniques initially considered for a test may not be able to measure the event.

The risk events can be described in a flow diagram as follows:

Given that a risk has been identified, then there are two fundamental ways to determine a decision metric for risk: a two-factor system that considers the probability of a failure event occurring and the impact of the failure event on the test. Given the probability of the failure and given that the impact of the failure is measured in currency, a value outcome can be considered as the product of the probability and the impact. The calculated "predicted risk factor" value would have currency units that could be added to the unit under test cost for cost-estimating purposes and/or management of the test design (Figure 26.1).

The second method of determining a metric for risk cost includes the probability of a failure event, the impact or cost of a failure event and the probability of being able to measure the failure event. For this three-metric calculation, the probability of measuring/sensing the error is divided into the predicted risk factor. The reality is that the more difficult the failure even is to measure, the higher the over-risk factor becomes and higher the impact to the test. In many automatic test scenarios, the test manager designs testing protocol and test devices to ensure that a failure event

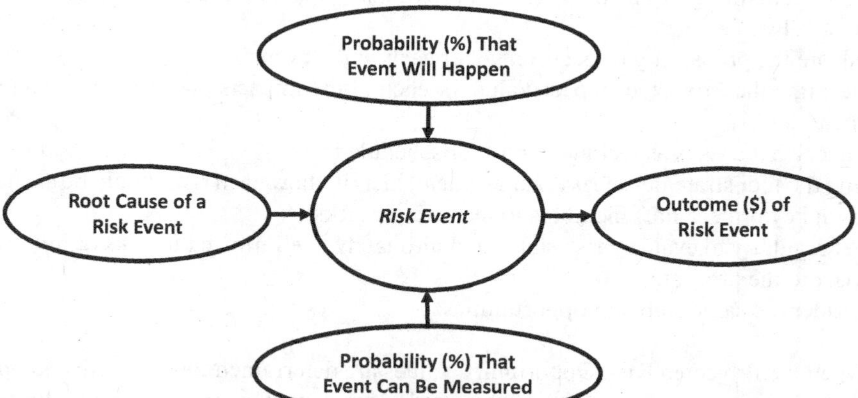

FIGURE 26.1 Risk events.

can be accurately sensed and measured, so the probability that the event can be measured is unity for practical purposes.

After the list of risks and opportunities has been decided and assigned to team members, a reporting system is necessary to communicate periodic risk/opportunity status. One popular method to do so is by using a matrix diagram and color scheme to offer a visual representation of the issues.

26.6 RISK REPORTING MATRIX

Often 4 by 4 or 5 by 5 matrices are used to show the impact and probability for each of the risk/opportunity issues. These matrices often use colors such as red, yellow, and green for risks and blue, violet, and blue-green for opportunities to show the severity of the impact and probability of occurrence. For example, considering risk for a test design project each risk would be identified as a cost, technical, or schedule risk and be shown on the matrix as C, T, or S. The most severe occurrence/probability events would be red, the medium events would be yellow, and the low value events would be green. The risk reporting periodic discussion would be focused on taking the appropriate actions to turn "red" issues to "yellow" and "yellow" events to "green." The risk meetings then settle on discussing the "return to green" plans. The diagram below shows an example of the risk matrix without colors but with the scale definitions for probability and impact for a fictitious project. As mentioned above, the matrix must be defined to be relevant to the project (Figure 26.2).

For some projects, managers have used a dollar value graphing technique that monitors the total project risk cost as a function of time. The goal of the graphing technique is to show how the projected risk values decrease in time through effective risk management. Hence, the graphical technique is often identified as the risk "waterfall." Often some more complex projects use both techniques described above: Individual risks are evaluated using the matrix in the periodic risk review meetings, and the "waterfall" plot would show the success of the overall project risk efforts.

To transition from the matrix to the waterfall, a risk management "plan" could be used by the risk team members to show progress toward retiring the risk or at least reducing the impact. The risk reporting protocol usually involves regular periodic meetings, with timing appropriate to the project schedule.

*Notional Probability Descriptions**

Ident	*Probability*	*Description of Occurrence*
P1	< 19%	Remotely probable
P2	20-39%	Probable but limited likelihood
P3	40-59%	Probable
P4	60-79%	Very probable or likely
P5	≥ 80%	Quite probable to certain to happen

*Relevant Probabilities and Descriptions *must* be defined for each test project.

Probability	I1	I2	I3	I4	I5
P5	L	M	H	H	H
P4	L	M	M	H	H
P3	L	M	M	H	H
P2	L	L	L	M	H
P1	L	L	L	L	M

Impact

*Notional Rating Guidance**

Risk Rating
High - Often shown Project; Major disruption likely. Different approach probably required. Priority Management Attention is required.
Moderate – Often yellow in the matrix. Some disruption to Test Project. Different approach probably required. Additional management attention required.
Low - Often green in the matrix. Minimum Impact to Test Project. Minimal/nominal oversite required to ensure risk remains low.

*Notional Cost, Technical & Schedule Impacts**

Ident	*Cost To Plan*	*Technical*	*Schedule*
I1	Little or no impact	Little or no impact to plan	Little or no impact to Test Plan
I2	≤ 5% increase	Nominal & Acceptable Reduction in Planned Test Scope	Planed end dates can be met with additional resources
I3	5-10 % Increase	Significant Reduction in Test Scope with performance still within plan	Test Plan Milestones will be missed; end date to slip
I4	10—12%	Significant Reduction in Test Scope with performance reducing final outcome	Key milestone delays; end date slip, borders on unacceptable
I5	> 12% Increase	Performance Reduced to unacceptable level	Critical path is effected; end date will slip

*Relevant Cost, Technical & Schedule Impacts *must* be defined for each test project.

FIGURE 26.2 Risk probability, consequence, assessment guide (Office of the Deputy Assistant Secretary of Defense for Systems Engineering, 2017; Vose, 2008).

A specific, suggested format will not be presented here because the format should be appropriate to the enterprise management project reporting system. Rather, a list of typical information for each risk is shown below:

- The Risk number or identification;
- Risk report number, usually sequentially numbered from the first meeting;
- The type of risk...Technical (scope), Schedule or Cost. [A single risk may have more than one type];
- Risk Management Strategy being followed (Avoid/Transfer/Control/Assume);
- The risk matrix shows the position in the matrix of the risk, including a reference to show the previous risk position and the new position in the most recent report;
- Risk team lead and/or action person name;
- Person responsible/authorized to sign off on the risk retirement;
- Documents that contain the requirements that have the risk elements;
- Specific actions being taken to bring the risk to retirement or closure;
- Specific dates for closer actions;
- A Schedule showing dates and action milestones.

Similar analyses and reports can and should be done for opportunities; below is an example of a matrix for an opportunity. While an opportunity report is not provided in this chapter, one can easily be generated for each opportunity with information to be reported similar to that for a risk report shown above. Note that the "consequence" scale for an opportunity is identified as a "benefit" and the scale values are reversed (Figure 26.3).

A final note about managing risk and opportunity for a project can be summarized by a matrix representation called the Arrow of Attention often highlighted by Dr. David Hillson in his writings on risk management (Hillson, 2001). The visual in Figure 26.4 shows that project (test design and execution) managers should focus on reducing risk *and* increasing opportunities to establish more effective projects.

*Notional Probability Descriptions**

Ident	*Probability*	*Description of Occurrence*
O1	< 19%	Remotely probable
O2	20-39%	Limited likelihood
O3	40-59%	Likely
O4	60-79%	Very probable or likely
O5	≥ 80%	Quite probable to certain

*Relevant Probabilities and Descriptions *must* be defined for *each* test project.

Probability	B5	B4	B3	B2	B1
O5	H	H	H	M	L
O4	H	H	M	M	L
O3	H	H	M	M	L
O2	H	M	L	L	L
O1	M	L	L	L	L

Benefit

*Notional Rating Guidance**

Opportunity Rating
High – Blue color. Very likely to happen with desirable benefits. Needs plan to implement with required resources.
Moderate – Violet color. Implementation Plan must show resourcing (staff and budget) plan appropriate to benefit.
Low - Light Green color. Minimum/nominal Benefit to Test Project. Resources required probably exceeds estimated benefit

*Notional Cost, Technical & Schedule Benefits**

Ident	*Cost (% of Plan ETC)*	*Technical*	*Schedule Reserve Remaining on Test Plan Critical Path*
B5	> 10	May have very positive business level opportunity	> 30%
B4	7-10	"Exceeds Customer Expectations" for performance	20 – 29%
B3	6-7	Significant Reduction in Test Scope with performance still within plan	10 – 19%
B2	<5	Significant Reduction in Test Scope with performance reducing final outcome	5 – 9%
B1	Minimum / None	Performance Reduced to unacceptable level	< 5%

*Relevant ETC, Critical Path and Technical Benefit metrics *must* be defined for *each* test project.

FIGURE 26.3 Opportunity probability, consequence, assessment guide (Office of the Deputy Assistant Secretary of Defense for Systems Engineering, 2017; Vose, 2008).

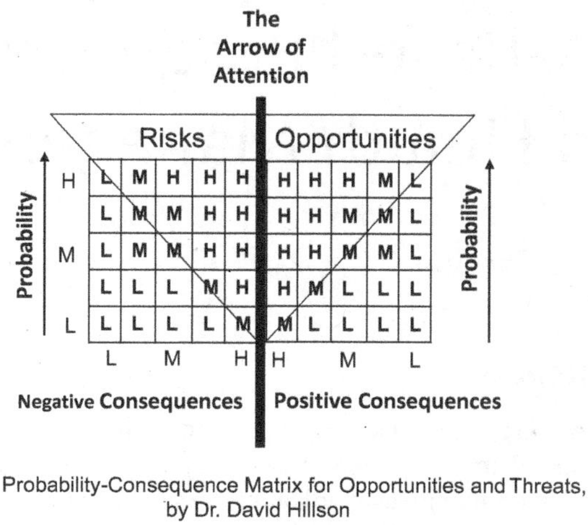

FIGURE 26.4 Probability consequences matrix (Hillson, 2001).

26.7 CHAPTER SUMMARY

ATS/TPS project managers must recognize that with every project something will go wrong. Typically, these problems are unforeseen and unplanned, not considered relevant to the project during the planning stage, and should not have occurred in the first place. The ATS/TPS project manager must recognize the fact that problems will occur. This chapter described various techniques to help project managers in risk analysis by considering and analyzing causal events. It presented four key strategies for mitigating risk: accept, control, transfer, and avoid. This chapter focused on the ATS/TPS design projects; however, these strategies and processes can apply to any engineering project.

Some final advice is offered to all test project managers; three rules to live by to produce less risk and more opportunities for your test projects.

- Not identifying a risk does not mean that it doesn't exist.
  - …the project is taking the risk without the team being aware of it!
- Bad news does not get better with age.…ignoring a risk does not make it go away or make it less of a risk!
  - Good things normally do not just happen:
- …Smart work and planning often leads to "Good Luck!"

WORKS CITED

Association for Project Management (2004). *Project Risk Analysis and Management Guide*. Buckinghamshire, UK: Association for Project Management.

Hillson, D. (2001). Extending the risk process to manage opportunities. In *Fourth European Project Management Conference*, London, UK.

Office of the Deputy Assistant Secretary of Defense for Systems Engineering (2017). *DoD Risk, Issue, and Opportunity Management Guide for Defense Acquisition Programs.* Washington, DC: US Department of Defense.

Vose, D. (2008). *Risk Analysis: A Quantitative Guide*, 3rd Edition. Chichester, UK: John Wiley & Sons.

27 Risk-Based Prognostics and Health Management

John W. Sheppard

27.1 INTRODUCTION

As engineering fields mature, new technologies are emerging that are beginning to serve as the foundation of many societal improvements. For example, modern medical diagnostic equipment provides valuable information that gives medical professionals a better understanding of a patient's needs and ultimately improves the quality of life (Cosmus & Parizh, 2011). Improvements to vehicle designs make transportation in cars or aircraft safer and more environmentally friendly (Farokhi, 2014). Military equipment continues to be developed that better supports and protects personnel in the field (Pullen & Khimeche, 2014). Manufacturing practices and robotic equipment improve work safety conditions and reduce a product's price point, making amenities available to a wider range of consumers (He & Xu, 2015).

One approach to maximizing system availability is to incorporate some means of health assessment into the system itself. Doing so is often referred to as "integrated system health management" (ISHM) or "prognostics and health management" (PHM), which has been applied successfully to many complex systems (Figueroa et al., 2009). By integrating health assessment into the very functioning of a system, more information can be obtained that provides a better understanding of the system as a whole, thus allowing system owners to become proactive in how they deal with system degradation. ISHM and PHM promise to focus on system conditions, thus supporting initiatives in what has become known as condition-based maintenance (CBM). This, in turn, enables maintenance events to be initiated based on specific system conditions rather than waiting until a failure occurs (Jardine et al., 2006).

One of the key ingredients of ISHM/PHM is diagnostics, which corresponds to the process of determining the health state of the system based on sets of observations (or tests). Such tests are designed specifically to track system behavior and determine whether or not a failure has occurred. In many cases, it is impossible to identify a single fault that explains the observations with certainty. Instead, candidate sets of faults are often indicated, and when using applicable models, probabilities or confidence values are associated with the faults to provide additional information. One historic approach to using test observations for diagnosis is to apply a decision tree – sometimes referred to as a fault tree[1] (Simpson & Sheppard, 1994). In a diagnostic fault tree, a path through the tree corresponds to a sequence of test results that end in identifying a fault, replaceable unit, or ambiguity group.

A companion process to diagnostics is prognostics. With prognostics, the goal is not only to determine the current health state of a system but also to predict the health state at some time in the future (Pecht, 2008). This prediction is key to the "P" part of "PHM." Similar to diagnostics, exact identification of this future health state is rare, and a probabilistic approach is desirable to quantify the level of uncertainty associated with the health state predictions more reliably. When moving beyond detecting incipient failures (another common task in PHM), a temporal component must be introduced into the models. A variety of model types have been proposed for this, including the Kalman filter (Peel, 2008; Lall et al., 2010), the particle filter (Orchard et al., 2008; Jouin et al., 2014), the dynamic Bayesian network (Sheppard & Kaufman, 2005; Codetta-Raiteri & Portinale, 2015), and recently the continuous-time Bayesian network (CTBN) (Perreault et al., 2015; 2016).

 DOI: 10.1201/9781003188377-32

In real-world, complex system support, PHM introduces significant challenges, not only with how to predict faults but also with how support organizations measure the effectiveness of maintenance processes that act upon such predictions. To address this issue, we propose coupling the prognostic process explicitly, rather than implicitly, with a risk-based framework. Specifically, we begin by observing that faults, by their very nature, produce undesirable effects in the system, which we refer to as hazards. With the ability to predict the health state of a system at some point in time, it is possible to anticipate faults and the risks associated with resulting hazards. This information can then be used to assess the severity or criticality of the risk and potentially mitigate that risk through preventative maintenance. Effectiveness measurement then shifts to be in terms of hazards averted. Alternative measures based on overall availability can also be used for such assessment.

27.2 RISK-BASED PHM

While much work in PHM has focused more on the health management aspects, a large part of PHM involves prognostics. Such approaches often use a prognostic model to predict when a system fault is likely to occur, based on an understanding of the current health of that system. Identifying the most critical faults is achieved by domain experts synthesizing the prognostic results, who then use these faults and existing reliability data to identify likely hazards (Mosallam et al., 2015). Finally, based on the determined likely effects, the domain expert makes a decision about the best course of action, given the predicted behavior.

Although this method has proven itself to be relatively effective, there are several shortcomings of this approach. The first is the heavy reliance on data synthesis by human experts. This can be a problem in that the process requires highly skilled domain experts to understand the details of the system whenever information needs to be collected and analyzed. Relying on the human expert, however, is problematic given the goals for real-time health assessment and prediction, and the steady increase in system complexity. Thus, minimizing the need for human involvement in PHM is critical to the success and integration of PHM into real-time system support.

Another concern regarding current PHM approaches is that the process often involves a pipeline of largely disjoint models. In particular, a diagnostic model is used to assess observations and test results to determine the health of the system. This health information is then provided to a prognostic model, which then predicts the future occurrence of potential faults. An entirely separate model is then used to identify which effects will be most impacted by those faults. Finally, yet another model, set of equations, or domain knowledge is used to evaluate the potential impact of emerging hazards on the system based on the likelihood of the different effects and other behaviors. Each step in the process depends on separate technologies, making implementation and integration difficult. Furthermore, information passed between technologies tends to be transcribed manually, unless a standard data exchange format is observed. This disjointed approach is inefficient and has the potential for information loss due to model incompatibilities. A better solution would seamlessly work with all tasks necessary to support PHM in responding to emerging hazards and their associated risks.

One approach to performing "risk-informed decision making" that was developed by the United States Nuclear Regulatory Commission is embodied in a tool called Systems Analysis Programs for Hands-on Integrated Reliability Evaluations (SAPHIRE) (Smith et al., 2008). The SAPHIRE system has been designed to support large-scale analysis of fault trees and event trees to determine how faults propagate to effects. SAPHIRE uses a process for generating minimal cut sets for the fault trees, thereby reducing their complexity to the most critical set of faults. The associated probabilistic analyses focus on deriving importance measures such as Fussell-Vesely importance and risk achievement, which are used to assess probabilistic properties of risk in a system (Van der Borst & Schoonakker, 2001). Ultimately, SAPHIRE's intended use is static, intended to be used to assess the risk of a design; it is not typically used for online risk assessment or PHM.

In this chapter, we introduce risk-based PHM (rPHM), a cohesive, online methodology and framework that supports a variety of PHM tasks using a single type of model. This framework supports both diagnostics and prognostics and incorporates effects or hazards using the same model semantics. By building hazards into the model itself, predictions can be made not only about likely faults but also about the effects that may occur as a result of those faults. The integration of faults, observations, and hazards is handled algorithmically and avoids the need for real-time human synthesis. Furthermore, the framework supports the definition of user-specified functions that place value on various system states, thus providing a means for quantifying system quality while still using the semantics of the underlying prognostic model. The framework also allows decisions (also referred to as "scenarios") to be modeled, which means that multiple system options can be compared to one another to facilitate risk mitigation. Finally, by incorporating performance functions into the model, a multiobjective approach to optimization can be used to assess possible courses of action and to facilitate more effective risk-informed decision-making.

This chapter provides an overview of techniques we have developed to define the rPHM framework. Section 27.3 provides background information on the models and data structures used in the rPHM framework. Sections 27.4 and 27.5 explain the procedures necessary to build the rPHM models using readily available reliability information. Section 27.6 goes into detail concerning how the rPHM model is intended to be used, drawing on our work in performance functions and continuous-time decision networks. The main points of the chapter are summarized in Section 27.7.

27.3 BACKGROUND

In this section, we review the models and data structures used throughout this chapter.

27.3.1 Continuous-Time Bayesian Networks

Many systems change state as a function of time, and the continuous-time Bayesian network (CTBN) is a framework that is capable of modeling such systems (Nodelman et al., 2002). The CTBN represents each variable in a system as vertices in a directed graph. These vertices are connected using directed edges such that the behavior of each vertex depends directly on the behavior of its parents in the graph.

To describe the behavior of a variable, the CTBN defines a conditional Markov process for the vertex. This consists of an initial distribution defined over the states of the variable, as well as a conditional intensity matrix (CIM) that describes how the variable changes its state over time. More formally, a CIM $\mathbf{Q}_X$ for some variable (vertex) X specifies a set of transition intensities $q_{i,j}\left(\mathbf{Pa}(X)\right)$, where $q_{i,j}$ is the entry in the ith row, jth column of $\mathbf{Q}_X$, which change as a function of the states assigned to the parents of X ($\mathbf{Pa}(X)$). These entries specify the main parameter of an exponential distribution, $f(t) = q_{i,j}\exp\left(-q_{i,j}t\right)$. When specifying the CIM, we also note that the entry on the diagonal is given as $q_{-} = \sum_j q_{i,j}$.

The resulting CTBN defines a distribution over the states of each variable at any given time $t \in [0, \infty)$. Approximate inference algorithms have been developed that are capable of answering queries about state probabilities or transition times tractably. These algorithms work with continuous-time evidence, which allows the states of variables to be set for specified time intervals (e.g., during maintenance). In this chapter, we use the CTBN as the basis for supporting PHM since it provides a model by which we can predict the probability of faults and hazards based on the evidence provided by the tests.

27.3.2 Notation

We begin our formal treatment of the CTBN model by providing an overview of notation we use through the remainder of this chapter. Capital letters F_i and T_j denote faults and tests respectively,

TABLE 27.1
Notation Used in this Chapter

Symbol	Interpretation
F_i	The ith fault in the network
T_i	The ith test in the network
$\mathbf{F}_i$	The vector of faults monitored by T_i (i.e., the parent set of T_i)
$\mathbf{T}_i$	The vector of tests that will detect the failure of F_i
$\mathbf{D}$	A D-matrix
$d_{i,j}$	The F_i, T_j entry of 0 or 1 in a D-matrix
λ_i	The failure rate for fault F_i
μ_i	The repair rate for fault F_i
ND_i	The non-detect probability for test T_i
FA_i	The false alarm probability for test T_i
$ND_{i,j}$	The non-detect probability for test-fault relationship T_i and $\mathbf{F}_i[j]$
$FA_{i,j}$	The false alarm probability for test-fault relationship T_i and $\mathbf{F}_i[j]$
$\mathbf{Q}_X$	The intensity matrix for variable X
$\mathbf{Q}_{X\mid Y}$	The conditional intensity matrix for variable X, conditioned on variable Y
$\mathbf{Pa}(X)$	The parents of variable X

where the subscript indicates the specific test or fault in the system. A bold $\mathbf{F}_i$ is a vector of faults detected by some test T_i, and $\mathbf{F}_i[j]$ indicates the jth fault detected by T_i. We use a bold $\mathbf{T}_j$ to denote a vector of tests that detect a fault F_j. This corresponds to what is known as the "diagnostic signature" for that fault. The symbols λ_i and μ_i specify the failure and repair rates respectively for fault F_i. Similarly, ND_i and FA_i denotes the non-detect and false alarm probabilities respectively associated with test T_i. $\text{ND}_{i,j}$ and $\text{FA}_{i,j}$ also describe non-detect and false alarm probabilities, but they do so for the specific test $(T_i, \mathbf{F}_i[j])$ pair. This allows us to capture the notion that a test may fail to detect a particular fault while remaining operational. A complete list of notation used in this chapter is provided in Table 27.1.

27.3.3 D-Matrices

In order to determine the health state of a system, whether for purposes of fault diagnosis or prognosis, we need certain information about the unit under test (UUT). This information is collected via tests performed on the UUT, whether through the use of built-in test, on-board health monitoring, off-board automatic testing, or manual testing. The outcome of these tests then indicates the presence or absence of faults in the system (Simpson & Sheppard, 1994). In a complex system, the way tests and faults interact with one another can make diagnosis and maintenance difficult since a single fault may be monitored by multiple tests, and a single test may monitor multiple faults. To manage this complexity, we represent the relationships between the faults and tests explicitly using a representation that corresponds to a graph adjacency matrix, known as a D-matrix. With this matrix, the columns correspond to tests and the rows correspond to the faults or failures observed by these tests. The D-matrix has been adopted by several modeling tools used to perform fault diagnosis, including the aerospace community (Sheppard & Butcher, 2007). In fact, the concept of a "fault dictionary," which is used in digital testing, can be regarded as a special case of a D-matrix (Abramovici et al., 1990; Sheppard & Simpson, 1996).

Constructing a D-matrix involves building diagnostic signatures for each fault within the UUT. Given a set of faults $\mathbf{F}$ and tests $\mathbf{T}$, we define the matrix D as follows:

$$d_{i,j} = \begin{cases} 1 & \text{if } T_i \text{ detects } F_j \\ 0 & \text{otherwise} \end{cases}.$$

For some fault F_i, the diagnostic signature corresponds to $\mathbf{F}_i = \left[d_{i,1},\ldots,d_{i,|\mathbf{T}|}\right]$ for that fault (Sheppard & Butcher, 2007; Strasser & Sheppard, 2013). This diagnostic signature is determined for each $F_i \in \mathbf{F}$, and each signature $\mathbf{F}_i$ becomes the ith row of the D-matrix.

The relationships between tests and faults described by a D-matrix provide information needed to diagnose the UUT, but probabilistic information is not included in this structure (Strasser & Sheppard, 2013). Similarly, the basic D-matrix does not provide information about fault-to-fault or test-to-test relationships; although, test-to-test relationships can be derived through a process known as "logical closure" (Simpson & Sheppard, 1994). For these reasons, while the D-matrix can be used to construct a network structure (Section 27.4.1), more information is required to parameterize a CTBN in order to perform diagnosis and prognosis (Section 27.4.2).

27.3.4 Fault Trees

Fault tree analysis (FTA) is a powerful method for evaluating system design in a reliability and safety-oriented context. In several critical domains, fault trees provide a graphical representation of how failures of one or more system components can lead to higher-order system failures. A fault tree is a directed acyclic graph (DAG) consisting of a set of events and a set of logic gates. Interior vertices of the graph represent effects, while the leaf vertices are faults or base-level failures. Leaf states propagate upward through the tree toward the root, capturing how faults contribute to possible system effects. Each interior vertex receives input from its children, which are then combined through a logic gate such as an AND gate or an OR gate.

Some fault tree representations allow for redundancies in the system that can cause the same component to appear in multiple parts of the tree, thus violating the strict tree structure of the underlying graph. When these redundant components are removed from the tree through a pruning process, the resulting fault tree becomes easier to store, update, and process since the pruned tree remains functionally equivalent to the original tree.

Note that many of these redundancies can be removed by finding the minimal cut sets of the fault tree. Since minimal cut set analysis is focused on identifying the cut sets affecting the top-level effect of the tree, we found that this type of analysis is not suitable for our purposes. This is because we are also interested in the occurrence of the intermediate effects as well (Barlow & Chatterjee, 1973).

A related diagnostic model is the Fault Isolation Manual (FIM), which often plays a significant role in maintaining large systems (Simpson & Sheppard, 1994). FIMs are derived from decision trees that are also typically referred to as fault trees.[2] The internal vertices of the FIM correspond to tests, that are represented in the D-matrix, and the leaves correspond to faults, also represented in the D-matrix. To use the FIM, the test specified at the root of the tree is performed, and the result of that test indicates which branch to follow, leading to the next vertex in the tree. This process continues until reaching a leaf corresponding to a diagnosed fault or ambiguity group. Thus, following the decision tree structure in a FIM leads to isolating faults or ambiguity groups (Strasser & Sheppard, 2013).

27.4 DERIVING CTBNs FROM D-MATRICES

Given that our work focuses on performing fault prognosis as a foundation, we begin by considering how information derived from a diagnostic model—namely the D-matrix—can be used as the first step in creating our rPHM model.

27.4.1 D-Matrix Network Structure

The non-zero entries in a D-matrix represent dependence relationships between tests and diagnoses (e.g., faults). Similarly, the directed edges in a CTBN indicate influence between variables. This observation can be used to derive the network structure of a CTBN from a D-matrix. Let $\mathbf{D}$ be an $m \times n$ D-matrix, and let $\mathcal{G}$ be a graph structure for a CTBN (initially containing no vertices). First, m fault vertices are added to $\mathcal{G}$, each one representing a row in $\mathbf{D}$. Then, n test vertices are added, representing the columns in $\mathbf{D}$. Finally, for each entry in the D-matrix where $d_{i,j} = 1$, an edge is inserted from F_i to T_j, where F_i is the fault vertex corresponding to the ith row in $\mathbf{D}$, and T_j is the test vertex associated with the jth column in $\mathbf{D}$. Each of these edges asserts that the behavior of a test depends on the state of a fault at some point in time.

The resulting network $\mathcal{G}$ has $m + n$ vertices and m_d edges, where $m_d = \sum_{i=1}^{m} d_{i,j}$. These directed edges are inserted from a fault vertex to a test vertex, with no edges inserted from tests to faults, between two test vertices, or between two fault vertices. The resulting graph $\mathcal{G}$ is therefore bipartite, with one layer consisting of the m faults and the other consisting of the n tests.

Typically, the variables contain relatively few states, meaning that the initial probability distributions and intensity matrices for each vertex are small. As a result, the total number of intensity matrices defined over the set of all vertices is the primary factor driving model complexity. None of the fault vertex will have parents, thus only a single intensity matrix is required to parameterize each fault. The test vertices, however, may have any number of fault vertices in their parent sets. As a result, a CIM for a test vertex will have $\left|\mathbf{Q}_{T_j}\right| = \prod_{i=1}^{m} d_{i,j} \cdot \left|F_i\right|$, matrices, where $\left|F_i\right|$ is the number of states for F_i, and $d_{i,j}$ equal to one indicates the presence of a dependence. In the worst case, a test may monitor all faults in the system, resulting in an upper bound $O(cm)$, where c serves as a bound on the number of states for each fault. In fault diagnosis, it is common for $c = 2$; however, this could increase multiple states, perhaps representing levels of degradation, be included.

27.4.2 D-Matrix Parameterization

The way one variable influences another, as well as default failure transition behavior, is captured in the CTBN by parameterizing each vertex. One way of doing such parameterization is by using reliability information; however, these parameters could be updated as data is collected during the operation and maintenance of the system. For the purpose of this chapter, we assume binary states for both fault and test vertices; however, the same ideas could be applied to variables with more than two states if supporting data is available.

1. **Parameterizing Fault Vertices**: The fault vertices in the network can be parameterized independently of the test vertices since they have no parents in the network. This leads to the requirement to specify only a single unconditional intensity matrix for each fault. Recall that the intensity matrix defines exponential transition distributions over the states. In other words, the CIM specifies the probability distributions associated with transitioning from a non-failed state to a failed state and vice-versa. Failure and repair rates are often available via either historical data or domain knowledge. Formally, a failure rate λ indicates the rate at which a failure will occur given that no failure currently exists. Note that the reliability analysis often captures information about mean time between failure (MTBF), and that with the exponential distribution assumption, is defined as $\text{MTBF} = 1/\lambda$.

   The repair rate μ indicates the rate at which a fault will transition back to having no failure and depending on the associated repair policy is often set to zero. In that case, the failure state will become an absorbing state. In general, when parameterizing a fault's intensity matrix, the failure rate λ is assigned to the entry corresponding to the transition from non-failure to failure, and the repair rate μ is assigned to the reverse transition (i.e., failure to non-failure). The diagonal values are set to be the negative sum of the remaining entries in the row (i.e., $-\lambda$ and $-\mu$ respectively). Let F_i be a fault vertex. Then the intensity matrix for F_i is

$$\mathbf{Q}_{F_i} = \begin{array}{c} f_i^0 \\ f_i^1 \end{array} \overset{\begin{array}{cc} f_i^0 & f_i^1 \end{array}}{\begin{pmatrix} -\lambda_i & \lambda_i \\ \mu_i & -\mu_i \end{pmatrix}}$$

where λ_i and μ_i are the failure and repair rates for F_i.

Note that in some cases, transition rates follow non-exponential distributions. In reliability modeling, the Weibull and lognormal distributions are two distributions that are commonly used to describe time-to-failure (Kalbfleisch & Prentice, 2011). While a CTBN is defined based on exponential distributions, it is possible to approximate non-exponential distribution using "phase-type distributions" (Nodelman & Horvitz, 2003). Furthermore, work has been done to obtain approximations of non-exponential parametric distributions automatically, thereby allowing for more complex failure distributions to be represented. The corresponding phase-type distribution is then learned and embedded directly into the existing matrix $\mathbf{Q}_{F_i}$.

2. **Parameterizing Test Vertices:** Parameterizing test vertices is more complex than parameterizing the fault vertices, in that test vertices have $n \geq 1$ parents in the graph, meaning that the number of intensity matrices required to specify a test CIM grows exponentially with respect to n.

The intensity matrices for a test vertex's CIM can be derived using false alarm (FA) and non-detect (ND) likelihoods for each test. FA and ND values can be estimated based on sensitivity and calibration information on the test. The relationship between FA and ND rates and the probability distribution for a test can then be modeled as line failures, described in detail by Perreault et al. (2015). The resulting probability distribution is given as

$$P(T_i = 0|\mathbf{F}_i) = (\text{ND}_i)P(p_i = 1|\mathbf{F}_i) + (1 - \text{FA}_i)P(p_i = 0|\mathbf{F}_i),$$

where T_i is a test in the network, F_i is the set of parent faults for T_i, and ND_i FA_i are non-detect and false alarm probabilities respectively, defined for T_i.

Note that the distribution incorporates two probabilities $P(p_i = 1|\mathbf{F}_i)$ and $P(p_i = 0|\mathbf{F}_i)$, which can be calculated as follows:

$$\begin{aligned} P(p_i = 1|\mathbf{F}_i) &= \prod_{\{f_{i,j} \in \mathbf{F}_i | f_{i,j} = 0\}} (1 - \text{FA}_{i,j}) \cdot \prod_{\{f_{i,j} \in \mathbf{F}_i | f_{i,j} = 1\}} (\text{ND}_{i,j}) \\ P(p_i = 0|\mathbf{F}_i) &= \left(1 - P(p_i = 1|\mathbf{F}_i)\right) \end{aligned}$$

Here, $\text{ND}_{i,j}$ and $\text{FA}_{i,j}$ correspond to the non-detect and false alarm probabilities for test T_i with respect to the jth fault in the parent set of T_i. The entire probability distribution $(p_i|\mathbf{F}_i)$ can be thought of as the probability that the true state of faults $\mathbf{F}_i$ are detected correctly (or incorrectly) by test T_i.

Given the previous equation, the number of non-detect and false alarm probabilities to be modeled is equal to the number of edges in the CTBN network, plus the total number of tests: $O(|E|+|V|)$. Although this number is quite manageable from a computational standpoint, it may be that no data exists to describe the needed probabilities for the individual fault-test relationships. Instead, specific fault-test values $\text{ND}_{i,j}$ and $\text{FA}_{i,j}$ may be assumed to be zero, and any false alarm or non-detect likelihoods can be summarized by ND_i and FA_i for test T_i.

With this assumption, the number of user-specified parameters is reduced drastically. The only non-detect and false alarm probabilities that need to be specified are for the test vertices: $O(|V|)$. If we assume a test fails in the presence of any failure of the monitored components, then the test's output will follow a deterministic OR gate. Then the intensity matrix corresponding to no faults in the system becomes

$$\mathbf{Q}_{\{T_i|\wedge_{F\in\mathbf{F}_i}\ F=0\}} \quad \begin{matrix} & t_i^0 & t_i^1 \\ t_i^0 & -(1-\text{FA}_i)^{-1} & (1-\text{FA}_i)^{-1} \\ t_i^1 & (\text{FA}_i)^{-1} & -(\text{FA}_i)^{-1} \end{matrix}.$$

and the intensity matrix for the cases with faults becomes

$$\mathbf{Q}_{\{T_i|\vee_{F\in\mathbf{F}_i}\ F=1\}} \quad \begin{matrix} & t_i^0 & t_i^1 \\ t_i^0 & -(\text{ND}_i)^{-1} & (\text{ND}_i)^{-1} \\ t_i^1 & (1-\text{ND}_i)^{-1} & -(1-\text{ND}_i)^{-1} \end{matrix}.$$

Then each matrix in CIM $\mathbf{Q}_{T_i}$ can be parameterized based on the high-level false alarm and non-detect probabilities. Again, this information can be learned based on historical maintenance data.

27.5 DERIVING CTBNS FROM FAULT TREES

The approach to prognostics espoused in this chapter is to model hazards that might arise in a system and capture the associated risks. The approach taken is based on the fault tree analysis discussed in Section 27.3.4. In this section, we present a method for deriving CTBNs from fault trees, resulting in a new predictive model with faults and hazard-based effects. We then show how to relate this information to the D-matrix and explain how to merge the two networks.

27.5.1 FAULT TREE NETWORK STRUCTURE

The structure of a hazard/effect-based CTBN can be derived directly using the fault tree developed through FTA as follows. First the set of vertices **V** is obtained by extracting the variables directly from the faults and hazards in the fault tree. Recall that the hazard/effect nodes correspond to logic gates. Note that either a fault or hazard may occur in multiple locations of the fault tree, but the corresponding vertex in the CTBN will occur only once. Thus, the resulting CTBN structure may not be defined strictly as a tree.

Next, directed edges are inserted between the vertices. Once again, these can be obtained directly from the structure of the fault tree. First recall that faults only occur as leaves in the tree. The corresponding vertices in the CTBN, as with the D-matrix-based CTBN, therefore have no parents. Next we consider the hazards in the tree whose states are determined as a logical combination of their inputs. We represent this dependence on the inputs by adding an edge from the vertices corresponding to each input u_i to the vertex corresponding to the effect. The resulting CTBN has the same underlying structure as the fault-tree after pruning. As mentioned, collapsing multiple occurrences of faults into a single vertex will result in a graph that is no longer a tree. The pruning process may yield substantially fewer edge insertions, and given that the space complexity of a vertex is exponential with respect to the number of parents, pruning is a critical preprocessing step.

27.5.2 Fault Tree Parameterization

In system reliability, the prior probability that each component starts in a failing state is often known, and in this chapter, we assume all variables start in a non-failed state with probability 1.0. In other words, this forms the prior probability for the fault vertices. As before, we assume the variables are binary, consisting of a failing state and a non-failing state. Then the intensity matrix for each vertex X requires $2^{(|\mathbf{Pa}(X)|+1)}$ rates, which can make identifying parameters for even relatively small models a difficult task. Just as in Section 27.4, the goal is to reduce the number of rates required to parameterize a vertex by taking advantage of available reliability information and by exploiting behavioral information obtained from the underlying fault tree. The fault vertices in the fault tree-based network are identical to those obtained from the D-matrix; therefore, the same parameterization process from Section 27.4.2 can be employed.

1. **Parameterizing AND Effects**: Let Φ_X be the discrete function corresponding to the gate from which variable X was derived. Then we expect $\lim_{t\to\infty} P\left(X(t) = \Phi_x\left(\mathbf{Pa}(X)\right)\right) = 1.0$. This behavior is guaranteed for each CIM $Q_{X|\mathbf{Pa}(X)}$ by forcing a non-zero transition rate *to* the state produced by $\Phi_X\left(\mathbf{Pa}(X)\right)$, and a zero rate *from* the state produced by $\Phi_X\left(\mathbf{Pa}(X)\right)$. In other words, this creates an absorbing state, at which point the variable will remain in this state, unless a change in its parent states occurs.

   For a vertex created corresponding to an AND gate, Φ_X becomes the logical AND function, where Φ_X will produce a value of 1 if and only if all inputs are 1. To accomplish this, the CIM is parameterized for a vertex X where all parents are 1 as follows:

$$\mathbf{Q}_{X|\mathbf{Pa}(X)} = \begin{array}{c} \\ x^0 \\ x^1 \end{array} \begin{array}{c} \begin{array}{cc} x^0 & x^1 \end{array} \\ \begin{pmatrix} -\lambda_X & \lambda_X \\ 0 & 0 \end{pmatrix} \end{array}.$$

   when $\Phi_X\left(\mathbf{Pa}(X)\right) = 1$ (i.e., all parent states are ones).

   Considering the cases where not all parents of the vertex are 1, the function Φ_X produces a value of 0, so we wish to describe the time it takes to transition back to a state of 0 for the vertex in the CTBN. For this, the CIM is parameterized as:

$$\mathbf{Q}_{X|\mathbf{Pa}(X)} = \begin{array}{c} \\ x^0 \\ x^1 \end{array} \begin{array}{c} \begin{array}{cc} x^0 & x^1 \end{array} \\ \begin{pmatrix} 0 & 0 \\ \mu_{X|\mathbf{Pa}(X)} & -\mu_{X|\mathbf{Pa}(X)} \end{pmatrix} \end{array}.$$

   when $\Phi_X\left(\mathbf{Pa}(X)\right) = \overline{1}$ (i.e., one or more of the parents is in state 0).

   Note that we need to specify a rate for transitioning back to 0 for each of the instances where not all parents are in state 1 since each of these cases has a separate intensity matrix.

Although this level of detail may be necessary in some cases, it may be possible to simplify the parameterization process by assuming the rate is the same, regardless of the parent set. This means that X transitions back to 0 at the same rate as long as Φ_X evaluates to 0. Given this, a variable X derived from an AND vertex in a fault tree can be specified using only two parameters, where λ_X defines the time it takes to transition to 1 in the event that all parents are in state 1, and μ_X is used in all other cases to indicate when X will transition back to 0.

2. **Parameterizing OR Effects:** When a variable X is created from an OR gate, Φ_X is defined as the logical OR function. Again, it is guaranteed that this CTBN vertex eventually reaches state 0 by parameterizing the intensity matrix such that state 0 is an absorbing state:

$$\mathbf{Q}_{X|\mathbf{Pa}(X)} = \begin{array}{c} \\ x^0 \\ x^1 \end{array} \begin{array}{c} \begin{array}{cc} x^0 & x^1 \end{array} \\ \begin{pmatrix} 0 & 0 \\ \mu_X & -\mu_X \end{pmatrix} \end{array}.$$

when $\Phi_X(\mathbf{Pa}(X)) = 0$ (i.e., all of the parent vertices are in state zero).

Next, consider the case where Φ_X evaluates to 1. Since Φ_X is an OR gate, this occurs whenever at least one parent is in state 1. In this case, vertex X is parameterized such that the matrices in the CIM eventually transition to state 1, ensuring that X. conforms to the correct output of the OR gate:

when $\Phi_X(\mathbf{Pa}(X)) = \bar{0}$ (i.e., one or more of the parents is in state one).

Once again, we need to specify a rate parameter for every possible state instantiation of the parents where at least one parent is in state 1. This means that $2^{|\mathbf{Pa}(X)|}$ parameters would required for vertex s in the case of the AND vertex, this number can be reduced by making a simplifying assumption. Given the causal interpretation of the model, the concept of disjunctive interaction, also referred to as the Noisy-OR model, may be employed (Perreault et al., 2016). This allows us to specify the rates λ_{X_i} for only the cases where a single parent is in state 1. This reduces the number of required parameters to be linear in the number of parents. The remaining parameters for the CIMs with multiple parents in state 1 are then accounted for by disjunctive interaction.

27.5.3 Merging the D-Matrix and Fault Tree Models

Now that we have a way to construct a CTBN that captures diagnostic and prognostic relationships and a way to create a separate CTBN that models effect/hazard propagation, the next step is to combine them into a single model. The key observation that makes this merge process possible is that both the D-matrix-derived CTBN and a fault tree-derived CTBN describe different aspects of the same single system $\mathcal{S}$; the set of faults is the same for both models.

Let $\text{CTBN}_\mathbf{D}$ be a CTBN constructed from the D-matrix for system $\mathcal{S}$, and let $\text{CTBN}_\mathbf{F}$ be a CTBN derived from the fault tree for the same system $\mathcal{S}$. As before, let $\mathbf{T}$ be the set of test vertices in $\text{CTBN}_\mathbf{D}$, $\mathbf{H}$ be the set of hazard vertices in $\text{CTBN}_\mathbf{F}$, and $\mathbf{F}$ the set of faults contained in both CTBNs. A new model $\text{CTBN}_\mathcal{S}$ can be constructed for system $\mathcal{S}$ by combining both networks to obtain a single CTBN with vertices $\mathbf{T}$, $\mathbf{H}$, and $\mathbf{F}$. The vertices in the sets $\mathbf{T}$ and $\mathbf{H}$ will have the same parents and parameters as in the original networks. The only difference is that the vertices in the set $\mathbf{F}$ in the combined CTBN have more children than they did before being merged, but this does not affect their parameterization. Furthermore, the fault vertices in $\text{CTBN}_\mathbf{D}$ and $\text{CTBN}_\mathbf{F}$ are parameterized using the same method.

27.6 USAGE AND DECISION-MAKING

A CTBN model can use observations collected from tests to perform diagnostics and prognostics. More specifically, we can regard the test results (which indicate whether or not the tests passed or failed as well as when the test was run) as evidence and then apply that evidence to the test vertices. We can then query the model to determine probability distributions over the states of faults and effects at any point in time (both future and past). These probability distributions can then be used to identify the most likely effects and how likely they are with an eye toward mitigating the risks associated with the hazards.

Falling under the umbrella of what is called "risk-informed decision making," the rPHM approach described here is intended to use the derived probability information about emerging risks to help operators and maintainers do their jobs more effectively. Specifically, our approach is to incorporate performance directly into the model and use standard query techniques to quantify the overall evolution of the state of the system, relative to health, risk, and potential impact on mission effectiveness. An elegant way of achieving this while still maintaining the benefits of the CTBN's factored representation is to make use of what we refer to as "performance functions," which extend the standard model by allowing more sophisticated, user-defined queries (Sturlaugson & Sheppard, 2014). Essentially, a value is assigned to the time associated with entering or remaining in a system state, and this value can be aggregated over multiple vertices and throughout the entire time of interest. The remainder of this section describes how the quantities from these user-defined functions can be used to help decide how best to respond to that evolving system state.

27.6.1 Handling System Scenarios

Here we will assume that we have constructed a CTBN consisting of a D-matrix and a fault tree. For this discussion, we will make use of a running example. Specifically, suppose we have modeled a vehicle that contains the subsystems shown in Table 27.2. Notice that this table also includes MTBF and MTTR, from which various parameters for λ and μ can be derived respectively. Suppose also that we seek to mitigate the following hazards.

TABLE 27.2
Summary of Subsystems in Vehicle Model

Subsystems	Subsystem Name	MTBF	MTTR	Repair Cost
AI	Air System	900	5	250
AL	Alternator	1,300	15	900
AX	Axles	1,800	12	2,000
BR	Brakes	1,200	5	950
CO	Cooling System	2,100	6	200
EL	Electronics	1,700	20	600
FU	Fuel System	800	16	1,200
IG	Ignition System	600	13	300
PR	Compression System	1,100	35	9,000
PW1	Power Source 1	400	3	250
PW2	Power Source 2	200	3	250
SU	Suspension System	1,500	8	850
TR	Transmission	2,500	30	6,500
WT	Wheels and Tires	700	3	700

1. Loss of power
2. Loss of electrical
3. Loss of chassis
4. Loss of engine
5. Loss of power train
6. Loss of vehicle
7. Loss of crew

In some cases, it may be useful to construct multiple versions of the same model. We refer to these model versions as "scenarios", examples of which could correspond to maintenance actions or different possible system operational modes. For instance, one possible situation might be to perform preventative maintenance on a vehicle at time $t = 500$ where the maintainer would replace the wheels and tires before this subsystem fails. Addressing this would require our model to include a subsystem (represented as a vertex) corresponding to "wheels and tires," which we denote as WT. This situation results in two possible scenarios: (1) do not perform a preventative action, and (2) replace the WT component. In the replacement scenario, downtime for the system will result in order to perform the maintenance, and there will be a cost associated with the maintenance process itself. But the goal of such proactive maintenance would be to reduce the likelihood that the subsystem fails in the near term, which might in turn prevent a more drastic Loss of Vehicle event. Indeed, it might even prevent a Loss of Crew event in the extreme case.

Another example of a scenario would be where we need to change the operational mode of the vehicle. For example, suppose the cooling system has a high probability of failing in the near future. Then rather than continuing to operate the vehicle as normal, a more conservative approach could be taken that reduces speed and conserves power consumption. This would then reduce the likelihood of failure of the cooling system (or adjacent subsystem), but it might have other costs associated with the action. For example, by reducing power, the time to get on target would likely be prolonged.

There are several options for modeling scenarios. As suggested above, the first option is to duplicate the base model and make the necessary changes to the network structure or parameters to account for the differences in various scenarios. The result is a set of networks (i.e., models) that vary in some way based on the specifics of the scenarios. Queries can then be performed over the set networks simultaneously, and the results can be compared. Although this works well in theory, scenarios often differ slightly, which means the corresponding networks will have several redundancies, thus potentially increasing the computational complexity associated with processing such similar models. Furthermore, it is possible that scenarios depend on other choices, which can lead to an exponential explosion in the number of models required. This would especially be a concern when there are many scenarios, when the models are very large, or when there are several complex interactions among the scenarios. Furthermore, in the event that a change must be made to the model, then this change would need to be propagated through all of the copied networks, raising the risk of the resulting models being inconsistent.

An alternative approach that addresses some of these issues is to introduce a new vertex to the model. We refer to this new vertex as a "decision vertex." This vertex acts identically to a standard CTBN vertex, but it is used in a specific way to handle various scenarios while preserving model commonality across the scenarios. Each state of a decision vertex corresponds to a different decision or scenario that may be "selected," and we assume the state of the decision vertex is known at all times. Multiple decision vertices can be included in the same model. Each of the decision vertices is defined with no parents, but any vertices that are affected by the scenario are added to the child set of the decision vertex. Then the child vertex's parameters (i.e., CIMs) are adjusted to account for different states of the scenario.

More specifically, the approach is to insert a new parent node on the fault, subsystem, or effect/hazard of interest, where the parent node reflects the conditions for that subsystem in that scenario.

For example, suppose we are considering the cooling system (CO) situation described above and we define a parent decision vertex labeled "Operation." For the purposes of this example, assume the vehicle has two operational modes: standard and conservative. We recognize that this violates the constraint imposed previously that each fault has no parents, which means the components of the vehicle no longer behave independently of external factors. Specifically, the external factors correspond to the decisions made to determine the appropriate scenario (i.e., setting the operational mode).

The model must now be reparameterized to account for this change to the network's structure. Note that subsystem CO originally had a single CIM with only one intensity matrix $\mathbf{Q}_{\mathrm{CO}}$. This assumed standard operation of the vehicle. The new CIM now needs two matrices to be defined, one for each state of its new parent.

$$\mathbf{Q}'_{\mathrm{CO|Operation}} = \left\{\mathbf{Q}'_{\mathrm{CO|Standard}}, \mathbf{Q}'_{\mathrm{CO|Conservative}}\right\}.$$

Since the original model assumed standard operation, the transition behavior in that case is already known and $\mathbf{Q}'_{\mathrm{CO|Standard}} = \mathbf{Q}_{\mathrm{CO}}$. The remaining intensity matrix $\mathbf{Q}'_{\mathrm{CO|Conservative}}$ then needs to be defined to describe the new cooling system behavior, given a conservative operational mode. For this scenario, the failure rate would likely decrease with the conservative driving style due to the reduced wear-and-tear on the vehicle. Note, however, that the repair rate would remain the same. Since the state of the decision vertex is known at all times, no inference will need to be performed to obtain this state. Thus, there is no need to parameterize this vertex.[3] The only requirement to maintain mathematical consistency is that all states of the decision vertex are controllable at any given time, meaning that the initial distribution must have non-zero entries, and a transition may not occur with infinite intensity. For example, we could use a uniform initial distribution over the states and zero transition rates.

To use the model extended to include decision vertices, inference can be run over the network or each possible decision, and the results are compared – that means fixing the operation vertex to be in the standard state and running inference and then repeating the process after setting the operation vertex to the conservative state. To aid in comparing the results, performance functions can be defined (Sturlaugson & Sheppard, 2014). Let π_{Op} be a performance function that assigns a positive value for every hour spent in the operational or partially operational state and no value when in a non-operational state. For example, we could define the following.

$$\pi_{\mathrm{Op}} = \begin{cases} 10t & \text{if } LOV = \text{false} \wedge Operation = \text{standard} \\ 5t & \text{if } LOV = \text{false} \wedge Operation = \text{conservative} \\ 0 & \text{otherwise} \end{cases}$$

where *LOV* is the abbreviation for Loss of Vehicle.

Inference can then be run by considering the two options available, and the value of each option can be quantified by querying over function π_{Op}. While running the vehicle in a conservative state is less valuable than standard operation, the decrease in the failure rate for the cooling system could be substantial enough to prevent a total LOV event. In fact, it could mitigate a significant risk. In particular, under standard operation, there would be lower cost (due to the shorter time required), and the mission might have a high probability of failure. On the other hand, under the conservative model, there would be an increase in cost (due to the increase in time required), but it would have a higher probability of success. It might therefore make more sense to take the partial value that is available rather than risk losing the vehicle entirely and thereby failing the mission.

Performance functions can be very complex; they can be defined to use more sophisticated equations or incorporate other risks that are of interest, such as the loss of crew hazard. Multiple performance functions can also be defined, each attempting to address different objectives or goals, thus setting up a multiobjective optimization problem. With these functions built directly into the rPHM framework, however, it is possible not only to support diagnostics and prognostics but also to compare scenarios using metrics that are relevant to the system's application. After inferring performance values for each possible scenario, choosing the best scenario reduces the task of optimizing over those values (Sturlaugson et al., 2017). In the multiobjective case, methods exist to obtain a Pareto front of nondominated decisions. This subset of Pareto-optimal options can then be presented to a domain expert who can decide what is best for the given context (Coello Coello et al., 2007; Peerlinck & Sheppard, 2023).

27.6.2 Performance-Based Logistics

In addition to supporting decisions for runtime operations, the rPHM models discussed in this chapter can be used during system design as well. This is especially useful when working under the objectives of performance-based logistics (PBL), which is a contracting strategy that aims to improve operational effectiveness in large organizations like the US Department of Defense. A more detailed treatment of what follows can be found in Perreault et al. (2015). Unlike other approaches that contract for resources, PBL contracts for performance according to a variety of prespecified criteria (Sols et al., 2007). These criteria are related directly to system performance and factors such as reliability, maintainability, and supportability (Kumar et al., 2007). In short, PBL contracts require contractors and suppliers to meet life-cycle performance criteria, thus requiring an objective method for evaluating performance against those criteria.

By building a rPHM CTBN during the design and development process, accurate predictions of system behavior can be made. Furthermore, performance functions can be constructed and included directly in the model that relates to the objectives laid out in the PBL contract. Often, these objectives incorporate the impact of failure to achieve mission objectives and are well suited to incorporating risk-based modeling. Finally, different design alternatives can be added to the model using the decision vertices discussed in the previous section. This allows design objectives (including in the multiobjective sense) to be evaluated with respect to the contract requirements prior to actual implementation, saving time and money for the contractor and improving the overall solution for the end user.

To demonstrate this idea, we continue to use our hypothetical vehicle model. During the design phase, there may be a choice between two different axle systems. Table 27.2 shows that the AX component has an MTBF of 1,800 hours, an MTTR of 12 hours, and a repair cost of 2,000. Suppose there is a second axle system AX' to be considered that has an MTBF of 2,100 hours, an MTTR of 20 hours, and a repair cost of 2,500. This data reflects that the new axle design is more complex (due to the higher repair cost), but this increase in complexity results in higher reliability and therefore a longer mean time between failures. Unfortunately, in addition to the higher repair cost, the additional design complexity also increases the mean repair time. Then the question that the contractor is left with is: which design alternative to choose to best meet the contract requirements?

The correct choice between these axle design alternatives is not immediately obvious, and it depends on a number of factors, including how changes to reliability impact the associated risks and what performance criteria are most important for the contract. From the rPHM's perspective, the vehicle CTBN model can be modified to incorporate various design options. Similarly, we inserted a parent vertex operation into our model, suppose we define a parent vertex AxleChoice, which we denote D_{AX}. We make this the parent of the AX vertex in the network. To parameterize the AX vertex, an intensity matrix is derived once for each design alternative and paired with the corresponding state in the D_{AX} decision vertex. Once again, the parameters in the decision vertex are largely unimportant since the state will be assigned deterministically. With this model, inference can be run once for each design alternative, allowing the performance values to be inferred for each objective. Alternatives whose performances do not meet the contract requirements can be eliminated, and a design choice can be obtained from the remaining valid solutions employing multiobjective decision-making methods (Peerlinck & Sheppard, 2023).

In addition to parameter changes, decision vertices can be used to support structure changes to the associated network as well. For example, there may be two design alternatives concerning the design of the power subsystem, where one option includes a redundant power source for fault tolerance while the other alternative proceeds from a single power source. For example, suppose instead of a single power source, we have two power sources, denoted PW1 and PW2, respectively. Here, we place a decision node D_{PW} as the parent of the loss of power event. Assuming PW1 and PW2 are both binary, then the CIM for loss of power originally consisted of four intensity matrices corresponding to the redundant power source scenario. These four intensity matrices are mapped to the first state of D_{PW}, and four new intensity matrices are mapped to the second state that covers the single power source scenario. With these four new matrices, loss of power is parameterized according to PW1 alone as if PW2 did not exist. When D_{PW} is in the first state (redundant design), loss of power is dependent on both PW1 and PW2. When D_{PW}. is in the second state (no redundancy), loss of power is dependent on only PW1, and the edge between PW2 and Loss of Power is effectively severed.

These models provide a mathematically sound and inexpensive method for evaluating design choices with respect to contract requirements prior to implementation. In addition, once the system is deployed, the model can be used for diagnostics, prognostics, and online risk mitigation. This is achieved by fixing the decision vertices representing design choices to the alternatives that were chosen during operation. Going forward, these models can then be used to identify fault and effect likelihoods as well as being used to evaluate specified performance functions. Additional decision nodes can be inserted to model additional operational decisions, which can serve as mitigation strategies that could minimize performance loss.

27.7 SUMMARY

In this chapter, we introduced the concept of risk-based prognostics and health management, with a specific focus on developing a CTBN framework for implementing this concept. To reduce the barrier to entry when building the CTBN models, we introduced a number of procedures that naturally generate the structure and parameters of the model using commonly available diagnostic and reliability information. By explaining how both structure and parameters can be obtained using this data, we are able to eliminate the need to apply expensive learning algorithms or manual modeling processes, while providing a modeling framework that could make use of machine learning methods, if desired. We then explained how to extend the modeling framework to support capturing different scenarios and demonstrated how to apply them with a hypothetical vehicle model.

In addition to detailing processes for building CTBN-based prognostic models, we discussed how the resulting models can be used to perform diagnostics and prognostics by applying evidence corresponding to test results. Furthermore, we presented the methods for representing and reasoning over operational decisions to support making design decisions and to incorporate strategies for mitigating risk. Specifically, we argued for incorporating decision vertices and performance functions into the CTBN, which enable representing various scenarios and utilities, and can be optimized to evaluate alternatives and make risk-informed decisions during runtime. Finally, we showed that these models were not only useful for making operational decisions, but could also be applied to design time decisions as well. By providing a framework for optimizing various design decisions, these models can support tasks like PBL while also continuing to serve as prognostic models capable of describing runtime behavior after deployment.

NOTES

1 For most of this chapter, we will use an alternative version of a fault tree that characterizes how faults propagate to effects and associated hazards in a system.
2 Note that FIM-based fault trees are different from fault trees developed through FTA.
3 It is possible to have adaptive scenarios, in which case such parameterization and inference would be required. Note that the CTBN model already supports this, but it is beyond the scope of the discussion here.

REFERENCES

Abramovici, M., Breuer, M., & Friedman, A. (1990). *Digital Systems Testing and Testable Design*. Hoboken, NJ: Wiley-IEEE Press.

Barlow, R., & Chatterjee, P. (December 1973). Introduction to Fault Tree Analysis, DTIC Technical Report ORC 73-30. Berkeley: University of California.

Codetta-Raiteri, D., & Portinale, L. (2015). Dynamic Bayesian networks for fault detection, identification, and recovery in autonomous spacecraft. *IEEE Transactions on Systems, Man, and Cybernetics: Systems*, 45(1), 13–24.

Coello Coello, C., Lamont, G., & Van Veldhuizen, D. (2007). *Evolutionary Algorithms for Solving Multi-Objective Problems*, 2nd Edition. New York: Springer.

Cosmus, T. C., & Parizh, M. (2011). Advances in whole-body MRI magnets. *IEEE Transactions on Applied Superconductivity*, 21(3), 2104–2109.

Farokhi, S. (2014). *Aircraft Propulsion*. Chichester, UK: John Wiley & Sons.

Figueroa, F., Schmalzel, J., Walker, M., Venkatesh, M., Kapadia, R., Morris, J., Turowski, M., & Smith, H. (2009). Integrated system health management: Foundational concepts, approach, and implementation. In *AIAA Infotech@ Aerospace Conference*, Seattle, WA (pp. 1915–1928).

He, W., & Xu, L. (2015). A state-of-the-art survey of cloud manufacturing. *International Journal of Computer Integrated Manufacturing*, 28(3), 239–250.

Jardine, A. K., Lin, D., & Banjevic, D. (2006). A review on machinery diagnostics and prognostics implementing condition-based maintenance. *Mechanical Systems and Signal Processing*, 20(7), 1483–1510.

Jouin, M., Gouriveau, R., Hissel, D., Péra, M.-C., & Zerhouni, N. (2014). Prognostics of PEM fuel cell in a particle filtering framework. *International Journal of Hydrogen Energy*, 39(1), 481–494.

Kalbfleisch, J. D., & Prentice, R. L. (2011). *The Statistical Analysis of Failure Time Data*. New York: John Wiley & Sons, ch. Failure Time Models.

Kumar, U. D., Nowicki, D., Ramirez-Marquez, J. E., & Verma, D. (2007). A goal programming model for optimizing reliability, maintainability, and supportability under performance based logistics. *International Journal of Reliability, Quality and Safety Engineering*, 14(3), 251–261.

Lall, P., Lowe, R., & Goebel, K. (2010). Prognostics using Kalman-filter models and metrics for risk assessment in BGAs under shock and vibration loads. In *Proceedings of the 60th IEEE Electronic Components and Technology Conference (ECTC)*, Las Vegas, NV (pp. 889–901).

Mosallam, A., Medjaher, K., & Zerhouni, N. (2015). Component based data-driven prognostics for complex systems: Methodology and applications. In *First IEEE International Conference on Reliability Systems Engineering (ICRSE)*, Beijing, China (pp. 1–7).

Nodelman, U., & Horvitz, E. (2003). Continuous time Bayesian networks for inferring users' presence and activities with extensions for modeling and evaluation. Microsoft Research Technical Report, MSR-TR-2003-97.

Nodelman, U., Shelton, C. R., & Koller, D. (2002). Continuous time Bayesian networks. In *Proceedings of the 18th Conference on Uncertainty in Artificial Intelligence*, Edmonton, Alberta (pp. 378–387). Morgan Kaufmann Publishers Inc.

Orchard, M., Kacprzynski, G., Goebel, K., Saha, B., & Vachtsevanos, G., (2008). Advances in uncertainty representation and management for particle filtering applied to prognostics. In *IEEE International Conference on Prognostics and Health Management*, Denver, CO (pp. 1–6).

Pecht, M. (2008). *Prognostics and Health Management of Electronics*. Chichester, UK: John Wiley & Sons.

Peel, L. (2008). Data driven prognostics using a Kalman filter ensemble of neural network models. In *IEEE International Conference on Prognostics and Health Management*, Denver, CO (pp. 1–6).

Peerlinck, A., & Sheppard, J. (December 2023). Managing objective archives for solution set reduction in many-objective optimization. In *Proceedings of the Symposium Series on Computational Intelligence*, Mexico City, Mexico.

Perreault, L., Strasser, S., Thornton, M., & Sheppard, J. (2016). A noisy-or model for continuous time Bayesian networks. In *Proceedings of the 29th International Florida Artificial Intelligence Research Society Conference*, Key Largo, FL (pp. 668–673).

Perreault, L., Thornton, M., & Sheppard, J. (2015). Valuation and optimization for performance based logistics using continuous time Bayesian networks. In *IEEE AUTOTESTCON Conference Record*, National Harbour, MD.

Perreault, L., Thornton, M., & Sheppard, J. W. (2016). Deriving prognostic continuous time Bayesian Networks from fault trees. In *Proceedings of the Annual Conference of the Prognostics and Health Management Society*, Denver, CO (pp. 1–10).

Perreault, L., Thornton, M., Strasser, S., & Sheppard, J. W. (2015). Deriving prognostic continuous time Bayesian networks from D-matrices. In *IEEE Autotestcon Conference Record*, National Harbor, MD (pp. 152–161).

Pullen, J. M., & Khimeche, L. (2014). Advances in Systems and Technologies Toward Interoperating Operational Military C2 and Simulation Systems, DTIC Document, Technical Report.

Sheppard, J. W., & Kaufman, M. A. (2005). A Bayesian approach to diagnosis and prognosis using built-in test. *IEEE Transactions on Instrumentation and Measurement*, 54(3), 1003–1018.

Sheppard, J., & Butcher, S. (2007). A formal analysis of fault diagnosis with D-matrices. *Journal of Electronic Testing*, 23(4), 309–322.

Sheppard, J., & Simpson, W. (1996). Improving the accuracy of fault diagnostics provided by fault dictionaries. In *Proceedings of the 14th VLSI Test Symposium*, Princeton, NJ (pp. 180–185).

Simpson, W. R., & Sheppard, J. W. (1994). *System Test and Diagnosis*. Norwell, MA: Springer.

Smith, C., Knudsen, J., Kvarfordt, K., & Wood, T. (2008). Key attributes of the SAPHIRE risk and reliability analysis software for risk-informed probabilistic applications. *Reliability Engineering & System Safety*, 93(8), 1151–1164.

Sols, A., Nowick, D., & Verma, D. (2007). Defining the fundamental framework of an effective performance-based logistics (PBL) contract. *Engineering Management Journal*, 19(2), 40–50.

Strasser, S., & Sheppard, J. (March 2013). An empirical evaluation of Bayesian networks derived from fault trees. In *Proceedings of the IEEE Aerospace Conference*, Schaumburg, IL (pp. 1–13).

Sturlaugson, L., & Sheppard, J. (May 2014). Factored performance functions with structural representation in continuous time Bayesian networks. In *Proceedings of the Florida Artificial Intelligence Symposium*, Pensacola Beach, FL (pp. 512–517).

Sturlaugson, L., Perreault, L., & Sheppard, J. (2017). Factored performance functions and decision making in continuous time Bayesian networks. *Journal of Applied Logic*, 22, 28–45.

Van der Borst, M., & Schoonakker, H. (2001). An overview of PSA importance measures. *Reliability Engineering & System Safety*, 72(3), 241–245.

28 Structural Health Monitoring

Jamie Blanche, Ranjeetkumar Gupta, Daniel Mitchell, Sam Harper, and David Flynn

28.1 INTRODUCTION TO STRUCTURAL HEALTH MONITORING FOR OFFSHORE WIND SECTOR COMPOSITES

The operational, maintenance, and servicing requirements of offshore wind farms represent economic, logistical, and technical challenges that require significant developments to maintain efficiency, competitiveness, and safety (ORE Catapult, and BVG Associates 2020). The cost of Operations and Maintenance (O&M) is significant, where 80% of the associated costs can be attributed to the transport of engineers and crew to offshore sites (Barnes et al. 2018). Following the commitment of the UK government to increase offshore wind energy production to 40 GW by 2030, with 1 GW via floating foundations, an estimate provided by the Global Wind Energy Council states that the UK input to EU offshore wind capacity will be 34%, with total EU capacity expected to reach 89 GW in the same period (HM Government 2020; Lee and Zhao 2020; The Crown Estate 2019; Crown Estate Scotland 2020). Continuing developments in asset integrity and fault detection using novel sensing technologies, post-processing of data, and prognostics have played a major role in reducing the risk of injury to offshore energy workers. This lessens the need for expensive and time-consuming intervention strategies carried out by a human workforce in the offshore environment (Antonopoulos et al. 2020; Blanche, Mitchell, et al. 2020b).

The necessity for a sustainable and safe global energy economy is a major driver in the design and development of the next generation of offshore wind turbines. Where new wind farm array sites are situated farther offshore, the need to monitor the structural health and integrity of WT assets accurately and safely will require advances in sensing methods, deployment methods and developments in composite engineering. This will result in a combined approach that utilizes human rope access teams and resident robotics performing automated inspections (Mitchell et al. 2021).

The low weight-to-modulus of strength ratio of composite structures is a major factor in their cost effectiveness and has led to a wide range of applications in onshore and offshore renewables, civil infrastructure engineering, the automotive and aerospace sectors (Ashby and Cebon 1993; Shafiee et al. 2021; Kong et al. 2020; Blanche, Mitchell, et al. 2020b; Barnes et al. 2018). The key application areas for load-bearing composite structures are shown in Figure 28.1, where their utilization requires integrity evaluation and monitoring of the manufacturing process structural health during the in-service lifetime of an asset. The production cycle of composite structures involves multiple variables, with a high likelihood of some form of defect occurrence during the service life of a composite-based asset, such as a WT blade. The consequent need for SHM during the manufacture and service life of a composite structure stems from the requirement to accurately assess the remaining useful life (RUL) of that structure and to avoid critical failure by informing timely intervention strategies (Amenabar, Lopez, and Mendikute 2012; Wang et al. 2014). As a result, there exists a need for non-destructive evaluation (NDE) methods to assess composite asset health at all stages of the asset lifecycle.

This chapter is structured as follows: Section 28.2 discusses WT blade defects and failure modes. Section 28.2.1 and Section 28.2.2 outline manufacturing defects and operationally induced defects, respectively. Section 28.3 provides a comparative analysis of SHM for WT blade integrity inspection methods. Section 28.3.1 evaluates physical contact methods. Section 28.3.2.1 and Section 28.3.2.2 evaluate non-contact methods in the visual spectrum and non-contact methods outwith the visual

DOI: 10.1201/9781003188377-33

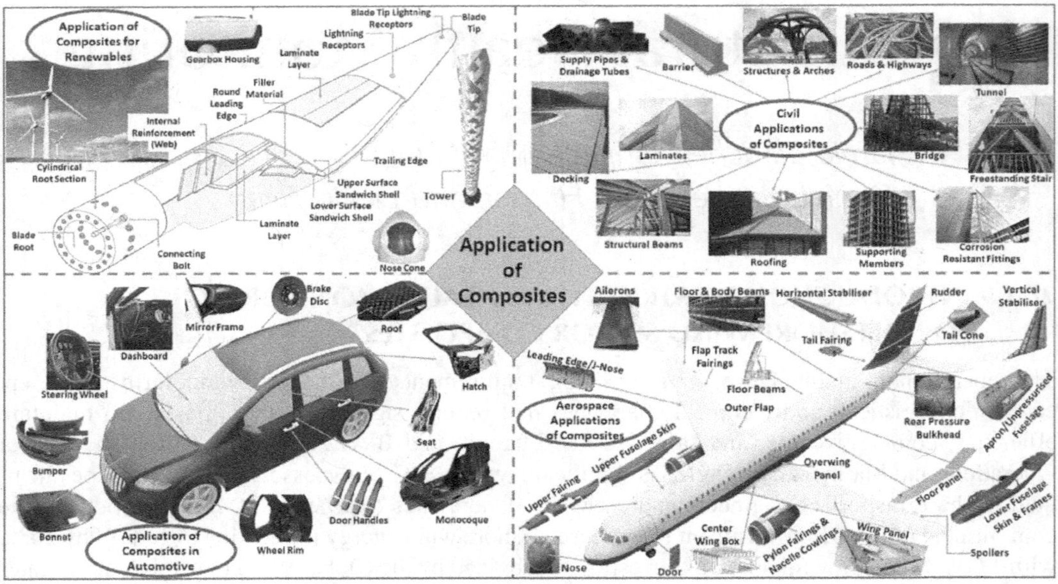

FIGURE 28.1 (Clockwise from left) Case examples of composite application in four key sectors: wind energy, civil engineering, aerospace, and automotive. (Gupta et al. 2021.)

spectrum, respectively. Section 28.4 introduces frequency-modulated continuous wave radar for WT SHM. Section 28.4.1 provides the numerical theory of FMCW. Section 28.4.2 outlines equipment parameters and data acquisition workflow. Section 28.4.3 provides a tutorial via the implementation and setup for the evaluation of two key defects. Experiment A demonstrates FMCW sensitivity to internal structural delamination and Experiment B evaluates water ingress within a delamination. Section 28.5 provides a discussion of the results from Section 28.4. Section 28.6 concludes. Section 28.7 outlines the future scope of FMCW for WT SHM and asset integrity inspection.

28.2 WIND TURBINE DEFECTS AND FAILURE MODES

The range of defects within WT blade structures can be divided into two main categories: defects resulting from errors in the manufacturing process and defects that result from operational loading and exposure to the operating environment.

28.2.1 Manufacturing Defects

Several factors can affect the cost of production during the manufacturing stage of composite structures. These can be typified by non-uniform curing time, non-uniform resin distribution, and other variables that affect consistency and reproducibility across all composite batches on a production line. Additionally, manufacturing process parameters can affect the overall quality of the composite structure. Filling defects, often manifested as foreign object material inclusions and/or porosity, and cure-induced defects, such as residual stresses and under-cured resin, represent significant risks to the required structural properties of the composite material. To mitigate this, applied quality control sensing is required to eliminate any defective products from use beyond the manufacturing stage. Process inputs such as raw-material type, geometry, and relevant thermodynamic parameters are also dominant mechanisms that affect the filling and curing processes during composite manufacture (Konstantopoulos, Fauster, and Schledjewski 2014). Ensuring consistent structural integrity during the manufacturing and operations phases of a composite structure is reliant on the reduction of intrinsic signal noise during the detection of defects, which may often be characterized as anisotropic and heterogeneous materials (Flynn et al. 2014; Fagogenis, Flynn, and Lane 2015; Miguelañez-Martin and Flynn 2015). The broad nature of defect and damage modes within a composite structure also requires sensing methods that may be applied at a wide range of length scales, representing a challenge for the detection of dynamic damage or defect progression over time, and with the risk of non-detection, leading to accumulation of significant damage prior to reliable detection. In all cases, continued material degradation leads to a reduction in structural integrity and RUL of the composite-based components of a system (Wang et al. 2014).

Figure 28.2 shows a range of typical manufacturing defects found in the monolithic composite structures used for the main structural elements of a large scale (~90 m) WT blade, such as the spar cap, internal reinforcement web and central support structure for the flanges (see top left Figure 28.1). As such, common defects relate to variations in material thickness, such as the adhesive layers that bond the monolithic elements to the sandwich composite structures (such as balsa strengthened with glass-fiber veneer), as detailed in Figure 28.2a. Figure 28.2b shows the effect of non-uniform resin saturation of the glass-fiber layering. This leads to regions of poor penetration of the resin prior to hardening, resulting in a weakened structure. The effect of wrinkles forming in the fiber-glass layering prior to resin saturation are varied; however, two examples are given: Figure 28.2c shows air voids and material discontinuities resulting from improper layering of glass-fiber and Figure 28.2d shows the variations in material thickness that can result from a small wrinkle early in the glass-fiber layering process.

FIGURE 28.2 Examples of variations in manufacturing tolerances and defects. (a) Difference in adhesive thickness at flange joint. (b) Poor resin distribution resulting in fiber areas unexposed to resin. (c) Lateral effect of wrinkles within fiber/resin mix causing cavities and material discontinuities. (D) Effect on material thickness caused by wrinkle during the early fiber-laying process.

28.2.2 Operational Defects

The static and dynamic forces WT blade structures are exposed to during their 20-year operational lifecycle are significant and include aerodynamic lift, drag, and gravitational loads (Al-Khudairi and Ghasemnejad 2015). The design and construction of WT blades utilize the strength and weight optimization afforded by composite materials, and construction utilizes a range of composite materials such as epoxies, adhesives, and resins. The optimization of these materials results in decreased cost of manufacture, while also maximizing the operational lifespan of these valuable offshore assets (Gupta et al. 2019). The categorization of common WT failure modes is well documented and can be described as one of seven distinct types of defect or failure, where each type is distinguished by the nature of the fault, in addition to the location on the structure, as shown in Figure 28.3 and Table 28.1 (Branner and Ghadirian 2014; Crespo 2016).

One common defect in WT blade structures is the formation of areas of delamination within the airfoil structure. These regions can be due to operationally induced cyclic stresses, or defects extant from the point of manufacture. These are the result of interlaminar sheer stresses creating cracks and distortions in the composite that may be identified as both longitudinal and transverse features; associated with a significant increase in localized porosity. The presence of frozen water within the WT blade structure is a unique defect resulting from the combined effect of increased porosity and structural permeability resulting in water ingress following leading edge erosion. The air flow over the WT blade during operation lowers the temperature of the leading edge, resulting in the freezing of any trapped moisture within, further weakening the structure.

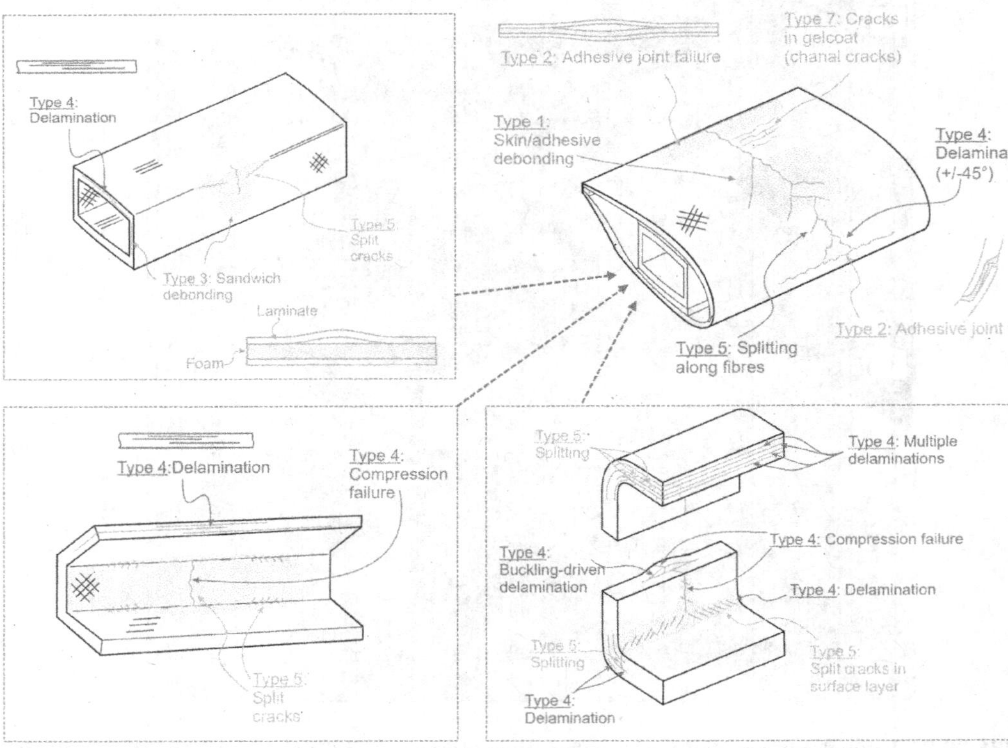

FIGURE 28.3 Wind turbine failure modes and associated categories with subassemblies inset. (Modified from Sørensen 2004; Branner and Ghadirian 2014.)

TABLE 28.1
Description of Wind Turbine Failure Modes, as Classified in Figure 28.3

Type	Description
1	Adhesive layer damage formation and growth in bond joining skin and main spar flanges (skin/adhesive debonding)
2	Adhesive layer damage formation joining the up- and downside skins along leading and trailing edges (adhesive joint failure)
3	Damage formation and growth at the interface between face and core in sandwich panels in skins and main spar web
4	Internal damage formation and growth in laminates in skin and/or main spar flanges, under tensile or compression load
5	Laminate skin and main spar splitting and fracture of separate fibers (fiber failure in tension; laminate failure in compression)
6	Buckling of skin due to damage formation and growth in the skin and main spar bond under compressive load (skin/adhesive debonding induced by buckling, a specific type 1 case)
7	Formation and growth of cracks in the gel-coat; debonding of the gel-coat from the skin (gel-coat cracking and gel-coat/skin debonding)

Source: Sørensen (2004); Crespo (2016).

28.3 COMPARATIVE ANALYSIS OF STRUCTURAL HEALTH MONITORING METHODS FOR WIND TURBINE BLADE INTEGRITY

The increase in demand for NDE of composite structures is concurrent with the continuing development of sustainable energy generation and WT design. This analysis provides a comparison of key asset integrity inspection methods applicable for use on WT blade structures and will outline the working principles of each method. Each subsection will provide a commentary on benefits and limitations via an interim conclusion.

28.3.1 Physical Contact Non-Destructive Evaluation Methods

28.3.1.1 Ultrasonic

Ultrasonic testing equipment typically comprises a transducer tool, transmitter, and receiver units and is designed to detect the size, location, and orientation of cracks in the test material and other material discontinuities. Figure 28.4 illustrates a test material with bistatic transmitter and receiver transducers. The emitted ultrasonic signal is reflected by an internal flaw and is displayed via the user interface. This mode of sensing is applicable to assembly line quality control testing, where defect detection and tolerance measurements must be frequently applied (Katunin, Dragan, and Dziendzikowski 2015).

Notable advantages of the ultrasonic testing method are good resolution, a short scan time for rapid data acquisition and the system is generally portable for field deployment. Disadvantages are a complicated technical setup requiring a skilled operator. The ultrasonic method is optimal with medium thickness materials and a smooth contact interface is required to couple the transducer. Some limitations of this method are the detection of multiple material interfaces, reducing the contrast in signal used for the detection of a defect or material discontinuity. This is due to multi-reflections and high wave scattering effects in sandwich panels and laminated composites. The accuracy of this method is limited by ultrasonic wave attenuation effects in porous materials and, as a result, it is not suitable for composites of thickness greater than ~50 cm (Adams and Cawley 1988).

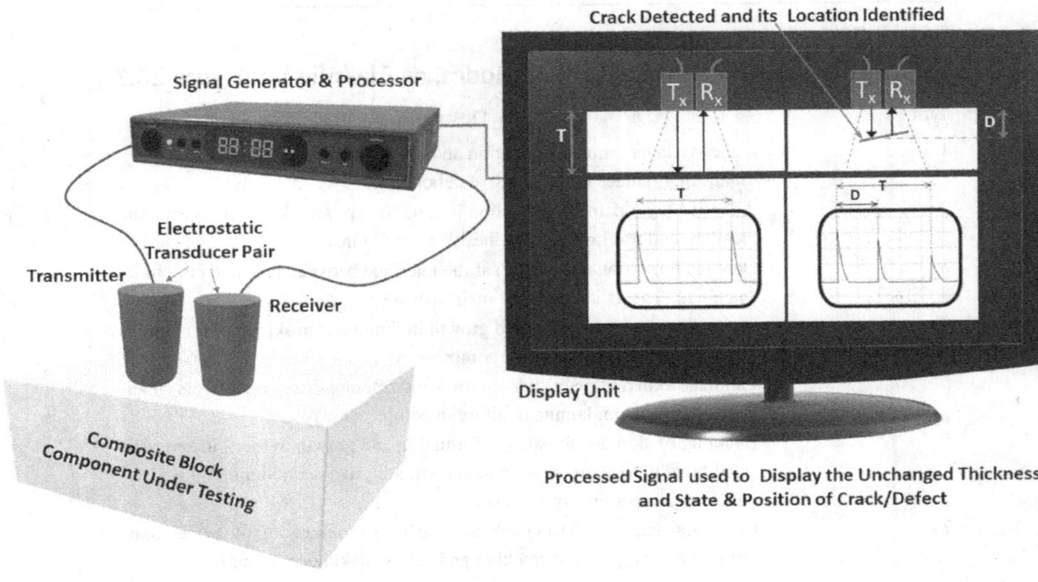

FIGURE 28.4 NDE of composites via ultrasonic testing. (Gupta et al. 2021.)

28.3.1.2 Piezoelectric Transducers

The piezoelectric method is a useful form of inspection for non-porous and homogenous composites, with high efficiency in defect detection in laminated structures (Mal and Bar-Cohen 1988). The manufacture of laminated structures requires quality control methods with sensitivity to delamination features and concentration of inspection in small areas. Piezoelectric methods include active and passive sensing, where the reversibility of the piezoelectric effect has relative merits for reflection and transmission parameters, which may indicate the direction of a detected defect (Tanasoiu, Miclea, and Tanasoiu 2002). Figure 28.5 illustrates the most basic configuration of a piezoelectric transducer, which uses a cylindrical, monostatic piezoceramic element aligned perpendicular to the layering of the material under test.

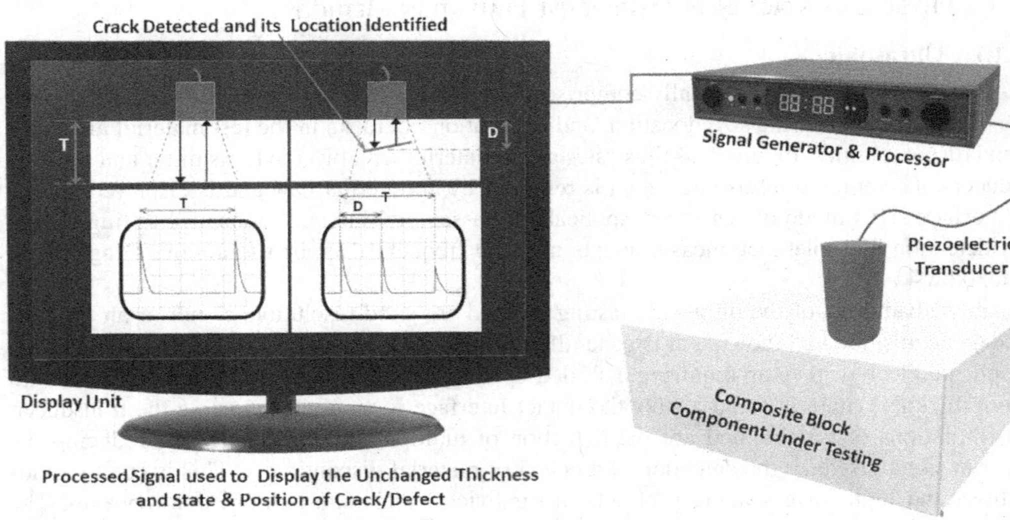

FIGURE 28.5 NDE of composites via piezoelectric transducer. (Gupta et al. 2021.)

The advantages of piezocomposites relate to the flexibility of the composite materials used to make the sensor, where the inspection of curved surfaces is made easier and where the traditional use of non-flexible materials results in wave loss and a poor signal-to-noise ratio. A key challenge to the application of piezoelectric materials is the limitation for use in high-temperature environments. This is primarily due to the high thermal expansion of polymer fillers, low mechanical quality factors, and low thermal conductivity associated with these materials, which are known to cause debonding and cracking in the composite structure under test. The low thermal conductivity inhibits thermal dissipation to adjacent areas, resulting in highly localized heating near the point of contact of the sensor and results in melting of the test structure (Kažys, Voleišis, and Voleišienė Baršauskas 2008; Charchuk et al. 2012; L. Li et al. 2013b).

28.3.1.3 Acoustic Emission

As defects occur in deforming and loaded materials, mechanical vibrations are emitted from the source of the defect. These are often the result of fiber breakage, cracking of the matrix material and/or material or fiber debonding (Arumugam et al. 2011; Gholizadeh, Lemana, and Baharudinb 2015). The resulting material failure results in the emission of stress waves from the highly localized failure zone, which is detected via an array of sensitive piezoelectric sensors. Acoustic Emission (AE) is particularly effective in imperfection analyses in composite structures. Figure 28.6 illustrates a schematic of a typical AE setup. The AE method is a passive mode of sensing that differs from most NDE methods, where the primary source of data is the collection and processing of emitted sound waves from within the material under test. A strong advantage of this sensing modality is the ability to track, via triangulation, the evolution of a dynamic defect, in addition to inactive and non-critical defects (Gholizadeh 2016).

AE has various advantages, such as large area testing by means of multiple sensors, high sensitivity, and no requirement for disassembly prior to reuse or sample preparation (Lu 2010). In addition, the technique is helpful in detecting various defect types resulting from fatigue loading. Fatigue damage type defects detected by AE are fiber-matrix debonding, fatigue cracks, matrix micro-cracks, delamination, and fiber fractures. A disadvantage of AE testing is that it requires high skill for the correlation data collected to specific damage mechanism type (Gholizadeh 2016).

28.3.1.4 Eddy Current

An eddy current sensor measures the electromagnetic (EM) properties of a conductive sample, typically metal. A current is passed through a coil to generate a magnetic field to induce a circular electric current in the conductive sample. Discontinuities and variations in the metal structure, such as cracks, differing metal interfaces, corrosion damages, coating thickness, material thickness, and

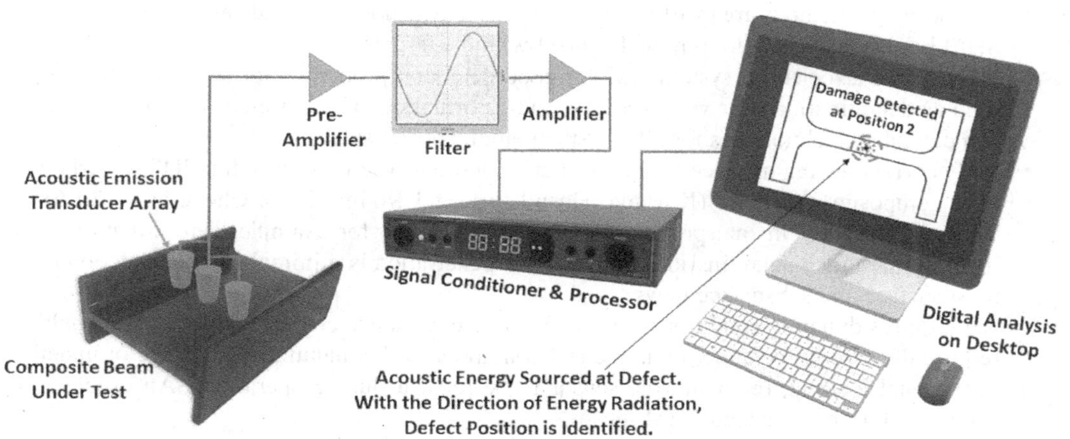

FIGURE 28.6 NDE of composites via acoustic emission. (Gupta et al. 2021.)

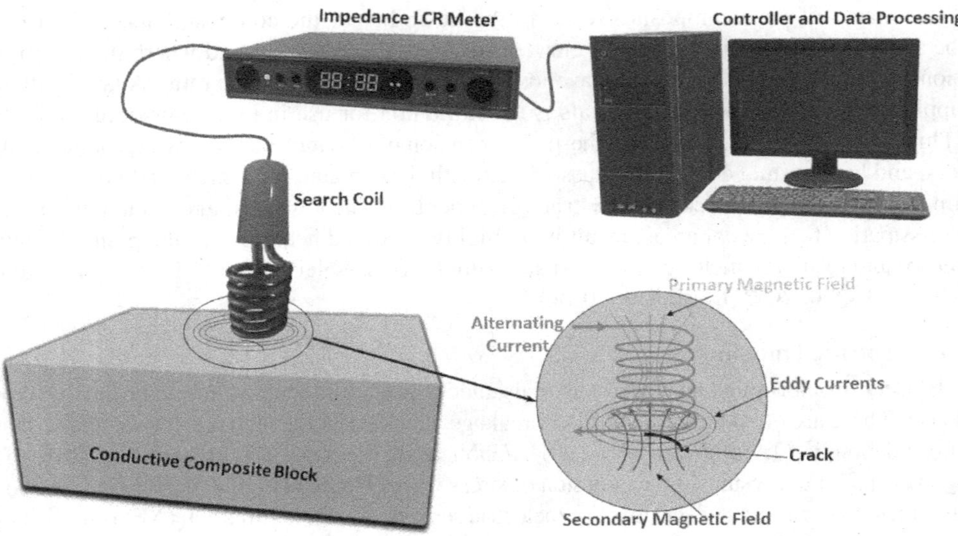

FIGURE 28.7 NDE of composites via eddy current testing. (Gupta et al. 2021.)

material conductivity are detected as variations in the electric field generated within the sample (de Goeje and Wapenaar 1992; He et al. 2014). Figure 28.7 illustrates the working principle, where the coil can be scanned over the surface area of the target material, allowing for concentration over a potentially defective area and increasing the likelihood of detection. The eddy current method has superior sensitivity in comparison to other NDE methods, such as ultrasonic or acoustic sensing. However, this only applies to sensing in conductive materials, such as metals or carbon fiber reinforced polymer composites. This method is capable of mapping subsurface contours in layered, conductive structures and requires minimal sample preparation time, resulting in a highly portable NDE method (He et al. 2014; Oral 2019). Eddy current testing relies on electric conductivity, which is not always a property of composite structures, resulting in difficulties interpreting the return signal from the noise floor and a minimal penetration depth within low conductivity materials.

28.3.1.5 Interim Conclusion

The following conclusions can be made for this section:

- All the above methods are insufficient for the task of offshore WT blade asset integrity and SHM because they require physical contact with the target.
- Ultrasonic/Piezoelectric systems exhibit good detection properties for some key defects, such as delaminations, air voids, cracks, and deformation. These are consequently excellent candidate technologies for NDE at the manufacturing stage.
- Eddy current testing has faced some further limitations when it is used for NDE of carbon fiber composite structures (Koyama, Hoshikawa, and Kojima 2013; Cheng et al. 2016). There is difficulty in interpreting the measured signals, for example, determination of interlaminar crack delamination. The depth of penetration is minimal for the detection of most surface and subsurface flaws.
- AE requires dynamic conditions, such as load, to elicit a detectable response. This would require the asset to be under dynamic test conditions at the manufacturing stage or under operational loading, requiring the sensor to be present during operations. An excellent candidate for an embedded SHM sensor.

The following sections focus on non-contact methods with an appreciable standoff range.

28.3.2 Non-Contact Non-Destructive Evaluation Methods

28.3.2.1 Visual

The use of unmanned aerial vehicle (UAV) platforms for visual inspection of WT blades is a well-established practice and forms the basis of the aerial inspection sector, often used as contractors for wind farm operators. Multiple modes of photography are used and include visual spectrum, infrared, digital image correlation (DIC), and 3D photogrammetry. This section discusses the applications of these sensing modes and signal image processing techniques for NDE.

28.3.2.1.1 Photography

Visual spectrum photography inspection of wind farm assets is a mature field of non-contact SHM and utilizes high-resolution digital cameras to provide standoff imagery of all areas of a WT exterior structure. The acquired photographs can be assessed by skilled photographic interpreters or used as input for AI-driven analysis software packages designed to identify specific defect types. Figure 28.8 shows a range of common defects on WT structures, with Figure 28.8a showing regions of peeling from the WT skin, Figure 28.8b showing transverse cracking on a WT tower structure and Figure 28.8c showing the effects of leading edge erosion due to fine particulates in the air stream, such as sand and dust (Yang et al. 2021). Advantages of this method are a significant standoff distance from the asset, with camera systems often deployed at ground level, or ease of deployment via aerial platforms, such as UAVs. Disadvantages of this method are the limitation to inspect only the exterior of a structure, with internal defects being undetectable until significantly advanced, reducing the likelihood of a timely intervention strategy. Visual spectrum photography is also reliant on ambient lighting conditions, which change significantly during the course of a day.

28.3.2.1.2 DIC and Photogrammetry

DIC is a common method of defect identification used in onshore and offshore wind farm integrity inspection and is an optical technique that utilizes pattern matching and image recognition methods. This method can provide two- and three-dimensional calculations and any contrasts in shape caused by damage or deformation. This results in data useful for finite element analyses, as well as material and structural analyses (McGinnis 2006; McGinnis, Pessiki, and Turker 2005; Schmidt, Tyson, and Galanulis 2003a; 2003b). Figure 28.9 illustrates the DIC method, which combines photogrammetric images to correlate and identify contrasts in material position, orientation, or size, for example. Photogrammetry uses triangulation to map key features of a structure in 3D (McGinnis, Pessiki, and Turker 2005). The photogrammetric method has some limitations, including the requirement for exact knowledge of the orientation and position of the cameras to ensure accurate triangulation. These typically require two cameras mounted at opposing ends of a mounting bar, resulting in a

FIGURE 28.8 Examples of externally detectable WT blade defects. (a) Peeling, (b) cracking, and (c) leading edge erosion. (Yang et al. 2021.)

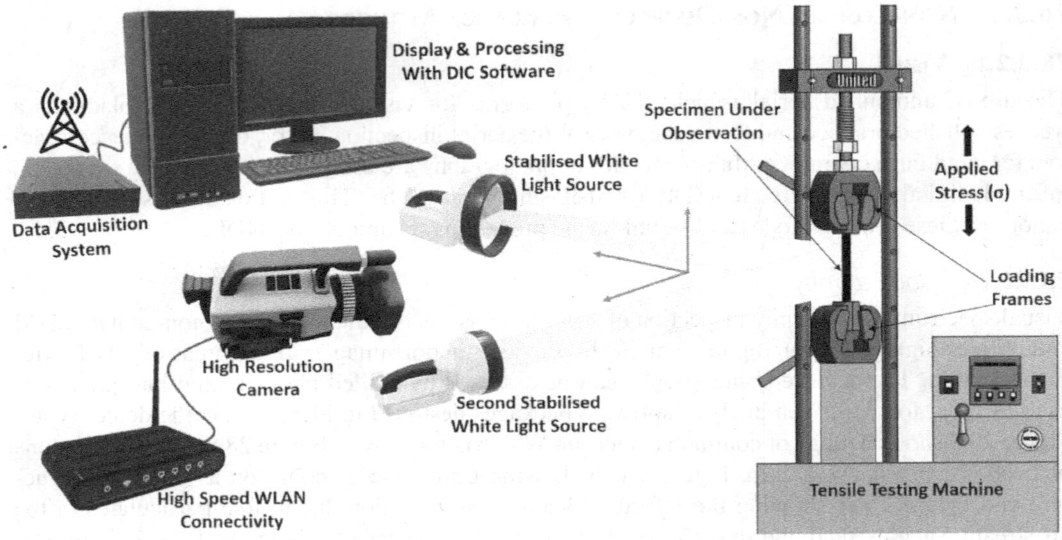

FIGURE 28.9 NDE of composites via digital image correlation. (Gupta et al. 2021.)

stereoscopic image with the same path length to the target for both cameras (Hohmann et al. 2012; Poozesh et al. 2017; Baqersad et al. 2017).

28.3.2.1.3 Interim Conclusion

The application of visual spectrum photogrammetry to WT SHM methods has been shown to focus on blade displacement from a baseline position to monitor the operational dynamics of the composite structures. As a result, visual spectrum photogrammetric methods are most effective in the measurement of large displacements from the norm and to detect low-frequency vibrations within the WT structure; inferring a defect within the mechanical systems. Most literature sources consequently address developments in vibrometry and analysis techniques (Ozbek et al. 2010; Ozbek and Rixen 2011; Poozesh et al. 2017; Baqersad et al. 2017).

The combination of visual spectrum photogrammetry with DIC techniques provides a powerful analytical tool for non-contact NDE. The application of this technique to UAV operations both on and offshore has created a specialist industry in asset integrity inspection. However, visual inspection can only provide evidence of external defects, with internal failures only detectable at a late stage.

To conclude this section, visual spectrum non-contact EM SHM methods are useful and applicable for external assessment of composite structures; however, these methods are insufficient for the accurate assessment of internal structural integrity, internal defect formation, or propagation. The next section will evaluate non-contact EM SHM methods outwith the visual spectrum.

28.3.2.2 Non-Visual

28.3.2.2.1 Infrared Thermography

Infrared (IR) thermography (or thermal imaging) measures the thermal conductivity of a target material. Contrasts in conductivity indicate areas of latent damage within the structure. This method of inspection is particularly effective when applied to laminated composite structures that have undergone cyclic loading and deformation. The defects that result from these conditions are typified by interruptions in the material thermal conductivity, indicating the presence of a delamination, air gap or cracks in the structure that may not be observable in the visible spectrum. Figure 28.10 illustrates the working principle of IR thermography, where defects and flaws in the test material

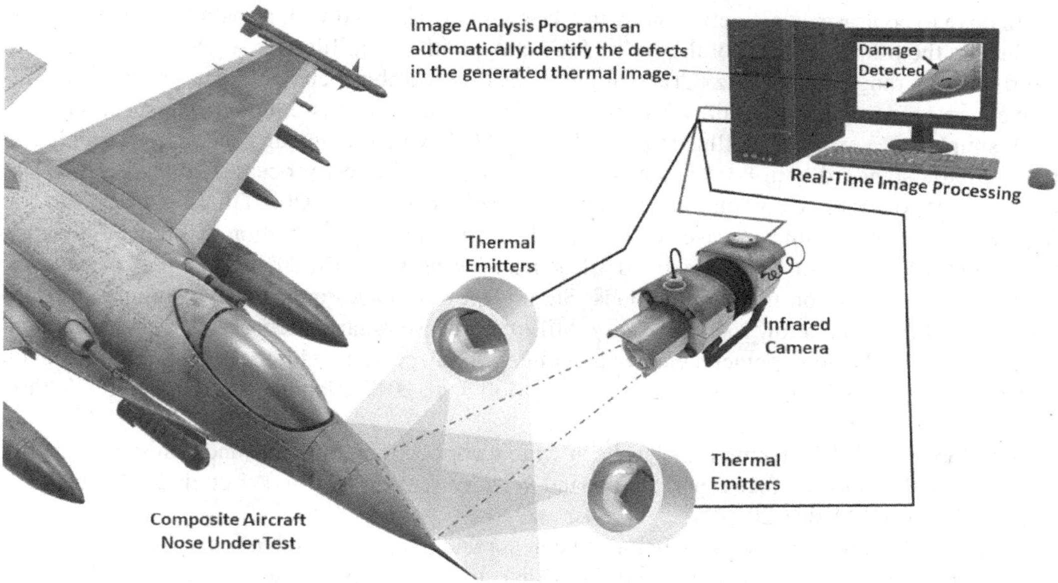

FIGURE 28.10 NDE of composites via thermography. (Gupta et al. 2021.)

lead to highly localized variations in the reflected IR waves of the sample (Vollmer and Mollmann 2018; Meyendorf, Nagy, and Rokhlin 2004).

A key advantage of this method is a wide sensor field of view and the lack of a requirement for physical contact with the target material. One major disadvantage of this method relates to the defect size and depth from the target surface, where a defect that is located deep within the structure must be of sufficient size to affect the thermal signature at the target-air interface. Consequently, the IR thermography method is only likely to pick up large defects at any depth within the sample. This method is also sensitive to surface irregularities and environmentally influenced temperature variations (Grinzato, Vavilov, and Kauppinen 1998; Kashif Ur Rehman et al. 2016; Mulaveesala and Tuli 2005).

IR thermography can be deployed in both active and passive modes, where the active mode utilizes thermal emitters to generate a signal, from which contrasts may be detected. In the passive mode, the thermal contrasts are observed using ambient thermal energy. This mode is particularly useful for instances where the target is not thermally balanced (Montesano, Fawaz, and Bougherara 2013). However, this is impractical for field deployment offshore.

28.3.2.2.2 Frequency-Modulated Continuous Wave Radar

Frequency-modulated continuous wave (FMCW) radar utilizes a continuous frequency sweep and represents a significantly lower powered inspection system, when compared to standard pulsed wave systems. The size of an FMCW system is defined by the operating frequency band, the antenna dimensions and the intended measurand of the system. The continuing trend of electronic miniaturization has led to systems with much improved power economics and higher frequency bands, resulting in high-resolution systems capable of functioning in harsh and sensitive environments.

The standard FMCW system consists of a monostatic horn or patch antenna acting as both the transmitter and receiver and offers significant standoff distance from the target when used in conjunction with a highly directional horn antenna. The signal emitted from a millimeter-wave system has been demonstrated as capable of characterization of single layer dielectrics with the latest FMCW units able to provide kilohertz data acquisition rates (Barowski, Zimmermanns, and Rolfes 2018). This method, with a sensitivity that is directly related to the bandwidth of the emitted signal, is sensitive to contrasts in refractive index in a layered material, and with a spatial resolution that is defined by the frequency band of the emitted radiation, with sub-millimeter resolution capability in the frequency range 40–90GHz (Cristofani et al. 2014). The above characteristics provide significant advantages for the application of FMCW systems in SHM and asset integrity inspection roles.

A simple case example is illustrated in Figure 28.11, where a signal generator is coupled to a monostatic standard gain horn antenna and acquired data is post processed following wireless transmission to a remote computerized display unit (Blanche et al. 2020b; Gupta et al. 2021). The limitations of the millimeter-wave method are limited depth of penetration at higher frequency bands and a lower resolution than standard benchmarking methods, such as ground penetrating radar, X-ray and neutron imaging methods. Significant advantages are low-power demand, bandwidth tuneability, and a low risk of toxicity. Millimeter-wave radar systems represent an emergent NDE method, and a non-contact means of asset integrity inspection, with the potential to inform on internal properties within a composite structure (Moll et al. 2017; Blanche, Mitchell, et al. 2020b).

Automotive: FMCW technology has been widely applied to automotive and industrial sensing, with a specific focus on distance and velocity sensing (Kaminski et al. 2014; Yadav, Dahiya, and Mishra 2016).

Security: This sensing modality has also been applied to multiple defense and security roles, such as concealed weapon and explosive detection, perimeter security, surveillance, and roadside device detection (Beasley 2008).

Medical: The use of FMCW for medical diagnostics and research is a maturing field, with the capability to detect displacement at the sub-micron level for heartbeat and respiration monitoring in patients too vulnerable for physical contact assessment of vital signs, such as premature births requiring incubation and the immunosuppressed (Vinci et al. 2012; Aardal et al. 2013; C. Li et al. 2013a).

Snow Research: Research into snow and ice dynamics has also used FMCW in the X-band for the detection of ice and snow stratigraphy and provides information relating to geometrical layering, snow and ice depth, density, water equivalence and moisture content in void spaces (Gubler and Hiller 1984; Holmgren et al. 1998; Marshall and Koh 2008).

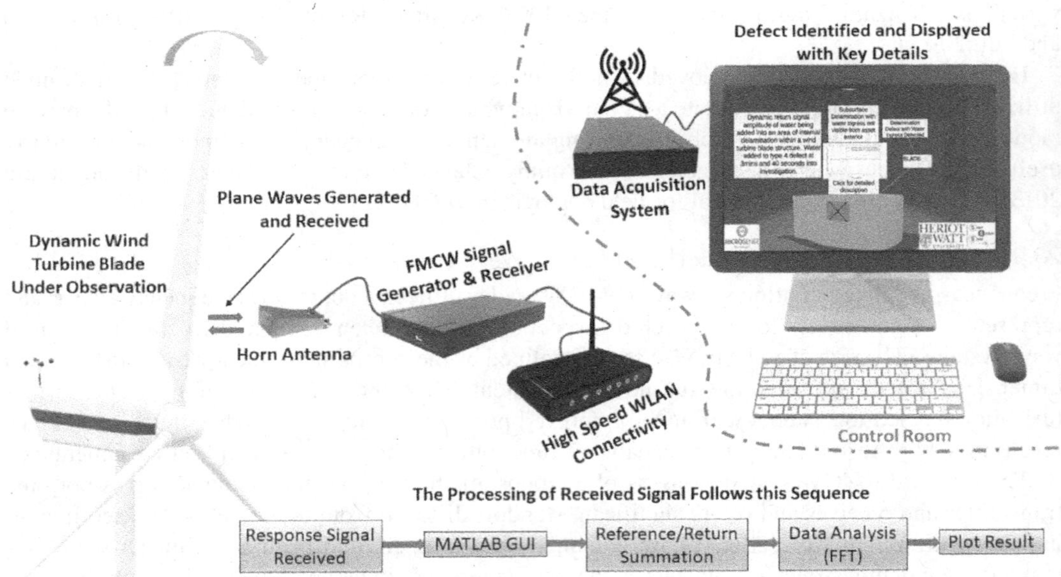

FIGURE 28.11 NDE of composites via frequency-modulated continuous wave radar. (Gupta et al. 2021.)

Corrosion under Insulation: The detrimental effects of Corrosion Under Insulation (CUI) represent a multi-billion-dollar loss to the offshore energy sector annually, where CUI represents the degradation and failure of steel infrastructural assets due to the aggressive operating environment. This compromises insulation systems, and the resulting ingress of water shortens the RUL and functional integrity of an asset. Recent research has also demonstrated the applicability of FMCW radar to both locate and assess the severity of CUI features in pipeline and other offshore assets, both civil and military (Jones 2012; Jones et al. 2012; Flynn et al. 2014; Herd 2016).

Geological: Research into the sensitivity of FMCW to geomaterial pore content and material properties has also successfully identified the presence of water, oil or air within sandstone samples, with a view to providing an NDE sensor for energy sector deployment. This is in addition to providing a sensor to compliment the full-field toolbox used by geomaterial specialists; neutron beam tomography, X-ray Energy Dispersive X-ray (EDX) and Scanning Electron Microscopy (SEM). The low-power requirement, low-toxicity risk, and high portability of the FMCW sensor offer a useful alternative to the limitations imposed by the high-cost and fixed location of industry standard geomaterial benchmarking and characterization tools, such as heavy equipment including neutron, X-ray, EDX and SEM (Blanche et al. 2018; Blanche, Lewis, et al. 2020a).

28.3.2.2.3 Interim Conclusion

This section has provided a brief literature survey on NDE techniques commonly used in the offshore renewables and associated sectors. Based on this review, we show that IR thermography, though capable of remote large area monitoring of a structure, requires costly and sensitive instrumentation that needs highly skilled workers to operate and interpret the acquired data, in addition to the likelihood of non-detection of a defect deep within the structure.

Frequency-Modulated Continuous Wave (FMCW) radar sensing represents a strong candidate technology for WT blade integrity that is uniquely suited to non-contact and non-destructive analysis of composite structures, with the capability to map internal, as well as external, asset integrity features.

The FMCW sensing modality exhibits strong advantages when compared to contemporary methods for asset integrity inspection, as illustrated in Figure 28.12. A key advantage of K-band

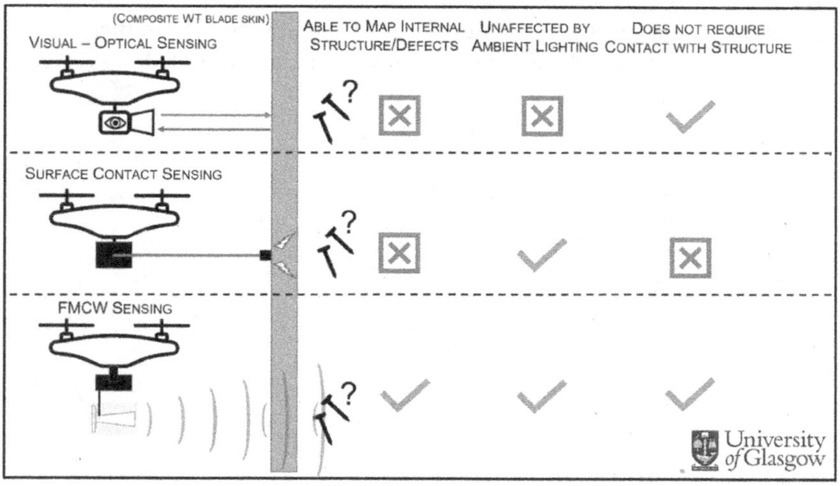

FIGURE 28.12 Comparative analysis of visual, contact and non-contact sensing methods for WT asset integrity inspection and SHM.

FMCW is the capability of the sensor to penetrate low to medium dielectric materials and return data pertaining to the internal features of a composite material. The system is:

- Non-contact, allowing for a significant standoff distance to the target (1–10 m or more).
- Non-visual and consequently unaffected by lighting conditions.
- Low power and lightweight, allowing for integration with handheld or robotic asset integrity inspection systems.
- Demonstrably unaffected by harsh environments and tunable depending on the measurand required and operating frequency.

The next section will provide an outline of FMCW theory and the data acquisition workflow before detailing the experimental setup for the detection of key contrasts relating to defects within a decommissioned WT blade structure.

28.4 FREQUENCY-MODULATED CONTINUOUS WAVE RADAR FOR WIND TURBINE BLADE STRUCTURAL HEALTH MONITORING

Following the identification of FMCW sensing as a strong candidate for non-contact NDE with a capability of SHM of asset integrity for WT internal defects, this section will provide an overview of key numerical theory, equipment parameters and the data acquisition workflow. This section will then demonstrate the use of FMCW in the K-band for the detection of key operationally induced defects on a decommissioned section of WT blade.

28.4.1 Numerical Theory

The rich and varied history of RAdio Detection And Ranging (RADAR) applications is beyond the scope of this tutorial, however, the key function of radar involves an active transmission and target interaction resulting in a signal reflection containing specific measurands for information extraction via signal processing. The most common measurands are direction, range and velocity, however, amplitude and phase variations of the reflected signal also provide critical information on the material properties of the target material, and it is these properties that will be the focus of this tutorial.

FMCW is characterized by a continuous change in signal frequency, as output from the sensor as a low-power transmission. This contrasts with more commonly used pulsed radar systems, which emit a single, pulsed frequency at a much higher power output to travel significant distances in air. The modulation of the output results in a sawtooth waveform, as illustrated in Figure 28.13.

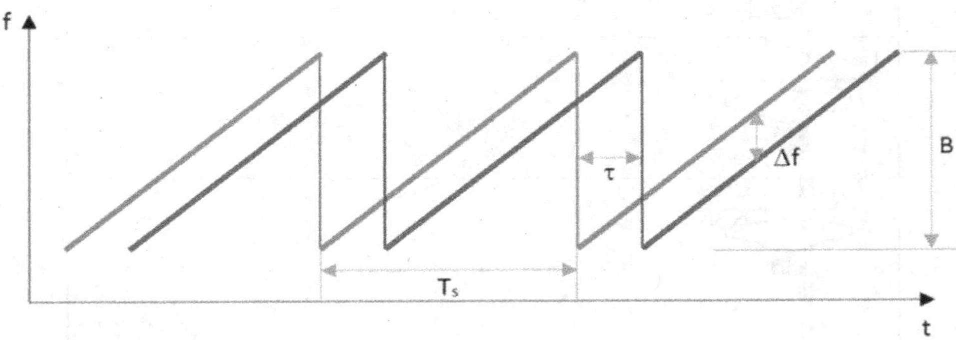

FIGURE 28.13 Transmitted and received signal properties, where f=frequency, t=time, T_s=frequency sweep duration, τ=two-way travel time, B=sweep bandwidth and Δf= intermediate frequency. (Blanche, Lewis, et al. 2020a.)

The output of an FMCW sensor is the intermediate frequency (IF) in the time domain, which is the summation of the transmitted and received waveforms. Post-processing of the IF, via a Fast Fourier Transform (FFT), provides amplitude and phase data in the frequency domain, which can be interrogated for distance and the target material properties. The difference in IF data (Δf) is transferred to a data logger as a column vector for each data acquisition.

The derivation of the IF signal is as follows (Blanche, Lewis, et al. 2020a):

$$f_{\mathrm{RF_{out}}} = f_{\mathrm{RF_0}} + k_f \times t \tag{28.1}$$

where, T_s is the frequency sweep time, $f_{\mathrm{RF_0}}$ is the initial frequency and k_f is the sweep rate, where $0 \leq t \leq T_s$

$$k_f = \frac{B}{T_s} \tag{28.2}$$

where B is the frequency sweep bandwidth and the two-way time, $\Delta t = \tau$, of the signal is:

$$\tau = 2\frac{d}{c} \tag{28.3}$$

where d is the distance between the target and antenna, and c is the speed of light in air.

Due to the delay in signal return, the reflected frequency compared to the emitted frequency is:

$$f_{\mathrm{RF_{received}}} = f_{\mathrm{RF_0}} + k_f * (t - \tau) \tag{28.4}$$

where $\tau \leq t \leq T_s + \tau$. The intermediate frequency (Δf) between the emitted and received signal is therefore:

$$\Delta f = k_f * (-\Delta t) \tag{28.5}$$

If the negative time of flight is taken as a magnitude:

$$\Delta f = \frac{B}{T_s} * 2\frac{d}{c} \tag{28.6}$$

To ensure the IF represents measurands that may describe the target material properties, it can be seen in (6) that the distance, d, to the target must remain a constant. Thus, signal variances in data acquired from a static target at a fixed distance, can be attributable to the intrinsic properties of the composite material under test, which may contain contrasts in asset integrity due to the presence of fluid ingress or operationally induced delamination.

28.4.2 Frequency-Modulated Continuous Wave Equipment Parameters and Data Acquisition Workflow

A block schematic of the data acquisition and processing stages is shown in Figure 28.14 and may be described in six key stages (Blanche, Lewis, et al. 2020a):

a. A signal generator provides a sawtooth wave with a 1.5 GHz (24–25.5 GHz) frequency sweep lasting 300 ms (see Figure 28.13). Data was acquired once per second (experimental parameters shown in Table 28.2),

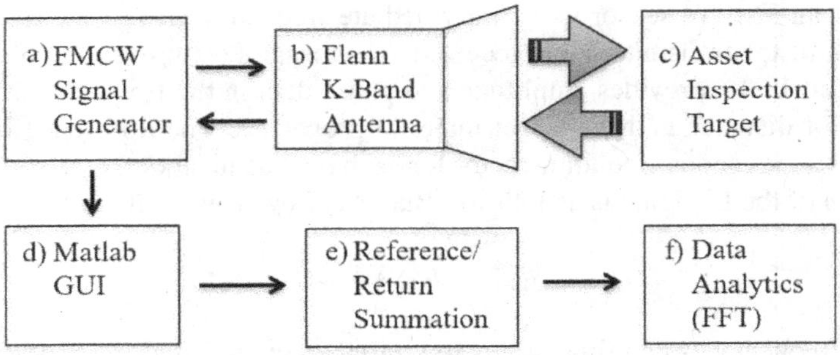

FIGURE 28.14 Key stages of experimental procedure including hardware elements and post-processing steps. (Blanche, Lewis, et al. 2020a).

TABLE 28.2
Key FMCW Parameters during WT Blade Tests

Parameter	Value
Band	K – Band (24–25.5 GHz)
Chirp duration	300 ms
Bandwidth	1,500 MHz
Acquisition frequency	1 Hz (0.2–3 Hz tunable)

Source: Blanche, Mitchell, et al. (2020b).

b. A Flann Microwave model #21240-20 (rated 17.6–26.7 GHz) standard gain horn antenna (serial #219405) was used with a nominal gain of 20 dB at 22.15 GHz. This antenna was aligned to directly face the WT section at a constant distance of 10 cm.
c. The inspection target. In this asset SHM use case, a WT section with a type 4 delamination was evaluated (Table 28.1).
d. A graphical user interface (GUI) and post-processing algorithm was programmed via MATLAB.
e. Reference and return signal waveforms were summed, providing the IF in the time domain, as shown in (28.6).
f. Data processing consisted of an FFT, from which amplitude extractions in the frequency domain were analyzed and presented. The time taken to process each data acquisition was 31 ms.

28.4.3 Implementation and Setup

The purpose of this tutorial is to demonstrate the detection of delamination and water ingress events within a decommissioned WT blade section. The experimental setup is shown in Figure 28.15, where a type 4 defect is clearly visible on the skin structure (right hand side). The antenna can be seen in the right image of Figure 28.15. It may be noted that the delamination has occurred on the inside edge of the defective skin section and therefore would not be visible from the asset exterior. An antenna to target distance of 10 cm was employed, giving a Field of View (FoV) on the target of $4.2 \times 10^{-3}\,m^2$ (Blanche, Mitchell, et al. 2020b).

For Experiment A, a baseline measurement was taken to characterize a healthy section of the WT blade structure. Care was taken to ensure that the baseline area was a healthy analog of the

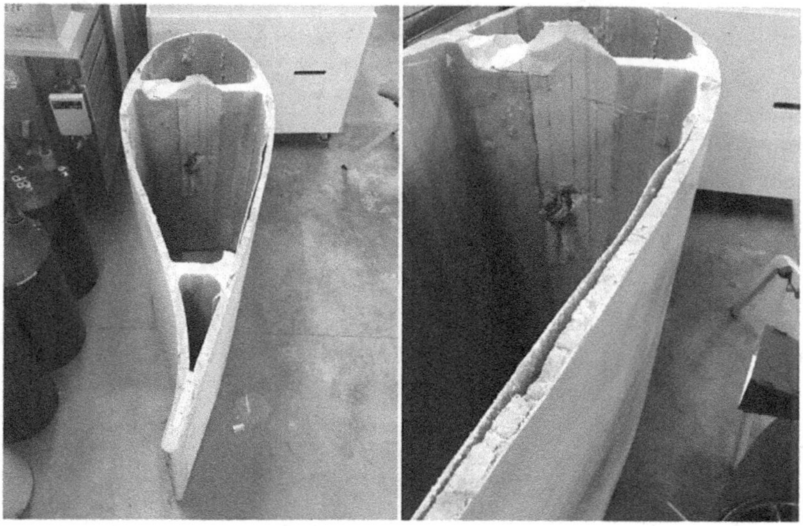

FIGURE 28.15 Section of decommissioned WT blade used for FMCW detection of type 4 delamination and water ingress. The K-band FMCW antenna can be seen 10 cm to the right of the structure. Note that the delamination defect is located on the inner surface of the composite skin and would not normally be visible to an observer. (Blanche, Mitchell, et al. 2020b).

defect area, with similar initial thickness and composition. A measurement was then taken with the defect in the sensor FoV. Each measurement was repeated ten times to ensure consistency.

For Experiment B, the same delamination feature as experiment A was used. FMCW data was acquired continuously for 15 minutes at a rate of 1 acquisition per second (1 Hz). During this experiment, a volume of 3 mL of water was applied to the delamination within the FoV of the sensor.

Experiment A: Delamination

The decommissioned WT blade exhibiting a type 4 delamination defect, as fully described in Table 28.1 is shown in Figures 28.15 and 28.16, where a contrast in return signal amplitude (RSA) may be observed, when compared to the baseline measurement for an adjacent healthy region. For both measurements, a peak exists at an IF of 8 Hz, which is consistent with the constant time of flight to the target of the emitted and received signals. The contrast in RSA is evident, with the defect response at 106 dB a lower amplitude than the healthy baseline at 114 dB.

This distinct and repeatable contrast in the RSA for healthy and defective sections of WT skin structures validates the use of K-band FMCW in this role, where the signal response is evident at both the peak IF associated with the air/composite boundary of the WT skin and the higher IF regions associated with internal volumes of the asset structure. These delayed returns provide insight into the material properties of the internal composite structure and any further interfaces within that structure, which may be numerous given the porous nature of the balsa and fiber-glass laminate construction.

Experiment B: Water Ingress within Delamination Defect

The results shown in Figure 28.17 demonstrate the ingress of 3 mL of water added at 3 minutes and 40 seconds into the experiment. For this experiment, the data shown represents the dynamic signal at IF 8 MHz at the signal interface and within the WT skin structure. To inform on internal properties of the WT skin structure, an IF of 14 MHz was selected; a value representative of material beyond the target-air interface at 8 MHz.

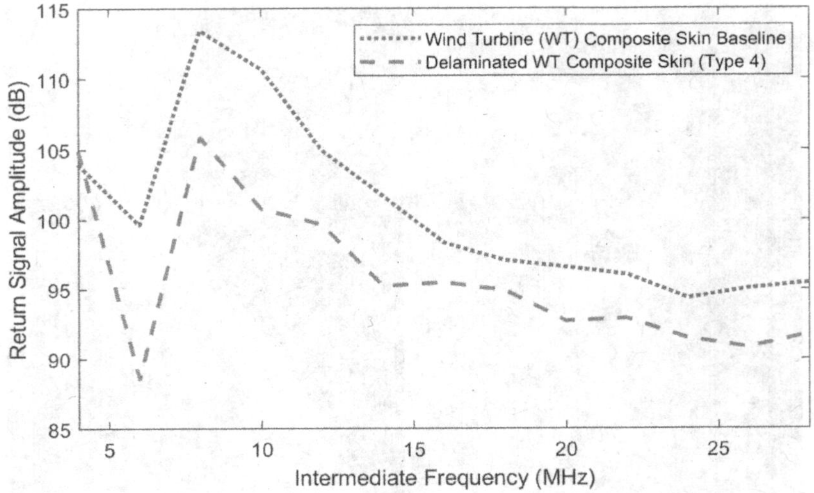

FIGURE 28.16 Comparison of RSA in the frequency domain for a healthy section of wind turbine blade wooden core structure versus a section of wooden core exhibiting a type 4 defect (see Table 28.1 and Figure 28.15). (Blanche, Mitchell, et al. 2020b.)

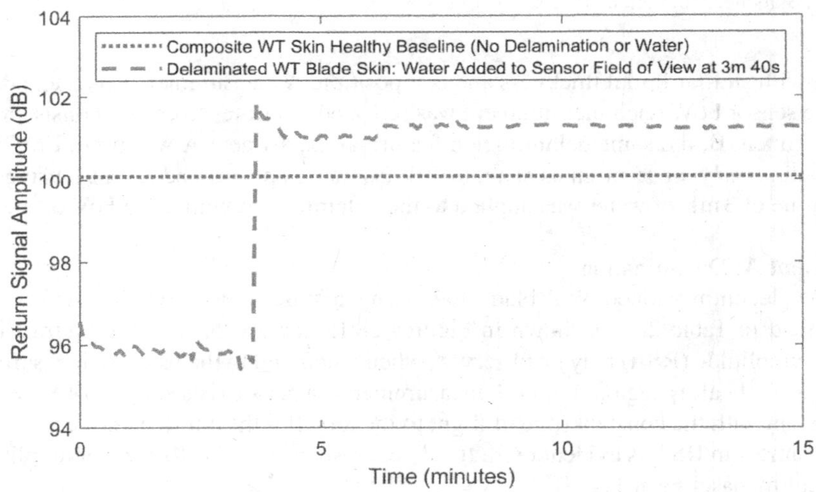

FIGURE 28.17 Signal contrast versus baseline for a type 4 delamination and dynamic water ingress at 3 minutes and 40 seconds. (Blanche, Mitchell, et al. 2020b.)

From the acquired data, the following may be observed:

- The RSA of the healthy baseline region of the WT blade structure is ~100 dB. This represents an undamaged low dielectric material of constant thickness and composed of porous polymers reinforced by laminate carbon fiber and resins.
- The RSA of the type 4 delaminated region of the WT blade structure ranges between 95.5 and 96.5 dB, representing a consistently lower dielectric than the healthy baseline value. This is due to increased porosity in the damaged volume and inconsistent skin thickness.

- The deposition of 3 mL of water within the type 4 defect area, though visually obscured from the sensor by ~2 cm of sandwich composite material, shows a distinct and consistent contrast in RSA, with a post water deposition range between ~101 and 102 dB.

28.5 DISCUSSION

Following the identification of millimeter-wave sensing as a high potential emergent candidate technology for detection of operationally induced WT defects, this chapter presents the novel application of FMCW radar technology to the detection of surface and internal structural analysis of a section of damaged WT blade. This has resulted in the identification of specific defects and their key measurands. Significant contrasts in the return signal properties have been correlated with delamination and water ingress within the composite interior structure of a WT blade, highlighting the suitability of the FMCW sensing modality as a deployable sensor to act as a discriminator for the presence of defects within vulnerable and compromised WT blade assets. The detection and mapping of WT structural degradation at an early stage of defect formation will increase the operational effectiveness of a vital renewable energy resource and carbon reduction asset at a time when there exists a need to deliver a net zero carbon economy. Millimeter-wave sensing represents a significant advance for NDT sensing, allowing for the near-to-real-time characterization of defects via a sensing mode that is unaffected by impediments to visual sensing, such as mist, fog or smoke. This represents a suitable sensor for structural embedding within the fabric of a WT structure and enabling use even when optical sensing methods are unable to operate due to local weather conditions. Consequently, the versatility of FMCW sensing will improve defect detection strategies and add to the state-of-the-art in sensing modalities in this role, such as visual photography and photogrammetry, DIC and IR thermographic inspections.

28.6 CONCLUSION

The need to meet climate targets to reduce global warming is a major driver in the continued expansion of the offshore renewables sector, where asset optimization is required to ensure that carbon reducing technologies operate at peak performance. Achieving this goal, while also driving significant global expansion of the sector, requires improvements in both the capability of offshore renewable asset integrity sensing and SHM. This is in addition to the reduction of asset downtime associated with inspection and defect intervention strategies. This chapter has evaluated the effectiveness of current NDT/SHM technologies for the offshore renewables sector and demonstrates the advantages of FMCW sensing in this role. A decommissioned WT blade section was used as an exemplar to successfully evaluate the sensitivity of FMCW in the K-band to detect the presence of delamination and water ingress faults internal to the composite blade structure.

The millimeter-wave FMCW system used in this tutorial has several advantageous features:

- It is simple to use and allows rapid operation with resilience to challenging ambient conditions where optical sensors are likely to be impeded, such as fog, mist, or smoke.
- It is a non-contact, NDE system with a low radiation output, maximum 900 mW, representing a safe to use and non-toxic sensor with a standoff capability, requiring no couplant.
- It is tunable in terms of the frequency sweep, chirp duration and output power, allowing for variable sensitivity to operational conditions and parameters.
- It is a small and lightweight payload and is suitable for operation in confined spaces.

Applied to WT blade SHM practices, the FMCW sensor deployed in this role has the potential to reduce the cost of O&M, allowing for earlier defect tracking and leading to previously inaccessible data regarding the internal features of a WT structure. This will lead to timely intervention strategies, reducing asset downtime and improving the efficiency of asset integrity management for the offshore renewables sector.

28.7 FUTURE SCOPE

Sensor Development: The development of the FMCW systems applied to this role is aimed at improving strategies for the characterization of defects and target material identification. This will include developments in the Terahertz (THz) band, where the higher spatial resolution is offset by a reduction in target material penetration and overall sensor range in air. The application of dielectric lensing to collimate the emitted FMCW radiation pattern, allowing for the optimization of the radiation footprint on the target material and will facilitate a greater standoff distance to the target for the same output power. Tunable dielectric lensing allows alterations to the radiation spot size on the target to provide beam focus or a wider area sweep of an asset under inspection.

Robotic Integration: Removing the need for human presence and intervention in the challenging offshore environment remains a major driver in the pursuit of a low-carbon footprint for the offshore renewables sector. Integration of sensing suites with robotic platforms to detect, mitigate or repair defects will lessen the need for human intervention, and the associated costs and risks that result from human deployment offshore. Increasing the practical range to the inspection target represents a key step in the operational deployment of FMCW sensing and will facilitate integration with robotic sensor suites, capitalizing on the platform agnostic nature of the FMCW modality and the full SHM sensing toolset.

Machine Learning: Continuing research into machine learning for FMCW SHM data has demonstrated that AI-based algorithms aid in the classification of defect types with over 95% accuracy and can identify composition and diameter differentials of up to 3 mm. The development of this method presents a promising solution to the efficient classification of surface and internal defect characterization of composite structures (Tang et al. 2021).

Resident Robotics: The continuing development of wind farm arrays far offshore presents a logistical problem for the current paradigm of robotic deployment, where human-robot teams are transported to the inspection site and are supported by crew vessels and other ancillary services. The cost effectiveness of this method reaches a tipping point the further offshore a wind farm array is situated. The solution is the development of resident robotic systems that are operated remotely from onshore facilities and/or increasingly autonomous robotic systems able to operate independently beyond visual line-of-sight conditions. The presence of resident robotic systems also serves to provide enhanced operational overviews and integration of acquired asset integrity data to significantly improve situational awareness of these remote assets via a digital twin of the offshore environment. Key advances in the development of these cyber-physical systems and the required levels of trusted autonomy as a service are the subject of active research in this space (Zaki et al. 2020; Kong et al. 2020; Mitchell et al. 2021; McConnell et al. 2022).

REFERENCES

Aardal, Ø., Paichard, Y., Brovoll, S., Berger, T., Lande, T., & Hamran, S. (2013). Physical working principles of medical radar. *IEEE Transactions on Biomedical Engineering*, 60(4), 1142–1149. doi: 10.1109/TBME.2012.2228263.

Adams, R. D., & Cawley, P. (1988). A review of defect types and nondestructive testing techniques for composites and bonded joints. *NDT International*, 21(4), 208–222. doi: 10.1016/0308-9126(88)90333-1.

Al-Khudairi, O., & Ghasemnejad, H. (2015). To improve failure resistance in joint design of composite wind turbine blade materials. *Renewable Energy*, 81 (September), 936–951. doi: 10.1016/J.RENENE.2015.04.015.

Amenabar, I., Lopez, F., & Mendikute, A. (2012). In Introductory Review to THz Non-Destructive Testing of Composite Mater. doi: 10.1007/s10762-012-9949-z.

Antonopoulos, I., Robu, V., Couraud, B., Kirli, D., Norbu, S., Kiprakis, A., Flynn, D., Elizondo-Gonzalez, S., & Wattam, S. (2020). Artificial intelligence and machine learning approaches to energy demand-side response: A systematic review. *Renewable and Sustainable Energy Reviews*, 130(September), 109899. doi: 10.1016/J.RSER.2020.109899.

Arumugam, J., Suresh Kumar, C., Santulli, C., Sarasini, F., & Joseph Stanley, A. (2011). A global method for the identification of failure modes in fiberglass using acoustic emission. *Journal of Testing and Evaluation*, 39(5), 954–966. doi: 10.1520/JTE103730.

Ashby, M. F., & Cebon, D. (1993). Materials selection in mechanical design. *Le Journal de Physique*, IV 03(C7), C7–C1. doi: 10.1051/JP4:1993701.

Baqersad, J., Poozesh, P., Niezrecki, C., & Avitabile, P. (2017). Photogrammetry and optical methods in structural dynamics: A review. *Mechanical Systems and Signal Processing*, 86(March), 17–34. doi: 10.1016/J.YMSSP.2016.02.011.

Barnes, M., Brown, K., Carmona, J., Cevasco, D., Collu, M., Crabtree, C., Crowther, W., et al. (2018). Technology Drivers in Windfarm Asset Management. *HOME Offshore*. doi: 10.17861/20180718.

Barowski, J., Zimmermanns, M., & Rolfes, I. (2018). Millimeter-wave characterization of dielectric materials using calibrated FMCW transceivers. *IEEE Transactions on Microwave Theory and Techniques*, 66(8), 3683–3689. doi: 10.1109/TMTT.2018.2854180.

Beasley, P. (2008). Advances in millimetre wave FMCW radar. In *2008 Proceedings of Microwaves, Radar and Remote Sensing Symposium, MRRS 2008* (pp. 246–249). Kiev: IEEE. doi: 10.1109/MRRS.2008.4669588.

Blanche, J., Flynn, D., Lewis, H., Couples, G., Buckman, J., Bailey, C., & Tilford, T. (2018). Analysis of sandstone pore space fluid saturation and mineralogy variation via application of monostatic K-band frequency modulated continuous wave radar. *IEEE Access*, 6(August), 44376–44389. doi: 10.1109/ACCESS.2018.2863024.

Blanche, J., Mitchell, D., Gupta, R., Tang, A., & Flynn, D. (2020b). Asset integrity monitoring of wind turbine blades with non-destructive radar sensing. In *11th Annual IEEE Information Technology, Electronics and Mobile Communication Conference, IEMCON 2020* (pp. 498–504). doi: 10.1109/IEMCON51383.2020.9284941.

Blanche, J., Lewis, H., Couples, G., Buckman, J., Lenoir, N., Tengattini, A., & Flynn, D. (2020a). Dynamic fluid ingress detection in geomaterials using K-band frequency modulated continuous wave radar. *IEEE Access*, 8, 111027–111041. doi: 10.1109/ACCESS.2020.3002147.

Branner, K., & Ghadirian, A. (2014). Database about Blade Faults. DTU Library. https://backend.orbit.dtu.dk/ws/portalfiles/portal/118222161/Database_about_blade_faults.pdf.

Charchuk, R., Werstiuk, C., Evans, M., & Sjerve, E. (2012). High temperature guided wave pipe inspection. In *4th International CANDU In-Service Inspection Workshop and NDT*, Canada.

Cheng, J., Qiu, J., Xu, X., Ji, H., Takagi, T., & Uchimoto, T. (2016). Research advances in eddy current testing for maintenance of carbon fiber reinforced plastic composites. *International Journal of Applied Electromagnetics and Mechanics*, 51(3), 261–284. doi: 10.3233/JAE-150168.

Crespo, B. H. (2016). Damage sensing in blades. In W. Ostachowicz, M. McGugan, J.-U. Schroder-Hinrichs, & M. Luczak (eds.), *MARE-WINT: New Materials and Reliability in Offshore Wind Turbine Technology* (pp. 25–52). Switzerland: Springer Open. doi: 10.1007/978-3-319-39095-6.

Cristofani, E., Friederich, F., Wohnsiedler, S., Matheis, C., Jonuscheit, J., Vandewal, M., & Beigang, R. (2014). Nondestructive testing potential evaluation of a terahertz frequency-modulated continuous-wave imager for composite materials inspection. *Optical Engineering*, 53(3), 031211. doi: 10.1117/1.OE.53.3.031211.

Crown Estate Scotland (2020). ScotWind Leasing Seabed Leasing For New Offshore Wind Farms, Edinburgh, UK. https://www.crownestatescotland.com/our-projects/scotwind.

de Goeje, M. P., & Wapenaar, K. E. D. (1992). Non-destructive inspection of carbon fibre-reinforced plastics using eddy current methods. *Composites*, 23(3), 147–157. doi: 10.1016/0010-4361(92)90435-W.

Fagogenis, G., Flynn, D., & Lane, D. (2015). Novel RUL prediction of assets based on the integration of auto-regressive models and an RUSBoost classifier. *2014 International Conference on Prognostics and Health Management, PHM 2014*. doi: 10.1109/ICPHM.2014.7036373.

Flynn, D., Herd, D., Lofting, D., Record, P., & Skinner, N. (2014). Health and usage monitoring systems: Enabling the future prediction of remaining useful life of submarines. In *International Naval and Engineering Conference*.

Gholizadeh, S. (2016). A review of non-destructive testing methods of composite materials. *Procedia Structural Integrity*, 1(January), 50–57. doi: 10.1016/J.PROSTR.2016.02.008.

Gholizadeh, S., Lemana, Z., & Baharudinb, B. T. H. T. (2015). A review of the application of acoustic emission technique in engineering. *Structural Engineering and Mechanics*, 54(6), 1075–1095. doi: 10.12989/SEM.2015.54.6.1075.

Grinzato, E., Vavilov, V., & Kauppinen, T. (1998). Quantitative infrared thermography in buildings. *Energy and Buildings*, 29(1), 1–9. doi: 10.1016/S0378-7788(97)00039-X.

Gubler, H., & Hiller, M. (1984). The use of microwave FMCW radar in snow and avalanche research. *Cold Regions Science and Technology*, 9(2), 109–119. doi: 10.1016/0165-232X(84)90003-X.

Gupta, R., Huo, D., White, M., Jha, V., Stenning, G., & Pancholi, K. (2019). Novel method of healing the fibre reinforced thermoplastic composite: A potential model for offshore applications. *Composites Communications*, 16(December), 67–78. doi: 10.1016/J.COCO.2019.08.014.

Gupta, R., Mitchell, D., Blanche, J., Harper, S., Tang, W., Pancholi, K., Baines, L., Bucknall, D., & Flynn, D. (2021). A review of sensing technologies for non-destructive evaluation of structural composite materials. *Journal of Composites Science*, 5(12), 319.

He, Y., Tian, G., Pan, M., & Chen, D. (2014). Impact evaluation in Carbon Fiber Reinforced Plastic (CFRP) laminates using eddy current pulsed thermography. *Composite Structures*, 109(1), 1–7. doi: 10.1016/J.COMPSTRUCT.2013.10.049.

Her Majesty's Government (2020). The Ten Point Plan for a Green Industrial Revolution, London. https://assets.publishing.service.gov.uk/government/uploads/system/uploads/attachment_data/file/936567/10_POINT_PLAN_BOOKLET.pdf.

Herd, D. S. (2016). Microwave Based Monitoring System for Corrosion Under Insulation, Edinburgh.

Hohmann, B., Bruck, P., Esselman, T., & Schmidt, T. (2012). Digital Image Correlation (DIC): An advanced nondestructive testing method for life extension of nuclear power plants. In *International Conference on Nuclear Power Plant Life Management (PLiM) for Long Term Operations (LTO)*. https://inis.iaea.org/Search/search.aspx?orig_q=RN:43070819.

Holmgren, J., Sturm, M., Yankielun, N. E., & Koh, G. (1998). Extensive measurements of snow depth using FM-CW radar. *Cold Regions Science and Technology*, 27(1), 17–30. doi: 10.1016/S0165-232X(97)00020-7.

Jones, R. E. (2012). Use of Microwaves for the Detection of Corrosion under Insulation, London. https://core.ac.uk/download/pdf/9833196.pdf.

Jones, R. E., Simonetti, F., Lowe, M. J. S., & Bradley, I. P. (2012). Use of microwaves for the detection of water as a cause of corrosion under insulation. *Journal of Nondestructive Evaluation*, 31(1), 65–76. doi: 10.1007/S10921-011-0121-9.

Kaminski, P., Staszek, K., Wincza, K., & Gruszczynski, S. (2014). K-band FMCW radar module with interferometic capability for industrial applications. In *Proceedings International Radar Symposium*. IEEE Computer Society. doi: 10.1109/IRS.2014.6869237.

Kashif Ur Rehman, S., Ibrahim, Z., Memon, S., & Jameel, M. (2016). Nondestructive test methods for concrete bridges: A review. *Construction and Building Materials*, 107(March), 58–86. doi: 10.1016/J.CONBUILDMAT.2015.12.011.

Katunin, A., Dragan, K., & Dziendzikowski, M. (2015). Damage identification in aircraft composite structures: A case study using various non-destructive testing techniques. *Composite Structures*, 127(September), 1–9. doi: 10.1016/J.COMPSTRUCT.2015.02.080.

Kažys, R., Voleišis, A., & Voleišienė Baršauskas, B. K. (2008). High temperature ultrasonic transducers: Review. *Ultragarsas/Ultrasound*, 63(2), 7–17. https://www.ultragarsas.ktu.lt/index.php/USnd/article/view/17063.

Kong, L., Harper, S., Mitchell, D., Blanche, J., Lim, T., & Flynn, D. (2020). Interactive digital twins framework for asset management through internet. *2020 IEEE Global Conference on Artificial Intelligence and Internet of Things, GCAIoT 2020*. doi: 10.1109/GCAIOT51063.2020.9345890.

Konstantopoulos, S., Fauster, E., & Schledjewski, R. (2014). Monitoring the production of FRP composites: A review of in-line sensing methods StorHy view project MoVeTech view project. *EXPRESS Polymer Letters*, 8(11), 823–840. doi: 10.3144/expresspolymlett.2014.84.

Koyama, K., Hoshikawa, H., & Kojima, G. (2013). Eddy current nondestructive testing for carbon fiber-reinforced composites. *Journal of Pressure Vessel Technology, Transactions of the ASME*, 135(4). doi: 10.1115/1.4023253/379034.

Lee, J., & Zhao, F. (2020). Global Offshore Wind Report 2020, Brussels, Belgium. www.gwec.net.

Li, C., Lubecke, V. M., Boric-Lubecke, O., & Lin, J. (2013a). A review on recent advances in Doppler radar sensors for noncontact healthcare monitoring. *IEEE Transactions on Microwave Theory and Techniques*, 61(5), 2046–2060. doi: 10.1109/TMTT.2013.2256924.

Li, L., Zhang, S., Xu, Z., Geng, X., & Shrout, T. R. (2013b). 1-3 ceramic/polymer composites for high-temperature transducer applications. *Physica Status Solidi (a)* 210(9), 1888–1891. doi: 10.1002/PSSA.201228787.

Lu, Y. (2010). Non-Destructive Evaluation on Concrete Materials and Structures Using Cement-Based Piezoelectric Sensor, Hong Kong.

Mal, A., & Bar-Cohen, Y. (1988). Ultrasonic NDE of bonded solids. In *Proceedings of the International Workshop on Nondestructive Evaluation for Performance of Civil Structures* (pp. 299–308). University of Southern California, Los Angeles.

Marshall, H. P., & Koh, G. (2008). FMCW radars for snow research. *Cold Regions Science and Technology*, 52(2), 118–131. doi: 10.1016/J.COLDREGIONS.2007.04.008.

McConnell, A., Mitchell, D., Donaldson, K., Harper, S., Blanche, J., Lim, T., Flynn, D., & Stokes, A. (2022). The future workplace: A symbiotic system of systems environment. In T. Semwal, & F. Iqbal (eds.), *Cyber-Physical Systems: Solutions to Pandemic Challenges* (vol. 1(in print)). Routledge. https://www.routledge.com/Cyber-Physical-Systems-Solutions-to-Pandemic-Challenges/Semwal-Iqbal/p/book/9781032030371.

McGinnis, M. J. (2006). Experimental and Numerical Development of the Core-Drilling Method for the Nondestructive Evaluation of in-situ Stresses in Concrete Structures, Ann Arbor, MI. https://www.proquest.com/docview/305316147?pq-origsite=gscholar&fromopenview=true.

McGinnis, M. J., Pessiki, S., & Turker, H. (2005). Application of three-dimensional digital image correlation to the core-drilling method. *Experimental Mechanics*, 45(4), 359.

Meyendorf, N. G. H., Nagy, P. B., & Rokhlin, S. I. (2004). *Nondestructive Materials Characterization: With Applications to Aerospace Materials*, 1st Edition. Berlin, Germany: Springer-Verlag. doi: 10.1007/978-3-662-08988-0.

Miguelañez-Martin, E., & Flynn, D. (2015). Embedded intelligence supporting predictive asset management in the energy sector. In *IET Conference Publications* (vol. 2015). Institution of Engineering and Technology. doi: 10.1049/CP.2015.1752/CITE/REFWORKS.

Mitchell, D., Blanche, J., Zaki, O., Roe, J., Kong, L., Harper, S., Robu, V., Lim, T., & Flynn, D. (2021). Symbiotic system of systems design for safe and resilient autonomous robotics in offshore wind farms. *IEEE Access Reliability Society Section*, 9(October), 141421–141452. doi: 10.1109/access.2021.3117727.

Moll, J., Arnold, P., Mälzer, M., Krozer, V., Pozdniakov, D., Salman, R., Rediske, S., Scholz, M., Friedmann, H., & Nuber, A. (2017). Radar-based structural health monitoring of wind turbine blades: The case of damage detection. *Structural Health Monitoring*, 17(4), 815–822. doi: 10.1177/1475921717721447.

Montesano, J., Fawaz, Z., & Bougherara, H. (2013). Use of infrared thermography to investigate the fatigue behavior of a carbon fiber reinforced polymer composite. *Composite Structures*, 97(March), 76–83. doi: 10.1016/J.COMPSTRUCT.2012.09.046.

Mulaveesala, R., & Tuli, S. (2005). Implementation of frequency-modulated thermal wave imaging for non-destructive sub-surface defect detection. *Insight: Non-Destructive Testing and Condition Monitoring*, 47(4), 206–208. doi: 10.1784/INSI.47.4.206.63156.

Oral, I. (2019). Characterization of damages in materials by computer-aided tap testing. In *IOP Conference Series: Materials Science and Engineering* (vol. 707, p. 012019). IOP Publishing. doi: 10.1088/1757-899X/707/1/012019.

ORE Catapult, & BVG Associates (2020). Wind Farm Costs: Guide to an Offshore Wind Farm. https://guidetoanoffshorewindfarm.com/wind-farm-costs.

Ozbek, M., & Rixen, D. J. (2011). Optical measurements and operational modal analysis on a large wind turbine: Lessons learned. In *Conference Proceedings of the Society for Experimental Mechanics Series* (vol. 5, pp. 257–276). New York: Springer. doi: 10.1007/978-1-4419-9428-8_21.

Ozbek, M., Rixen, D. J., Erne, O., & Sanow, G. (2010). Feasibility of monitoring large wind turbines using photogrammetry. *Energy*, 35(12), 4802–4811. doi: 10.1016/J.ENERGY.2010.09.008.

Poozesh, P., Baqersad, J., Niezrecki, C., Avitabile, P., Harvey, E., & Yarala, R. (2017). Large-area photogrammetry based testing of wind turbine blades. *Mechanical Systems and Signal Processing*, 86(March), 98–115. doi: 10.1016/j.ymssp.2016.07.021.

Schmidt, T., Tyson, J., & Galanulis, K. (2003a). Full-field dynamic displacement and strain measurement using advanced 3D image correlation photogrammetry: Part II. *Experimental Techniques*, 27(4), 22–26. doi: 10.1111/J.1747-1567.2003.TB00118.X.

Schmidt, T., Tyson, J., & Galanulis, K. (2003b). Full-field dynamic displacement and strain measurement using advanced 3D image correlation photogrammetry: Part I. *Experimental Techniques*, 27(3), 47–50. doi: 10.1111/J.1747-1567.2003.TB00115.X.

Shafiee, M., Zhou, Z., Mei, L., Dinmohammadi, F., Karama, J., & Flynn, D. (2021). Unmanned aerial drones for inspection of offshore wind turbines: A mission-critical failure analysis. *Robotics*, 10(1), 26. doi: 10.3390/ROBOTICS10010026.

Sørensen, B. F. (2004). Improved Design of Large Wind Turbine Blade of Fibre Composites Based on Studies of Scale Effects (Phase 1) - Summary Report. Risø National Laboratory.

Tanasoiu, V., Miclea, C., & Tanasoiu, C. (2002). Nondestructive testing techniques and piezoelectric ultrasonics transducers for wood and built in wooden structures. *Journal of Optoelectronics and Advanced Materials*, 4(4), 949–957.

Tang, W., Mitchell, D., Blanche, J., Gupta, R., & Flynn, D. (2021). Machine learning analysis of non-destructive evaluation data from radar inspection of wind turbine blades. In *International Conference on Sensing, Diagnostics, Prognostics, and Control (SDPC)* (pp. 122–128). Weihai, China: IEEE. doi: 10.1109/sdpc52933.2021.9563264.

The Crown Estate (2019). Introducing Offshore Wind Leasing Round 4 Information Memorandum, London, UK. https://www.thecrownestate.co.uk/media/3321/tce-r4-information-memorandum.pdf.

Vinci, G., Lindner, S., Barbon, F., Hofmann, M., Fischer, G., Kissinger, D., & Koelpin, A. (2012). 24GHz six-port medical radar for contactless detection and heartbeat monitoring. In *9th European Radar Conference* (pp. 75–78). Amsterdam: IEEE.

Vollmer, M., & Mollmann, K.-P. (2018). *Infrared Thermal Imaging: Fundamentals, Research and Applications*, 2nd Edition. Weinheim, Germany: Wiley-VCH.

Wang, B., Zhong, S., Lee, T. L., Fancey, K. S., & Mi, J. (2014). Non-destructive testing and evaluation of composite materials/structures: A state-of-the-art review. *Advances in Mechanical Engineering*, 12(4), 1–28. doi: 10.1177/1687814020913761.

Yadav, R., Dahiya, P., & Mishra, R. (2016). A high performance 76.5GHz FMCW RADAR for advanced driving assistance system. In *3rd International Conference on Signal Processing and Integrated Networks, SPIN 2016* (pp. 383–388). Institute of Electrical and Electronics Engineers Inc. doi: 10.1109/SPIN.2016.7566724.

Yang, X., Zhang, Y., Lv, W., & Wang, D. (2021). Image recognition of wind turbine blade damage based on a deep learning model with transfer learning and an ensemble learning classifier. *Renewable Energy*, 163(January), 386–397. doi: 10.1016/j.renene.2020.08.125.

Zaki, O., Flynn, D., Blanche, J., Roe, J., Kong, L., Mitchell, D., Lim, T., Harper, S., & Robu, V. (2020). Self-certification and safety compliance for robotics platforms. In *Offshore Technology Conference* (vol. 2020). Houston, TX: OnePetro. doi: 10.4043/30840-MS.

Part 6

System Security

29 Risk Management Framework

Derek Reimanis

29.1 INTRODUCTION

In Chapter 32, "Software Security", we look at software-specific issues in software vulnerability and software security, and how attackers can target the code base as well as the underlying data itself to achieve malicious intent. Understanding the code and data level vulnerabilities in a software system is critical to preventing attacks, yet such actions must be paired with organization-level standards and process to ensure that security protections are properly and consistently applied while being cognizant of stakeholder priorities (which might not align with security issues) (Figure 29.1).

This process gap is where the Risk Management Framework (RMF) situates itself. The RMF outlines a seven-step process to manage security and risk that any organization can use. The RMF was developed by the National Institute of Standards and Technology (NIST) and is public and freely available under the NIST Special Publication 800-37 Revision 2 (NIST 800-37r2) document (NIST [rmf], n.d.).[1] The RMF has been adopted by the United States Department of Defense (DoD) as the de-facto standard for strengthening risk management-related process in IT organizations. The RMF is a framework, not a regulation. That means it provides high-level guidelines and process to follow that results in a more secure organization. It does not specify the technical details that an organization needs to achieve compliance. Each organization implements the RMF differently; some organizations, such as the DoD, adhere to the framework using strict organizational and compliance-based requirements. The DoD outlines their RMF implementation using several formal documents, notably the DoDI 8510.01 Instruction document (DOD INSTRUCTION 8510.01, n.d.).

FIGURE 29.1 The risk management framework

DOI: 10.1201/9781003188377-35

Such level of formal adherence is not implicitly required by the RMF, yet it is not uncommon for critical organizations to formalize their RMF implementation as such. While the RMF does allow for varying degrees of implementation, there is a clear structure that defines the RMF, with necessary elements and process.

This chapter describes the seven major steps of the RMF. After reading this chapter, it is expected the reader understands of the different stages of the RMF, has a clear understanding of how the RMF helps organizations mitigate and accept risk, understands the responsibility roles of RMF officials in the organization, and has the tools in place to implement the RMF. As a caveat, because every organization must manage risk with respect to their own mission, there is no boilerplate RMF implementation to follow. This means there is no NIST or RMF-sponsored standard format for most of the required documentation. Skeleton formats for most documentation can be found online but is largely catered to the organization that hosts it, and a particular format might not be applicable in another context. This trade-off is what allows the RMF to maintain such a strong and effective presence in the field; too structured of a format will make the RMF difficult to adapt at organization levels, and not enough structure will remove the standardization properties that the RMF offers. The RMF is very effective at finding this balance.

29.2 PREPARE

The first step toward implementing the RMF is a pre-curser exercise that prepares the organization with groundwork before the actual RMF implementation begins. This step, aptly named Prepare, involves identifying tasks at the organization, and system levels to help prepare the organization to manage its security and privacy risks using the RMF (Figure 29.2).

29.2.1 Objectives

The purpose of the Prepare step is to align the organization to enable easier and more cohesive adaption of the RMF. This involves preparing the information and resources (both managerial and

FIGURE 29.2 The first step of the Risk Management Framework, Prepare.

technical) that are necessary to complete RMF adaption. The tasks within the Prepare step are conceptually divided into two categories, *organization* and *system*.

29.2.1.1 Organization

From the organization's perspective, the Prepare step involves two subtasks: (1) identifying key personnel to assume responsibility for the implementation, completion, and future support of each RMF task and (2) identification of the organization's high-value assets that require increased levels of protection. The format of these deliverables will vary from organization to organization, but generally, they are relayed as text-based tables. The completion of these two tasks ultimately facilitates communication between leaders and executives at the organization, in addition to providing a unified strategy at management levels.

29.2.1.2 System

From the system's perspective, the Prepare step involves two subtasks: (1) The collection or generation of system-wide concepts in the form of technical diagrams, such as is covered in Part 1 of this series [Data Frameworks] and (2) identification and collection of system and personnel controls across the organization. These controls range from system-level controls, such as root access into a computer system, to physical controls such as card-key scanners or biometric access points. NIST provides a supporting standard (NIST 800-53) (NIST [53], n.d.) that solely covers security and privacy controls within an organization, which is intended to be used in tandem with the RMF. Complementary to the collection of controls is the identification of information type; information type refers to the type of information that will be processed, stored, and transmitted by the system to support organizational missions. NIST provides a supporting standard (NIST SP 800-60) (NIST [60], n.d.) that provides guidelines for the selection of information types across organizational types; for example, government organizations can be broken down into the following categories:

- Mission-based information types:
  - Purpose of government *(citizen services)*
  - Mechanisms that the government uses to achieve its purpose *(delivery)*
- Management and support information types:
  - Support functions necessary to conduct government operations *(delivery support)*
  - Resource management support of government *(resource management)*

Information types are specific to organizations, so the supporting standard (NIST SP 800-60) (NIST [60], n.d.) might only be used as a guideline, yet the collection and proper documentation of information types facilitates conceptual breakdowns of important aspects to protect in an organization. The completion of these tasks provides the resources needed from a systems level to enable a correct and efficient RMF implementation.

29.2.2 Outcomes

Following the Prepare step, the organization will have the groundwork laid to begin implementing the RMF. From the business perspective, managerial members will have clear responsibilities, and RMF implementation strategies will be cohesive. Furthermore, an assessment of the value of the organization's high-level assets will be performed so that risk mitigation efforts can be prioritized. From the system perspective, important system-wide and control-based technical documents will be organized and current. These technical documents are necessary to complete the next step in the RMF process, Categorize.

29.3 CATEGORIZE

The second step of the RMF is the Categorize step. Categorization entails the structuring of system-critical information, typically related to system inputs, processing, storage, and output. Categorization seeks to capture the adverse impact of these system-critical functions with respect to each facet of the CIA triad, consisting of confidentiality, integrity, and availability. The outcome of the categorization step is a security category for the system, which presents the severity of adverse impacts to an organization broken into each impact's effect on key organizational concerns (Figure 29.3).

29.3.1 System Description

The first subtask of system categorization consists of creating a documented system description. A system description describes the characteristics of a system and is constructed from all available system design or requirement artifacts. These artifacts may include system identifiers, controls, regulations, network rules, supply-chain information, contractor responsibilities, etc.; the key point is that the documents are as extensive as possible. Included in the system description is a list of the information types pertinent to the system. Many of these artifacts would have already been collected in the Prepare step of the RMF (see Section 29.1). The RMF does not specify the format for these artifacts, yet a list is sufficient.

29.3.2 Security Categorization

Following the generation of a documented system description, each security objective (confidentiality, integrity, and availability) is ranked with a single provisional impact value (low, moderate, high) for each information type, to produce a Security Category (SC). That is, each information type has a SC that consists of ranks for each security objective. The RMF provides a general format for documenting each SC:

FIGURE 29.3 The second step of the Risk Management Framework, Categorize.

$$SC_{information\_type} = (\text{Confidentiality}, X), (\text{Integrity}, X), (\text{Availability}, X),$$
$$\text{where } X \text{ is } \{\text{low, moderate, high}\}$$

As an example, consider the information type of employee passwords in a company system. In RMF format, this SC would be described as:

$$SC_{employee\ passwords} = (\text{Confidentiality, high}), (\text{Integrity, moderate}), (\text{Availability, low})$$

This describes that when considering employee password security, security objective Confidentiality is ranked high, Integrity is ranked moderate, and Availability is ranked low. These rankings are made with respect to the SC, not with respect to one another. Note that there is no restriction to duplicate rankings; for example, if the organization deemed it appropriate, this SC could also be described as:

$$SC_{employee\ passwords} = (\text{Confidentiality, moderate}), (\text{Integrity, moderate}), (\text{Availability, low})$$

This suggests that the impact of confidentiality factors is only moderate when considering employee passwords. There is no correct method to generate SCes, and each organization will decide upon information types as well as their security objective rankings based on aspects that are deemed important to the organization. Ultimately, this ranking system helps to determine how to implement controls to best protect data, based on the balance between the impact and probability of an event occurring.

29.3.3 Security Categorization Review

The final step in RMF Categorization is a Security Categorization review. In an SC review, senior members at an organization review all the SCes to ensure that the concrete SCes align with the organization's greater strategy of managing risk. As part of this process, SCes with similar characteristics are grouped together and the highest impact values are taken as the information type group. For example, given two SCes that cover password authentication such as:

$$SC_{employee\ passwords} = (\text{Confidentiality, moderate}), (\text{Integrity, low}), (\text{Availability, low})$$

$$SC_{customer\ passwords} = (\text{Confidentiality, high}), (\text{Integrity, moderate}), (\text{Availability, moderate})$$

a generalized grouping of these two would be the following SC:

$$SC_{passwords} = (\text{Confidentiality, high}), (\text{Integrity, moderate}), (\text{Availability, moderate})$$

where each security object is the maximum impact value from the sub-SCes. Categorization produces a list of SCes with rankings of risk impact. The next RMF task is the Select step, in which controls are assigned to each SC.

29.4 SELECT

The third step of the RMF is the Select step. The purpose of the Select step is to select, tailor, and document the controls that will be used to protect the system and organization (Figure 29.4).

FIGURE 29.4 The third step of the Risk Management Framework, Select.

29.4.1 Control Selection

The initial subtask within Select is the selection of controls documentation. This subtask details the *initial* selection of document(s) pertaining to controls, based on the SCes generated in the previous step. There are two approaches that can be used for the initial selection: *baseline* or *organization-generated*. In a baseline controls selection approach, pre-defined sets of controls are identified and selected based on their application to the organization's SCes. Generally, the NIST SP 800-53r5 [800-53] publication is used for baselining. In federal organizations, it is required to use the NIST SP 800-53 baseline of controls. In an organization-generated approach, an organization uses its own selection process to select controls, which is advantageous when an organization has a highly specialized use-case that would not be captured in baseline controls documents such as the NIST SP 800-53. In either controls selection approach, the organization develops a set of security and privacy requirements that guide and inform the itemized selection of controls. These requirements can be gathered using a systems engineering process, similarly to the matter in which an organization gathers product requirements from a customer.

29.4.2 Control Tailoring

Controls, especially baseline controls such as are presented in the NIST SP 800-53r5 [800-53], are left general enough to maintain a high degree of applicability across domains. However, this generalization often means separating them from the concrete implementation details. That is why, following control selection, the controls must be tailored to suit the organization's concrete implementations of their systems.

Control tailoring is a general process that involves applying more detail to each control so that they better fit into an organization's systems. Because of the wide variety of different controls and control systems, there is no universal approach to tailoring controls. Each organization must decide the level of tailoring for themselves. This can take the form of combining baseline controls to match a complex system, applying scope considerations to existing baseline controls to capture specific use cases, or assignment parameters to baseline controls. The controls tailoring process is heavily dependent on the implementation of the systems that an organization has implemented.

29.4.3 Control Allocation

Following control tailoring, an organization proceeds to allocate controls to the system by designating each individual control as a system-specific, hybrid, or common. Each designation carries inheritance-related meaning; common designations refer to controls that satisfy security and privacy requirements that are inherited by one or more systems. In other words, the scope of a common control extends to other systems. A hybrid designation refers to controls that are partially inherited by other systems, and a system-specific designation refers to controls that are specific to one system only. These designations allow an organization to allocate controls to system elements in a manner that provides an adequate amount of control coverage per system element, as well as identify the controls that do not need to be applied to a system element. It is the organization's decision to decide how much control coverage and potential control overlap they implement.

29.4.4 Documentation of Control Implementations

Furthered documentation surrounding control implementations is performed after control allocation. This specific documentation subtask specifies that an organization builds a security and privacy plan that outlines the correct implementation of each control as well as how to assess the effectiveness of the control. Assessing the effectiveness of the control is as important as the implementation of the control, because the RMF is a continuous process and identifying controls that are outdated or ignored is important to the organization security. The control documentation constructed in this subtask describes planned inputs, expected behaviors, and expected outputs for hardware, software, and firmware. For mechanisms, such as key-card entry devices, the plan should include the functional specifications obtained from the manufacturer or vendor. There is no standard format for this plan, it is expected that the organization design the format of the plan to make it easy to implement in the future.

29.4.5 Continuous Monitoring Strategy

The RMF is not only concerned with the assessment and implementation of security controls but also concerned with continuous monitoring to gauge the effectiveness of those controls. Specifically, the continuous monitoring strategy should specify guidelines regarding frequency and the approach used to assess each control. At a system level, many of the expected control monitoring strategies may be consistent with an organization's existing continuous monitoring strategy, in which case the continuous monitoring strategy of the controls should simply be added to the existing continuous monitoring strategy. In cases where the organization does not have an existing continuous monitoring strategy, a document should be constructed that lists each control and specifies the frequency at which that control should be monitored, as well as the approach that should be used to perform the assessment. The nature of the approach will differ based on the details of the control. This document forms the continuous monitoring strategy.

29.4.6 Plan Review

As a final subtask in the Select step, the RMF requires a plan review. This plan review takes the form of a series of meetings with executive representatives from the organization, for the purpose of identifying if the plan satisfies security and privacy requirements for the system. If there are disagreements with respect to the plan, the appropriate changes need to be made. If all parties agree with the plan, the plan is approved, the set of controls and system requirements are finalized, and the RMF proceeds to the Implement step.

29.5 IMPLEMENT

As the name suggests, the Implement step within the RMF details the implementation and documentation of controls. This step is very simple in explanation, but will likely take the longest to complete, especially if many of the controls are being implemented from scratch (Figure 29.5).

29.5.1 Control Implementation

Control implementation refers to applying the control to its context. The RMF expects that best practices are followed per control implementation. Many times this takes the form of default configurations for each control, yet controls should be re-configured as necessary to be properly implemented. Some controls, such as commercial off-the-shelf (COTS) products, may not have the level of configurability at a system level, so in these cases, it is important to adequately test that the control is properly implemented. Several assessment facilities, such as the NIST Cryptographic Module Validation Program (CVMP) (NIST [CVMP], n.d.) can be used to assist with validating that certain modules are conformant.

During control implementation, system engineers should consider the adequacy of each control. Specifically, it may be determined that the planned control is insufficient to meet the security and privacy requirements for the system. This can happen if the control is poorly documented, or the act of implementing the control is more challenging than initially thought. In such a case, engineers can identify compensating or supplementary controls that will be implemented alongside the initial controls, or ignore the control and accept greater risk to their system, with acknowledgment and approval of the organization. It is important to remember that the RMF is a tool to manage and mitigate risk, not to eliminate it, and thus in some circumstances the benefits of not implementing a control might outweigh the cost of having it be compromised.

FIGURE 29.5 The fourth step of the Risk Management Framework, Implement.

29.5.2 Update Control Implementation Information

As controls are implemented, they should be documented. This documentation should take the form of updating the controls plan with any changes to inputs, behavior, or output of the control. The form of this documentation may look different for different controls, but it is vital to ensure that documentation is completed. Documenting the state of the implemented control ensures that the organization can track security and privacy impacts on the system accurately, while also providing visibility for cases where the control needs to change in the future.

29.6 ASSESS

The fourth step of the RMF is the Assess step. The purpose of the Assess step is threefold; to validate that the controls are implemented correctly, to ensure that they are operating correctly, and to verify that they are producing the desired outcome for the organization (Figure 29.6).

29.6.1 Assessor Selection

The first subtask within Assess is Assessor Selection, which is concerned with selecting the appropriate assessment team for a controls assessment. This team must be selected by considering the technical expertise with respect to control subject matters and the level of independence of potential team members. Technical expertise should be judged on skills applicable to control types (system-specific, hybrid, or common), knowledge of hardware, software, or firmware components, and general knowledge of risk management within the organization. These skills are important to consider because it is the determination of the assessment team members to judge the degree to which a control has been implemented correctly; frequently the successful implementation of a control will not be a binary outcome. The second team criterion is level of independence, which refers to the ability to conduct an impartial assessment. Assessments should be impartial, which

FIGURE 29.6 The fifth step of the Risk Management Framework, Assess.

means they are free from conflicts of interest with respect to the development and operation of the system, as well as the application of the controls. The RMF allows for independent assessors from third-party organizations, who provide excellent assessments at a high price tag.

29.6.2 Assessment Plan

The second subtask within Assess is to develop, review, and approve the assessment plan. An assessment plan consists of roles, responsibilities, control objectives, and assessment procedures for each control. The RMF does not specify the format for the assessment plan; organizations may choose a format that aligns to their internal documentation format. Regardless of format, items in the assessment plan must reflect the type of assessment the organization is conducting, per control. Assessment plans are then reviewed and approved by organizational authority to ensure that the plan is aligned with security and privacy goals of the organization. Items in the plan must be comprehensive while balancing resource investments; plans that feature total coverage of controls might be infeasible to implement because of time and cost resources. The goal of developing a plan is to ensure that the appropriate resources and directions are used with control assessment.

29.6.3 Control Assessments

After an assessment plan has been approved, the process of assessing all controls begins. Control assessment consists of determining the extent to which controls are implemented correctly, operating as intended, and producing the correct output. This assessment can occur in two phases; an initial assessment prior to system operation that validates the control, and a periodic assessment that may be performed frequently enough to satisfy the security and privacy requirements of the control. Control assessments may occur early in the system's development process, with preference being in the development phase. This allows control assessments to be incorporated (or modified) quickly to provide the greatest value to the organization; controls that are assessed as being incorrectly implemented or infeasible can be modified quickly with minimal cost. Control assessments that occur in the development phase are referred to as developmental testing and evaluation. In the case where the organization employs an agile development process, iterative controls assessment may be a lucrative path forward (i.e., a percentage of each sprint can be used to assess controls). Automation is desirable for control assessments, but many times it is infeasible to do so because of the complexity of the control or the nature of the control. For example, hardware controls pertaining to physical objects such as key-card scanners might be impossible to automate, while software controls pertaining to testing might be worthwhile to automate.

29.6.4 Assessment Reports

The RMF specifies how control assessments should be reported in the next subtask, assessment reports. The individual, or assessor, of the control assessment is primarily responsible for documenting the assessment report. The report should contain assessment results, which may take the form of pass/fail/custom based on the control criteria, as well as commentary on the effectiveness of the control. Ideally, each control has an associated assessment result in the report, yet controls that span the organization's lifecycle may be withheld until the assessment is complete. From a report management perspective, the RMF suggests that an organization can maintain one executive assessment report document while managing more granular control assessment results on a team or interim-based standing. However, an executive report document is preferable as it provides a single point of reference for the organization to use.

29.6.5 Remediation Actions

The next subtask in the RMF is remediation actions, which details how organizations fix any deficiencies recognized in the previous subtask, or any deficiencies that are discovered post-development. Largely, control deficiencies are divided into three categories; significant/unacceptable risk, quick-and-easy fixes, or inconsequential. In the least impactful category (inconsequential findings), if the organization finds the control assessment deficiency to be inconsequential, no remediation is suggested and the control is further evaluated in the next RMF step, Monitoring. In the cases where the control deficiency falls into the quick-and-easy fix, assessors can reassess the controls and apply the correct remediation strategy. The assessors can re-evaluate the control but should keep the findings from the initial assessment. The greater assessment plan should be updated if it needs to be. While it is difficult to provide a universal strategy for dealing with quick-and-easy fixes, organizations are expected to document everything and handle everything on a reasonable basis. Finally, control deficiencies that are deemed significant and have the potential to cause unacceptable risk are highlighted and the assessment plan is updated with new control plans to reflect the deficiency. Significant deficiencies do not always originate from oversight; new attack vectors and 0-day can cause a stable control to become deficient, in which case the organization needs to retain its documentation while providing new strategies to remediate it.

29.6.6 Plan of Action and Milestones

The final subtask within the Assess step is the development of a plan of action and milestones. The plan of action and milestones is a documented plan that details how the report findings are to be remediated. This plan describes actions that are planned to correct deficiencies from the previous subtask and includes (1) resources that are required to complete the deficiency correction, (2) the establishment of milestones that complete the task, and (3) the scheduled completion dates of the task. It is expected that the organization agrees upon the specifics of the controls' deficiencies, the criticality of the deficiencies, and the proposed risk mitigation strategy. This requires alignment from senior members of the organization; a new risk mitigation strategy may include new personnel, hardware, software, third-party support, or other expensive resources.

29.7 AUTHORIZE

At this point, the controls have been planned, implemented, and assessed, which means they are in place and working correctly. This leads to the sixth step of the RMF, Authorize, in which the organization selects a senior management official to authorize the system and formally accept the risk of implementing the system. As it is resource-infeasible and technologically impossible (because of unknown vulnerabilities such as 0-day) to protect the entire system, the organization must be prepared to accept the outcome of an attack on an unprotected asset; this is what accepting the risk means (Figure 29.7).

29.7.1 Authorization Package

As a first subtask in Authorizing a system, the organization constructs an authorization package document. This document includes security and privacy plans, assessment reports, plans of actions and milestones, and an executive summary. While these topics are required for the package, an organization can add additional information to the package if it is deemed worthwhile. Proper versioning and change controls should be incorporated into the package so that the package is up-to-date for audit or reauthorization actions. Automated tool support is highly encouraged for maintaining the package as it can become cumbersome to keep all supporting documents organized. When the

FIGURE 29.7 The sixth step of the Risk Management Framework, Authorize.

authorization package is prepared, the organization presents the package to authorizing officials, who make the informed decision to accept the risk of implementing the system.

29.7.2 Risk Analysis and Determination

Once the authorization package is prepared, it is presented to security officials within the organization who are tasked with providing a determination of risk. The determination of risk is a breakdown of risk with respect to organizational operations and assets, individuals, and other organizations and groups. The RMF does not specify a format for the risk determination, as each organization will have different systems as well as different expectations of risk. The determination of risk might include organizational risk tolerance, system/control dependencies, and impacts on system-critical components. Risk determination is used to assist authorizing officials with authorizing the system.

29.7.3 Risk Response

After risk has been determined, a risk response document is constructed. A risk response document details how organizations choose to respond to risk items from the risk determination subtask. An organization can choose to accept the risk, meaning they accept the outcome of such an attack, or they can choose to mitigate the risk. The risk mitigation route involves documenting planned mitigation actions, which are tracked using the plan of actions and milestones. Risk mitigation might include adding new controls, which requires a re-assessment of controls. When controls are re-assessed, the initial findings are not changed but rather updated in the same process as Control Assessment. Choosing to accept or mitigate a risk is a matter of organizational resources; it is impossible to protect against every threat so risks should be mitigated based on priority or potential damage to the organization.

29.7.4 Authorization Decision

After the authorization package has been prepared, risks have been documented, and risk responses elicited, an authorizing official of the organization chooses to accept the risk. This is not a light decision as the authorizing official takes responsibility for any damages due to attacks that occurred because of accepted risk. Such responsibilities include residual risk, which impacts dependent systems, organizations, and groups. The official must use all documentation available as well as any consultation with other organizational members and technical personnel for this decision. The authorizing official conveys terms and conditions for the authorization to operate, such as termination date (when the authorization expires) or time-driven control updates. The termination date must be agreed upon by the system owners or control owners. While the system is operating under ongoing authorization, the authorizing official is continually responsible for impacts of risk. The RMF does not specify a format for the authorization decision; instead it as well as the process of communicating the decision with the rest of the organization is left to the organization's choice. Finally, the authorizing official may deem that the risk is no longer acceptable for the organization. In such a case further risk response or full denial of the authorization is performed, in which case the RMF iterates to the required step;' for example, a new control might be necessary for the authorizing official to be comfortable accepting risk, in which case a new control would be added to the control plan.

29.7.5 Authorization Reporting

The authorization decision may be reported after it has been made. Reporting occurs only when the authorizing official is not in a senior position in the organization, which is not desirable from the RMF perspective but might be preferable for the organization. When this is the case, the authorizing official must build an authorization report that details each risk item, risk decision, and known system vulnerabilities. This report is presented to senior members within the organization to make organization decisions.

29.8 MONITOR

The final, yet vital, step in the RMF is Monitor, in which continuous monitoring is performed on the systems and organization to provide a continual feedback cycle to keep the organization situationally aware of its security and privacy position. Risk changes over time; threats, or 0-day, that were non-existent 1 day might become a critical threat the next. Furthermore, as business goals change and markets shift, products and systems change, which can cause controls to be outdated. Monitor seeks to organize and actionize the process of keeping the controls effective (Figure 29.8).

29.8.1 System and Environment Changes

Cyber systems and environments are in constant states of change. System elements, such as the hardware, software, or firmware of a cyber system, require updates to keep them functionally correct and patched. Furthermore, changes to the environment, such as changes to the human elements or changes to the physical facility in which the system resides, might cause certain controls to become irrelevant. While the Select step seeks to capture system and environmental changes and integrate them into the RMF plan, unforeseen changes might occur from a variety of sources, such as new technologies, inadvertent errors, or even purposeful attacks. When these changes occur, the organization must establish and document the root cause of the change. The organization's response to the root cause is addressed in subtask c, Ongoing Risk Response, yet because of the nature of this problem, there is no universal solution. If the change is caused by staff error, for example, training

FIGURE 29.8 The seventh step of the Risk Management Framework, Monitor.

courses might be mandated. If the change is caused by an attack, the nature of the attack must be established and configuration or system changes must follow.

29.8.2 Ongoing Assessments

The organization must establish a schedule for ongoing assessment of controls. The frequency of these assessments is not formally specified by the RMF, and organizations can choose the frequency based on their organization's needs and business concerns. In practice, many organizations perform ongoing assessments annually, but many argue for more frequent assessments. As part of ongoing assessments, organizations should use as many assessment results as possible; the assessments that occur during the authorization step, or even general system testing, are valuable in ongoing assessments. The organization is expected to generate a report for ongoing assessments; in classic RMF form, the RMF does not specify a format for a report of ongoing assessments, but the report might consist of control item, assessment performed, control effectiveness, and control relevance.

29.8.3 Ongoing Risk Response

As controls assessments are continuously performed to evaluate for relevancy, so are risk assessments. A risk assessment, or risk response, is the decision made with respect to risk, following a controls assessment update or when determined by the authorizing official. As risk is different from controls, a risk response does not take the form of hardware, software, or firmware, but rather a plan of action following an attack. In similar context to Authorization, risk responses can have the labels of acceptance or mitigation. An acceptance risk response means the risk is not worth the resources to mitigate, which means the organization understands and accepts the impacts such an attack would bring. Accepted risks are documented and continuously monitored, especially across changing risk factors. A mitigated risk response means that the risk is being reduced to acceptable levels through the implementation of controls. Mitigated risks are tracked via documentation, and any controls that are modified or extended are re-assessed to ensure that they have been implemented correctly and provide the correct output.

29.8.4 Authorization Package Updates

As ongoing assessments and risk responses are performed, the organization should update the authorization package that was prepared from the Authorize step. It is the responsibility of the organization to decide the frequency at which this package is updated. In an ideal world, these updates would occur as close to real time as possible, though organization management and resource utilization make this task difficult. Additionally, it is difficult to automate because many assessments and especially risk response items are manual tasks that change dynamically based on the system and new threats. However, automation should be preferred when possible because of the difficulty of tracking all assessments and risk response items.

29.8.5 Security and Privacy Reporting

The results from all monitoring activities should be documented and reported to chief officers within the organization, as well as pertinent individuals outside of the organization if applicable. The format of these reports is not specified by the RMF, but they should contain information alluding to the effectiveness of the controls, any ongoing monitoring activities, any risk information discovered during audits or controls assessments, and how system owners plan to address the risks. The greater goal of reporting is not necessarily to secure audit trails, but rather to facilitate communication between authorizing officials and senior members of the organization. The rate at which reporting is done is determined by the organization.

29.8.6 Ongoing Authorization

Separate from reporting, it is critical that the organization review the security and privacy posture of the system(s) to continually assess whether the accepted risk is still acceptable. This task is the responsibility of the authorizing official, who also must provide direction to the system owner or controls technician if the risk is no longer acceptable. If the risk is no longer acceptable, the authorizing official should issue an authorization denial, which ceases the operation of the system elements under consideration. These elements can be re-instantiated when their controls have been updated to a significant level. Automation is highly preferred for ongoing authorization, as automated tooling and dashboards can assist the authorizing official by providing visibility into the system.

29.8.7 System Disposal

In the event that a system is removed from operation, whether because it is deemed obsolete or de-authorized, the organization should have controls in place that address system disposal. Such plans must cater to the system; hardware systems may require destruction such as hard drive destruction, software systems may require modifying configuration management control, physical entry systems may require destruction, etc. When disposed of, the organization should record that the system has been removed from service. Because organizations have different systems, there is no one-size-fits-all solution to disposal, and disposal controls should be relied upon for direction. System disposal is addressed in the Monitor step of the RMF because it is monitoring that will reveal when a system needs to be removed.

29.9 CONCLUSION

Correctly implementing and maintaining the RMF is a difficult task, but the rewards are well worth the effort. The RMF is not a 'push button deployment' solution, meaning that a successful RMF iteration is much more complicated than the issuance of a single command to fire off a chain of events that completes it. Instead, the RMF requires continuous diligence and astuteness from all aspects of an organization; executives, officials, and engineers or technical users. The RMF requires continual updates and monitoring to track system updates, system obsolescence, new attack vectors, dynamic marketplace ventures, etc. Because of these reasons, the RMF may not be suitable for all organizations. Smaller organizations such as start-ups or organizations that value flexibility while settling on identifying a market niche might find that the overhead of the RMF is more costly than the value it provides. However, the value of the RMF does not only fall in its literal outcome (an extensive risk management plan) but also in its ability to assemble an organization. The extensive documentation requirements of the RMF coupled with its requiring communication and agreement from members across all levels of an organization are great boons. For established organizations that require cyber aware decision-making, especially nation-critical organizations, the RMF is an excellent framework to manage security and privacy.

NOTE

1 https://csrc.nist.gov/publications/detail/sp/800–37/rev-2/final

REFERENCES

DOD INSTRUCTION 8510.01 (n.d.). [dodi-8510]. https://www.esd.whs.mil/Portals/54/Documents/DD/issuances/dodi/851001p.pdf.

NIST [53] (n.d.). https://csrc.nist.gov/Projects/risk-management/sp800-53-controls.

NIST [60] (n.d.). https://csrc.nist.gov/publications/detail/sp/800-60/vol-1-rev-1/final.

NIST [CVMP] (n.d.). https://csrc.nist.gov/projects/cryptographic-module-validation-program.

NIST [rmf] (n.d.). https://nvlpubs.nist.gov/nistpubs/SpecialPublications/NIST.SP.800-37r2.pdf.

30 Information Assurance, Vulnerability Analysis, and Remediation

William R. Simpson

30.1 INTRODUCTION

The sheer volume of regulations, recommendations, and analyses to reduce cyberspace threats is astounding (US Department of Defense, 2014 and 2021, NIST, 2015, Common Criteria, 2017), are just a few. Threat mitigation is undertaken to reduce the attack space and to minimize losses due to cyber activities either malicious or accidental. From Hacks to Nation-States, these threats continue to attempt to penetrate networks every day. These threats are growing, evolving, and sophisticated. Loss of information capability and information integrity as well as the loss of intellectual property is a significant security risk.

The goal of this chapter is to describe a process to mitigate many known vulnerabilities and to be responsive in mitigating new or newly discovered vulnerabilities. Vulnerability identification is a continuous process whereby threats are reviewed and prioritized with mitigation through either the flaw remediation process, modifications to the operating environment, or the reliance on multiple products in mutual support of one another. The analyses are reviewed either periodically or on demand as new threats evolve or are identified.

30.2 VULNERABILITY NUMBERS AND SEVERITY

Cases of vulnerabilities are numerous. The Common Weakness Enumeration (CWE), Mitre (2021b), was developed to track vulnerabilities as a guide to their elimination. Vulnerabilities are of concern because they may be turned into exploits that in turn can be used to disrupt IT systems or ex-filtrate their assets. The number of vulnerabilities has increased dramatically and continues to increase over time, showing the importance of patching continuously. Figure 30.1 shows the trend of numbers as derived from NIST (2021). Figure 30.1 is provided with permission from The NIST National Vulnerability Data Base.

Figure 30.2 shows the more common causes of vulnerability among software development categories for the top 15 categories from MITRE (2021a). Figure 30.2 is derived from data provided with permission by the Mitre Common Vulnerability and Exposures Database.

The data in Figure 30.2 indicates that for software development vulnerabilities are fairly equally distributed among the top 15 areas. From a risk standpoint, everything is fair game. Vulnerabilities do not necessarily lead to exploits. According to MITRE, "Categories (which are not technically weaknesses) are special CWE entries used to group weaknesses that share a common characteristic. Classes are still very abstract, typically independent of any specific language or technology. Base-level weaknesses are used to present a more specific type of weakness. A chain is a set of weaknesses that must be reachable consecutively in order to produce an exploitable vulnerability. While a composite is a set of weaknesses that must all be present simultaneously in order to produce an exploitable vulnerability."

DOI: 10.1201/9781003188377-36

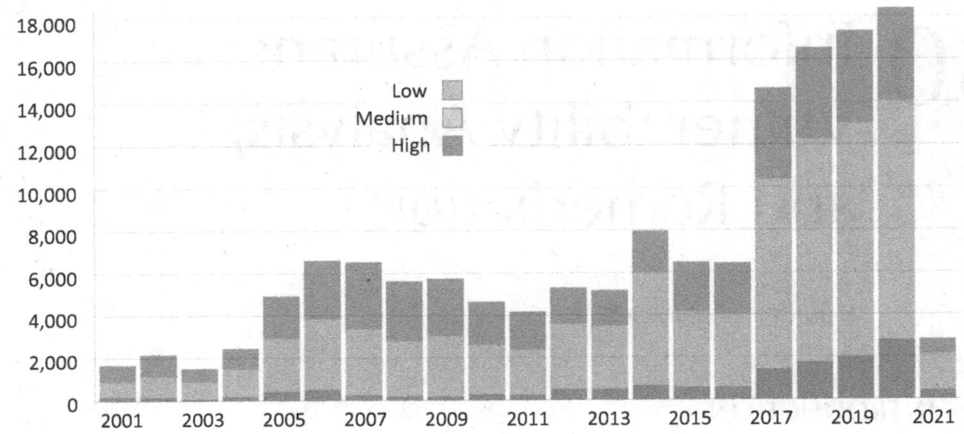

FIGURE 30.1 Vulnerabilities numbers over time.

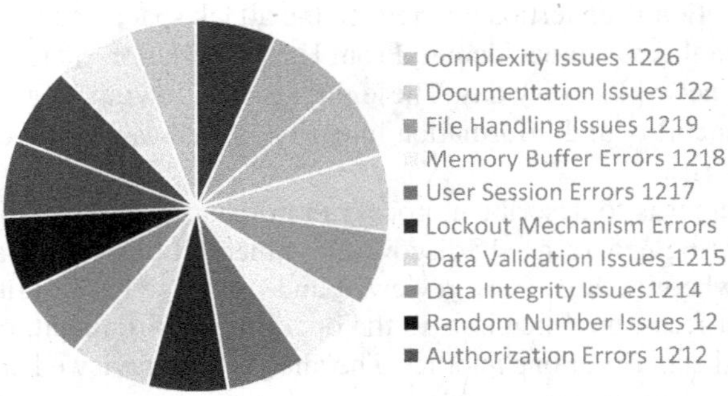

FIGURE 30.2 Software development vulnerabilities by type.

FIGURE 30.3 Vulnerabilities word frequency chart.

Figure 30.3 shows the more common causes of vulnerability among vulnerability descriptions as derived from NIST (2021). Figure 30.3 is provided with permission from The NIST National Vulnerability Data Base.

The identification of vulnerabilities as they relate to software allows software development practices and tools to look for structures and processes in software that may be exploited. Many of these tools will check for susceptibility to specific exploits also. Proper software development practices will eliminate many of these vulnerabilities. Formal design practices offer additional capabilities to avoid buffer overflows, improper data acceptance, and other software flaw-based vulnerabilities, Jones (2010). Nonetheless, tools must be used to verify that they have been properly done, or remediation must be applied. These tools are not perfect, and the analysis may miss some vulnerabilities. Their use should be followed by penetration testing. Even then, other security measures are needed within the enterprise. Those additional measures will not be discussed in this chapter. These analyses, however, will reduce the attack space and eliminate some common attack methodologies for the enterprise.

30.3 BASIC EXPLOIT CLASSES

Malware comes in various forms. Viruses and Worms are downloaded programs that can copy themselves and spread those copies including, in some cases, modified copies. To be called a virus or worm, inserted exploit programs (often called malware) must have the ability to reproduce itself. The difference is that a worm operates without outside support, whereas a virus needs a host program to spread itself. Ransomware captures and encrypts data from the targets data stores and either threatens to publish it or destroy it if demands are not met (usually money – ransom). We will discuss these and other exploits in the following paragraphs.

30.3.1 Viruses and Anti-Virus Programs

Viruses are attached to a program or application (WebRoot, 2021). The most common exploit is the human one where clicking on a link or opening an attachment is an entry point. Viruses propagate by inserting a copy of their code (or a modified copy of their code) into and becoming part of a specific program or application. Viruses get activated by the execution of the host program or application. The virus moves from one computer to another, leaving infections where it can find the host program or application. Viruses may cause mild effects, like pop-up ads, or spinning off tracking cookies, or they may maliciously damage software or data and may cause denial-of-service (DoS) conditions. When the host program or application is executed, the viral code may be executed as well. One defense is to block active elements in the code of the host program. For example, Microsoft Office Programs will block active code, but allow the user to override this blocking. Many viruses are spread because individuals do not understand the consequences of overriding these active elements. The host program or application may keep functioning after it is infected by the virus. Some viruses overwrite other programs and applications with either themselves or other material that is not executable, which renders the program or application unusable. Viruses spread when the program or application or a related document is transferred between computers. This can be done by attachments in emails. Infecting a disk that is moved to another computer, transferring files within the network, or, infecting a file that is moved to another computer. Keeping virus scanners up to date and frequently scanning files is a defense against unknown viruses. Monitoring whom you receive files from is also a defense.

30.3.2 The Life-Like Worms

Computer worms also copy themselves. A worm enters a computer through an exploit by using one or more weaknesses and vulnerabilities in the system (Fruhlinger, 2019). Worms will not be obvious and if you pull up a list of programs and applications running, they may masquerade as a familiar application in that list. They can cause damage similar to viruses and are often hard to distinguish from viruses. However, they do not require a program or application. Worms are entire programs. To move, worms either exploit a weakness or vulnerability in the system or use a trick on the human operator. Often these tricks are termed social engineering. The trick gets the human to execute the program by clicking on an attachment or opening a website. It may use file or information transport commands on the system, allowing it to travel without assistance. Worms may be configured to use encryption, wipers, and ransomware technologies to harm the system. Many virus scanners will find and identify known worms. Keeping virus scanners up to date and frequently scanning files in a defense against unknown viruses. Avoiding links in emails will improve your defenses also. Links should be saved in favorites, or typed in by hand when needed.

30.3.3 The Many Classes of Ransomware

Ransomware is a whole class of exploit software (McAfee, 2021). Ransomware may be a worm or a Trojan. The software may be a Trojan that communicates with a hacker. The hacker can make threats such as releasing your information or destroying your information. The Ransomware may have encrypted your data or made it hard to find. Paying the ransom may or may not get your data released. Vendors sell end point defense systems to protect against ransomware. One defense is the frequent backup of files to another system. This way files that are compromised are replaceable.

30.3.4 Trojans

A Trojan is a software that communicates with a hacker (Avast, 2021). The hacker can then direct the software to do various things such as mapping the system in preparation for a wider scale attack. The Trojan may also assist the hacker in downloading additional malicious tools. It derived its name from the wooden horse that the Greeks used to infiltrate Troy, and it is often downloaded because it looks attractive or comes with something that looks interesting or attractive. It is a dangerous piece of software that looks legitimate. Once it is executed, it will set up a communication with the hacker. The hacker achieves numerous attacks on the system, from trivial display changes to exfiltration of data, destruction of data and files, and importing other viruses and Trojans. Trojans do not copy themselves, they do not infect other programs or applications. Trojans are initiated by interacting with a user including downloading a file from the Internet or clicking on an email attachment. A good defense is to be very careful when it comes to email attachments or unknown websites. Careful control of ports and protocols through two-way firewalls will reduce the likelihood of a Trojan being able to set up a link with the hacker.

30.3.5 Phishing

Phishing is an attacker sending emails that appear to be from a reputable sender with the goal of getting you to share sensitive information (Bisson, 2020). The sender's address may be spoofed, or it may come from a compromised computer. It may simply ask you to reply with a piece of information. Often, an attachment or a link may invite you to execute. The attacker seeks information that can be used in further exploits. This may be bank account details, credit card information, login credentials, or other personal information. They may want your contact list to send phishing attempts to your contacts. Phishing attacks are usually from a generic mass email. A more personalized form is called spear-phishing, a spear-phishing attack is personalized with specific details about the message's recipient. The attackers may have gathered personal information from someone else's contact list, the dark web or your social media accounts to craft messages that appear to come from trusted sources—like your own bank, or credit card, or even individuals that you know. They use these personal details to trick you into taking an action that could cause you to share even more personal information.

The best defense is to never open an attachment or click a link unless you have a separate communication from a trusted source indicating that they are sending the email you see. If it appears from a trusted source and you were not expecting it, contact the apparent sender to verify they sent the email. This extra step can save a lot of grief.

30.3.6 Robots, Bots and BotNets

A robot is a piece of software or an automated process that will execute a work flow or a script, often in concert with other services on the network or in the system. Bot is an abbreviation for robot. Bots are used to automate repetitive tasks retrieving information or exercising services that a human would be doing in sequence. A web crawler is a bot that gathers information, from websites, or emails and places it somewhere to be picked up by a user or another bot. The bot is not intrinsically a threat. It can be used for good or bad. A malicious bot will copy itself and occupy as many systems as possible (SectigoStore, 2020). Often they sit there awaiting a trigger, but they can simply be used to do background processing on a system. Such things as hosting nefarious sites, or generating spam emails. When a large number of bots in various systems and networks act in concert, we have a BotNet. The BotNet has the power of numbers and bandwidth and can be used to outperform the masses on social media or shut down target websites by flooding them with traffic. The latter is often termed Distributed Denial of Service (DDOS). The compromise of many systems is used to attack a few systems. Defenses include traffic filtering and using cloud resiliency to expand bandwidth for the targeted systems. These are expensive. The long-term solution is to hunt down and destroy malicious BotNets and avoid their congregation by increasing the overall hygiene of computer systems. BotNets may be established in small individual devices on the network, termed Internet of Things (IoT). This would include commercial HVAC devices or home electronics. The IOT devices often do not have enough computational resources to provide adequate security defenses and an incoming bot will displace what resources are available.

30.3.7 Defending Against These Exploits

The malicious entity that inserts a virus, worm, Trojan, or ransomware must gain control of your system with enough privilege to insert the virus into an application or place the Trojan on your system. Often the human is the weak link. However, there is a large catalog of techniques or vectors that do not use human weaknesses and are used to gain control by exploiting weaknesses in the underlying systems and programs. The known weaknesses are cataloged and made available by MITRE Corporation, Mitre (2021b). There are three basic defenses:

1. Eliminate human weaknesses through training and education.
2. Analyze for weaknesses and vulnerabilities. These may be combined to become exploits. This vulnerability analysis will be described in detail.
3. During operations, constantly be vigilant against new weaknesses and vulnerabilities. They may be discovered by researchers or reported by users. This is called Flaw Remediation and will be discussed in detail. After flaw remediation, many vendors issue patches to the software or recommend system configuration changes. Pay careful attention to these. Many software patches are automatically sent out, but that is no guarantee as we have seen with the *SolarWinds* which co-opted the supply chain and gave false updates to system software. The average user has no defense against supply chain vulnerabilities.

30.3.8 Zero-Day Exploits

A zero-day exploit is an exploit that combines weaknesses and vulnerabilities that have not been previously seen or anticipated (Norton, 2019). Attackers are always innovating and trying new methods. Complexity is the problem, see below. While defense needs to be all-inclusive, the bad guys only need to find one way to enter your system. There is no defense if you have the ingredients for the zero-day exploit. The ingredients may be a combination of software or operating systems, interfaces, and security methods. If you are a victim of a zero-day exploit, you can only be resilient and recoverable. Both of these factors are discussed below. Once a zero-day exploit is used and reported, several things happen. The organizations that track vulnerabilities and weaknesses include these in their reporting. Often a work around is provided that breaks up the combination of factors that allow the exploit. This is why registering your software purchases is important so you can be notified of a work around and/or an update. Next, software developers work on fixes to patch the problem. This is one reason patches are so important. The exploit that is patched will be prevented, But caution is advised here because if the software is very complex, it may open yet other unforeseen vulnerabilities and weaknesses, see complexity below. In general, a not too complex piece of software, such as your accounting program should always be patched. More complex software and hardware systems, such as your operating system are more problematical. Many Industrial Control Systems – ICS – refuse to patch operating systems for this reason. The best defense is simplicity and assure every software system in your environment is needed for a real purpose, and sometimes convenience alone is not a real purpose.

30.3.9 Supply Chain Exploits

We have recently seen a level of exploitation that is beyond our ability to control. Many software packages today, use a build process that brings together software from open repositories and creates a software package that provides common services. Recently we have seen that those repositories may be compromised – SolarWinds (FireEye, 2020). These supply chain issues make a much larger concept of the software system to include dependencies that are related to software builds and updates. It is clear that these processes need to be well-defended and where they are not widespread exploits, across many systems will happen. Defense is two-fold, the first is the reduction of complexity, and the second is establishing a resiliency process.

30.3.10 Complexity as a Factor

The software landscape is becoming increasingly connected. The interconnections pose vulnerability problems that cannot be controlled by the casual user or even the enterprise. This complexity opens us all to uncontrolled vulnerabilities, weaknesses, and exploits. The best defense is to minimize this complexity and limit interconnectivity where possible. Each element in your software system must be analyzed by you on its necessity. If it is deemed necessary, you must further analyze the update process and decide whether or not an automated update is a requirement. Each update may bring security fixes and security flaws. At some point, this is a losing proposition. The minimal system is the safer system. Even then, exploits will happen, software will steal data and render data useless, so we must be resilient and recoverable.

It is important that in both cases, backup services be stored separately from the network and be encrypted and protected.

30.3.11 System Resilience and Recovery

In information technology, the resilience approach focuses on protecting core services and preventing issues before they occur, if possible (Stoever, 2018). This can involve identifying the risks and vulnerabilities associated with the services that support critical business processes and performing detailed risk assessments of the impact of an outage. It may also include backup systems put into place when primary systems are damaged. In cases where breeches occur, services outside of the core services are disabled and quarantined while they are being examined, analyzed, and repaired. Core services, if suspect may be reverted to earlier or alternate versions.

In information technology recovery is the act or process of regaining or repossession of something lost (Imperva, 2021). This includes capability as well as data. This latter is achieved by frequent, if not continuous backup of data, maintenance of previous versions of critical software for reverting to a previous state, and a set of tools that allow the system to be restricted to its previous state and backup data to be brought in to overwrite damaged or missing data.

30.4 RELATED WORK

Vulnerability analyses are usually coupled with software quality assurance and fall into three main categories:

1. Static code analysis
2. Dynamic analysis or execution tracing
3. Penetration testing

30.4.1 Static Code Analysis

Static program analysis is the analysis of computer software that is performed without actually executing programs. In most cases, the analysis is performed on some version of the source code and in other cases some form of the object code. The term is usually applied to the analysis performed by an automated tool, with human analysis being called program understanding, program comprehension, or code review. The sophistication of the analysis performed by tools is rule-driven which provides the intellectual property associated with the tool set. A growing commercial use of static analysis is in the verification of properties of software used in safety-critical computer systems and locating potentially vulnerable code.

These types of analyses are discussed in Ruiz-Rube et al. (2020), Bavishi et al. (2019), and Marcilio et al. (2019).

30.4.2 Dynamic Code Analysis

Dynamic program analysis is the analysis of computer software that is performed by executing programs on a real or virtual processor. For dynamic program analysis to be effective, the target program must be executed with sufficient test inputs to produce interesting behavior. Care must be taken to minimize the effect that instrumentation has on the execution (including temporal properties) of the target program. The sophistication of the analysis performed by tools is both rule-driven and instrumentation (debug monitors) which provides the intellectual property associated with the tool set. Some dynamic code checker tools are given below:

HP Security Suite is a suite of Tools at various stages of development (HP, 2021).

IBM Rational is a suite of application security solutions targeted for different stages of the development lifecycle. The suite includes two main dynamic analysis products – (IBM, 2021).

Intel provides a complete line of security products, (Intel, 2021).

Additional dynamic code analysis data is given in Coulter et al. (2020) and Schnoor and Hasselbring (2020).

30.4.3 Penetration Testing

A penetration test is a method of evaluating computer and network security by simulating an attack on a computer system or network from external and internal threats. The process involves an active analysis of the system for any potential vulnerabilities that could result from poor or improper system configuration. The analyses include both known and unknown hardware or software flaws or operational weaknesses in process or technical countermeasures. This analysis is carried out from the position of a potential attacker using a threat model and can involve active exploitation of security vulnerabilities.

Jevtic (2019) provides a definition of type of penetration testing (White Box Tests, Blind Tests, Double-Blind Tests, External Tests, and Internal Tests) and an evaluation of a number of tools (Netsparker, Wireshark, Metasploit, BeEF, John The Ripper Password Cracker, Aircrack, Acunetix Scanner, Burp Suite Pen Tester, Ettercap, W3af, Nessus, Kali Linux, SQLmap, (SET) Social Engineer Toolkit, Zed Attack Proxy, Wapiti, Cain & Abel).

In the analysis of requirements penetration testing and both static and dynamic analyses to set up the penetration testing are required. Further, the work is only begun with these analyses as the discovery of vulnerabilities and exploits once the software is fielded must also be submitted to the flaw remediation system and resolved as quickly as possible. To get ahead of this cycle many software vendors now offer rewards for vulnerability and exploit discovery (Bugcrowd, 2021).

Current literature on penetration testing is included in Friesen and Vonk (2021), Ghanem and Chen (2020), Creasey and Glover (2019), Almubairik and Wills (2016), Applebaum et al. (2016), and Sarraute et al. (2019).

30.4.4 Resilience Testing

Resilience testing is a type of software testing that conducts test on applications under stress and measures how applications perform in chaotic conditions without a loss of core functions or data; it ensures a quick recovery after unforeseen, uncontrollable events (Mamnani, 2018).

Resilience testing includes (among others); conducting a failure mode analysis, validating data resiliency, configuring and testing health probes for load balancing and traffic management, conducting fault injection tests for every application, and validating that there is an automated rollback mechanism for code in production. Additional resources are at Vogels (2016), Firesmith (2020), and Burns (2020).

30.4.5 Data and Software Recovery

The best way to recover corrupted software and data is to reinstall the software from a backup copy, preferably the previous version saved in a separate file. Many operating systems offer a restore function to facilitate this type of recovery. If the applications are in the cloud, it is the responsibility of the cloud provider to restore software systems.

For data recovery, the same is true that a copy of the data is backed up to a secondary site that is unlikely to be corrupted at the same time the primary system is corrupted. Cloud recovery of data is essentially by the backup method. Absent that type of recovery, several firms will offer a way to attempt to recover corrupt file. "In computing, data recovery is a process of salvaging inaccessible, lost, corrupted, damaged or formatted data from secondary storage, removable media or files, when the data stored in them cannot be accessed in a usual way" – Wikipedia

Further reading can be found at Mendelson (2018), Turner (2021), and Wiltshire (2013).

30.5 VULNERABILITY ANALYSIS

The vulnerability analysis requirement is levied on all software procured for high-assurance environments. This is true for commercial products, as well as legacy systems to be ported to the high-assurance environment, and assumes that the developer (either product producer or the enterprise representative through a custom developed contract) will perform these analyses and provide the flaw remediation [an adaptation of Common Criteria, 2017].

While formal methods in the development process and the obtaining of software development credentials such as Capability Maturity Model (CMM), (White, 2018), are valuable, they cannot be used to replace the vulnerability analyses. In an adversarial procurement environment, where the developer either cannot or will not perform these analyses, the developer must agree to allow the procuring agency to perform these analyses or hire a competent laboratory to do the analyses.

These costs, if not included in a competitive bid will be added to the price for the evaluated product when evaluating alternative products for procurement. The consumer should purchase only products that are so evaluated. They will probably cost a little more than competitors, but will be well worth it in the long run.

The level of effort will depend on the product complexity. Execution of the vulnerability analyses does not require the execution of a full Common Criteria evaluation, although any such Common Criteria evaluation should include these analyses. The vulnerability analysis described here is based on the product in its operational environment. The flaw remediation process cannot be performed by anyone other than the software developer. It is assumed that the license or purchase of a product will include a support contract that includes the flaw remediation provisions if these are not provided with the product.

Within each country, a number of competent laboratories exist to do these analyses. These laboratories have tools and testing techniques for penetration testing and will provide a level of assurance on the threat mitigation that will lead to informed decisions. Many are able to conduct these analyses at higher security classification levels and have the capability to sign non-disclosure, non-complete arrangements with the software developers. Contracting for these analyses would include a competitive procurement where the developer is unable or unwilling to perform such tasks. The consumer should make the purchase of such software and/or equipment depending upon the successful completion of the vulnerability analysis. Prior vulnerability analysis reports may be used to the extent that the operational environments are similar or modifications to the prior analysis are completed and executed.

In the US there are seven laboratories that are licensed under the National Voluntary Laboratory Accreditation Program (NVLAP) Program, (NIST, 2021). NVLAP provides third-party accreditation to testing and calibration laboratories. NVLAP's accreditation programs are established in response to Congressional mandates, administrative actions by the US Government, and requests from private-sector organizations and government agencies.

NVLAP operates an accreditation system that is compliant with ISO/IEC 17011, Conformity assessment — General requirements for accreditation bodies accrediting conformity assessment bodies, which requires that the competence of applicant laboratories be assessed by the accreditation body against all of the requirements of ISO/IEC 17025, General requirements for the competence of testing and calibration laboratories.

30.5.1 Objective of the Vulnerability Analysis

The objectives of a vulnerability analysis are:

1. To determine whether the product(s), in its operational environment, has easily identifiable exploitable weaknesses and vulnerabilities.
2. Identify those weaknesses and vulnerabilities.
3. Begin a flaw remediation process that will close those vulnerabilities.
4. Excessive vulnerabilities may disqualify the product for enterprise use and may compromise marketability of commercial products.

Figure 30.4 shows the basic vulnerability analysis flow.

30.5.2 Required Information for Vulnerability Analyses

The information required for vulnerability analyses includes:

- The Product(s) themselves.
- The guidance documentation provided with the products.

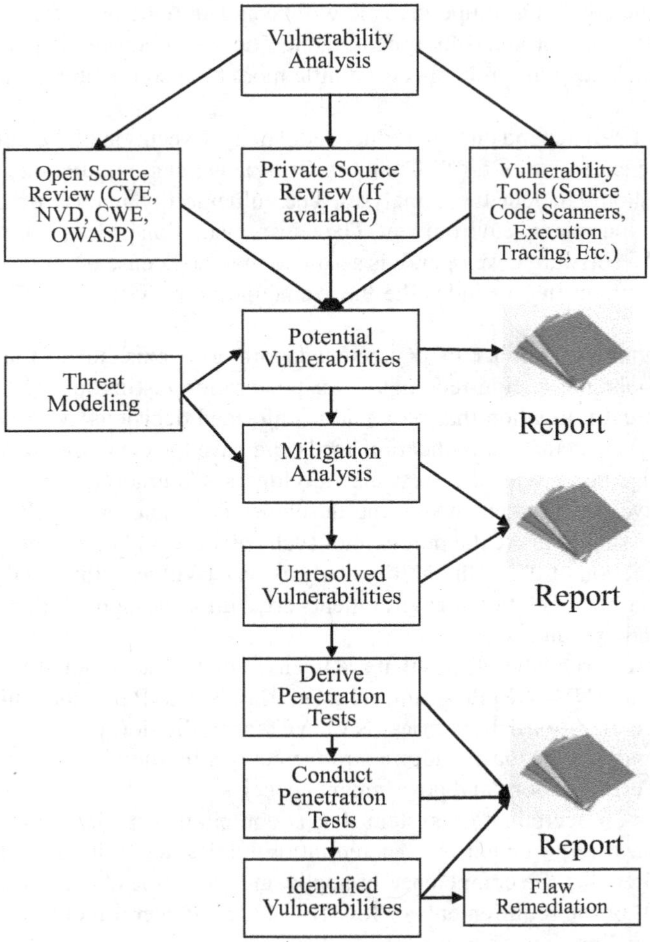

FIGURE 30.4 Vulnerability analysis process.

- Identification of all interfaces to the product, physical and functional (software).
- Harnesses and software instrumentation are necessary for testing the product.
- Information publicly available to support the identification of potential vulnerabilities. Information that may be used in these analyses is listed below:
  - Common Vulnerabilities and Exposures (CVE®) (Mitre, 2021a) is a dictionary of common names (i.e., CVE Identifiers) for publicly known information security vulnerabilities. CVE's common identifiers make it easier to share data across separate network security databases and tools and provide a baseline for evaluating the coverage of an organization's security tools. If a report from one of the security tools incorporates CVE Identifiers, the tool may quickly and accurately access remediation information in one or more separate CVE-compatible databases to remediate the problem.
  - National Vulnerability Database (NVD) (NIST, 2021) is the U.S. government repository of standards-based vulnerability management data represented using the Security Content Automation Protocol (SCAP). This data enables automation of vulnerability management, security measurement, and compliance. NVD includes databases of security checklists, security-related software flaws, misconfigurations, product names, and impact metrics.

- Common Weakness Enumeration (CWE™) (Mitre, 2021b) International in scope and free for public use, CWE provides a unified, measurable set of software weaknesses that enables more effective discussion, description, selection, and use of software security tools and services that can find these weaknesses in source code and operational systems as well as better understanding and management of software weaknesses related to architecture and design.
- Others include the Open Web Application Security Project (OWASP, 2021).

The product developer or evaluator performs additional tests as a result of potential vulnerabilities encountered during the conduct of other parts of the evaluation. The use of the term guidance in this process refers to the operational guidance and the preparative guidance. Potential vulnerabilities may be in information that is publicly available, or not, and may require skill to exploit, or not. These two aspects are related, but are distinct. It should not be assumed that, simply because a potential vulnerability is identifiable from information that is publicly available, it can be easily exploited.

30.5.3 Obtaining Vulnerabilities

The product developer or evaluator examines sources of information publicly available to identify potential vulnerabilities in the product. There are many sources of publicly available information, which should be considered, but a minimum set is listed above as input to the analysis. The product developer or evaluator should not constrain his consideration of publicly available information to the above, but should consider any other relevant information available.

The product developer or evaluator records in an evaluation report the identified potential vulnerabilities that are applicable to the product in its operational environment. The product developer or evaluator may use manual methods or source code scanning and execution tracing tools, or all in this analysis, but minimal coverage of the above listed sources of vulnerabilities and weaknesses are 100%. Each of the known vulnerabilities identified as appropriate to the product in its operational environment is assigned a category and rationale as follows:

- No further consideration of the potential vulnerability is required if for example the product developer or evaluator identifies mitigations in the operational environment, either IT or non-IT that prevent exploitation of the potential vulnerability in that operational environment. This may include mitigations within the product itself.
- For any reason, the potential vulnerabilities may be excluded from further consideration if the potential vulnerability is not applicable in the operational environment.
- Otherwise, the evaluator records the potential vulnerability for further consideration. This list of potential vulnerabilities applicable to the product in its operational environment, which can be used as an input into penetration testing activities, is reported in the evaluation report.

30.5.4 Deriving Penetration Tests

The product developer or evaluator derives penetration tests that are based on the search above for potential vulnerabilities, threat modeling activities, and other analysis methods. The product developer or evaluator prepares for penetration testing as necessary to determine the susceptibility of the product, in its operational environment, to the potential vulnerabilities identified during the search of the sources of information publicly available.

The product developer or evaluator produces penetration test documentation for the tests based on the list of potential vulnerabilities in sufficient detail to enable the tests to be repeatable. The test documentation includes:

- Identification of the potential vulnerability the product is being tested for;
- Instructions to connect and setup all required test equipment as required for conducting the penetration test;
- Instructions to establish all penetration test prerequisite initial conditions;
- Instructions to initially set up the product, instruction sequences, and exploit attempts for the product;
- Instructions for observing the expected behaviors and actual behaviors of the product;
- Descriptions of all expected results and the necessary analysis to be performed on the observed behavior for comparison against expected results;
- Instructions to conclude the test and establish the necessary post-test state for the product.

The product developer or evaluator conducts penetration testing based on the list of potential vulnerabilities identified above and prepares an evaluation report on these tests. Where, as a result of evaluation, the product developer or evaluator discovers a potential vulnerability, this is reported in the evaluation report as a residual vulnerability. The vulnerability is reported as a flaw in the flaw remediation system with a priority commensurate with its potential for exploitation and the consequences of a successful exploit of the vulnerability.

30.5.5 Continuous Updating

The product developer or evaluator re-examines sources of information publicly available to identify potential vulnerabilities in the product either periodically or on demand. The analysis may be on demand when critical vulnerabilities or damaging exploits in similar products have been identified, changes in the operational environment are made, or other changes require further analysis. This continuous updating should proceed for the active life of the product in the commercial market.

30.5.6 Review and Approve

The enterprise or product representative examines the results of the evaluation report of actions above and of all penetration testing to determine that the product, in its operational environment, is resistant to an attacker appropriate to the high-assurance environment. If the results reveal that the product, in its operational environment, has vulnerabilities exploitable by an attacker appropriate to the high-assurance environment, then remedial action must be taken by the product developer. The enterprise or product representative ensures that periodic re-evaluation as provided above is undertaken.

For critical and trusted software such as the Secure Token Servers or services within the Enterprise Attribute Stores, the enterprise or product representative may, at his option, conduct or procure independent penetration testing as described above, including repeating some of the testing described and independently derived tests. The results of these tests may lead to the identification of vulnerabilities that are subject to the flaw remediation processes described below.

30.6 FLAW REMEDIATION

The flaw remediation process is the responsibility of the product developer [an adaptation of Common Criteria, 2017, Evaluation Methodology]

30.6.1 Flaw Remediation Objectives

The objective of this flaw remediation is to demonstrate that the product developer has put in place procedures that:

- provide for the tracking of exploitable weaknesses, exploitable vulnerabilities, and/or other security flaws,
- the identification of corrective actions,
- the distribution of corrective action information to product users.

Additionally, this process demonstrates that the product developer's procedures provide for:

- the corrections of security flaws,
- for the receipt of flaw reports from product users,
- for assurance that the corrections introduce no new security flaws,
- for the establishment of a point of contact for each product user, and
- for the timely issue of corrective actions to product users.

The administrative tracking and reporting of flaws may be outsourced, but the developer must be involved in the corrective actions and must be included in the developer's quality control system.

In order for the developer to be able to act appropriately upon security flaw reports from product users, product users need to understand how to submit security flaw reports to the product developer, and product developers need to know how to receive these reports. Flaw remediation guidance addressed to the product user ensures that product users are aware of how to communicate with the developer; flaw remediation procedures describe the product developer's role in such communication.

The Flaw Remediation process as applied to security issues is shown in Figure 30.5.

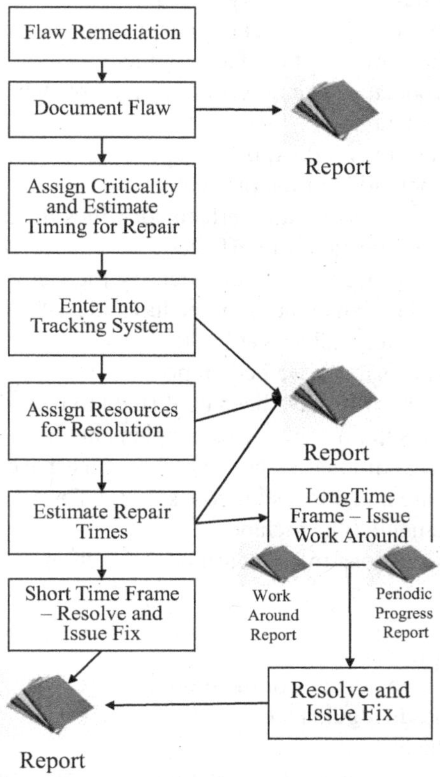

FIGURE 30.5 Flaw remediation process.

30.6.2 Information Required for Flaw Remediation

The required information for the evaluation of flaw remediation processes is:

- The flaw remediation procedures documentation;
- Flaw remediation guidance documentation.

The product developer may use automated flaw remediation tracking systems such as trouble ticketing software or may manually track such flaws. However, it is expected that any flaw not remedied with 10 working days are referred to the quality assurance tracking system for formalized assignment of actions and priorities, and timely reviews to assure progress in resolution of open items. Security flaws normally have the highest assigned priority for remediation in high-assurance systems. Remediation may be code changes and patch update, configuration changes and recommended changes to configurations, or other remedies outside of the product in the operational environment. The enterprise or product representative may reject the last approach depending on its impact.

30.6.3 A Flaw Remediation Process

The product developer provides a flaw remediation process:

- The procedures describe the actions that are taken by the product developer from the time each suspected security flaw is reported to the time that it is resolved. If a flaw is discovered not to be security-relevant, there is no need (for the purposes of the Flaw Remediation requirements) for the flaw remediation procedures to track it further; only that there be an explanation of why the flaw is not security-relevant.
- The procedures identify the actions that are taken by the product developer to describe the nature s of each security flaw and the effects on the operation of the product in sufficient detail to be able to reproduce it. The description addresses whether it is:
  - an error in the documentation,
  - a flaw in the design of the product, or
  - a flaw in the implementation of the product, etc.

  The description of the security flaw's effects identifies the portions of the product that are affected and how those portions are affected.
- That when applied these procedures would identify the status of finding a correction to each security flaw. The flaw remediation procedures identify the different stages of security flaws. This differentiation includes at least:
  - suspected security flaws that have been reported,
  - suspected security flaws that have been confirmed to be security flaws, and
  - security flaws whose solutions have been implemented.

  It is permissible that additional stages (e.g. flaws that have been reported but not yet investigated, flaws that are under investigation, security flaws for which a solution has been found but not yet implemented) be included.
- That requires corrective actions to be identified for each of the security flaws. Corrective action may consist of:
  - a repair to the hardware,
  - a repair of the firmware, or
  - a repair of the software portions of the product,
  - a modification of product guidance, or
  - a combination of the above.

  Corrective action that constitutes modifications to product guidance (e.g. details of procedural measures to be taken to obviate the security flaw) includes both those measures serving as only an interim solution (until the repair is issued) as well as those serving as a permanent solution (where it is determined that the procedural measure is the best solution).

If the source of the security flaw is a documentation error, the corrective action consists of an update of the affected product guidance. If the corrective action is a procedural measure, this measure includes an update made to the affected product guidance to reflect these corrective procedures.

- That describes the methods used to provide flaw information, corrections, and guidance on corrective actions to product users. The necessary information about each security flaw consists of:
  - its description,
  - the prescribed corrective action, and
  - any associated guidance on implementing the correction.

  Product users may be provided with such information, correction, and documentation updates in any of several ways:
  - posting to a website,
  - sent to product users, or
  - arrangements made for the product developer to install the correction.

  In cases where the means of providing this information requires action to be initiated by the product user, product guidance must be adequate to ensure that it contains instructions for retrieving the information.

  The only metric for assessing the adequacy of the method used for providing the information, corrections and guidance is that there is a reasonable expectation that product users can obtain or receive it.

  For product users who register with the passive availability of this information is not sufficient, product developers must actively send the information (or a notification of its availability) to registered product users.
- The procedures ensure that product users have a means by which they can communicate with the product developer. By having a means of contact with the developer, the user can report security flaws, enquire about the status of security flaws, or request corrections to flaws. This means of contact may be part of a more general contact facility for reporting non-security-related problems.
- The issue of timeliness applies to the issuance of both security flaw reports and the associated corrections. The product developer processes include a procedure requiring timely response and the automatic distribution of security flaw reports and the associated corrections to registered users who might be affected by the security flaw. It is recognized that flaw reports should be generated and issued as soon as an interim solution is found, even if that solution is as drastic as turn off the product. Likewise, when a more permanent (and less drastic) solution is found, it should be issued without undue delay. It is unnecessary to restrict the recipients of the reports and associated corrections to only those product users who might be affected by the security flaw; it is permissible that all product users be given such reports and corrections for all security flaws, provided such is done in a timely manner.
- Automatic distribution does not mean that human interaction with the distribution method is not permitted. In fact, the distribution method could consist entirely of manual procedures, perhaps through a closely monitored procedure with prescribed escalation upon the lack of issue of reports or corrections. It is unnecessary to restrict the recipients of the reports and associated corrections to only those product users who might be affected by the security flaw; it is permissible that all product users be given such reports and corrections for all security flaws, provided such is done automatically.

30.6.4 Flaw Remediation Quality System

The product developer provides reporting processes:

a. The product developer ensures reported security flaws are remediated and the remediation procedures are issued to product users.
   - The flaw remediation procedures cover not only those security flaws discovered and reported by developer personnel but also those reported by product users.
   - The flaw remediation procedures are sufficiently detailed so that they describe how it is ensured that each reported security flaw is remediated.
   - The flaw remediation procedures contain reasonable steps that show progress leading to the eventual, inevitable resolution.
   - The flaw remediation procedures describe the steps that are taken from the point at which the suspected security flaw is determined to be a security flaw to the point at which it is resolved.

b. The product developer ensures that the product users are issued remediation procedures for each security flaw. The procedures describe the process that is taken from the point at which a security flaw is resolved to the point at which the remediation procedures are provided. The procedures for delivering remediation procedures should be consistent with the security objectives.

c. The product developer provides safeguards that any corrections to these security flaws do not introduce any new flaws. Through analysis, testing, or a combination of the two, the developer may reduce the likelihood that adverse effects are introduced when a security flaw is corrected.

30.6.5 Flaw Remediation Reporting

The product developer provides remediation guidance:

a. The product developer describes a means by which product users report to the developer any suspected security flaws in the product. By having a means of contact with the developer, the user can report security flaws, enquire about the status of security flaws, or request corrections to flaws.

b. The product developer describes a means by which product users may register a point of contact with the developer, to be eligible to receive security flaw reports and corrections. This point of contact is to be used to provide the product user with information related to security flaws that might affect that product user, along with any corrections to the security flaw. Registering the product user may be accomplished as part of the standard procedures that product users undergo to identify themselves to the developer, for the purposes of registering a software license, or for obtaining update and other useful information.

There need not be one registered product user per installation of the product; it would be sufficient if there were one registered product user for an organization. It should be noted that product users need not register; they must only be provided with a means of doing so. However, users who choose to register must be directly sent the information (or a notification of its availability).

30.6.6 Review and Approve

The enterprise or product representative examines the results of the above actions to determine that the product, in its operational environment, has sufficient flaw remediation. If the results reveal that the product, in its operational environment, has insufficient flaw remediation, then remedial action must be taken by the product developer or the product may be replaced.

30.7 SUMMARY

This chapter has provided a set of processes by which the enterprise can field relatively vulnerability-free software, at least to the extent of known vulnerabilities and exploits. New exploits are subject to temporary solution and become part of the flaw remediation system where they are reported and monitored until a satisfactory solution is achieved. It is not anticipated that the enterprise will be 100% exploit free from this process alone. Additional measures, including firewalls, ports, protocol restrictions, and some communication scanning, may be required. This process should, however, reduce the attack space significantly. The implementation will have to be monitored and evaluated as it proceeds. This includes the tracking of exploits, the response of the software remediation system, and the degree to which the vulnerability was knowable before the exploit as well as newly discovered vulnerabilities. It is expected that the analyses listed in this chapter will need refinement based on that feedback. Portions of this architecture are described in (Simpson, 2016, Simpson and Foltz, 2020).

REFERENCES

Aftandilian, E., Sauciuc, R., Priya, S., & Krishnan, S. (2012). Building useful program analysis tools using an extensible Java compiler. In *2012 IEEE 12th International Working Conference on Source Code Analysis and Manipulation* (pp. 14–23). doi: 10.1109/SCAM.2012.28.

Almubairik, N. A., & Wills, G. (2016). Automated penetration testing based on a threat model. In *Proceedings of the 11th International Conference for Internet Technologies and Secured Transactions, ICITST*, Barcelona, Spain, December 5–7, 2016.

Applebaum, A., Miller, D., Strom, B., Korban, C., & Wolf, R. (2016). Intelligent, automated red team emulation. In *Proceedings of the 32nd Annual Conference on Computer Security Applications (ACSAC'16)*, Los Angeles, CA, December 5–8, 2016 (pp. 363–373).

Avast (2021). Avast, Security, What is Trojan Malware? The Ultimate Guide. https://www.avast.com/c-trojan. (Accessed March 17, 2021).

Bavishi, R., Yoshida, H., & Prasad, M. R. (2019). Phoenix: Automated data-driven synthesis of repairs for static analysis violations. In *Proceedings of the 2019 27th ACM Joint Meeting on European Software Engineering Conference and Symposium on the Foundations of Software Engineering*, ESEC/FSE 2019 (pp. 613–624). New York: ACM. doi: 10.1145/3338906.3338952.

Bisson, D. (2020). Tripwire, 6 Common Phishing Attacks and How to Protect against Them. https://www.tripwire.com/state-of-security/security-awareness/6-common-phishing-attacks-and-how-to-protect-against-them/. (Accessed March 17, 2021).

Bugcrowd (2021). *Public Bug Bounty Program List.* San Francisco, CA. https://www.bugcrowd.com/bug-bounty-list/. (Accessed March 8, 2021).

Burns (2020). The Newstack, James Burns, the Importance of Resilience Testing and Observability. https://thenewstack.io/the-importance-of-resilience-testing-and-observability/. (Accessed March 19, 2021).

Common Criteria (2017). Common Criteria for Information Technology Security Evaluation, 2017 (All Version 3.1, Revision 5, April 2017): "Part 1: Introduction and General Model, Part 2: Functional Security Components, Part 3: Assurance Security Components, Common Methodology for Information Technology Security Evaluation". https://www.commoncriteriaportal.org/cc/. (Accessed March 3, 2021).

Coulter, R., Han, Q. L., Pan, L., Zhang, J., & Xiang, Y. (2020). Code analysis for intelligent cyber systems: A data-driven approach. *Information Sciences*, 524, 46–58. doi: 10.1016/j.ins.2020.03.036.

Creasey, J., & Glover, I. (2019). *A Guide for Running an Effective Penetration Testing Program.* Slough, UK: CREST Publication. https://www.crest-approved.org. (Accessed December 18, 2019).

Finifter, M., Akhawe, D., & Wagner, D. (2013). An empirical study of vulnerability rewards programs. In *22nd USENIX Security Symposium (USENIX Security 13)* (pp. 273–288).

FireEye (2020). Threat Research, Highly Evasive Attacker Leverages SolarWinds Supply Chain to Compromise Multiple Global Victims with SUNBURST Backdoor, December 13, 2020 | by FireEye. https://www.fireeye.com/blog/threat-research/2020/12/evasive-attacker-leverages-solarwinds-supply-chain-compromises-with-sunburst-backdoor.html. (Accessed March 16, 2021).

Firesmith, D. (2020). System Resilience Part 6: Verification and Validation, Softwrare Engineering Institute, SEI LBOG. https://insights.sei.cmu.edu/sei_blog/2020/04/system-resilience-part-6-verification-and-validation.html. (Accessed March 19, 2021).

Friesen, L., & Vonk, K. (2021). *Honing Your Poker Face: Pentesting & Project Management in Cybersecurity*. Evolve Security, EvolveSec Online. https://www.evolvesecurity.com/resource/honing-your-poker-face-pentesting-project-management-in-cybersecurity. (Accessed March 19, 2021).

Fruhlinger, J. (2019). *What Is a Computer Worm? How This Self-Spreading Malware Wreaks Havoc*. CSO. https://www.csoonline.com/article/3429569/what-is-a-computer-worm-how-this-self-spreading-malware-wreaks-havoc.html. (Accessed March 17, 2021).

Ghanem, M. C., & Chen, T. M. (2020). Reinforcement learning for efficient network penetration testing. *Information*, 11, 6. doi: 10.3390/info11010006.

Habib, A., & Pradel, M. (2018). How many of all bugs do we find? A study of static bug detectors. In *Proceedings of the 33rd ACM/IEEE International Conference on Automated Software Engineering, ASE 2018* (pp. 317–328). New York: ACM. doi: 10.1145/3238147.3238213.

HP Security Tools (2021). https://www8.hp.com/us/en/solutions/computer-security.html. (Accessed March 10, 2021).

IBM (2021). *Introduction to Rational Products*. IBM Knowledge Center. https://www.ibm.com/support/knowledgecenter/en/SSCLKU_7.5.5/com.ibm.help.common.rational.doc/topics/c_overview_rational.html. (Accessed March 10, 2021).

Imperva (2021). *Availability, Disaster Recovery Solutions*. Imperva, Learning Center. https://www.imperva.com/learn/availability/disaster-recovery/. (Accessed March 16, 2021).

Intel (2021). *The Latest Security Information on Intel® Products*. Security Center. https://www.intel.com/content/www/us/en/security-center/default.html. (Accessed March 10, 2021).

Jevtic, G. (2019). 17 Best security penetration testing tools the pros use. In *Security Strategy, Data Protection, Devops*. PhoenixNAP Global IT. https://phoenixnap.com/blog/best-penetration-testing-tools. (Accessed March 8, 2021).

Jones, P. (2010). *A Formal Methods-Based Verification Approach to Medical Device Software Analysis*. Embedded Systems Design. https://www.embedded.com/design/prototyping-and-development/4008888/A-Formal-Methods-based-verification-approach-to-medical-device-software-analysis.

Mamnani, D. (2018). *Testing Strategies to Build Resilience in a Chaotic World*. Capgemini. https://www.capgemini.com/2018/07/testing-strategies-to-build-resilience-in-a-chaotic-world/. (Accessed March 19, 2021).

Marcilio, D., Furia, C. A., Bonifácio, R., & Pinto, G. (2019). Automatically generating fix suggestions in response to static code analysis warnings. In *19th IEEE International Working Conference on Source Code Analysis and Manipulation, SCAM 2019*, Cleveland, OH, September 30–October 1, 2019 (pp. 34–44). IEEE Computer Society. doi: 10.1109/SCAM.2019.00013.

McAfee (2021). What Is Ransomware? https://www.mcafee.com/enterprise/en-us/security-awareness/ransomware.html. (Accessed March 17, 2021).

Mendelson, E. (2018). *The Best Data Recovery Software*. PCMag. https://www.pcmag.com/picks/the-best-data-recovery-software. (Accessed March 19, 2021).

Mitre (2021a). Common Vulnerability and Exposures. https://cve.mitre.org/. (Accessed March 3, 2021).

Mitre (2021b). Common Weakness Enumeration. https://cwe.mitre.org/. (Accessed March 3, 2021).

Moore, B., Trapnell, B., Fox, J. F., & French, C. (2020). National Voluntary Laboratory Accreditation Program (NVLAP) Cryptographic and Security Testing (CST), Handbook (NIST HB)-150-17. National Institute of Standards. doi: 10.6028/NIST.HB.150-17-2020. (Accessed March 3, 2021).

NIST (2015). Recommended Security Controls for Federal Information Systems and Organizations, Revision 5: FIPS PUB 800-53. Gaithersburg, MD: National Institute of Standards.

NIST (2021). National Vulnerability Database. https://nvd.nist.gov/vuln/data-feeds. (Accessed March 3, 2021).

Norton (2019). Zero-Day Vulnerability: What It Is, and How It Works. https://us.norton.com/internetsecurity-emerging-threats-how-do-zero-day-vulnerabilities-work-30sectech.html. (Accessed March 17, 2021).

OWASP (2021). The Open Web Application Security Project (OWASP). https://www.owasp.org/.

Ruiz-Rube, I., Person, T., Dodero, J. M., Mota, J. M., & Sánchez-Jara, J. M. (2020). Applying static code analysis for domain-specific languages. *Software Systems Model*, 19, 95–110. doi: 10.1007/s10270-019-00729-w.

Sarraute, C., Buffet, O., & Hoffmann, J. (2019). POMDPs Make Better Hackers: Accounting for Uncertainty in Penetration Testing. https://arxiv.org/abs/1307.8182. (Accessed December 20, 2019).

Schnoor, H., & Hasselbring, W. (2020). Comparing static and dynamic weighted software coupling metrics. *Computers*, 9(2), 24. doi: 10.3390/computers9020024.

SectigoStore (2020). Botnet Attacks: What Is a Botnet & How Does It Work? Sectigo, InfoSec Insights. https://sectigostore.com/blog/botnet-attacks-what-is-a-botnet-how-does-it-work/. (Accessed March 17, 2021).

Simpson, W. R., & Foltz, K. E. (2020). *Enterprise Level Security 2: Advanced Techniques for Information Technology in an Uncertain World* (p. 338). Boca Raton, FL: Taylor & Francis Group, CRC Press.

Simpson, W. R. (2016). *Enterprise Level Security: Securing Information Systems in an Uncertain World* (p. 397). Boca Raton, FL: CRC Press.

Stoever, K. (2018). *IT Resilience is the New Disaster Recovery*. DC Blox. https://www.dcblox.com/it-resilience-new-disaster-recovery/. (Accessed March 16, 2021).

Turner, et al. (2021). *Brian Turner, Cat Ellis, Darcy French, Best Data Recovery Software of 2021: Paid and Free File Recovery Solutions*. TechRadar. https://www.techradar.com/best/best-data-recovery-software. (Accessed March 19, 2021).

US Department of Defense (2014). "Security Categorization and Control Selection for National Security Systems" Categories for Moderate or High Risk Impact as delineated in NIST 800-53 Committee on National Security Systems Instruction (CNSSI) No. 1253. https://www.cnss.gov/CNSS/issuances/Instructions.cfm. (Accessed March 3, 2021).

US Department of Defense (2021). Enterprise Connections Document Library, Computer Network Defense (CND). https://public.cyber.mil/connect/ec-document-library/. (Accessed March 3, 2021).

Vogels (2016). Usernap, Rebecca Vogels, A Guide to Software Resilience Testing. https://usersnap.com/blog/resilience-testing/. (Accessed March 19, 2021).

WebRoot (2021). What is a Computer Virus and How Can I Protect My Computer? WebRoot, Cybersecurity Resources, Tips/Articles. https://www.webroot.com/us/en/resources/tips-articles/computer-security-threats-computer-viruses. (Accessed March 17, 2021).

White, S. K. (2018). *Standard CMMI Appraisal Method for Process Improvement (SCAMPI) Version 1.3a: CMMI Certifications: "A Complete Guide"*. CMMI Institute. https://www.cio.com/article/3308676/cmmi-certifications-guide.html#tk.cio_rs. (Accessed March 3, 2021).

Wiltshire, S. (2013). *2 Types of Data Loss that Cloud Backup Services Don't Cover & 5 Tips on How to Protect Your Data*. Ontrack. https://www.ontrack.com/en-gb/blog/two-forms-data-loss-cloud-backup-services-dont-cover-5-tips-protect-data. (Accessed March 19, 2021).

Zampetti, F., Scalabrino, S., Oliveto, R., Canfora, G., & Di Penta, M. (2017). How open source projects use static code analysis tools in continuous integration pipelines. In *2017 IEEE/ACM 14th International Conference on Mining Software Repositories (MSR)*. IEEE. doi: 10.1109/msr.2017.2.

31 Cryptographic Systems

Keith M. Martin

31.1 INTRODUCTION: THE ROLE OF CRYPTOGRAPHY

All electronic systems process digital information. It is extremely important that the security of this information is considered carefully at the design stage. One of the easiest ways of compromising a system is through manipulation or falsification of the information on which it relies. This is true of any system, not just those deemed mission-critical. Some of the greatest threats to systems based around everyday technologies such as mobile phones and web browsers come from attacks on the information that they process.

At the heart of information security lies the need for a range of atomic data security services, from which the security of systems can be constructed. These include:

- **Confidentiality**: The obfuscation of data in such a way that only authorized entities are able to view the original data.
- **Data Integrity**: The assurance that data has not been altered in an unauthorized manner since the time that it was created by an authorized entity.
- **Data Origin Authentication**. The assurance that a given entity was the original source of data (a stronger notion of data integrity).
- **Non-Repudiation**. The ability to prove to a third party that a given entity was the original source of data (a stronger notion of data origin authentication).

Cryptography provides a toolkit of mathematically based mechanisms for the provision of these atomic security services. It also provides techniques for combining these mechanisms in order to provide more complex services, such as *entity authentication*, which is the assurance that a given entity is present and active in a communication session, and *anonymity*, which is the ability to disguise the originator of data. Cryptography also concerns the design of protocols between two or more entities in order to achieve a combination of precise security goals.

Cryptographic tools do the security job specified, but nothing more. This is why it is particularly important to take a system-wide view when assessing digital security. After reviewing the most important cryptographic tools, we will consider the other layers of a cryptographic system, namely, implementation, key management, and endpoints. If cryptographic tools are chosen wisely, these other layers are the most likely sources of security failure. We then briefly consider three case studies of the use of cryptography and consider future developments.

31.2 CRYPTOGRAPHY FUNDAMENTALS

The main component of most cryptographic tools is a cryptographic *algorithm*. The standard security assumption is that this algorithm is public knowledge. In fact, most uses of cryptography deploy algorithms that are published in open standards. Most cryptographic algorithms have at least two inputs – the data requiring protection and a cryptographic *key*. This key is a variable, generally much shorter than the data being protected, which ensures that the output of a cryptographic algorithm is not fixed each time the same data in input. The output of a cryptographic algorithm is normally pseudorandom, rendering it meaningless to anyone without knowledge of the key.

 DOI: 10.1201/9781003188377-37

Since cryptographic algorithms are assumed to be publicly known, the security provided by cryptography is dependent on the security of cryptographic keys. Indeed cryptography could be considered a set of techniques for transferring the security of (potentially large amounts of) data onto the security of (much shorter) cryptographic keys.

Although cryptography can be used to protect data in local storage, it is most compelling to consider cryptography being used to protect data in a communications setting. In this case, we can think of a sender applying cryptography to data, which is then sent over an insecure channel to a remote receiver. While both sender and receiver will deploy the same cryptographic algorithm, a crucial cryptographic classification arises from the relationship between the sender's and the receiver's keys:

- *Symmetric* cryptography concerns algorithms where the sender and receiver need to use the same key. Since knowledge of this key is what distinguishes the sender and receiver from a potential adversary, a symmetric key is necessarily secret. A critical challenge in applications of symmetric cryptography is thus how to ensure the sender and receiver both obtain the same secret key.
- *Asymmetric* cryptography concerns algorithms where the sender and receiver use different (mathematically related)l keys. The primary advantage of having different keys is to restrict just one of them to being secret (often referred to as the *private* key), allowing the other to be public (hence, asymmetric cryptography is sometimes known as *public-key* cryptography). This, in turn, enables cryptography to be used in applications where it is not realistic for the sender and receiver to agree a (symmetric) secret key ahead of use.

In fact, by far the majority of uses of cryptography involve symmetric algorithms. This is because symmetric algorithms are significantly faster from a computational perspective than asymmetric algorithms. In closed environments where it is relatively straightforward to distribute secret keys to their required points of use, symmetric cryptography tends to be deployed exclusively. This is the case for applications such as card payments, Wi-Fi networks and mobile call protection, as well as many corporate networks.

Asymmetric cryptography tends to be used for just two purposes. The first is to support the exchange of symmetric keys in open environments, where senders and receivers may have no prior relationship that facilitates prior symmetric key exchange. This is the case for many internet applications, such as secure browsing, email protection, and secure messaging services. Once asymmetric cryptography has been used to securely transfer secret keys, symmetric cryptography based on these keys is used for bulk data protection. The other primary use of asymmetric cryptography is digital signature schemes.

Modern cryptographic algorithms are subject to rigorous analysis and testing, making it unlikely they contain serious flaws. The most obvious attack strategy on paper is thus to determine the secret cryptographic keys. Both symmetric keys and asymmetric private keys must appear to be chosen randomly from a large set of possibilities, thus preventing an attacker from simply trying out every possible key until they come across the correct one by chance. Symmetric keys are typically at least 128-bits long, and asymmetric private keys are even longer. Finding keys of such lengths simply through a brute-force search is well beyond the capabilities of any conventional supercomputer today, or imaginably in the future.

31.3 CRYPTOGRAPHIC TOOLS

We now review some of the most important tools provided by cryptography. It should be noted that cryptography provides many more tools than those we address here, but the majority of cryptographic systems are built using the tools we discuss.

31.3.1 Symmetric Encryption

The most common symmetric encryption mechanism is the *block cipher*, so-called because it encrypts data in *blocks* of bits. The size of this block is fixed by the encryption algorithm. A block cipher algorithm thus takes as input a block of *plaintext* bits and a symmetric key and outputs a block of *ciphertext* bits. Only anyone possessing the symmetric key can decrypt the block of ciphertext bits to obtain the original block of plaintext bits.

By far the most widely deployed block cipher algorithm is the Advanced Encryption Standard (AES) (NIST, 2001). This has a block size of 128 bits and supports three different key lengths: 128, 192, and 256. The AES is well-respected and was designed through an open `competition' in the late 1990s, led by the US National Institute for Standards and Technology (NIST). It has since been adopted in numerous technical standards and deployed in a wide range of everyday technology. The AES is by no means the only block cipher in widespread use. Indeed, its predecessor the Data Encryption Standard (DES) is still deployed throughout the financial sector, albeit in the form of Triple DES.

Encrypting a long plaintext involves first breaking the data into blocks (the final block will typically need to be filled to a full block length by means of *padding*). Encrypting each block independently is regarded as insecure, since two identical plaintext blocks will be encrypted into the same ciphertext block, revealing data patterns. There is thus a wide range of standardized *modes of operation* of a block cipher, which dictate how to encrypt multi-block plaintexts securely. Modes of operation need to be selected carefully, based on the nature of the application. There are, for example, modes of operation designed specifically to support parallel processing, modes that additionally provide authentication, and modes that optimize the security of full-disk encryption. Secure deployment of a block cipher thus involves not just the selection of a block cipher algorithm, but also choosing an appropriate mode of operation.

A special class of symmetric encryption mechanisms known as *stream ciphers* offer bitwise, rather than blockwise, encryption. These are fast and have good error handling, making them well-suited to noisy networking environments where ciphertext bits are readily prone to corruption. Stream ciphers have traditionally been deployed for applications such as military radios and mobile telecoms, and more recently to protect streaming services. We are now seeing less use of stream ciphers since block ciphers can be deployed in modes of operation that essentially convert them into stream ciphers.

31.3.2 Public-Key Encryption

A *public-key encryption* algorithm enables a sender to use a public key to encrypt some plaintext data (typically a symmetric key). Only the holder of the matching private key can decrypt the resulting ciphertext to retrieve the plaintext. Public-key encryption algorithms are built around computational processes that are easy to perform but hard to reverse. Very few such processes are understood well enough to build reliable public-key encryption algorithms around.

There are only really two types of public-key encryption algorithms in widespread use. The *RSA* public-key encryption algorithm is based on the computational hardness of factoring large composite numbers. *Elliptic-curve* public-key encryption algorithms are based on the computational hardness of versions of the *discrete logarithm problem*.

It is worth re-emphasizing that public-key encryption is much slower than symmetric encryption. It also involves larger keys, with the equivalent security of a 128-bit AES key requiring over 3,000 bits for RSA and around 256 bits for elliptic-curve public-key encryption (BlueKrypt, 2022). For these reasons, public-key encryption tends to be deployed only surgically, if needed at all.

31.3.3 Hash Functions

Cryptographic *hash functions* are unusual in that they do not feature a key. A hash function takes as input some data and outputs a *hash* of the data, which is essentially a compressed digest representation of the data. Hash functions are carefully designed to provide a number of security properties. These include *preimage resistance*, which means it is hard to find an input that hashes to a given output, and *collision resistance*, which means it is hard to find any two inputs that hash to the same output.

Because there is no key involved, hash functions are publicly computable. This means that hash functions are of little use on their own as cryptographic tools. They can be used to provide data integrity, but only in situations where the hash output can be protected from tampering, However, they are extremely useful building blocks for other cryptographic mechanisms, including message authentication codes and digital signature schemes as well as providing a useful cryptographic `glue' for binding data together within cryptographic protocols. Indeed, the entire blockchain structure underpinning many cryptocurrencies and other distributed ledgers is constructed using hash functions.

The length of the hash output is an indicator of the security offered by a hash function. A general rule of thumb is that this output length should be double the length of an equivalent symmetric key. There are several hash functions in widespread use, including the SHA-2 family (NIST, 2015a), which includes hash functions whose output is double the length of each of the three AES key lengths (SHA-256, SHA-384, and SHA-512), and SHA-3 (NIST, 2015b).

31.3.4 Message Authentication Codes

Message authentication code (MAC) algorithms are similar to hash functions except that they also input a symmetric key. This means that MAC algorithms can be used to provide data origin authentication. Suppose a sender and receiver share a symmetric MAC key. The sender can input some data and the key to a MAC algorithm, the data and the output of the MAC algorithm can now be transmitted to the recipient, who performs their own local computation of the MAC output and then compares this with the MAC output sent to them. If they match then the recipient has good grounds to believe that the data is correct and originated with the genuine sender.

While it is possible to design dedicated MAC algorithms, the most common MAC algorithms in use are built from other cryptographic tools. The MAC algorithms widely deployed in financial networks are built from block ciphers. Possibly the most common MAC algorithm is HMAC (Krawczyk et al., 1997), which is a secure way of taking a hash function and using it to create a MAC algorithm.

31.3.5 Authenticated Encryption Schemes

There are many applications where it is important to provide both confidentiality and data origin authentication of data. This includes almost all secure networking protocols, including Transport Layer Security (TLS) and IPSec, secure Wi-Fi, and secure file transfer services. Using cryptography to support such applications traditionally involved symmetrically encrypting data using a block cipher and then providing the output of a MAC algorithm as an integrity check. There are different ways in which these two processes can be combined and, unfortunately, some of these have turned out to be insecure. In addition, providing both using two separate cryptographic tools involves two separate keys and double the processing.

A best practice increasingly adopted by new applications is to use *authenticated encryption schemes*, which input data and a symmetric key and then output both a ciphertext and an integrity check on the data. Authenticated encryption schemes only require a single key, are more efficient than applying two separate cryptographic tools, and are carefully designed to ensure the combination of the two cryptographic services is secure. Another advantage of authenticated encryption schemes is that they support the processing of data in situations where only part of the data needs confidentiality, but all the data needs data origin authentication (for example, network data may have routing headers that must not be encrypted).

Most authenticated encryption schemes are special modes of operation of a block cipher and include GCM mode (NIST, 2007). Others are dedicated designs, including ChaCha20-Poly1305 (Nir and Langley, 2018a). The use of authenticated encryption schemes is fast becoming the default means of implementing symmetric encryption.

31.3.6 Digital Signature Schemes

The main cryptographic tool for providing non-repudiation is a *digital signature*. A digital signature algorithm takes input a message and a private key and outputs a digital signature. This is an integrity check that can be verified by anyone who possesses the equivalent public key. A digital signature provides non-repudiation because only the signer has access to the private key used to generate the digital signature. This contrasts with symmetric MAC algorithms, where both the sender and the receiver have equal capability to generate a MAC algorithm output, making it hard to resolve disputes between them.

The requirements for generating and verifying digital signatures somewhat mirror those for public-key encryption. Only the holder of a private key can create a digital signature while anyone can verify, whereas for public-key encryption anyone can encrypt a message that only the holder of a private key can decrypt. It should thus not be surprising that there is a close relationship between digital signature algorithms and those for public-key encryption. The main digital signature algorithms deployed are those based on RSA, such as in (Moriarty et al., 2016), and the Digital Signature Algorithm (DSA) (NIST, 2013), which relies on the hardness of the discrete logarithm problem. The elliptic-curve version of DSA, known as ECDSA, is also deployed widely.

31.3.7 Key Exchange

Prior to using symmetric cryptography, it is necessary for communicating parties to agree on a common symmetric key. In many applications of cryptography, this is resolved through a manual process (for example, through the pre-installation of keys on physical devices). Cryptography provides two general approaches for exchanging a symmetric key in situations where manual key agreement is not possible.

The first general approach is *key encapsulation*, which uses public-key encryption to transfer securely a symmetric key from one party to another.

The main alternative is *Diffie-Hellman-based key exchange* (DHKE) (Diffie and Hellman, 1976). This technique involves both parties generating some private and public information (essentially private and public keys) and then exchanging public information. The design of DHKE is such that each party can use their own private information and the public information of the other party to compute a common secret. They can then both compute a common symmetric key by inputting the common secret locally into a *key derivation function*.

Key encapsulation is the more traditional approach, but DHKE is preferred by most contemporary applications. The reason is that DHKE offers *perfect forward secrecy*, a property that preserves the security of communications in the event that a long-term cryptographic key is compromised.

31.4 IMPLEMENTING CRYPTOGRAPHY

Because cryptography provides such a sensitive role in electronic systems, there is no room for error when it comes to implementation. Very small deviations from the intended specification can sometimes render a cryptographic algorithm insecure. While the focus of this chapter is not on the implementation of systems, several aspects of cryptographic implementation are worth highlighting.

31.4.1 Cryptographic Standards

For reasons of compatibility and wide scrutiny, most deployed cryptography uses algorithms that are published openly, descriptions of which can be found in a range of different sources. Many of these descriptions may be simplified, some even incorrect. It is imperative that relevant standards are followed when implementing cryptography in any system. Standards bodies oversee not just the selection of algorithms but also review many of the details of implementation.

The three most prominent standards bodies overseeing cryptographic specifications are ISO/IEC, the National Institute for Standards and Technology (NIST), and the Internet Engineering Task Force (IETF). Some sector-specific standards bodies are also active in cryptography, such as the European Telecommunications Standards Institute (ETSI).

31.4.2 Cryptographic Libraries and APIs

Because cryptography is challenging to implement, most developers are likely to rely on cryptographic libraries accessed by means of an Application Programming Interface (API). This is, in general, recommended; however, care should be taken when selecting both the cryptographic library and the associated API since some well-publicized attacks have exploited weaknesses in both libraries and APIs.

Issues to consider when selecting libraries and APIs include functionality, compatibility, validation (some such as Bouncy Castle (Bouncy, 2022) are validated against the FIPS 140-2 standard) and whether they are open source.

31.4.3 Side Channels

One of the most surprising developments in the past few decades of cryptographic research has been the plethora of attacks against cryptographic implementations through the leakage of information (particularly about cryptographic keys) by means of *side channels*, which are sources of information other than those anticipated in the theoretical design of the cryptographic algorithm itself. These include, but are by no means restricted to, deducing information based on the differences in the time taken to perform specific computations, differences in the power consumed when performing specific computation, and analyzing the responses of an implementation to the deliberate introduction of erroneous data.

Software and hardware defenses against side channels are beyond the scope of this chapter, but developers are advised to familiarize themselves with best practice in this area. See, for example, (Piessens, 2021)) and (Verbauwhede, 2021).

31.4.4 Good Implementation Practices

The most important takeaway is that great care must be taken in all aspects of the implementation of cryptography in digital systems. Potential problems are often unintuitive. For example, there are examples of two cryptographic systems that are both secure in isolation becoming insecure when used in composition. There is more and more general guidance over time about how to develop secure software and hardware, which should certainly be followed when building cryptography into a system. Another welcome trend is the strengthening of libraries and APIs. With luck, we should see more off-the-shelf secure cryptographic implementations, which should help to reduce the many implementation weaknesses in current deployments.

31.5 KEY MANAGEMENT

Cryptography relies crucially on the keys that underpin it. In any complex system, managing those keys is arguably the most challenging aspect of making cryptography work. Weaknesses in key management processes are responsible for a significant percentage of attacks against systems relying on cryptography. This is because key management represents the interface between cryptographic technology itself and the complexity of the wider systems in which it is deployed and the humans who use these systems.

31.5.1 The Key Lifecycle

Almost all cryptographic keys have a finite lifetime, after which they are no longer valid and should be changed and/or destroyed. Long-term keys may be valid for a number of years, such as those embedded in hardware tokens, while short-term keys are often valid for single session, such as the duration of a mobile phone call. When considering key management, it thus helpful to imagine a *key lifecycle*, from the point at which a key is generated through to the point at which it is withdrawn and/or destroyed. The management of each stage of this key lifecycle must be considered carefully. Weaknesses in any single stage can undermine the entire security that the use of the key was intended to provide.

31.5.2 Key Generation

Cryptography relies on the randomness of keys, which in turn rely on an ability to generate random bits. Relevant standards should be consulted when selecting a suitable means of key generation, since there have been many real-world attacks where strong cryptographic algorithms are deployed but weak processes have been used to generate the underlying cryptographic keys. Particular care needs to be taken when generating asymmetric key pairs since these require special mathematical properties.

Ideally, random bits should be extracted from physical hardware sources and various technologies can be used to do this. However, using physical sources is expensive and slow. It is also impossible to use such techniques to generate the same key in two different places, which is something we require when using symmetric cryptography.

A commonly deployed compromise is to generate keys using a *pseudorandom number generator* (PRNG), an algorithm specially designed for this purpose. A PRNG takes a seed as input, and outputs pseudorandom bits that can be used to form a cryptographic key. While the seed can be unique to a specific source, such as system information extracted from the computer on which the PRNG is implemented, it is also possible to use a seed that is known to more than one entity, including an already-shared cryptographic key. This latter approach to generating keys is often called *key derivation.*

Mobile phones provide a good example of the compromises made when generating keys. The SIM card inserted in a mobile phone contains a long-term key shared between the SIM and a mobile network operator. This key was likely generated from a source that relied on a secure hardware generator. Each mobile call, however, is protected by a short-term session key that is generated from the long-term key using a key derivation function. This same session key can be generated locally at point of use by both the SIM and the network operator.

31.5.3 Key Establishment

An important aspect of key management is making sure that the right keys end up in the correct places, often referred to as *key establishment.* This is particularly challenging for symmetric cryptography, where we typically require two entities in different places to have access to the same secret symmetric key.

We have already mentioned several approaches to key establishment, including manual pre-installation of keys before deployment, deriving keys locally from existing shared information, and key exchange techniques based on the use of asymmetric cryptography. Indeed these three broad key establishment approaches are by far the most commonly used.

It is worth noting a couple of more specialized approaches. Large complex digital systems, such as those governing the security of multinational organizations, often rely on a sophisticated *key hierarchy* to facilitate key establishment. At the top of this hierarchy, a manual key ceremony involving various senior officials from the organization governs a process whereby a central security module generates a system master key. Each key in the hierarchy can then be used to securely transfer keys lower in the hierarchy to the places they are needed in the wider system. Another specialist key establishment approach is *quantum key distribution* (QKD), which allows a key to be sent in the clear over a quantum channel in such a way that any attempt to view it on the channel should, with high probability, be detected. Expensive and reliant on special equipment, QKD is a niche technology, but is an approach to key establishment that may gain traction in some high-security environments.

31.5.4 Key Storage

Many cryptographic keys, particularly long-term keys, need to be stored. This makes keys a prime target for anyone attacking the system, hence great care needs to be taken to store keys securely.

By far the safest option is to store keys in secure hardware. Indeed, there are many different hardware security modules (HSMs) available whose primary task is to generate, store, and govern access to cryptographic keys. Technologies such as mobile phones and personal computers increasingly have their own secure hardware components on which keys can be stored. At the very least, keys should be stored on secure chips such as those on smartcards.

An alternative option to using secure hardware is to store keys in encrypted form. This requires a second cryptographic key to be used to access the first one. One approach to avoiding storage of this second key is to derive it from a user password. This password is input to a key derivation function, which generates the key that can be used to decrypt the stored key. Of course, the security of the stored key is now reliant on the protection and strength of the user password.

31.5.5 Key Change

There are many reasons why cryptographic keys may need to be changed. The simplest scenario is one that can be anticipated, namely the end of an existing key's lifetime. However, there are many reasons why an unplanned key change may be necessary. These include the compromise of an existing key through loss, damage, or theft, the potential loss of a key following an attack on the system, or the unexpected departure of a member of staff who has access to the key.

Key change typically involves a new generation and establishment process, but this could be on a substantial scale should, for example, an unplanned key change be required for a master key on which many other keys depend. Since this could involve, at worst, the entire reissuance of hardware tokens to a user community (for example, if the master key of a bank was compromised this might necessitate new cards being issued to all customers). It is thus essential that in any complex system, the potential impact of an unplanned key change is anticipated, and policies and procedures are put in place as part of business continuity planning.

31.5.6 Public-Key Management

It is a fallacy to assume that managing asymmetric public keys is simpler than managing secret keys. There are two aspects of key management that are particularly problematic for asymmetric public keys.

Firstly, while it is theoretically easier to establish a public key (since it is not a secret), the very fact that a public key is *not* a secret means that is typically widely available. This, in turn, accentuates a problem that exists for all cryptographic keys – how can any entity be sure that a given key is the right key to be used for a specific purpose? For secret keys, this problem is normally addressed by the wider key management aspects associated with it. For example, the symmetric key in the chip on a bank card is delivered to the customer through wider banking processes, and cannot be used by anyone other than the bank and the cardholder. The key is utilized when the customer presents their bank card to a card reader. A public key, on the other hand, is essentially public knowledge. Anyone relying on this key in a specific situation needs some way of being sure that this is the right key to be used.

The most common solution to this problem is certification. A public key is made available within a *public-key certificate*, which is a data string that includes the public key alongside relevant data such as with whom the public key is associated, its expiry date, and valid uses. Importantly, this data string is digitally signed by a trusted *certificate authority* (CA), which should have checked the validity of this associated data before signing the certificate. The public key can only be relied on if a reliant party checks the various fields of the certificate, verifies the CA's digital signature, and holds the CA in sufficient regard to have conducted the issuing process correctly.

The second particular key management challenge for public keys is key change. Since public keys are public, nobody can be sure who has knowledge of them. This means that it is hard to change a public key since it is almost impossible to know how to inform potential users of them.

The solution to this issue is to maintain either white lists of valid certificates or black lists of those that have been changed. The latter approach, which is most common, involves CAs publishing signed *certificate revocation lists* (CRLs) that identity revoked certificates. Anyone wishing to rely on a public key is then obliged not just to check the apparent validity of the associated public-key certificate as previously defined, but also check the latest CRLs to see whether the particular certificate is listed there.

31.6 ENDPOINT SECURITY

When assessing the security of a digital system, it is easy to overlook an obvious point. Cryptography offers security services to data only at the point of use. In the case of encryption, data is confidential once converted into ciphertext, but could exist as plaintext at the endpoints of a digital system both before encryption and after decryption. This is not just a concern relating to confidentiality. Incoming data can be integrity-checked on receipt within a digital system but then modified subsequently to the check being performed, Cryptography can support entity authentication of a network connection, but then the session could be hijacked by a malicious entity attacking the endpoint by other means.

This is a stark reminder that the security of digital systems requires much more than good use of cryptography. Securing endpoints requires a broader range of cyber security controls, including effective security management governing the users and procedures followed by any organization.

31.7 WIDER CRYPTOGRAPHIC ISSUES

Having reviewed the function and security aspects of using cryptography, we now briefly consider two more general issues that could affect the deployment of cryptography within any digital system.

31.7.1 Cryptography as an Overhead

From a system engineering perspective, cryptography is an overhead. The use of cryptography involves numerous ongoing costs, beyond those of development, maintenance, and management. Fundamentally, cryptography uses resources relating to computation, communication, and bandwidth. Cryptography burns up energy and introduces delays. This is particularly acute when deploying asymmetric cryptographic techniques, which are much more computationally heavy and thus slower than symmetric ones.

When designing any system supported by cryptography, there is thus an inevitable trade-off to be made between security and efficiency. It is natural to seek to deploy a bare minimum of cryptography. However, using insufficient cryptography can lead to devastating security weaknesses. Finding the balance is never easy.

The good news is that the costs associated with deploying cryptography do not seem to scale with improvements in technology. Modern computers, even resource-constrained ones, are more capable of handling cryptography than they ever have been. We are thus seeing more systems engineers defaulting to greater security as the relative cost of overheads abates. Today, there seems less need for the types of cryptographic shortcutting we have seen in past decades.

31.7.2 Societal Aspects

An even broader issue that affects the use of cryptography in digital systems relates to law enforcement and national security. Modern encryption is highly effective and can thus be used to protect the data of potential targets of investigations. This creates a significant tension between a desire to have strong encryption to protect data and to have, in certain circumstances, means of accessing data that has been encrypted.

This issue makes use of cryptography politically, with very different competing societal views on how to navigate such issues. The history of attempts to address this issue reveals the importance of taking a system-wide view of cryptography. Past attempts to overcome the protection offered by encryption by governments have included the deployment of weak algorithms, restricting key lengths, and exploiting weaknesses in implementations, key management, and endpoints.

Designers of digital systems using cryptography thus need to take into account any restrictions arising from jurisdictions within which the system may be deployed. They should also consider the extent to which they desire to comply or resist pressures to overcome the use of encryption and take a system-wide view when doing so.

31.8 CASE STUDIES

We now briefly consider three case studies of the use of cryptography, highlighting some of their most significant cryptographic features.

31.8.1 Mobile Telecoms

Mobile telecoms was the first mass technology to deploy cryptography, doing so to protect calls between a mobile phone and the nearest mobile base station.

The basis for mobile telecom security is a unique symmetric key that is shared between a mobile user's SIM card and their network operator. Since users need to purchase a SIM card from an operator, possibly via a high-street store, this key can be pre-installed in a secure facility and manually delivered to the user at the point of purchase.

When a user makes a call, a cryptographic protocol is run between the mobile phone and the base station, which establishes entity authentication by initially checking that both parties know the pre-installed key. The shared key is then used to derive further symmetric keys for securing the call. Mobile calls are normally encrypted using a stream cipher since speed of encryption and minimal impact of channel-induced errors is a priority. Only a few cryptographic algorithms are supported by the mobile telecoms standards, to facilitate worldwide compatibility. Mobile calls are only encrypted as far as the base station, after which they typically enter the traditional telecom networks. Many jurisdictions place a legal requirement on network operators to enable access to call data after decryption at the base station, should this be required by investigative authorities.

31.8.2 Transport Layer Security

The *TLS protocol* can be used between any two devices to secure traffic over the internet but is most classically deployed to enable secure web browsing via sessions headed by https.

Since we cannot anticipate every web server we might want to connect to in advance, it is impossible to base TLS security on a pre-shared symmetric key. Asymmetric cryptography is thus required to establish a common secret between browser and server, which is then used to derive the symmetric cryptographic keys needed to secure a TLS session. This session authenticates the web server, and eventually both encrypts and offers data origin authentication of exchanged data.

Because TLS is deployed in a wide range of different environments, earlier versions supported a large number of different cryptographic algorithms. The precise choice is negotiated during the initial TLS handshake. The most recent version of TLS (1.3) (Rescorla, 2018b) has significantly reduced this choice, enforcing the use of DHKE to ensure perfect forward secrecy and authenticated encryption to secure the subsequent session.

The security of TLS is thus underpinned by the need to manage the TLS public key certificates that support the initial symmetric session key establishment. By far the most likely exploitation of TLS is through weaknesses in certificate management.

31.8.3 Bitcoin

Bitcoin is the most well-known example of a cryptocurrency, which is a means of digital payment offering anonymity with no requirement for a trusted central bank.

Bitcoin addresses are essential asymmetric public keys, and a transaction is a digital signed statement transferring value from one Bitcoin address to another. Bitcoin value is linked to the equivalent asymmetric private keys. This means that safeguarding Bitcoin value is a key management issue, since anyone getting hold of a private key has the ability to spend the associated coinage.

The primary role of a conventional bank is to act as a trusted witness of transactions in and out of an account. Bitcoin facilitates this role through an open distributed ledger, where all transactions made are publicly witnessed and added to millions of copies of this ledger on computers around the world. It is vital that copies of this ledger are reconciled and that data integrity is maintained at all times. Bitcoin deploys a data structure called a *blockchain*, which is essentially a binding using hash functions of recent transactions to all those of the past.

New currency is minted through a process known as *mining*, which involves exhaustively searching for preimages of a hash function. The difficulty of this can be adjusted to ensure that currency is minted at a steady rate.

31.9 FUTURE DEVELOPMENTS

We close by considering some future developments likely to have an impact on the use of cryptography in digital systems.

31.9.1 Quantum Computing

The greatest future threat to the cryptography we deploy in current systems arises from the development of quantum computers. To an extent, it does not matter exactly when and how quantum computers of real power emerge, since cryptography is used today to secure some data which will still need protection far into the future. We thus need to consider the threat posed by an adversary with a quantum computer now, even if such an adversary may not exist for a considerable time.

Quantum computers are not believed to pose a significant threat to symmetric cryptography. While they will be able to undertake exhaustive key searches much faster, it is believed that roughly doubling current symmetric key lengths should suffice. Since AES-128 is the most deployed symmetric encryption algorithm today, upgrading to AES-256 should keep data safe from quantum computers. This is an upgrade that should be seriously considered now if data is being protected which requires longevity of confidentiality. Of course, much of the data that we protect data has only short-term confidentiality requirements, so AES-128 is absolutely fine.

Similarly, it is believed that increasing hash output lengths should suffice to protect hash functions against a quantum adversary. This issue is not fully understood, but doubling, and possibly tripling, will probably suffice. Hash functions such as Keccak, which is the basis behind SHA-3, can support arbitrary hash output lengths, so should be fine.

The biggest threat posed by quantum computers is asymmetric cryptography. The two hard problems on which we base the main asymmetric algorithms used today, factoring and the discrete logarithm problem, can effectively be solved by a quantum computer. This means that we need alternative hard problems on which to base new asymmetric algorithms for public-key encryption, digital signature schemes and to support key exchange. A major design competition run by NIST is expected to lead to standardization of new asymmetric algorithms that are believed resistant to quantum computers. On completion of this process, a significant challenge to many organizations will be how to plan migration from existing algorithm suites.

31.9.2 New Cryptographic Tools

While we have discussed the most fundamental cryptographic tools, there are many more out there. Indeed there is an ongoing process of designing specialist cryptographic mechanisms for precise tasks that are required as technology evolves.

It can be quite hard to predict which cryptographic ideas will transfer from theory into widespread practice. Promising concepts such as *lightweight cryptography* have yet to see wide deployment, probably because many of the resource-constrained devices for which it is designed have proved capable of supporting conventional algorithms. Various tools for computing on encrypted data, such as *homomorphic encryption*, have computational costs that make them unattractive for widespread use. Others, such as mechanisms for facilitating electronic voting are constrained by the political issues surrounding them. One thing we can probably expect to see is greater emphasis not just on cryptographic tools that have been through a formal security analysis, but also implementations of these tools that have also done so.

31.10 FURTHER READING

Most of the concepts discussed in this chapter are general and widely treated in the cryptographic literature. An excellent starting point for a short `birds-eye' view of applied cryptography is (Paterson, 2021). For a comprehensive (non-mathematical) introduction to applied cryptography, including key management, use case studies, and discussion of societal aspects, see (Martin, 2017). There are numerous books discussing the more technical aspects of cryptography in detail, including (Aumasson, 2017) and (Paar and Pelzl, 2010). In general, an excellent source of guidance documents and advisories on specific aspects of cryptography is (NIST, 2022).

REFERENCES

Aumasson, J.-P. (2017). *Serious Cryptography*. San Francisco: No Starch Press.

BlueKrypt (2022). BlueKrypt: Cryptographic Key Length Recommendation. https://www.keylength.com.

Bouncy (2022). The Legion of the Bouncy Castle. https://bouncycastle.org.

Diffie, W., & Hellman, M. E. (1976). New directions in cryptography. *IEEE Trans. Inf. Theory*, 22(6), 644–654. doi: 10.1109/TIT.1976.1055638

Krawczyk, H., Bellare, M., & Canetti, R. (1997). HMAC: Keyed-Hashing for Message Authentication, RFC 2104 (Informational), Internet Engineering Task Force, Feb. 1997, Updated by RFC 6151. https://www.ietf.org/rfc/rfc2104.txt

Martin, K.M. (2017). *Everyday Cryptography: Fundamental Principles and Applications*, 2nd Edition. Oxford: Oxford University Press.

Moriarty, K., Kaliski, B., Jonsson, J., & Rusch, A. (2016). PKCS #1: RSA Cryptography Specifications Version 2.2, RFC 8017 (Informational), RFC Editor, IETF, Fremont, CA, USA. https://www.rfc-editor.org/rfc/rfc8017.txt.

Nir, Y., & Langley, A. (2018a). *ChaCha20 and Poly1305 for IETF Protocols, RFC 8439 (Informational)*, RFC Editor, IETF, Fremont, CA, USA. https://www.rfc-editor.org/rfc/rfc8439.txt.

NIST (2001). Advanced Encryption Standard (AES). https://nvlpubs.nist.gov/nistpubs/FIPS/NIST.FIPS.197-upd1.pdf.

NIST (2007). Recommendation for Block Cipher Modes of Operation: Galois Counter Mode (GCM) and GMAC. https://nvlpubs.nist.gov/nistpubs/Legacy/SP/nistspecialpublication800-38d.pdf.

NIST (2013). Digital Signature Standard (DSS). https://nvlpubs.nist.gov/nistpubs/FIPS/NIST.FIPS.186-4.pdf.

NIST (2015a). Secure Hash Standard (SHS). https://nvlpubs.nist.gov/nistpubs/FIPS/NIST.FIPS.180-4.pdf.

NIST (2015b). SHA-3 Standard: Permutation-Based Hash and Extendable-Output Functions. https://nvlpubs.nist.gov/nistpubs/FIPS/NIST.FIPS.202.pdf

NIST (2022). Cryptography. https://www.nist.gov/cryptography

Paar, C., & Pelzl, J. (2010). *Understanding Cryptography*. Heidelberg, Dordrecht, London, New York: Springer.

Paterson, K. G. (2021). *The Cyber Security Body of Knowledge, Applied Cryptography, Version 1.0.0*. University of Bristol. https://www.cybok.org/

Piessens, F. (2021). *The Cyber Security Body of Knowledge, Software Security, Version 1.0.1*. University of Bristol. https://www.cybok.org/

Rescorla, E. (2018b). The Transport Layer Security (TLS) Protocol Version 1.3, RFC 8446 (Proposed Standard), RFC Editor, IETF, Fremont, CA, USA. https://www.rfc-editor.org/rfc/rfc8446.txt

Verbauwhede, I. (2021). *The Cyber Security Body of Knowledge, Hardware Security, Version 1.0.1*. University of Bristol. https://www.cybok.org/

32 Software Security

John Marchesini and Sean Smith

32.1 WHAT IS SOFTWARE SECURITY?

When applied to software, the term *security* is challenging to define precisely. Some software vendors claim that their product is "Secure" or that it will "Protect You." Is security really a Boolean value, an adjective that can be used to describe a system? If so, why does any software lack this property? From whom or what are we being protected, under what conditions? In this section, we go behind the marketing slogans to define software security, discuss what makes it such an elusive property, and present several perspectives for thinking and reasoning about security.

32.1.1 Defining Software Security

What does it mean for a system to be "secure?" Informally, we think of something being insecure when a bad actor can somehow get the system to do something very wrong; a secure system is one that can resist such unpleasantness. This thinking thus brings up several concepts:

- The system doing something wrong.
- A bad actor doing something.
- Resistance to this action.

32.1.1.1 Functionality

Our approach to defining security centers on these concepts. To start with, if we talk about a system "doing something wrong," that means we have some notion of what it means for the system to behave correctly. For example, in (Smith and Marchesini 2008), we characterized the set of states a system can be in and asked "can the adversary lead the system into an incorrect state?"

Defining correctness requires looking at the system from different perspectives. First off, what is our software supposed to do? Early thinking about computer security focused on machines that stored information for groups like intelligence agencies, and so the initial correctness model for security (the "CIA" triad – see Section 32.2) focused on this stored information. However, the richer space of modern applications gives rise to many additional dimensions to consider.

- For one example, software that tracks real-world state (such as tracking the amount in a bank account) needs to stay in correspondence with the real-world state. If the software additionally interacts with the real world (e.g. by accepting cash deposits or telling an ATM to release a cash withdrawal), these interactions need to match the correct model.
- Depending on the application, we may also need to consider the physical aspects of the computation. For example, software controlling a relay in an electrical substation should not accept and act on an inauthentic command – but it may also need to act on authentic ones within a small time window. Taking too long to decide may be too late. Software controlling the operation of a rental car in northern New England may need to still operate correctly even when out of range of a cell signal.
- For software facing human users and connecting them to the Internet, we need to think about what the "correctness" of this interaction means. For example, in web browsers, the importance of ensuring that actions taken on behalf of a user originated with that user

DOI: 10.1201/9781003188377-38

leads to the "same origin policy." For PDF readers handling signed or certified content, it's important not just that the software correctly displays the status of this certification – it's also important that the human correctly perceive this display, and not be fooled by other content that gets rendered in a way that spoofs this output. (See (Rohlmann et al. 2021) for a recent examination of this problem.)

In summary: a programmer shouldn't just grab properties from a security textbook; the programmer should also think about what her or his software is supposed to be doing.

32.1.1.2 System Model

Software doesn't run on some abstract ethereal plane; it runs on physical machines in the real world. Most software runs on an operating system (OS) in either kernel mode or user mode. This OS may support other applications as well; potentially, the software may be designed to interact with them. As with the OS, we can ask the same questions about virtual machines (VMs): our software may be running inside one; other software may also be on the same host. Underlying all of this is the CPU, which may be single core, multi-core, or even heterogeneous multi-core; other software may be using these cores at the same time. Finally, the machine itself may be connected to a network and offer services on that network; our software may or may not be designed to use these services. (Section 32.3.3 expands on some of these topics.)

When considering software security, we also need to consider the rest of this system model. To the extent that we intend our software to interact with other software and with network interfaces, these other elements and their connection become part of the "correct operation" we wish to preserve. These other elements also may become vehicles for the adversary's actions: more avenues for the adversary to manipulate our system. Even if they are not part of our software's intended operations, these other elements are things our software can manipulate if the adversary compromises our software, and then uses it as a springboard to go further.

32.1.1.3 Threat Model

Having identified what the software is supposed to do and the system context in which it runs, we now need to think about the bad things the adversary might do. A standard way to break this down is to first think about the assets: what's valuable about this software? This consideration has several dimensions. We can think from the perspective of the defender, and ask: what aspects of correct behavior does our enterprise critically depend on? However, it's also important to think about what the adversary's objectives might be. Sometimes, these may have little to do with the software designer's ideas – maybe the adversary just wants to use our system to hide her tracks when attacking something else. Many IoT device compromises (so far) have been not motivated by breaking household appliance behavior, but rather by finding another foothold for a botnet.

We then must carefully consider the ways (the *threats*) that those assets can be compromised by an attacker. Can an attacker supply malicious input that forces the system out of the set of good states? If so, under what conditions? What is the interface to the system and how can other applications in the environment affect security? One aspect of thinking about the threats is the *attack surface:* where might the adversary launch an attack? The official interface to the software is the obvious surface, but software running on a real-world computer may have many other avenues as well. In the Stuxnet attack, Brutal Kangaroo malware was used to hop over air gaps – should that sort of attack vector factor into the threat model? (One of us had an IT security officer colleague who worried exactly about such inadvertent jumps.)

When thinking both about threats and assets, it's important to consider the "regular exceptional cases," not just the regular regular behavior. "Correct operation" of EMR software in a hospital may be especially critical in the edge case when a major accident (or a pandemic) has increased the caseload to an unusual high. The software controlling which university ID cards enable a door to be opened may usually operate with a network connection – but will a network outage, albeit rare, make it easier for an adversary to launch a successful attack?

A final aspect to consider is how strongly we want our software to resist attacks. One standard approach here is to determine an upper bound on the strength of the adversary we worry about. For example, if the attack surface includes an adversary having physical access to the computing machine, do we need to consider an adversary who can open and probe chips? Another approach we've seen is to place a lower bound on the damage of attacks we worry about. For example, colleagues in the power grid have told us that vandals frequently will use rifles to shoot insulators on transmission lines – and that any cyber attacks we defend against need to have more significant consequences than the use of insulators as target practice.

32.1.2 Why is Software Security Hard?

To put it bluntly: software security is hard because software is hard and (as we discussed above) security is hard.

The field of software engineering exists (and continues to exist) because when teams of developers write code, bugs are endemic. Even at the top-level of abstraction – the programming language being used – the interactions between the various moving parts can be complex and subtle.

However, to make matters worse, most software applications run at the top of a tower of abstractions. The instruction set architecture (ISA) abstracts the details of the CPU, compilers and language-specific VMs like the JVM and CLR abstract the processor's ISA, device drivers abstract hardware details, hypervisors abstract the physical hardware, an OS abstracts the hardware (via drivers) and protects processes on the same (virtual) machine from each other, and containers like Docker abstract applications from their environment. (In subsequent sections we discuss the security aspects of these issues in more detail.)

There are also a host of tools that operate at each layer that must work together to turn a high-level program into a series of electrical signals that drive the hardware: (dis)assemblers, loaders, linkers, interpreters, and compilers. These abstractions are there to reduce the complexity – you couldn't write a modern enterprise Software as a Service (SaaS) application in assembly. The state space is too big to fit in anyone's head. So, we build these layers to suppress detail. That means we have to take things on faith about the layer below us and make some assumptions: What size are integers on the CPU and in the VM? How will the compiler optimizations affect our program – particularly parts we've tried to make safe? Do the OS APIs behave as advertised?

We can put numbers to this complexity by comparing software to 747s. Which is more complex, the Linux kernel or a 747? By the latest count, one would have to collect five 747 jets before the number of parts (even things like rivets) matches the number of lines of code in the kernel alone. But what's worse is that a wild pointer in one line of code may cause failure in some completely different part of the kernel, and it's hard to imagine how failure of a part in one jet could lead to trouble in a different jet. Now, if we factor in the lines of code in all the other tools in our software tower, we can begin to appreciate why this is hard.

Furthermore, the underlying mathematics of computing shows that a general solution to security testing is impossible: there cannot exist an algorithm A that

- Takes any program P and an output of interest y
- And answers whether or not there exists an input x such that P on x gives y

32.1.3 Risk Management

The art of engineering involves making appropriate tradeoffs given a set of problem constraints. Balancing opposing forces lies at the heart of engineering and requires judgment and experience. Some of the classic tradeoff decisions in software development involve time vs. space and cost vs.

scope vs. time; the saying goes: "good, fast, cheap - pick two" (e.g., (MIcrosoft 2019)). As we discussed in the previous section, there is a natural tension between complexity and security which implies that we must manage these elements carefully if we want to get security right.

In order to make these types of management and engineering decisions, we must first be able to measure the quantities concerned. The right measurements for many of these elements are obvious and relative to the given problem space. For example, to measure processing time we could use milliseconds or CPU ticks, for project schedule time we might use months or number of the Agile/Scrum methodology's sprints (Schwaber and Sutherland 2020). Measuring something like scope is more ambiguous. Should we use the number of features, lines of code, or story points? Quantifying security is even more difficult; if we have a hard time measuring it, we have no hope of managing it.

The security industry has increasingly adopted tools from poker players, options traders, and insurance companies to help quantify security, most notably the concept of *risk management*. *Risk* can be informally defined as the probability of an outcome occurring multiplied by the impact of that outcome. In the context of software security, we are interested in the probability of the system getting into a bad state and/or exercising a vulnerability, and the impact of the system entering that state (Joint Task Force Transformation Initiative, NIST 2012). For example, to understand the risk of some particular use case we are implementing, we might try to assess the likelihood that an attacker Alice can get malicious SQL code that she inputs in our application's web form to execute in the database. To fully understand the risk, we would also need to assess the impacts of this occurring: is the database merely corrupted in the worst case, or does Alice now have access to all of the stored credit card information?

A key attribute of the risk management approach is that the probability of an outcome and its impact can be measured, or at least approximated and compared to other outcomes. When this attribute is missing (i.e., probabilities and impacts are not estimable) we are dealing with *uncertainty* (Knight 1921). Several processes, frameworks, and tools have been created to aid software vendors and IT organizations measure and manage these risks. Risk-based threat modeling frameworks such as STRIDE and OCTAVE (discussed further in Section 32.4.1) provide a systematic approach for quantifying risk. Once we understand risk, we can begin to address some of the most prevalent software security questions such as the following:

- How should we manage the complexity inherent in shipping features against the risk that complexity introduces?
- How much time, budget, and effort should we dedicate to mitigate our threat model?
- How do we manage time to market pressure vs. thorough security testing?

32.2 SECURITY MODELS

Much of today's thinking around security models came from the development of time-sharing systems in the 1960s and 70s such as the Compatible Time Sharing System (CTSS), the Dartmouth Time Sharing System (DTSS), the Multiplexed Information and Computing Service (Multics) and Unix. Such systems gave each user the illusion that they were alone on the machine, and as such were forced to design and implement models for isolating user processes from each other. Additionally, since multiple users were sharing the same memory and filesystem, the designers spent considerable effort controlling access to these shared resources. This was especially true for time-sharing systems used by the Department of Defense (most notably Multics) to store and process classified information.

32.2.1 Fundamental Security Properties

32.2.1.1 History

As just noted, much of the early thinking about software security focused on what that era considered important applications: machines storing important military and defense information. As a consequence, these applications gave rise to what analysts then considered (and what many textbooks today still consider) as the three fundamental security properties (ironically called "*CIA*"):

- **Confidentiality:** is the important secret information protected from being revealed to the wrong parties?
- **Integrity**: can the wrong parties change this important secret information?
- **Availability**: is this important secret information available to the right parties when they need it?

The importance of the CIA properties led to a lot of thinking (and even formal mathematical models) of how to ensure they hold –or at least the C and I parts. This thinking is probably best summarized under the term *multi-level security (MLS)*. First, we think of a set of levels of security – e.g., "unclassified," "secret," "top secret". We then order these by sensitivity – e.g., "unclassified"<"secret"<"top secret" (although in the general case, this does not have to be a total order). We label each piece of information in the system with one of these labels and label each party with the level for which they are cleared.

If we're interested in confidentiality, then we need to ensure that the system enforces "no read up" and "no write down" rules: if a party is cleared for level X1, then it cannot read anything at level X2>X1, and if it has read something at level Y1, it cannot write to any item at level Y2<Y1. (In its full formality, this is known as the *Bell-LaPadula Model*.) Conversely, if we're interested in integrity, then the system needs to enforce "no read down" and "no write up" rules: if a party has read a lower level of information and writes higher, then that higher level might be contaminated (the *Biba Model*).

If a system is supposed to enforce access rules such as this, then it needs to have a module of software that does this – this came to be called the *reference monitor* or sometimes the *security kernel*. This module actually needs to work – hence the principle of *complete mediation*: checking the rules at every access; hence also the concept of the *trusted computing base (TCB)*: the core modules that must work in order for the software to be secure. The administrator of the system needs a way to specify what rules should be followed – hence the idea of an *access control matrix* specifying the *security policy*: what *subject* A is allowed to do to *object* X.

32.2.1.2 Today

Earlier, we noted that securing software requires identifying what its correct behavior should be, and identifying how strongly it needs to resist what kinds of adversaries are disrupting it.

It turns out that even for software intended to store information in an MLS setting, the prospect of fully implementing something like the complete Bell-LaPadula model turned out to be too cumbersome and expensive. It also turns out that the MLS model doesn't capture many types of secure information flow that arise in the real world. For example, consider an investment bank where clients X and Y may have a potential business connection. Analyst A is free to look at either X or Y's information – but once A looks at one of them, then A cannot look at the other without creating the appearance of insider trading. This is sometimes called the *Chinese Wall* model: once A jumps from the top of the continent-wide wall to one side, A can no longer get to the other side.

Nonetheless, this early thinking is a good foundation to build on.

- Are confidentiality, integrity, and availability important parts of the correct behavior the system should have? Each problem domain may weigh these differently – e.g., critical infrastructure and other real-time systems applications may consider availability more important than the other two.
- Who are the subjects and what are the objects in your system?
- Where is the system's reference monitor? Does it work?
- What else is part of the system's TCB? Identifying the TCB is a key component of a threat modeling exercise (discussed in Section 32.4).

Thinking about information flow – and how to reason about and express the access control policy in a system – also became part of the standard security canon. Frequently, the access control matrix is implemented as a set of *access control lists (ACLs)*, where each object carries a list of who can access it and how. The Unix family of OSs still features a permission system ("owner, group, other") that's essentially a simplified ACL. (Less often, systems would attempt *capabilities* – where each subject carries around tickets to what it can in touch. However, this style never caught on on any large scale.)

32.2.2 Beyond CIA

Having a system enforce some kind of access policies for various users requires that the system knows who those users are! This leads to the concept of *authentication*: how does a user prove who he or she really is?

Traditionally, this was done by having the user sit in front of the machine and enter a username and password, and even this traditional approach raises security challenges:

- Users need to pick strong, unguessable passwords, and not write them down.
- The system needs to be able to verify passwords in a way that keeps them secure. For example, having the system store a master list of usernames and passwords is a bad idea – what if this list gets leaked? Fortunately, there are standard textbook ways of doing this securely; unfortunately, we regularly see reports of systems compromised because they didn't follow these practices.
- The system needs to take precautions against an adversary repeatedly trying to guess a user's password. Here, conventional wisdom used to have systems lock a user out after three incorrect attempts, although later thinking decried this as arbitrary (and possibly locking out legitimate users trying to use the system in a mission-critical scenario). Forcing a delay after each incorrect attempt and then doubling it each time may be a better balance.

However, the computing paradigm has been through many revolutions since that original thinking; consequently, many new approaches to authentication have emerged.

First, the notion of authenticating a user via her knowledge of a password generalizes to *knowledge-based authentication:* authenticating via other forms of knowledge. Then, we can think about other ways of authenticating besides knowledge. For example, the system can use *biometric* authentication: measuring something about the user (e.g., a fingerprint, iris patterns, typing rhythm). For another example, the system can require the user to possess some physical *token* (e.g., an RFID-equipped identity badge, or a smartphone tied to a specific phone number). In many enterprises, *multi-factor authentication (MFA)* – using two of these approaches – is standard.

As users begin to access systems not directly in front of them – e.g., via a network shell connection, or over the web – this physical separation introduces new risks. The user can no longer be trivially sure the system she is telling her password to is authentic, not an impostor. One way to address this problem is for the user to use cryptographic techniques (e.g., `https` with a trusted certificate chain, or `ssh` with a pre-established host public key) to authenticate the server before giving away secrets. Another approach is to reduce or eliminate the secret information a user must give away (again, there are standard cryptographic techniques here) in the event the remote system is indeed an impostor.

When a user's worklife in a large enterprise requires repeated access to many machines, the burden of authenticating to each one repeatedly can be onerous. Consequently, many enterprises use *single sign-on (SSO)* techniques, where a user needs to authenticate only once, and then that authentication follows them to other machines.

As the number of machines a user must work with increases and as the importance to the user of any one machine's security decreases, system security must rely more heavily on user behavior that becomes unreliable. If a user picks an easy-to-guess password or shares it with friends, then an enterprise can no longer be sure it's that user. If a user re-uses passwords across machines, then the security of a machine inside an enterprise becomes dependent on the security of machines completely outside of the enterprise. Countermeasures here include using standard tools to scan user passwords for guessability, and also to require users to regularly change passwords. However, these countermeasures still have challenges. Users who do not regard changing passwords as important will find workarounds; furthermore, fieldwork by many researchers (including one of us) reveals that users often share passwords not to be evil, but to be good: it appears to them that this is the only way to get their jobs done. (And often, they may be right!)

As computing became more advanced, the notion emerged that user *identity* alone is not always sufficient to control access. Consequently, much work has arisen in *authorization,* where the system tracks not just who you are, but also what you are allowed to do. One aspect of this blossomed into *role-based access control (RBAC),* where the system's security policy moves beyond matching users directly to permissions, and instead on the dual problems of matching users to roles, and then roles to permissions. Separating identity from authorization also allows for *delegation.* Rather than manager *A* giving her password to assistant B in order to complete necessary work tasks, manager *A* can delegate her authority to do those tasks to *B.* The need for systems across different enterprises to communicate and reason about authorization has given rise to sets of rich standards, such as the *Security Assertion Markup Language (SAML)* and *Open Authorization (OAuth).*

A system can use authentication and authorization to ensure that, in real time, user A is permitted to take action *X*. However, the challenge of later determining that action *X* did in fact take place and of establishing it was indeed user A who did can also be important. Hence, in important software systems, the flipside of authentication/authorization is *audit:* collecting these logs – but using chain-of-custody and cryptographic techniques to provide assurance of *non-repudiation:* strong evidence that a logged event really happened.

32.2.3 Software Security Design Principles

In addition to security models, the time-sharing era introduced several security focused design principles that are still relevant today. The most notable set of such principles came from Jerome Saltzer and Martin Schroeder in their 1975 paper "The Protection of Information in Computer Systems" (Saltzer and Schroeder 1975). The paper covers a wide range of security topics that arose from their experiences working with Multics, but their design principles are likely the most relevant aspect to a modern software developer. Their principles are paraphrased as follows (we discussed several of these in Section 32.2.1):

1. **Economy of Mechanism**: keep the design as simple as possible, especially for security-relevant parts of the system. As we discussed in Section 32.1.2, there is a tension between complexity and security.
2. **Fail-Safe Defaults**: the default security action should be to deny access. The system should grant access to resources only when that access grant is explicit.
3. **Complete Mediation**: every access to every resource must be checked for authority. Saltzer and Schroeder call this the "primary underpinning" of the protection system.
4. **Open Design**: the design should be open and security mechanisms should not rely on the ignorance of the attacker. A modern slogan for the anti-pattern of closed design is sometimes referred to as "security through obscurity."
5. **Separation of Privilege**: security mechanisms that require multiple keys, pieces of information, and/or parties are stronger than mechanisms that rely on one. A successful attacker has to compromise and coordinate multiple elements of the system instead of just one.
6. **Least Privilege**: every program and user in a system should operate with the least amount of privilege necessary to accomplish their task. This minimizes the impact of a security-related error and thus reduces risk.

7. **Least Common Mechanism**: minimizes the sharing of data, variables, and state between users and processes because such sharing can lead to an accidental exchange of information between parties.
8. **Psychological Acceptance**: the user interface (UI) must encourage users to apply protection mechanisms correctly each time. If the UI leads users to accurate mental models of the system, security mishaps will be minimized.

In his 2003 book, "The Art of Unix Programming," Eric Raymond captured the Unix Philosophy that originated from Ken Thompson's first Unix version (after Bell Labs withdrew from the Multics project), and evolved as the Unix culture flourished (Raymond 2003). This philosophy, which covers building systems in general, has several overlapping principles with Saltzer and Schroeder:

1. **Rule of Separation**: separate policy from mechanism because they evolve on different time scales. In a security context, access controls may change frequently while the policy enforcement code should probably only change if we have a bug.
2. **Rule of Simplicity**: keeping programs small and simple makes them easier to reason about. Again, complexity makes security hard.
3. **Rule of Transparency**: design programs to make them easy to read and debug. Design, and implementation, should be open.
4. **Rule of Robustness**: write programs to hold up under unexpected assumptions which stress the initial assumptions. This is especially relevant for security-critical code that will likely be the target of attackers.
5. **Rule of Least Surprise**: avoid gratuitous cleverness in the interface design. As Saltzer and Schroeder note, an interface which makes it easy for the user to build an accurate mental model of the system reduces the risk of an accidental security mishap.

As software systems have become more distributed and heavily networked, some new principles have emerged. These are especially relevant to cloud-based SaaS platforms that are comprised of (micro)services:

- **Defense in Depth**: avoid the temptation to assume that, just because an application is running behind a firewall, its networked surfaces are safe from attackers. Instead, design systems with several layers of defense so that if some outer layer (e.g., the firewall) is compromised, additional security measures are still in place.
- **Zero Trust Architecture**: taken to the extreme, defense in depth can lead to a paradigm where no implicit trust is granted to a user or account based on network location. These architectures move the trust boundary back to each surface, service, or asset in the system (Rose et al. 2020).

It is important to underscore that none of the above are formal design methods that, if followed, promise to yield a secure system. Rather, they are a set of pragmatic principles that emerged from the experience and practice of building real systems. They are akin to folk traditions and serve as guidelines for those building secure systems today.

32.3 SOFTWARE VULNERABILITIES

In this section, we explore some of the common software vulnerabilities, why they arise, and what can happen when they do. In Section 32.4, we'll cover several strategies and tactics used to mitigate the risk of shipping code with these types of issues.

32.3.1 Breaching Trust

A large number of vulnerabilities stem from programmers trusting input that they shouldn't. This input can be supplied by users or by other systems in the form of commands or data. For a very long time now, a huge fraction of vulnerabilities stem from the basic anti-pattern of a programmer assuming that input is structured in a certain way, and then acting on that input without first verifying that assumption (or, more subtly, only verifying it partially). As a result, clever adversaries can craft input that violates this assumption but which the program accepts and acts on anyway – and then goes down execution paths the programmer never envisioned. (For a sampling of the latest such vulnerabilities, search for "crafted input" at the Common Vulnerabilities and Exposures (CVE) database, or see Ali et al. 2021.)

Indeed, the mathematical underpinnings of computation (think formal language theory) can give fundamental insight here (Sassaman et al. 2013). Recognizing whether an input string is properly formatted (or the dual problem of *parsing* the string into a machine-readable representation of its structure) can be surprisingly difficult or even fundamentally impossible if the language is sufficiently complex. What can be even more vexing is that for many classes of languages that are recognizable, the problem of verifying whether two recognizers actually do the same thing is uncomputable. Indeed, *parser differential attacks,* where an adversary crafts input that is seen by two different programs as saying two different things, are common occurrences – e.g., (Kaminsky, Patterson, and Sassaman 2010, Siguza 2020).

An ancient (but still common) example of this anti-pattern is stack-based *buffer overflow.* A program allocates a buffer as a local variable, and then copies externally-supplied input into that buffer without first checking whether the input fits. In the C programming language, this is very easy to do, as we will discuss in Section 32.3.2. If the user supplies input that is too long, the program dutifully copies the extra bytes into the space beyond the buffer, overwriting whatever was there. Although ancient, buffer overflow remains a real problem – e.g., it was the basis of one of the vulnerabilities in the high-profile Ripple20 compromise (JSOF 2020) in software used in hundreds of millions of computers. Buffers do not need to be on the stack to be at risk; heap-based buffer overflow (due in part to code not checking when heap allocation failed) was one of the issues in 2020's zero-click iOS trouble (ZecOps Research Team 2020).

A question one of us often asks in the classroom is "how big is an integer?" In the world of pure math, such a question sounds absurd – an integer, positive or negative, can have arbitrarily large magnitude. However, in software, an integer is usually represented as a fixed number of bits (e.g. 8, 16, 32), and the software needs to decide whether to interpret these bits as a non-negative integer or as a signed integer – implicitly putting some limits on the range of values that can be stored there. Input that violates those assumptions can trigger *integer overflow* or *integer underflow* attacks. For some simple examples, a railroad control system in Switzerland would completely fail to detect a train that had exactly 256 axles (8-bits unsigned, overflowing to zero); the Solaris OS used to track bytes of RAM used is a signed 32-bit integer, which was fine until machines starting having gigabytes of memory. For more recent security-specific vulnerabilities, both Snapdragon and SigRed had scenarios where lengths provided by the adversary led to integer issues internally which then led to buffer overflows.

Adversarial input tricking a program into writing data into the wrong place can lead to a range of program behavior useful to an adversary. At first glance, perhaps the most obvious is allowing the adversary to change critical internal data fields. However, buffer overflow via crafted input can also lead to data exfiltration (e.g. Heartbleed, Ripple20).

What can be more insidious is when the crafted "data" the adversary inserts then gets treated as "code" by the machine (e.g., Ansley et al, 2007). In the classic stack-based buffer overflow attack (Aleph One 1996), the user is asked for a short character string but actually provides a sequence of executable machine instructions, followed eventually by the address of those instructions overwriting where the program's current function had left its return address. When the function returns, the machine executes the adversary's code. Subsequent families of attacks (e.g., Shacham 2007) can perform essentially the same result via the adversary injecting sequences of addresses that trick the program into executing an adversarial-chosen sequence of *gadgets* inside the program's own code. Observing the blurry distinction between code and data, some researchers call this "programming the weird machine." Programmers interested in securing their software should continually ask themselves: "Does this interface contain any weird machines?"

32.3.2 Programming Language Security

For the last two decades of the twentieth century, most software was written in C/C++. As we discussed in the previous section, these languages were (and still are) particularly susceptible to classes of vulnerabilities involving memory (mis)management – e.g., buffer and heap overflows and format string vulnerabilities. What is it about these and other programming languages that makes them so prone to this sort of security trouble?

32.3.2.1 Memory Safety

In part, the answer lies in the concept of *memory safety.* A programming language that prevents programs written in that language from accessing memory locations outside their intended memory space is memory safe. Protecting a program's memory is a necessary, though not sufficient, condition for security. Languages that lack this property (such as C/C++) allow programs to read and write arbitrary memory locations, which implies that seemingly innocuous programming errors can result in an attacker having the ability to write malicious data or code into memory addresses of their choice. In terms of our state model of security from Section 32.1, a lack of memory safety allows attackers to inject new states into the system and/or transition the system into a bad state.

Many programmers have moved on from languages without memory safety. However, there are still a number of application domains that require the programmer to have complete control over the memory space. Applications where the hardware is unique or constrained (e.g., IoT devices and robotics), and/or where performance is critical (e.g., OSs, the JVM, web browsers, and databases) still tend to be written in C/C++, and thus still need to be concerned about memory safety.

A particularly nasty illustration of memory safety issues came in the form of a set of 19 vulnerabilities released in 2020 under the name of Ripple20 (JSOF 2020). These vulnerabilities were discovered in a low-level custom TCP/IP library (written in C) which was sold to device manufactures across a wide range of industries including medical devices, avionics, transportation, robotics, industrial automation, embedded and other IoT devices. Because this library was so widely used across the device supply chain, the impacts "rippled" to an estimated hundreds of millions of devices worldwide. In addition to the buffer and heap overflow vulnerabilities discussed in the previous section, languages without memory safety are susceptible to other types of memory (mis) management issues as well. As we will discuss throughout this section, several of the vulnerabilities in the Ripple20 set fall into one of these categories.

32.3.2.2 Memory-Related Vulnerabilities

One class of common vulnerabilities involves using a variable before it has been properly initialized – i.e., when its contents are undefined. Depending on the data that occupied the memory location before, using a variable before it is initialized can potentially leak interesting information. While these types of vulnerabilities are typically considered low-risk because of the difficulty of executing a repeatable exploit, several researchers at Arizona State University and Samsung Research published an approach that reliably uses uninitialized variable vulnerabilities in the Linux kernel to leak the location of kernel pointers (Cho et al. 2020). These pointers can in turn be used to defeat kernel stack overflow protection mechanisms like Kernel Address Space Layout Randomization, where the layout of the kernel stack is randomized to make it impossible for shellcode to deterministically locate memory addresses. (Ironically, reading undefined variables can be useful in establishing entropy for cryptographic keys; sadly, in 2006, a major security hole emerged in Debian `SSL` because developers, concerned that such access was bad, commented it out.)

Another class of vulnerabilities related to memory stems from using a chunk of heap-allocated memory after it has been freed, either directly or by freeing it again. Most often, these vulnerabilities arise because a single chunk of memory has several pointers to it throughout the program, especially if the chunk holds some data structure and the various pointers point to different parts of the data structure. One pointer may be set to null once the chuck has been freed, but others may not. These other pointers may lead to heap corruption and/or a heap overflow.

One of the 19 vulnerabilities in the Ripple20 set was the result of using memory after free. This critical vulnerability was in the IPv4 Tunneling portion of the affected TCP/IP stack and lets an attacker with specially crafted input cause a memory corruption issue. The impacts vary by vendor, but several network device vendors released advisories stating that this particular issue allows unauthenticated attackers to cause a remote reboot of network hardware.

Yet another class of memory vulnerabilities arises from "minor" errors in determining length when copying too much data into a buffer. These off-by-one (or off-by-a-few) errors write bytes out of the buffer bounds and are sometimes caused by incorrect null termination of strings. (Recall that C strings end with a null character.) Vulnerabilities can also arise when a programmer miscalculates the index of where the null character belongs and writes it outside of the buffer bounds. When these situations occur, the C string handling functions, which assume that strings are null terminated, may read past the end of a buffer into adjacent memory. This can disclose data or lead to potentially exploitable memory corruption issues.

Another one of the Ripple vulnerabilities was caused by improper null termination in the stack's DHCP component. Although less critical than many of the other vulnerabilities, this issue lets an unauthenticated attacker send a specially crafted packet (recall Section 32.3.1) to a vulnerable device and potentially read sensitive information. A more serious example was found in versions of the popular Linux command `sudo`, a command used to temporarily escalate user privileges. An off-by-one error in `sudo` could be used by an attacker to launch a heap overflow which resulted in the attacker bypassing `sudo`'s policy mechanism and escalating their privileges to root.

32.3.2.3 Type Safety

Starting in the nineties, languages like Java and C# were introduced and have since gained widespread adoption, particularly in (enterprise) application and web development. These languages possess a property closely related to memory safety: *type safety*. A type safe language prevents classes of errors that occur as the result of assigning elements of one type to another. For example, in a type safe language like Java we cannot assign a variable of type integer to another variable of type string. We may convert the integer value to its string representation and then perform the assignment to a string variable, but direct cross-type assignments and operations are prohibited.

We care about type safety from a security standpoint because type safe languages are also memory safe. That does not imply that using a type safe language means you will write secure software. There are still plenty of pitfalls such as mishandling cryptographic keys, passing sensitive information in the clear, and generally violating the principles discussed in Section 32.2. It does imply that using a type safe language will free you from the memory corruption vulnerabilities we have been discussing. In order to make language type safe, the compiler and runtime (e.g., the Java Virtual Machine (JVM)) must carefully check stack and heap memory accesses to ensure that objects are of the correct type. This type checking eliminates the ability to read and write arbitrary memory locations, which makes the language memory safe as well.

From a security standpoint, type safe languages are a better choice, provided your problem domain does not require low-level control of the hardware. If you must use a memory unsafe language like C/C++, memory issues must remain top of mind. We will discuss some concrete tactics in Section 32.4.2.

32.3.3 Runtime Privileges

Being able to reason about software security at runtime requires understanding the underlying system model. At a high level of abstraction, the developer might think of the program somehow translating down to a sequence of instructions executing on a machine it has all to itself, maybe occasionally requesting special services from the OS. At a lower level of abstraction, the developer may realize that there are other programs executing on the machine (and the developer's program may even interact with them, via OS services). However, the developer still has the expectation that, except for explicit interactions, each of these programs acts "as if" it had the machine to themselves – the developer has an expectation that his or her program will run in isolation from the others and that the system will protect the developer's program from inadvertently doing anything bad to the rest of the machines.

32.3.3.1 Privileges

Modern computing systems aspire to achieve these goals of isolation and protection via a conspiracy of the underlying OS working with some CPU features:

- **Privilege**: CPUs typically have a flag storing the current *privilege level*. The hardware uses this privilege level to allow or disallow certain operations. Informally, one can think of these as two levels – "privileged" and "unprivileged."
- **Exceptional Control Flow:** CPUs have ways (such as an external interrupt, an explicit software-requested trap, or a software exception) to suspend the program currently executing, change the privilege level to "privileged," and begin running special code at a special location.
- **Memory Management Units:** CPUs also typically have MMUs which translate the *logical* addresses issued by the CPU to *physical* addresses seen by memory. The mapping can also be configured to reject certain memory accesses (e.g., all access to some address, or only reads, or only instruction fetches). Controlling the details of the mapping the MMU currently uses requires privileged mode.

Userland software is what runs in unprivileged mode, with the expectations of isolation and protection. *Kernelland* software runs in privileged mode and typically consists of the bulk of the OS, which works with these exceptions and MMU tricks (and also deciding which userland processes get which CPUs and resources when) to aspire to achieve these expectations.

In the Intel x86 architecture family which came to dominate the universe, privilege levels are called *rings,* and there are four: ring 0 (kernelland), ring 3 (userland), and two intervening rings that do not see much use. Modern large CPUs may also provide support for *virtualization* (e.g., running many different OSs on the same machine at the same time); on such machines, the term "*ring -1*" may be used to refer to the code and hardware that manages these various ring 0 kernelland instances.

The security of a program thus can significantly depend on the details of this underlying system model. First, following the dictum of *least privilege* above, the developer should aim to run the target program in unprivileged mode (to limit the damage of direct compromise). If the developer's notion of "secure behavior" requires isolation or fairness of resource usage between the target program and others running on the machine, then the developer should try to be as specific as possible about what such isolation means – and also try to gain assurance that the underlying OS and CPU actually provide these properties. As subverting the OS can subvert the target program, the target program also inherits the attack surface of the underlying OS. Given the massive complexity of OS software, bugs are endemic – hence ensuring the secure operation of a target program, even if the program doesn't change, usually requires continual patching and updates to the underlying OS. If the OS can be compromised, the target program may also inherit the attack surfaces of the other programs running on that machine – thus many installations will end up running mission-critical programs each on their own machine.

In order to reduce security dependence on the OS (and its administrators), higher-end CPUs in the two decades have begun supporting *enclaves* (e.g., Intel SGX). To the multiple levels of privilege, the CPU adds another dimension: "ordinary" or "secure." In theory, userland software running in secure mode is cryptographically protected even from privileged software. In practice, we've seen a "cat and mouse" game of the continual discovery of and fixing of security bugs in enclave implementations, so we expect some churn here for a while.

32.3.3.2 Containers

With the move toward delivering software across a variety of environments (e.g., on premise or in a cloud environment) and OSes, many vendors now deploy their code in a user space sandbox called a *container.* Container technologies like Docker allow a piece of software and all of its dependencies to be packaged together and run on any OS and machine where the Docker runtime is installed. This approach simplifies the work of managing dependencies and configuration and thus provides a higher degree of process isolation since processes in different containers do not share dependencies.

Container security and the security of similar technologies (such as `chroot` jails on Linux) rely on the assumption that processes running in the container are sandboxed and running at the appropriate privilege level. However, if an attacker in the container can escalate to root privileges, then the attacker can often break out of the sandbox and keep the elevated privileges.

32.3.3.3 When the Foundation Fails

As we noted above, the security of software depends on the system underneath it. Sometimes, these components fail.

At the CPU level, a good example here is the family of transient execution side-channel attacks that recently emerged, the most notable of which were Spectre and Meltdown. As noted above, CPUs use MMUs to keep a rogue userland process from spying on data that belongs to another process. However, modern CPUs also use rather complex pipelining and caching techniques to improve performance – and it turns out that a perfect storm of subtle errors in memory exceptions, caching, and pipelining could indeed enable rogue processes to violate the developer's expectation of isolation. (Loughlin et al. 2021 provides a very thorough discussion of attacks and defenses here.) At the OS level, a quick look at the CVE database (discussed in Section 32.3.5) just over the previous two months shows several denial-of-service and privilege escalation bugs in Linux, and several privilege escalation and remote code execution bugs in Windows.

Another example of runtime mischief involves an attacker breaking out of a container environment and getting code to execute on the underlying host – likely running as root (The MITRE Corporation 2019). Container environments such as Docker rely on an underlying runtime called `runC` to establish and maintain the sandbox that a container provides. `runC` either establishes a new process in a container or attaches to an already running process by placing itself in the container and then making an execve system call to overwrite itself with the requested binary. A clever attacker can use the `proc` filesystem to trick `runC` into overwriting itself on the host. The next time that `runC` is executed on the host, the attacker's code will be executed instead of `runC`, and because `runC` runs as root, the container-based attacker has effectively root privilege code execution on the host.

32.3.4 Supply Chain Attacks

In Section 32.3.2 we introduced the Ripple20 set of vulnerabilities as an example of a *supply chain attack:* attacks that target components early in the supply chain with the intention that the impacts will get magnified as that component travels through the supply chain. The Ripple20 vulnerabilities targeted a TCP/IP stack that was sold to device manufacturers. Had the stack been something that was sold directly to consumers in the same manner as Microsoft's Excel, then the damage would have been limited to the direct users. Instead, because the stack is a component found early in the supply chain for several types of embedded devices, the vulnerabilities impacted the customers of all of the companies that use the stack in their device and allowed Ripple20 to affect hundreds of millions of devices.

Most real-world software applications sit at the end of a complex supply chain that includes the tools and libraries that developers use to build the system, as well as the underlying hardware and software upon which the system runs. The fact that these tools change frequently and on different schedules makes managing this complexity a difficult task.

32.3.4.1 Compilers

Ken Thompson, in his 1984 Turing Award acceptance speech, gave an excellent example of the type of havoc that a mischievous compiler developer can cause to anyone writing code in the compiled language (Thompson 1984). He starts with a simple C program that prints itself and works his way up to a C compiler with two interesting properties:

- When it recognizes that it is compiling the Unix `login` command, it modifies the code to allow a special password of Thompson's choosing.
- When it recognizes that it is compiling a C compiler, it modifies the code to insert the previous "feature" along with this one.

Such an attack so early in the software supply chain would have given Thompson access to every Unix system in operation. (We have heard from one Bell Labs veteran that Thompson really did write such a compiler, and it very nearly escaped into the real world.)

In addition to malicious compilers, we must also be concerned with security-relevant side effects that a compiler can have on our code. Modern secure coding practices encourage developers to scrub memory locations that contain sensitive information such as passwords and cryptographic keys. Depending on optimization settings, some compilers may optimize away the functions and/or function calls responsible for such memory hygiene because they (rightly) detect that the memory is not used again. While this does in fact improve performance, it leaves sensitive data in memory (e.g., D'Silva, Payer, and Song 2015). An adversary that can inspect memory can capitalize on this and potentially use the secrets in these memory locations to propagate their attack.

32.3.4.2 IDEs

Another staple of the developer toolkit is the Integrated Development Environment (IDE). Modern IDEs are complex pieces of software in their own right, often featuring embedded scripting engines (e.g., `emacs` Lisp), library management features, integration with other systems in the developer environment (e.g., source control systems and container repositories), and integrated networking features such as embedded web servers and support for remote pair programming.

With so much complexity, so much focus on developer productivity and such an enormous install base, IDEs present another target in the software supply chain. In his Defcon 29 talk, David Dworken outlined several vulnerabilities over a wide range of IDEs from `vim` to VSCode, many of which result in remote code execution (Dworken 2021). As David points out, the security boundary for these systems is unclear; the mental model that we are safe using an IDE to just read untrusted code is not accurate. Most IDEs have integrated linting support that starts when the file is merely opened. Simply opening a malicious project can force your IDE to execute code and potentially even leverage the IDE's native integrations to propagate itself.

Google's Threat Analysis Group issued an update outlining a campaign by a North Korean "government-backed entity" that uses these techniques and social engineering to target security researchers (Weidemann 2021). The attackers invite their victims to join forces in researching a particular vulnerability and send them a Visual Studio project with the exploit. The project also contains malware that begins executing once the project is open. Keep in mind that the victims here are security researchers which are, generally speaking, a pretty untrusting group.

32.3.4.3 Libraries

A vast majority of modern software is built with the aid of other software in the form of libraries. If we remove the obvious library overhead of running in a VM such as the JVM, we can get down to a C program which still relies on a considerable number of libraries. A typical C compiler such as `gcc` is really a collection of programs that includes an assembler and a linker which pulls in system and language libraries (e.g., `glibc`) for the compilation target. Your program begins execution long before its main function is called, and all of that execution occurs in libraries that you probably did not write (and which may even change over time, even if the "program" does not).

If we consider the higher-level languages that most modern web and enterprise applications are written in, the percentage of library code in an application explodes. There are the runtimes and core language libraries such as the JVM, and the associated libraries that make up the chosen language (e.g., JVM languages Java, Scala, Groovy, Kotlin or Clojure). There is also an entire compiler toolchain for the chosen language to consider, as well as any prevalent frameworks often used to build enterprise or web applications (e.g., Spring Boot, ASP.NET). Finally, there are the third party components that the application depends on, including open source and commercial components. Managing these types of dependencies is a large enough problem that most modern languages include a package management framework such as Node.js's rpm, the JVM's Maven and Gradle,. NET's NuGet, and Rust's Cargo.

Finally, in addition to the libraries themselves, the systems that manage, build, package and release your application are part of this equation as well. The modern incarnation of this collection of systems is called the *Continuous Integration and Continuous Delivery* (CI/CD) pipeline. This includes an organization's source control, build systems, containerization technology (such as Docker) and deployment systems (especially for SaaS applications).

It is both important and terrifying to realize that all of these elements comprise the software supply chain and a weak link in any of these components can have a significant impact on your system. Some parts of the supply chain, most notably OS vendors, have been targeted for a long time and have built processes to help downstream developers manage risk. For example, since 2003 Microsoft has released security fixes for its OSs and core software on the second Tuesday of every month. Many other large software vendors have adopted similar practices and cloud vendors such as Amazon Web Services are likely to patch even more frequently.

Organizations later in the supply chain have a more difficult time and a more pressing need to manage the risks inherent in the supply chain. There are commercial auditing tools that can help monitor dependencies and notify when new updates are available. In our experience with large commercial codebases, such tools are invaluable, provided that they were not compiled with Ken Thompson's compiler.

32.3.5 Knowledge Bases and Threat Databases

So far, we have discussed a wide range of software vulnerabilities in this section. We have covered programming errors at several levels, provided several examples to illustrate what happens when they are discovered by security researchers in the wild, and shown how their impacts can be amplified by their position in the supply chain. For a developer, researcher, or manager, the volume and variety of this information can be overwhelming. Without a framework or taxonomy to help categorize this security information, it is impossible to manage risk consistently.

Many vendors and large open source projects have their own risk and severity scale for the vulnerability advisories they release, and we encourage you to get familiarized with the formats and scoring systems for the vendors that affect you. Here, we introduce several open, authoritative sources that make up the backbone of the security information flow.

32.3.5.1 Common Weaknesses Enumeration (CWE)

Most or all of the programming errors we have discussed so far are covered by The Mitre Corporation's CWE list (The Mitre Corporation 2021a). The list is an organized catalog of nearly one thousand hardware and software weaknesses that are known to cause security trouble. Each weakness is assigned a CWE number and the catalog contains a description of the weakness, relation to other weaknesses, common consequences, and their likelihood, sample code in several languages to illustrate the issue, and several mitigation tactics. The site also scores these entries based on risk and releases an annual Top 25 Most Dangerous Software Errors.

For example, in Section 32.3.2, we covered several such weaknesses: improper input validation (CWE-20), use of uninitialized variable (CWE-457), double free (CWE-415), use after free (CWE-416), improper null termination (CWE-170), and off-by-one (CWE-193). This catalog is incredibly useful for software developers and auditors, and we definitely encourage you to read the CWE entries for the issues discussed in this chapter.

32.3.5.2 Open Web Application Security Project (OWASP)

Another useful resource is the OWASP Foundation's annual Top Ten Web Application Security Risks (Open Web Application Security Project 2021b). Unlike the definitive CWE catalog, the Top Ten is a list of the most common vulnerabilities affecting web-facing applications, although many of the items apply to applications in general. Each entry in the Top Ten has a description, sample exploit, and instructions on how to prevent the issue. It references all of the applicable CWEs as well as other OWASP documentation, such as their Testing Guides and Cheat Sheets (discussed further in Section 32.4) that present common patterns and best practices.

As we will discuss in Section 32.4, OWASP develops automated testing tools for these issues that can and should be integrated into the testing process. Many application penetration test firms use the Top Ten as the starting point for their assessment. If your application is web-facing, you should keep up with the Top Ten; attackers certainly do.

32.3.5.3 Common Vulnerabilities and Exposures

Once a security incident is discovered, the best one can hope for is responsible disclosure. In this case, the security researcher who discovered the issue will work with the vendor to inform customers and prepare a patch. Part of that process entails getting a unique CVE-ID assigned to the vulnerability.

Mitre's CVE database contains over 16,000 entries for reported vulnerabilities (The Mitre Corporation 2021b). These entries contain a description of the vulnerability, links to the vendor's or project's advisory, relevant CWE entries, and a link to the National Vulnerability Database (discussed next) for further detail. Importantly, the CVE-ID is a *de facto* unique identifier for the issue and is used in vendor documentation and press coverage of the incident. For example, the Ripple20 package of vulnerabilities discussed in Section 32.3.2 contained 19 CVEs: CVE-2020-11896 through CVE-2020-11914.

32.3.5.4 National Vulnerability Database

In 2005, the National Institute of Standards and Technology (NIST) released the National Vulnerability Database (NVD) to expand the information given in a CVE record (NIST 2021). The NVD is regularly synchronized with the CVE list and enhances the CVE content with topics such as patch availability and severity scores. The Common Vulnerability Scoring System (CVSS) is a mechanism that considers several aspects of the vulnerability – such as its complexity; whether privileges are required; impact on confidentiality, integrity and availability – and assigns it a score (FIRST 2021). This score can then be used to help individuals and organizations assess and manage risk.

32.4 BUILDING SECURE SOFTWARE

In the previous section, we discussed several types of vulnerabilities and gave examples of their impacts. It is natural to wonder why software companies ship products with such vulnerabilities in the first place; no other type of manufacturing industry has such a lousy quality record. At a high level, we see two key reasons: complexity and economics.

First, software companies typically compete on the number of features in their product, and these features generally increase the size of the codebase and thus increase complexity. Second, most commercial software companies are not financially liable for shipping bad code. A vulnerability might impact the company's brand, but does not carry the same magnitude of penalties as building a defective pipeline that pumps millions of gallons of oil into the Gulf of Mexico, for example. Taken together, these imply that commercial software companies are balancing the market pressure of shipping features (i.e., complexity) against the probability and impact (i.e., the risk) of shipping a vulnerability. When the impact for shipping vulnerabilities is low, complexity wins and the result is the security landscape we see today. This situation may be changing; legal issues and regulations, particularly around data and privacy, are increasing the cost of vulnerabilities and forcing companies to rebalance their weightings of complexity and risk.

On a practical level, features are built and risk is measured throughout a company's Software Development Lifecycle (SDLC). The SDLC is a commercial software organization's core manufacturing process that covers requirements gathering, design, implementation, testing, and operations and maintenance. There are numerous prescriptive methodologies for tuning the SDLC to produce features (e.g., Rational Unified Process, Extreme Programming, Agile/Scrum, Kanban). In this section, we outline some of the tools and techniques that can be used at each phase of the SDLC to measure and mitigate risk.

32.4.1 Security Requirements and Design

The SDLC typically begins with formulating requirements and feature design. Feature requirements (often called *functional requirements*) are usually formulated by a Product Management team and specify the use cases, interactions and user interface that users will experience as they use the feature. These requirements are then discussed with senior technical staff (such as an Architecture team) that will add *non-functional requirements* such as performance, interoperability, and security concerns. Once both types of requirements are understood and agreed upon, some sort of feature design and planning phase begins, the details of which will vary depending on the development methodology.

While many of the vulnerabilities we discussed in Section 32.3 involve implementation, we cannot stress enough how important it is to start thinking about security in these early requirements and design stages. Bugs in general (and security bugs in particular) get more expensive to fix as they go through the SDLC. Catching a design flaw at final testing time, or worse: in the field, is far more expensive than catching it at design time, before any production code has been written.

There are several tactics that designers employ to check for vulnerabilities in the design, such as verifying that the feature or system supports attack detection, resists attacks, and reacts to and recovers from attacks as appropriate. As system builders, we are always on defense and have limited resources to deploy in order to defend from a seemingly infinite number of potential attack vectors.

32.4.1.1 Threat Modeling

One best practice for determining which security mechanisms are best suited to the specific feature or system is Threat Modeling (Shostack 2014). Like the term "security", the term "threat modeling" is not easy to define; there are threat modeling frameworks, the threat modeling process, and the threat model itself. Complicating matters further is that there is no universally accepted or standardized meaning behind these terms. However, the essence of this process is straightforward and valuable: to take a risk-based approach to identify the most likely sources of attack and their likely impact on your specific feature or system under review, and use the limited resources at your disposal to protect against those attacks. This is much like a game of chess where each player is trying to anticipate the likely moves of their opponent and position themselves to protect against, or capitalize on, such moves. Here, we use the term threat modeling to describe this process, and the term threat model to loosely convey the abilities of our opponent (the attacker) as we discussed in Section 32.1.1.

Generally, threat modeling begins by examining the system or feature from an attacker's perspective in an effort to identify the *attack surface*: the mechanisms by which an attacker can get input into the system. This includes user-facing input mechanisms such as fields in the user interface, as well as any network facing APIs (common in SaaS applications). This can also include alternate ways to get data into the system such as the filesystem (e.g., configuration files), messaging technologies, databases, and the like.

Once the relevant input channels have been identified, the next step is to identify the *assets*: the system elements that an attacker would be interested in. These include obvious items such as credentials, cryptographic keys, and security policies; and more subtle items such as access to a piece of the system that will execute code, or that is responsible for enforcing security policy. As assets are identified, the necessary permissions to access each asset should also be recorded.

With entry points, assets, and access levels in hand, the next step is to think about the use case or system as it is executing. This entails mapping the flow of data from inputs to assets, noting where system and security boundaries are crossed. Returning to our chess analogy, this is the stage where threat modeling moves from looking at the current board position to imaging sequences of moves that an attacker and the system might make. The key here is to think critically about each scenario: challenge design assumptions, assume all input is malicious, and consider other parts of the system and/or external dependencies have been compromised.

Finally, as threats are identified, they can be categorized using one of several frameworks such as STRIDE. Under STRIDE, each thread is categorized as **S**poofing, **T**ampering, **R**epudiation, **I**nformation disclosure, **D**enial of service or **E**levation of privilege. These threats can then be used to refine the design so that it can protect against the ones deemed most likely to occur and/or would have the most impact if exploited.

While not a perfect methodology for generating a secure design, we have found through decades of practice that the exercise of thinking critically about the systems we are building – especially the security-relevant use cases – is the best available way to find security gaps in our designs. Threat modeling gives us a methodical way to reason about the capabilities of our most likely attackers, gives us guidelines for allocating our defense resources, and ultimately helps us manage risk.

32.4.2 Implementation Security

Once requirements and designs are settled, developers set out to turn designs (which often include prototype-quality code) into production-grade software, and it is in this phase of the SDLC where many security bugs are introduced. Much of the likelihood that developers can "get it right" without introducing a vulnerability into the code base is determined before they write their first line of code.

For starters, the design we are implementing could have a security flaw that made it through the design phase. In practice, we ourselves have been given requirements for interfaces and protocols that have had real security problems. Second, as we discussed in Section 32.3, the tools and toolchain may make it hard to get it right. For example, if we are building a network-connected application in a memory unsafe language like C, it will take extra effort to ensure our code does not contain a vulnerability. Finally, does the codebase we are working in have a high degree of what Martin Fowler calls *internal quality* (Fowler 2019)? I.e., does it have a large amount of technical debt that complicates decision-making; is it layered and loosely coupled so that developers can make changes without inadvertently breaking other parts of the system; does it have adequate test coverage so that refactoring is easy?

Such issues aside, there are several patterns and practices that developers can and should employ to reduce the chances that they get into the types of security trouble we discussed in Section 32.3.

32.4.2.1 Trust

In Section 32.3.1, we covered several security issues that result from the system trusting input from users and/or other components of the system (e.g., other microservices). Our key advice here: trust no one; treat all input as malicious. In design sessions (and sometimes design documents, depending on the feature), we insist that the team replace the terms "input" or "user input" with "attacker input." This simple change of language used in a design discussion helps establish a security mindset and puts everyone on the defensive – from the designers in the room, to the developers building the feature, to the managers who are allocating resources and making risk calculations.

For developers, the best tactic we can suggest is to validate input religiously. In practice, we see much validation occurring in places like UI form fields, which is a fine start. However, the attack surface is often much larger than that and APIs are a significant part of that surface, particularly for cloud-based software. We have seen many APIs that impose robust authentication (and possibly authorization) procedures, and then just assume that every authenticated user is a good actor.

Best practice is to treat all input with suspicion; be on the lookout for escape sequences and code injection and look for reputable libraries in the appropriate language to help with this task. Doing input validation with a ball of *ad hoc* regular expressions scattered throughout the codebase is an anti-pattern and ultimately reduces the internal quality of the code. If you must roll your own, build a reusable, testable component and employ fail-safe defaults as Saltzer and Schroeder suggest. Also, keep in mind that (de)serializers and other parsers could also be processing attacker supplied input, especially if they are processing data passed in via API. Those should be made robust enough to handle malicious input, which should also be part of the test suite. (There has been a lot of recent progress on *parser combinators* and other tools to make it easier for the security-minded developer to validate input.)

32.4.2.2 Programming Languages

In Section 32.3.2, we discussed how features like memory and type safety impact the risk of using a specific programming language. Much of today's enterprise and application software is written in a statically typed, type safe programming language like Java or C#. While these languages and their associated runtimes and libraries are a good start from a security standpoint, it is important to remember that they are not magic bullets. It is still possible to implement bad protocols and generic logic errors and to allow bad data into the system. Attacks that involve hooking the.Net CLR and breaking out of the Java sandbox are still being developed (Parata 2021, Eauvidoum and noise 2021); beware of the false sense of security that these languages can provide.

Working in a memory unsafe language increases the level of difficulty a programmer must overcome to get things right. The need for input validation increases since there are classes of overflow attacks that rely on confusing code and data. Additionally, the added burden of managing memory opens up the possibility for vulnerabilities caused by using uninitialized variables, double freeing, and null termination. Best practices here include using a well defined coding standard that is tailored to security, such as CERT C (CMU Software Engineering Institute 2018). Such standards not only embody security best practice but they lend themselves to be checked by running a static analysis tool over the codebase, which we highly recommend running as part of the build process (discussed further in Section 32.4.3). There are many commercial and open-source static analysis tools to choose from and although not a panacea, can help reduce risk, particularly when paired with a secure coding standard.

32.4.2.3 Runtime Privileges

Recall that our notion of security is rooted in the system's ability to remain in the set of correct states, even in the presence of a malicious actor. To this end, the developers implementing the system must build the system in such a way that maintains this property by preventing the system from transitioning into a bad state. This involves managing the state of the system in its runtime context and carefully considering the system's interactions with its environment. As we explored in Section 32.3.3, things like underlying hardware, OSs, and container technology can be threats themselves.

Often, these types of components offer services to an application and are therefore implicitly trusted. Perhaps the best advice we can give here is "trust, but verify." For example, when making system or Win32 calls, developers should thoroughly check return codes and handle them accordingly. Such codes often give an indication of the state of the environment, especially when considered as a set or sequence. Additionally, many modern workhorse languages also include structured exception handling facilities. Exceptions are incredibly useful to developers as they often carry more detailed information than simple return codes and give more information about the state of the system, such as a call stack and execution thread details. Finally, certain classes of bugs such as *race conditions* and *time of check time of use errors* are more subtle and difficult to detect because of their non-deterministic nature. In practice, in real-world production environments, there may be no easy or even systematic way to find these types of issues other than just reading the code; yet when these issues exist, they can make it very difficult to reason about the state of the system.

32.4.2.4 Supply Chain

In Section 32.2.1, we defined the TCB as the set of core modules that must work in order for the software to remain in a safe state. It is natural to focus on the parts of the system that comprise the TCB; it is less natural to consider outside influences such as libraries, IDEs, and compilers having an influence on the TCB. However, as Ken Thompson pointed out, these elements in the developer's supply chain can have real and catastrophic impacts on security.

As developers in the age of GitHub, Npm, Maven, NuGet, and Cargo, we are constantly pulling code we did not write into our code base. It is crucial to consider which of these libraries become part TCB, intentionally or not. If the security of your application relies on a library that your organization did not produce, extra scrutiny should be applied to the library. Cloud software has the additional burden of vetting outside services that it integrates with. For example, there are some companies that offer Authentication as a Service, where an application's authentication is outsourced to one of these service providers. At a minimum, designers should consider whether that is appropriate for the application's threat model and how such an integration affects confidentiality, availability and risk overall.

We also discussed how IDEs and compilers can subtly introduce code into your codebase and/or change security-relevant code at compile time. Checking settings and establishing developer guidelines around IDE and compiler settings is standard good practice. While it may be impractical to vet all of these items over an entire commercial size codebase, we strongly recommend paying extra attention to these issues for the code in the TCB and for the developers responsible for that code.

32.4.2.5 Cheat

"Life is too short. AppSec is tough. Cheat!' is the slogan for the OWASP Cheat Sheet series (Open Web Application Security Project 2021a). The articles in this series are great starting points for designers and implementers when building a security-relevant software feature, such as authentication, cryptographic storage, session state management, logging, and Docker security, to name a few. The articles are typically up to date with best practices and often contain sample code. These are not tablets of stone to be followed under all conditions, but they are chock full of practical advice that can help during design, threat modeling, and implementation. We recommend searching for relevant cheat sheets before writing security-critical features or code that will end up in the TCB.

32.4.3 Review and Testing

Modern best practices dictate that once code is implemented, it goes through review and then a matrix of test cases. The code review process and policies are different between organizations, teams, and projects, but they typically share the common idea that, once a developer is done with implementation, he or she constructs a *pull request* or a *patch* that is reviewed by other team members.

32.4.3.1 Review

While manual peer reviews cover a wide range of issues such as adherence to coding standards and proper implementation of algorithms that could impact performance, our advice is to pay close attention when security-relevant parts of the system are under review. The review should consider the impact of the change on the TCB and other key parts of the security model. Use a checklist as a reminder to look for code that touches authentication, cryptographic material, and the like, and put extra scrutiny on those parts of the code. The challenge with manual review is that the quality depends on the experience of the reviewer. Try to match pull requests for security-relevant code with reviewers that have security expertise; this is one of the last places to catch a security vulnerability before it ships.

Once the review has been successfully completed, the new code is merged into the code base, a build is generated and run through a battery of automated and ultimately manual tests to ensure that the feature meets the original requirements. The steps from merging to deployment make up the CI/CD pipeline, and modern best practices force things like unit test runs and coverage metrics to be achieved or a build will fail. From a security standpoint, we advocate adding static code analysis to this process as well. There are numerous free and commercial *Static Application Security Testing* (SAST) tools that are designed to be integrated into the CI/CD pipeline. In addition to giving detailed reports about security issues in the codebase, many can be configured to fail a build that introduces a new security vulnerability.

32.4.3.2 Testing

Once a build is generated, it can be installed and configured and we can begin testing a running application. A *quality assurance* (QA) team typically begins by ensuring that the feature meets the requirements, so-called "happy path" testing – i.e., ensuring that the code behaves as specified if users follow the happy path and use the system as the designers intended. From a security standpoint, we are more interested in *negative testing*: what happens when users stray from the happy path? In terms of our state-based definition of security, the goal of negative testing is to determine if and how we can get the system to leave the set of correct states.

32.4.3.2.1 Fault Injection and Fuzz Testing

One mechanism that is useful for negative testing is *fault injection*: a targeted introduction of errors into the system to measure the robustness and tolerance of the system to the specific type of fault. This concept is heavily used in hardware and standard manufacturing and is becoming more common in software. Software fault injection systems may change things like network traffic or user input in specific ways in an attempt to force the system into a bad state.

One specific type of fault injection is *fuzz testing*. As noted in Section 32.3.1 above, a common security anti-pattern is programmers making assumptions about the input and acting on that input without fully checking if it meets those assumptions. Fuzz testing hunts for such holes, by taking a corpus of correctly formatted input, randomly "fuzzing" it, and feeding it to the system being tested. A secure system should reject this fuzzed input; if the system does anything else, that indicates a failure to validate. If the system merely crashes, the developer might be tempted to dismiss that as somewhat harmless; however, as crashes usually are caused by memory exceptions, a system crashing on fuzzed input means that unchecked input affects memory addressing – so cleverly crafted input might be able to use this address corruption to do something useful (that is, useful to the adversary). Fuzz testing can be harder if the system already does some basic (but incomplete) checking of input formats – "fuzzed" input still needs to be sufficiently syntactically correct to get through the incomplete checking.

32.4.3.2.2 Symbolic Execution

Another approach to testing for inputs that lead to bad states is *symbolic execution*. To start with an extremely simple example, suppose a program does this:

- Inputs two unsigned 16-bit integers, u1 and u2
- Casts them to signed integers s1 and s2, respectively
- Then goes down code path A if s1 is greater than s2, and throws an error otherwise

(Yes, this example was inspired by real vulnerabilities when the programmer assumed A executes only when u1 is greater than u2 – and bad things happen if u1 is less than or equal to u2 but s1 is greater than s2.)

In fuzzing, we have to hope that we might randomly generate such a case – maybe we'll find it, maybe we won't. In symbolic execution, however, we run a tool that analyzes the code and calculates the input conditions required to go down each path, and then continues for the rest of the steps and input – and we are much more likely to discover the existence of input that violates key assumptions. A downside is that symbolic execution requires such a tool, and then requires extensive automated analysis of the code rather than simply running it on a few tests – but as upside, its coverage can potentially be much larger. (See Baldoni et al. 2018 for a good survey.)

32.4.3.2.3 Penetration Testing

Another valuable security testing strategy is *penetration testing*. Penetration testing involves testing a running system with the expressed intent of finding security vulnerabilities. Some organizations have the expertise to do this in-house, although it is common for software vendors to engage outside firms to perform this type of testing on a regular basis. In fact, some regulatory frameworks insist that an outside penetration test be done on an annual basis. Some level of automated penetration testing can be done internally by using tools such as OWASP's Zed Attack Proxy (ZAP) scan (OWASP 2021). ZAP scans can be incorporated into the release process to perform an automated test for the OWASP top ten vulnerabilities. We think that both manual and automated penetration tests have a place, as long as they are part of a larger security strategy that includes design and code review, static analysis, and various forms of testing.

32.4.4 Operations and Maintenance

Once the system is built and tested, it is at last ready to release. The system's deployment model guides what type of operational security and maintenance program is needed post release. For example, an embedded system for space flight, a medical device, an internal business application, and a commercial SaaS offering will have very different operational and maintenance requirements.

Code that is deployed and run in the customer's environment (i.e., on-premise software) will need to consider how to provide updates (including security patches) to all instances of the running software. At the extreme, OS vendors like Apple and Microsoft have developed sophisticated processes and systems to patch their software. As we discussed in Section 32.3.4, Microsoft releases security patches on the second Tuesday of each month. Enterprises with large numbers of servers have developed equally sophisticated processes to roll patches out to the organization in waves, so as to balance the risk of having unpatched servers against the risk of patching potentially thousands of computers on the same day. In addition to the process, companies that make on-premise software also need to consider how to securely distribute security patches. As we learned in the SolarWinds hack of 2020, there is perhaps no greater satisfaction for an attacker than having a security company distribute an exploit disguised as a patch (SolarWinds 2021).

Software companies that offer their software as a cloud service have a different set of challenges. Since they control the environment where the code is running, they need not be concerned with patching every instance of their product in the field. However, a security incident that takes the production environment offline can potentially take all customers out at once. As a result, SaaS companies typically employ rigorous segregation of duties across their environments – e.g., allowing only operations personnel (which does not include developers) into production environments. They also rely heavily on monitoring technologies to keep a real-time view on the health of the environment. It is important to ensure that these systems are permissioned appropriately; a log aggregator or network traffic analyzer, for example, may contain sensitive information. These environments usually make extensive use of automation throughout the infrastructure, a practice called *infrastructure as code*. From a security standpoint, this code typically contains high-level administrative account information as well as cryptographic keys and is sometimes treated with less care than the production code that developers write. In our view, it should go through at least the same level of review and testing and should have strict access controls in place. For a cloud-based software company, this code is as much a part of the product as the code that implements the features. Security problems here can have catastrophic results for customers and the company.

32.5 THE TAKE HOME MESSAGE

Software security is difficult. Most of the software we write sits on top of layers of abstractions, each one attempting to suppress details and complexity for the layer above. These details and this complexity are what make security hard. In our view, the best approach is to view the system as a collection of states and try to keep the system out of the bad states. This means thinking about the system's security model in several different ways and ultimately framing software security as a risk management problem.

This field is an arms race with new attacks and defenses cropping up daily. There is an inherent asymmetry in this game: attackers have to be successful once, and system builders have to defend against everything. The only hope for such a defensive posture is to employ effort at every stage of the life cycle, from design to deployment and beyond. Although perfection in this regard is unobtainable, our aim in this chapter has been to introduce some of the tools and techniques that can help manage complexity and help reduce the risk of shipping a vulnerability.

REFERENCES

Aleph One (1996). Smashing the stack for fun and profit. *Phrack*, 7(49). https://phrack.org/issues/49/14.html#article.

Ali, S., Anantharaman, P., Lucas, Z., & Smith, S. W. (2021). What we have here is failure to validate: Summer of langsec. *IEEE Security and Privacy*, 19(3), 17–23.

Anley, C., Heasman, J., Linder, F., & Richarte, G. (2007). *The Shellcoder's Handbook*, 2nd Edition. Indianapolis, IN: Wiley Publishing, Inc.

Baldoni, R., Coppa, E., D'elia, D. C., Demetrescu, C., & Finocchi, I. (2018). A survey of symbolic execution techniques. *ACM Computing Surveys (CSUR)*, 51(3), 1–39.

Cho, H., Park, J., Kang, J., Bao, T., Wang, R., Shoshitaishvili, Y., ... & Ahn, G. J. (2020). Exploiting uses of uninitialized stack variables in linux kernels to leak kernel pointers. In *14th USENIX Workshop on Offensive Technologies (WOOT 20)*. Virtual: USENIX Association.

CMU Software Engineering Institute (2018). SEI CERT C Coding Standard. https://wiki.sei.cmu.edu/confluence/display/c.

D'Silva, V., Payer, M., & Song, D. (2015). The correctness-security gap in compiler optimization. In *IEEE Security and Privacy Workshops*.

Dworken, D. (2021). *Worming Through IDEs*. Las Vegas, NV: Defcon 29. https://www.youtube.com/watch?v=pzqu_qaoNuY&ab_channel=DEFCONConference.

Eauvidoum, I., & noise, d. (2021). Twenty years of escaping the java sandbox. *Phrack Magazine* 16(70). https://phrack.org/issues/70/7.html#article.

FIRST (2021). Common Vulnerability Scoring System. https://www.first.org/cvss/.

Fowler, M. (2019). Is High Quality Software Worth the Cost? https://martinfowler.com/articles/is-quality-worth-cost.html.

Joint Task Force Transformation Initiative, NIST (2012). *SP 800-30 Rev. 1: Guide for Conducting Risk Assessments*. Computer Security Resource Center. https://csrc.nist.gov/publications/detail/sp/800-30/rev-1/final.

JSOF (2020). Ripple20: 19 Zero-Day Vulnerabilities Amplified by the Supply Chain. https://www.jsof-tech.com/disclosures/ripple20/.

Kaminsky, D., Patterson, M. L., & Sassaman, L. (2010). PKI layer cake: New collision attacks against the global X.509 infrastructure. In *International Conference on Financial Cryptography and Data Security* (pp. 289–303).

Knight, F. H. (1921). *Risk, Uncertainty, and Profit*. Boston and New York: Houghton Mifflin Co., The Riverside Press.

Loughlin, K., Neal, I., Ma, J., Tsai, E., Weisse, O., Narayanasamy, S., & Kasikci, B. (2021). {DOLMA}: Securing speculation with the principle of transient non-observability. In *30th USENIX Security Symposium (USENIX Security 21)*.

Microsoft (2019). The Project Triangle. https://support.microsoft.com/en-us/office/the-project-triangle-8c892e06-d761-4d40-8e1f-17b33fdcf810.

NIST (2021). National Vulnerability Database. https://nvd.nist.gov/.

Open Web Application Security Project (2021a). OWASP Cheat Sheet Series. https://cheatsheetseries.owasp.org/.

Open Web Application Security Project (2021b). OWASP Top 10. https://owasp.org/Top10/.

OWASP (2021). Zed Attack Proxy. https://www.zaproxy.org/.

Parata, A. (2021). NET instrumentation via MSIL bytecode injection. *Phrack Magazine*, 16(70). https://phrack.org/issues/70/6.html#article.

Raymond, E. S. (2003). *The Art of UNIX Programming*. Boston, MA: Pearson Education.

Rohlmann, S., Mladenov, V., Mainka, C., & Schwenk, J. (2021). Breaking the specification: PDF certification. In *IEEE Symposium on Security and Privacy (SP)* (pp. 1485–1501).

Rose, S., Borchert, O., Mitchell, S., & Connelly, S. (2020). *Zero Trust Architecture*. Computer Security Resource Center. https://csrc.nist.gov/publications/detail/sp/800-207/final.

Saltzer, J. H., & Schroeder, M. D. (1975). The protection of information in computer systems. *Proceedings of the IEEE*, 63(9), 1278–1308. doi: 10.1109/PROC.1975.9939.

Sassaman, L., Patterson, M. L., Bratus, S., & Locasto, M. E. (2013). Security applications of formal language theory. *IEEE Systems Journal*, 7(3), 489–500.

Schwaber, K., & Sutherland, J. (2020). The 2020 Scrum Guide. https://scrumguides.org/scrum-guide.html.

Shacham, H. (2007). The geometry of innocent flesh on the bone: Return-into-libc without function calls (on the x86). In *Proceedings of ACM CCS*.

Shostack, A. (2014). *Threat Modeling: Designing for Security*. Indianapolis, IN: John Wiley & Sons, Inc.

Siguza (2020). Psychic Paper. https://github.com/Siguza/psychicpaper.

Smith, S., & Marchesini, J. (2008). *The Craft of System Security*. Boston, MA: Pearson Education, Inc.

SolarWinds (2021). SolarWinds Security Advisory. https://www.solarwinds.com/sa-overview/securityadvisory.

The MITRE Corporation (2019). CVE-2019-5736. https://cve.mitre.org/cgi-bin/cvename.cgi?name=CVE-2019-5736.

The Mitre Corporation (2021a). Common Weakness Enumeration. https://cwe.mitre.org/.

The Mitre Corporation (2021b). Common Vulnerabilities and Exposures. https://cve.mitre.org.

Thompson, K. (1984). Reflections on trusting trust. *Communications of the ACM*, 27(8), 761–763.

Weidemann, A. (2021). *New Campaign Targeting Security Researchers*. Google Threat Analysis Group. https://blog.google/threat-analysis-group/new-campaign-targeting-security-researchers/.

ZecOps Research Team (2020). You've Got (0-Click) Mail! https://blog.zecops.com/vulnerabilities/youve-got-0-click-mail/.

Part 7

System Usage

33 Human System Interaction – Interface Design

Laura Stanley and Vishnunarayan Girishan Prabhu

33.1 INTRODUCTION: HUMAN SYSTEM INTERACTION

Human System Interaction (HSI) is an interdisciplinary field that integrates human factors, psychology, computer science, and design technology to improve how humans interact and collaborate with work systems (Booher 2003). The earliest application of HSI dates back more than half a century ago, during the Second World War, when psychologists were asked to investigate aircraft accidents that involved well-functioning aircraft (Fitts and Jones 1947). However, the predominance of HSI as a field of research coincides with the launch of personal computers in the 1980s. Since then, along with the growth of technology and computing, the application of HSI has expanded into areas including business, transportation, defense, manufacturing, health care, etc. (*Handbook of Human-Computer Interaction* 1997). The broad field of application has also led to designing and developing user interfaces of varied complexities, from simple two-dimensional desktop interfaces to complex aircraft cockpits and human-robot interaction interfaces that utilize three-dimensional augmented, virtual, and mixed reality displays (Prabhu et al. 2019; Girishan Prabhu et al. 2023).

Although the application areas and complexities of the user interfaces have diversified since the inception of HSI, its primary focus has always been to design, evaluate, develop, and implement interfaces with the end-user in mind, allowing for safe, functional, and efficient interaction between the user and the work system (Booher 2003). However, designing and developing such interfaces are not simple and require a systematic approach. Over the years, researchers have developed design methodology, design principles, tools for data collection, interface usability evaluation metrics, and error prevention methods that are updated with the progress of technology. This chapter aims to provide insights into some of these methodologies and tools. The sections are organized as follows, where we first start by discussing systems-level thinking for interface design, the pervasiveness of interfaces, and the criteria for developing usable and functional systems. The next section discusses human cognitive abilities and limitations, decision-making, error classifications, and their relation to interface design considerations. Finally, in the last two sections, we discuss the design framework that considers users as the center of focus during the design process, known as User-Centered Design (UCD), and a few tools used to develop such interfaces.

33.2 SYSTEMS-LEVEL THINKING IN HSI INTERFACE DESIGN

As briefly mentioned above, HSI is an interdisciplinary field, and as a result, designing and developing an HSI interface requires systems thinking approach. Systems thinking is a thought process that approaches a system as a whole rather than looking at individual factors to understand barriers and user needs and develop solutions (Checkland 1999). Additionally, this approach helps understand the root cause of problems rather than temporary fixes and identifies the stakeholder needs across the entire organization. In terms of the HSI design process, systems thinking does not separate interface, work system, stakeholders, user, etc., as individual components; instead, it considers the interplay and dependencies of each piece. Further, in each of these individual components, various smaller factors are considered. For example, functions including spatial components, audio-visual feedback, display modality, human cognitive resources, user schemas, etc., are

DOI: 10.1201/9781003188377-40

considered in interface design. Lack of systems levels thing in the design process could significantly affect the usability and functionality of the system. In some cases, the system could be functional, but it could lead to failure at a system level.

A practical model known as the Human-Tech Ladder that considers the human relationship with technology from many perspectives was developed to avoid such failures (Vicente 2004). Although the model explicitly uses the word technology, the model can be applied to any system, including interface design. This model considers various aspects starting from an individual physical level to organizational and political factors that could affect technology functionality. Figure 33.1 represents the human-tech ladder model.

Starting from the bottom, the first step of the human-tech ladder calls to evaluate how users interact with technologies physically and identify the physical challenges associated with technology. Here, the user-technology mismatch can be because of the technology's size, shape, weight, etc. For example, while designing and developing a Virtual Reality or Augmented Reality headset, physical features, including weight, size, etc., are essential factors in making it functional. Additionally, the interface's physical features, including positioning, field-of-view, etc., should be appropriate and meet the user's needs, allowing them to interact effectively with the system.

The next step in the ladder highlights the importance of aligning technologies with the user's cognitive abilities. Humans have limited cognitive resources, and technologies that do not account for this limitation could lead to a severe mismatch. For example, assume an interface where completed tasks are not marked, car display without fuel level, etc. This would lead to cognitive resource expenditure on unnecessary effort to remember these details, increasing the chances of errors. The third step in the ladder identifies the importance of technologies on a team level. At the team level, technologies should facilitate communication and enable stakeholders to work toward a common goal. For example, in a flight cockpit interface, the display should be available to the pilot and co-pilot, allowing for collaboration and better decision-making.

The next rung of the ladder discusses the alignment of technologies with organizational characteristics such as culture, staffing levels, incentives, and disincentives. Finally, technologies must also be assigned to people at a political level. These include points like public opinion, laws, and budgets – for example, Data privacy and compliance approval, FDA approval for medical devices, etc. In essence, for a technology to be useful and functional, it should align with all the five levels in the human-tech ladder. A mismatch in either one of these levels could result in the technology failing to perform its function. For example, a user interface could have all the physical features that meet the requirements, but the interface has too much information leading to cognitive overload and creating a mismatch at the psychological level. The next section of the chapter discusses these cognitive capabilities, limitations, and decision-making processes.

33.3 UNDERSTANDING THE USER

Understanding the user is one of the preliminary and crucial steps toward developing a good, functional, intuitive user interface. Hence, irrespective of application area, the complexity of the work system, and the interface design, it is essential for designers to understand human capabilities, limitations, information processing, and the decision-making process. Figure 33.2 represents how humans process information along with the series of processing stages while interacting with a system, including interface (Wickens et al. 2012). As shown on the left of Figure 33.2, any changes in the interface are initially processed by the senses, including sight, sound, touch, etc., which may be held in the short-term sensory store (STSS) for no more than about a second. Although the changes in the interface are sensed, only a brief amount of changes are perceived depending on the user's prior experiences, which are stored in their long-term memory of facts, knowledge, and understanding of how a system works. Upon perception, the user process information based on either (or both) of two paths. In the bottom scenario, as observed in Figure 33.2, perceiving a situation will trigger an immediate response from a set of possible actions. Once the response is selected, it is then

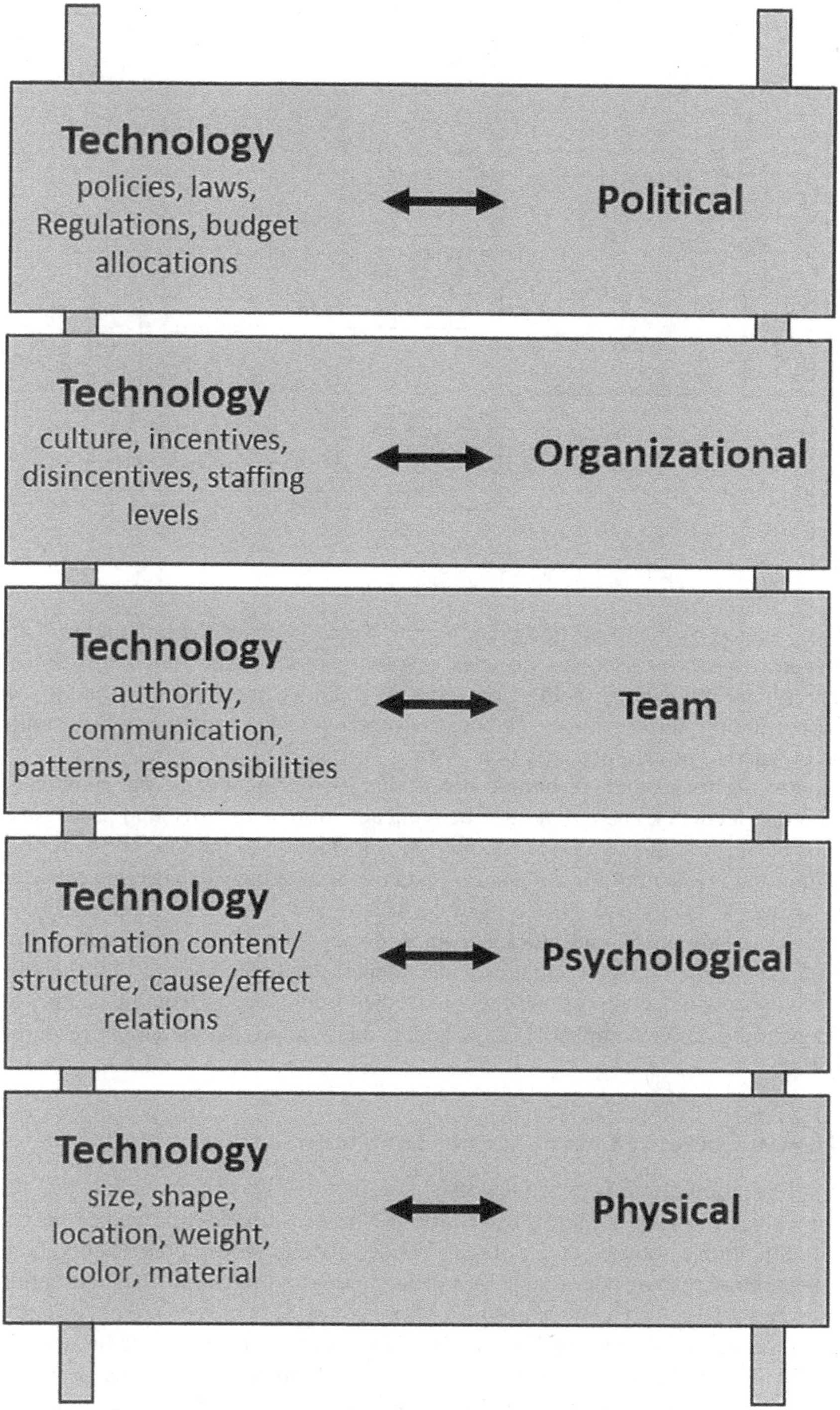

FIGURE 33.1 Human-Tech Ladder.

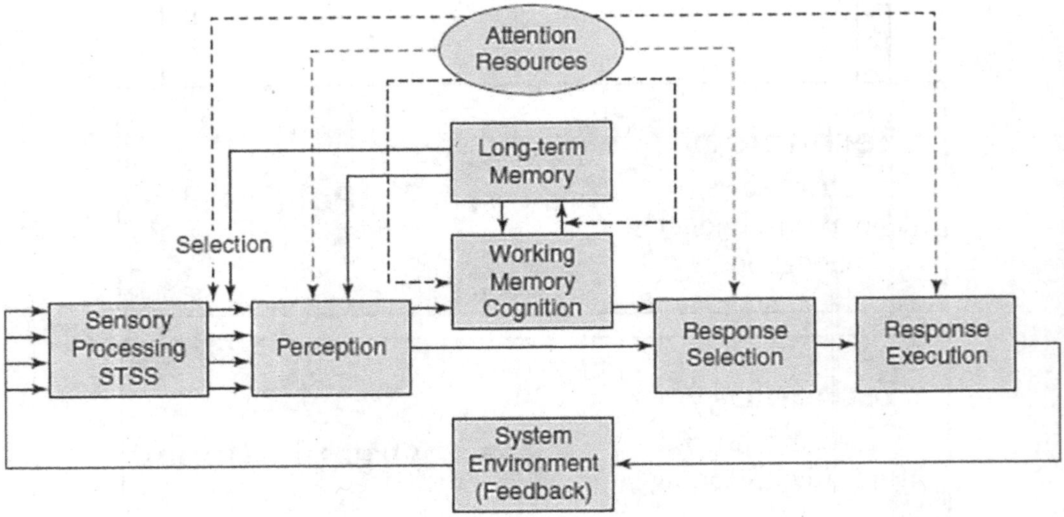

FIGURE 33.2 Human Information Processing.

immediately executed by the mind-body interaction, which involves the human brain and muscles. However, in some instances, users do not select responses immediately, especially during new situations as represented in the upper path where working memory may be interrupted to learn the new task and store it in long-term memory. Hence, the function of working memory is not just to store information but also to process the situation.

In addition to the four stages of human processing, two critical elements complete the human processing model: feedback and attention. The feedback loop helps learn and process new changes due to actions performed in stage four, adding new information to the long-term storage for future reference. Attention is essential for information processing playing two different roles during information processing (Wickens et al. 2007). Its first role acts as a filter to sieve and pass required information perceived by the users and block out unnecessary background information. In its second role, attention acts as the repository that provides mental resources required for the various stages of information processing, as represented by the dashed lines flowing from the supply of resources at the top. Depending on the complexity of the action, a few stages demand more resources for some tasks than others.

33.3.1 Human Cognitive Capabilities and Limitations

The human information processing framework shows that the two critical factors that play a role in how users perceive an act during a given scenario are based on their experience (memory) and the availability of attention resources. According to Wickens Multiple Resource Theory, humans have a limited set of attention resources available for mental processes (Wickens 1984). Attention resources can be considered a pool of resources used for various mental operations, from sensory-level processing to action execution. Additionally, this pool of resources is also utilized by any other competing activity that requires human attention. This theory explains why human performance decreases while multitasking on tasks that require similar resources compared to those requiring different ones. A classic example of this scenario is a decrease in driver performance when texting and driving.

Another theory that builds upon the human information processing framework is the Cognitive Load Theory (Sweller 2011). Cognitive Load Theory assumes that human memory can be divided into working memory and long-term memory, as seen in the human information processing model, where long-term memory stores information in the form of schemas based on our prior experiences.

However, new sensory information (task) results in a 'cognitive load' on working memory, affecting learning outcomes (Anderson et al. 1977). Working memory only has a limited capacity, so overloading task with information leads to redundant activities that hinder the learning process and developing schemas for future use. Hence the cognitive load associated with each task in the user interface should be well balanced such that the users can learn the task and make accurate decisions. In the context of Cognitive Load Theory, there are three different loads: (1) intrinsic load, which is the mental workload asserted by the task to be learned, (2) germane load, which is the demand for resources necessary to learn the task itself, and (3) extraneous load, the resource demands unrelated to either of the above.

Based on Wickens Multiple Resource Theory and the Cognitive Load Theory Hence while designing any systems, the overall goal is to minimize extraneous load and try (by adjusting the task) to keep the intrinsic load from being too high, such that sufficient resources are available to allocate to germane load. Although this concept might appear intuitive and trivial, it is not easy to develop systems and interfaces that meet these criteria. We discuss a few bad design ideas in the upcoming sections of the chapter to demonstrate how products have failed when not accounting for these factors. A few validated techniques that help in reducing cognitive workload and attention resource requirements include:

1. **Providing Short-Term Memory Support**: According to Miller's Law, a human brain can hold only 7±2 objects in short-term memory (Miller 1956). Hence to the cognitive load, users should be provided with information they would otherwise have to store in their short-term memory. These can be simple things like marking a task as done from the list of tasks, making a visited link look different from a non-visited one, etc. Another approach suggested by Miller to reduce cognitive load is known as 'chunking,' where objects are broken into smaller chunks. A classic example of chunking is how we memorize a phone number XXX-XXX-XXXX. From an interface standpoint, chunking is used along with other concepts such as Gestalt principles of display organization to develop a page with multiple items and easily chunk them into different items, for example, online shopping websites.
2. **Provide Multimodal Feedback**: Multimodal feedback involves some combination of audio, text, pictures, haptics (Mayer 2014). During cognitively demanding situations, multimodal feedback helps in redundancy gain and quickly identifying a change. A few subprinciples under this idea include highlighting, filtering unnecessary information, temporal and spatial contiguity (Mayer 2014). Some of the day-to-day applications of this include haptic and visual feedback while selecting an option on displays, hospital displays changing color with alarms during adverse events, etc.
3. **Avoiding Visual Clutter**: Interface design should be simple. Cluttering a display hinders the attention of users to sense and perceive information from the display. According to Wickens, the source of clutter can be numerosity clutter (information overloading), proximity clutter (multiple items close together), disorganizational clutter (similar items not grouped together), and heterogeneous clutter (heterogeneity of the background features). A way to overcome visual clutter is by using Gestalt principles of display organization. Figure 33.3 represents two similar displays where one uses Gestalt's principles of design (Wickens et al. 2012).
4. **Understanding Mental Models**: Mental models are users' internal representation and understanding of how a system works based on their experience. Hence while designing an interface, it is critical to understand the mental model of end-users to develop an intuitive and functional design. Researchers have observed that humans are proficient in recognizing things from the past, and this activity requires less processing in the human brain, avoiding the risk of error or failure. For example, a floppy disk icon on display interfaces represents saving action, and any changes to these functions would be against the mental model and could lead to errors.

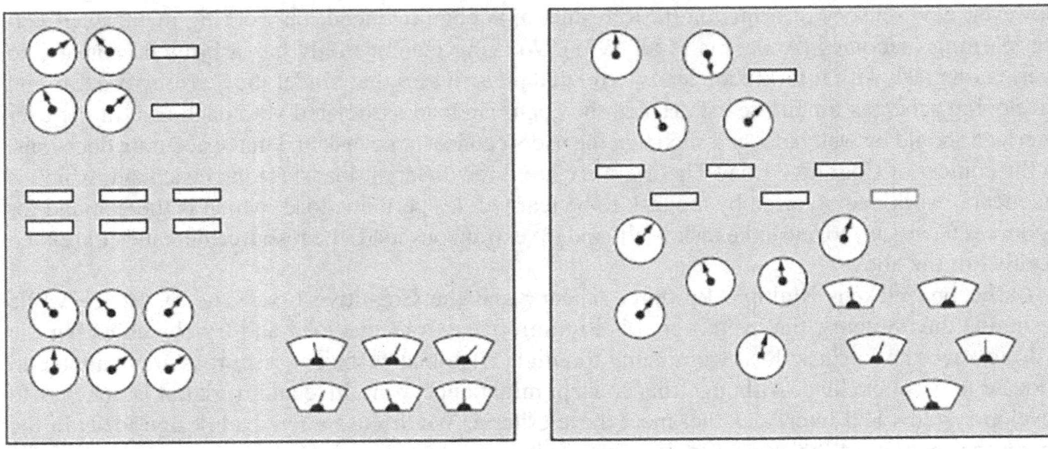

(a) (b)

FIGURE 33.3 (a) Designed based on Gestalt principles. (b) Unstructured display design.

These do not represent a comprehensive list of strategies for reducing the cognitive workload and attention resources, but these can be considered as guidelines for developing functional interface designs that reduce the risk of errors and failure. In the next section of the chapter, we briefly discuss a model that describes human behavior and decision-making while users interact with a system (including interface) and the resulting errors.

33.3.2 Decision-Making and Errors

Human decision-making is a complex process where users evaluate the value of the outcomes associated with alternative options. Various vital factors affect decision-making, including uncertainty, time, familiarity, and expertise. The degree of uncertainty of a decision is directly associated with risk. Hence, the user's confidence in selecting a decision decreases as uncertainty increases. Time influences human decision-making in at least two ways: first, by contrasting one-shot decisions with evolving decisions, and second, time pressure has a significant impact on the decision-making process (Svenson and Maule 1993). Finally, as familiarity and expertise increase, humans make decisions naturally and instantly than novices who may spend additional time and make poor choices. Although these are the key factors, various cognitive factors that are involved in information processing affect the decision-making process. To understand this, the human information processing model discussed in the previous section can be slightly modified and extended to the human decision-making process, as represented below (Wickens et al. 2012).

To begin with, the user first seeks cues or information from the environment, which could include uncertainty and could be ambiguous or interpreted incorrectly. Here user's selective attention (their focus) plays a critical role in decision-making, as it steers the user to pick up vital cues and filter out irrelevant cues. As seen from Figure 33.4, this is based on the user's experience stored in long-term memory and demands attentional resources. The user then perceives the selected cues and form the basis of understanding and situational awareness for the diagnosis process. During the diagnosis, the user generates hypotheses regarding current and future state based on the selected cues and experiences (long-term memory) and availability of attentional resources. As seen in the Figure 33.4, the diagnosis process is iterative, where the user generates a hypothesis and then rejects or accepts it based on all those resources stacked above the process and the selected cues. The set of activities from selecting the cues until diagnosis represents the front end of the decision-making.

The back end of decision-making starts with choosing an action (response selection). Although the user generates a set of possible courses of action from the long-term memory, there will be some

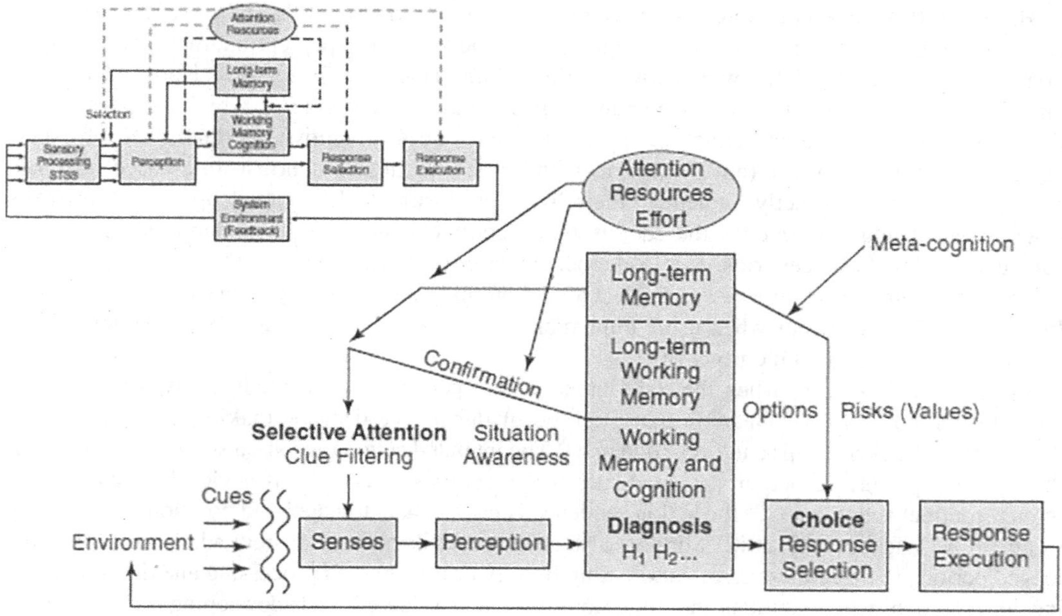

FIGURE 33.4 Human Decision-making Process.

risk or uncertainty associated with it, as mentioned above. The user then estimates the likelihood that outcomes will come to pass and the degree to which it will be good or bad. Further, time pressure, expertise, and familiarity also play a critical role in choice selection. The final step is executing the response and feedback as observed in the human information processing model, which helps develop long-term schemas.

During the decision-making process, failures could stem from an error during the front-end or back-end processes (Hoffman et al. 1998). In case of failures resulting from the front-end process, poor cues and situation assessment are the drivers that generate an incorrect action. In contrast, failures from the back-end process occur due to unsuitable choices in the face of a well-diagnosed situation (Wiegmann, Goh, and O'Hare 2002). Here the failures are caused by uncertainty, time pressure, experience, familiarity, etc.

Since errors are likely to occur during decision-making due to various causes, a framework was developed to classify these errors. The classification system known as the Skill, Rule, Knowledge-based (SRK) framework classifies human behavior and decision-making into three levels based on the cognitive demand required for decision-making (Rasmussen 1983). The skill-based classification refers to routine tasks and represents the lowest level in decision-making. In this scenario, cues are rapidly identified, perceived, and a decision is made quickly with high confidence and low uncertainty. Here the cognitive demand and attentional resources required for decision-making are low.

Rule-based classification refers to tasks that are not routine where decisions are made based on specific rules. Here, users access their working memory and other attentional resources to develop various "if-then" scenarios, and upon identifying a cue, they proceed to take action. Here the cognitive demand and attentional resources required for decision-making are comparatively higher than skill-based levels. Finally, knowledge-based classification corresponds to novel scenarios that users have never encountered in the past. Here, a user initially has to expend a lot of attentional resources to identify the right cue and perceive them. After identifying and perceiving the cues, users try to generate new hypotheses and map them to scenarios related to the situation by accessing working memory, long-term memory, and attentional resources. In these cases, users make their choices slowly as much uncertainty (risk) is associated with the decision, and cognitive demand and attentional resources required are significantly high.

Based on these classifications, errors can be classified into two: slips and mistakes. Mistakes occur when users fail to implement the right actions due to lacking perception, memory, or cognition. Mistakes are associated with knowledge-based and rule-based scenarios. Knowledge-based mistakes can be traced back to activities in the front-end decision-making, which includes failure to identify the right cue and situation. This can occur due to cognitive overload, lack of expertise to identify the right cue (novel scenarios), and time pressure. In contrast, rule-based mistakes occur when users incorrectly implement their "if," "then," rule. Unlike knowledge-based mistakes where users struggle to identify the cue, users are confident that they have identified the correct cue during rule-based scenarios. A mistake occurs here when users incorrectly identify cues to be correct based on their experience and select an action that does not fit the scenario and vice-versa. In essence, mistakes occur when users implement incorrect actions because they misinterpret the scenario or implement an incorrect rule.

In contrast, slips occur when the right intentions are performed incorrectly. Slips are associated with skill-based scenarios meaning that these occur during the decision-making of routine tasks. According to Wickens, slips can occur due to: (1) The intended action involves a slight departure from the routinely performed action, (2) some cues or the action sequence itself is closely related to the wrong (routine) action, and (3) the action sequence is primarily automated and not monitored closely by attention (Wickens et al. 2012). Slips are also observed while performing redundant tasks for a prolonged period. To avoid these errors originating from poor information processing and decision-making, human cognitive capabilities and limitations must be considered while developing user interfaces.

33.4 USER EXPERIENCE AND CONCEPTUALIZATION OF HSI INTERFACE DESIGN

The core of developing a good, usable interface design is to create interfaces that enhance the user experience. User experience represents how a user interacts and engages with an interface, including perceptions of utility, satisfaction, value, and ease of use. Although user experience entails many factors, the core is to empathize with users to understand their needs and develop an interface that provides a valuable user experience. According to Morville (2004), for any system to provide a valuable user experience, there are six facets that should be considered during the design process (Morville 2004). These six aspects with the valuable user experience at its core are known as the UX Honeycomb. Figure 33.5 represents the structure of UX Honeycomb (Morville 2004).

Although it is impossible to maximize all aspects of the honeycomb as each system would have various constraints, including time, budget, etc., UX Honeycomb helps set priorities for each area during the design process. Each aspect of the UX honeycomb can be briefly defined as follows:

1. **Usable**: The system is designed to be intuitive, simple, and easy to use. The goal is to minimize the learning curve.
2. **Useful**: The designed system satisfies the user's needs and addresses the current problem. A system that fails to address the needs cannot be considered useful.
3. **Desirable**: The system design is appealing, easy to operate, and minimal in design. The design should not confuse the user.
4. **Findable**: The system design allows for accessing and finding information quickly and easily. The system design should follow the user's mental model.
5. **Accessible**: The design considerations of the system should ensure that users with disabilities could have a similar user experience as others.
6. **Credible**: The system is credible and gains user trust.

These six facets can be considered design considerations, and maximizing these aspects leads to developing a system that enhances user experience. Conversely, failure to consider even one of these

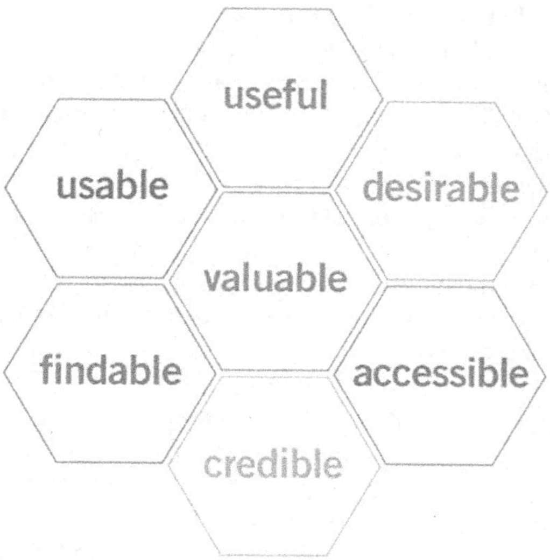

FIGURE 33.5 UX Honeycomb.

aspects during the design process can create a system with a bad user experience. Additionally, to develop a usable and functional interface validated design considerations specific to interface design should also be considered during the design process. Although there are various validated design considerations principles, two commonly used design consideration principles in practice are (1) Usability Heuristics for user interface design (Nielsen 1994) and (2) Eight Golden Rules of Interface Design (Shneiderman 1987). The first set of principles are considered as rules (heuristics) rather than design considerations. It is used extensively to evaluate an interface—this process, known as heuristic evaluation, is discussed in detail in the next section. Eight Golden Rules of Interface Design is one of the earliest interface design considerations developed approximately three decades ago and occasionally refined to meet specific design domains. Following are the Eight Golden Rules of Interface Design:

1. **Strive for Consistency**: The design, the sequence of actions used in similar scenarios should be consistent. For example, the prompts, menus, color, logo, fonts, etc., used should be consistent throughout. Any exceptions in the design that are unavoidable should be limited and easily comprehensible. For example, confirmation while using the delete command.
2. **Seek Universal Usability**: The user interface should consider the needs of a diverse set of users, including novice and expert users. Novice users would need details and explanations compared to expert users who prefer shortcuts.
3. **Offer Informative Feedback**: All user actions should have informative feedback. However, the feedback should not be redundant, and for frequent minor actions, these should be modest, whereas, for rare major actions, the feedback should be significant.
4. **Design Dialogs to Yield Closure**: Provide feedback to users after completing a set of actions to prevent users from keep guessing. For example, after completing an online purchase, providing messages such as "Transaction is processed" or "Transaction could not be completed" helps the user confirm the status of the transaction.
5. **Prevent Errors**: The user interface should be designed such that users cannot make serious errors. However, in case unavoidable errors occur, the interface should provide step-by-step action to correct the error. An example of error prevention would be limiting input to numbers for fields such as age etc.

6. **Permit Easy Reversal of Actions**: The user interface should allow for a quick reversal of actions. This feature should be included for simple and complex actions, and this allows users to explore unfamiliar options and reduces anxiety about making errors.
7. **Keep Users in Control**: The user interface and feedback to user actions should follow the user's expectations. This feature allows users to be in control of all actions and helps in gaining the trust of the users.
8. **Reduce Short-Term Memory Load**: As discussed in earlier sections, human attention and short-term memory are limited. Hence the user interface should be designed such that users are not required to remember information from another display and use it in the later stages. For example, for information with a lot of details, the display should be compacted to fit all details in a single view.

Although these do not include all the design considerations for developing a user interface and have their limitations, they assist in creating a good foundation for a usable and functional user interface. Incorporating these along with the design thinking process helps refine and develop an interface that considers all the facets of the UX Honeycomb that meets the user goals. The Design Thinking process is an iterative five-step process that focuses on developing solutions that address user needs by enhancing the user experience. The five-step Design Thinking process includes:

1. **Empathize**: This is the core of the UCD process as it helps to understand the problem through the user's standpoint without being influenced by individual biases. The primary goal in this phase is to observe, engage, and empathize with users to identify their problems and needs.
2. **Define**: In this phase, the goal is to aggregate the findings from the first phase to identify the user's core problems that need to be solved. This problem is then defined as a user-centered problem statement that needs to be addressed.
3. **Ideate**: This is a creative phase where designers think outside the box to come up with solutions that can address the problem statement defined in the second stage.
4. **Prototype**: In this phase, the design team will produce inexpensive scaled-down prototypes, including wireframes, post-it notes, etc. The goal is to evaluate the problem solutions generated in the prior stage and identify the best solutions. This is an iterative process where sometimes the solutions are combined, eliminated, or re-examined. As the iteration progresses, the fidelity of the prototype is also increased.
5. **Test**: This is the final phase of the design thinking process, where the actual users evaluate the prototyped solutions. Inputs collected from users during testing are then used to redefine the solutions and better understand user needs and problems.

Although the design thinking process may appear sequential, unidirectional, and a one-time process, usually, the process is iterative in practice. For example, after creating a prototype and/or testing, the design team might go back to the ideation phase to develop new ideas based on prototype evaluation and user testing. Moreover, sometimes designers might complete the testing phase and return to the empathize phase to evaluate if the interface meets the user's needs. As the design thinking process is iterative and time-consuming, each step should be carried out systematically to maximize the output. To assist in this process currently, there exist various UX research methods to address the goals of each step and improve the interface design. Broadly UX research methods can be classified into quantitative and qualitative methods based on the type of data collected and reported. Both methods are crucial and used during the iterative design process. Primarily qualitative methods provide insights for interface design, and quantitative methods aid in benchmarking against standards and existing products. Discussing the various qualitative and quantitative methods that could be used during each step of the design process is beyond the scope of this chapter.

Hence, a detailed discussion on some important techniques that help identify user needs, prototype development, and testing are discussed.

33.4.1 Empathy Mapping

An empathy map is a qualitative method that helps in visualizing and understanding user needs. This is one of the first steps in the design thinking process and part of the empathize phase. Ideally, an empathy map is generated right after the customer interviews and prior to developing the needs statement and persona generation. The earliest form of empathy map was developed many years ago and included four quadrants: (1) says, (2) thinks, (3) does, and (4) feels with the user in the center of the map. However, a few years ago, the empathy map was updated and split into seven sections including, (1) goals (two sections), (2) sees, (3) say, (4) does, (5) hears, and (6) thinks and feels put inside the head of the user as the center of the map. Further, to assist designers in completing various sections of the empathy map, the updated version has numbered the sections in the intended sequence of activities. This tool provides a high-level overview of the user's thought process. Figure 33.6 represents the updated empathy map canvas (Gray 2017).

The first step in completing the empathy map is starting with the *goal*. Within the *goal*, first, the user who we want to understand and empathize with is identified. Then their situation role is also summarized. After completing this, the second section of the *goal* section should be completed, which includes what we hope users will do, what success looks like, etc. This helps in setting the context for the activity. Following the completion of the goal section, the next step is to complete 3, 4, 5, and 6 in sequential order. These sections capture the impact of the outside world on the user. *The sees* section represents what the users experience in their daily life. These can be people, activities, and products they interact with. The key here is to identify and think from the user's standpoint and not your standpoint. The *says* section represents what a user says during a user interview. It is important to collect what the user mentions during the interview in its raw form and represent it as

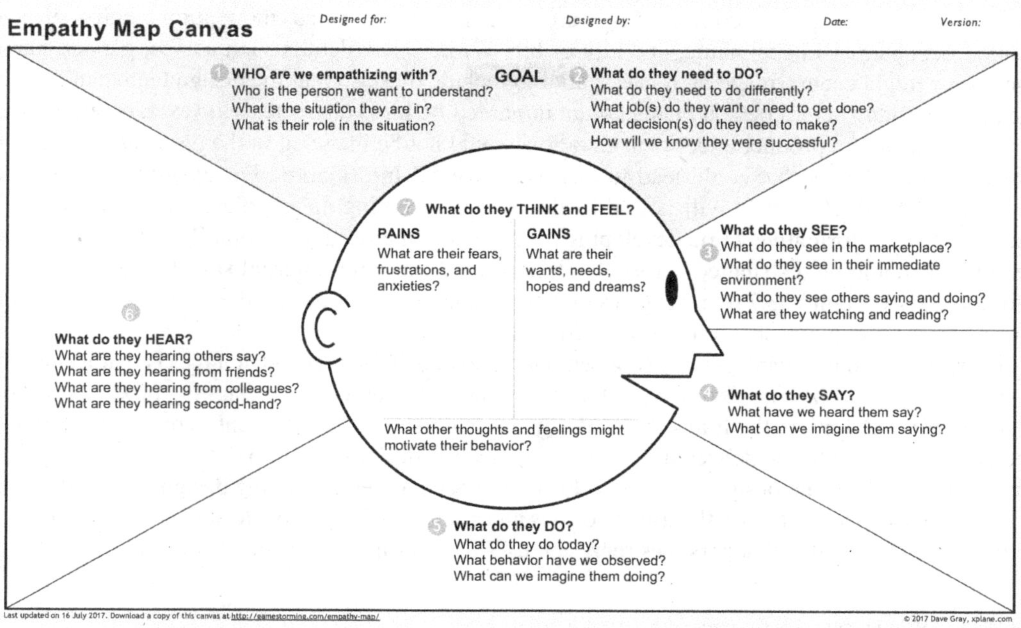

FIGURE 33.6 An Empathy Map Structure.

quotes in the section to capture what the user exactly says. An example of what the user says would be, *"I am not sure what I should do."* The *does* section captures the action completed by the user. Further, it also explores how the user goes about completing the action, etc. Finally, the *hears* section represents what the user hears from their friends, colleagues, public, social media, etc. and how it impacts user actions.

The last section, which includes what users *think* and *feels*, represents what users have in their mind. The *think* section captures what the user's thoughts are, what matters to the user etc. The *feels* section captures the emotional state or behavior of the user, how the user feels about the experience, what gets the users excited, what causes anxiety for the users, etc. Finally, the pains and gains section capture what success and failure look like for the user, their frustrations and challenges, dreams, hopes, etc.

Ideally, empathy maps are generated for individual users, but in certain cases, individual empathy maps are aggregated to represent a particular use segment. This should be utilized if the users exhibit similar behavior and goals. One of the main advantages of aggregated empathy maps is that it helps in creating personas representative of customer segments.

33.4.2 Persona Generation

Personas can be defined as fictional yet realistic characters that represent the target users. Users with similar characters and behaviors are usually gathered under a single persona, and usually, multiple personas are generated to map various user archetypes. Personas are based on user research and should be created after collecting information from actual users. Studies recommend interviewing between five to thirty users of the same archetype before creating a persona to avoid biases and misinformation. From the design thinking process perspective, personas are generated after conducting interviews and empathy mapping during the define phase of the design process. After collecting the data from interviews, the first step is to meet as a team and identify attributes that overlap emerged during the interview and merge these or eliminate the less valuable attributes. Although there is no standard persona template, a persona should be comprehensive and common details include the fictional character's needs, concerns/frustrations, goals, gains, pains, motivations, behavior, age, gender, occupation, image, name, and a brief bio. While a few details such as age, gender, name, and image might seem irrelevant, it is essential to include these so that the design team can develop the product for an actual person instead of an unnamed fictional character. However, details that do not add value to the product under consideration should not be included in the persona as it leads to overwhelming details that could lead to forgetting critical information. For example, understanding a user's food preference will not add value while designing an interface for a flight cockpit, but it could be informative while developing a new menu in a restaurant. Finally, after creating the persona, it should be distributed across the design team and other internal stakeholders for review and familiarity. Figure 33.7 represents a sample persona that helps design the *"About and Careers"* content for a company website (Harley 2015).

There are various advantages of using personas, and one of the main advantages is that it helps the designers to empathize and design a product for actual users rather than designing products without users in mind. Additionally, it also helps designers understand that different people have different needs and expectations, which should be considered while designing to avoid developing the product based on self-referential design potentially. Further, when having conflicting design ideas among the team, personas help to identify the user priorities and come to a design conclusion. However, a critical piece to keep in mind is that personas represent user patterns and not different user roles.

33.4.3 Prototype Development

Prototyping is a critical step in the design thinking process where a representation of the product or design is created for visualization, evaluation, and validation. Similar to the entire design thinking

Company "Investigator"

Rosa Cho
Content Strategist, Freelance

Age: 34
Location: Seattle, WA

"I'm looking to join the right company that challenges me and allows me to grow and develop my skills."

About Rosa

Rosa does not believe in settling. She won't settle for a job with a company that isn't as innovative and cutting edge as she believes she deserves. She wants to get the most out of every professional experience, and before moving to a new position, Rosa investigates every angle of aligning herself with a company.

Behavioral Considerations

+ Expects the site experience to reflect the business's culture and values
+ Interested in career opportunities within the organization that fit her career goals
+ Thoroughly compares multiple companies with similar opportunities
+ Is interested in the unique benefits of working at a company, including cultural elements, mentoring programs, and continuing education policies
+ Needs to be confident the company has innovative products that will be interesting to work on
+ Needs to know company has reputable partners and customers

"I crave variety in the types of industries and goals of each content project I work on. I need to ensure I won't get bored."

Frustrations

+ Thinks that too many companies have career sections that just talk about open positions but not why she would actually want to work there
+ Would like to challenge herself and have a more stable job, but is comfortable as a freelancer and wouldn't stop for just any job

Goals

+ Needs to see reasons why a company is interesting: has it won awards, had intense growth, won big contracts?
+ Wants to figure out how to get in touch with someone at the company to explore opportunities further

Tasks

+ Learn about current customers and success stories
+ Read press releases about recent big contract wins and other accolades
+ Read about culture, benefits and perks, and the people that work there
+ View job openings and apply

FIGURE 33.7 A Sample Persona.

process, prototyping also follows an iterative approach where multiple prototypes are developed based on feedback during the design phase. Although there are no set standards of prototyping methodologies, the general approach is to progress from a low-fidelity prototype to a high-fidelity prototype over the course of the design process. Low-fidelity prototypes are simple, basic representations of the product or design that are developed fairly quickly to get a tangible representation of the product/design under consideration. The goal of a low-fidelity prototype is to lay a foundation for the final design by prototyping early-stage ideas to visualize multiple design alternatives. Additionally, it helps the designers in developing the design flow (how things are organized), understanding functionality, and testing usability based on user needs and design principles for filtering early-stage ideas. For example, during the early stages of an interface design, there could be multiple ideas for buttons, menu, layout, etc., and some of these can be easily filtered out using the design principles and user priority of needs which helps in saving time and costs. The simplest low-fidelity prototyping method is to sketch, which are often free-form pen-and-pencil drawings. As mentioned earlier, the design team uses sketches to represent early-stage design ideas to brainstorm and review design alternatives. The main advantage of sketching is that they are quick to generate (on demand), inexpensive, collaborative, disposable, and minimally detailed. One of the main disadvantages of sketching is that it makes a high-level prototype and is not helpful as design complexity increases.

Few other low-fidelity prototyping techniques include wireframe models and paper prototypes. Similar to sketching, wireframe modeling is also a visual representation of the design ideas. However, these prototypes capture high-level and complex ideas as opposed to sketches. In the iterative design process, wireframe modeling can be considered the second stage of prototyping following sketching. Wireframe modes develop design ideas that come out of sketching that the design team wants to pursue further during the design process. For example, while developing a design interface, wireframe models help develop an overall structure and general flow to explore various ideas based on inputs from sketches. Although slightly advanced and refined, wireframe models can be quickly developed and inexpensive, but costlier and time-consuming than sketches. Similar to sketches, wireframe modeling cannot represent complex ideas which sometimes can be challenging for evaluation.

High-fidelity prototypes are more advanced than low-fidelity prototypes, representing design and functions closer to the final product. Further, these prototypes are aesthetically close to the actual potential design allowing for better feedback and user evaluation. One of the most common high-fidelity prototypes used for interface design is digital prototypes. Digital prototypes can be

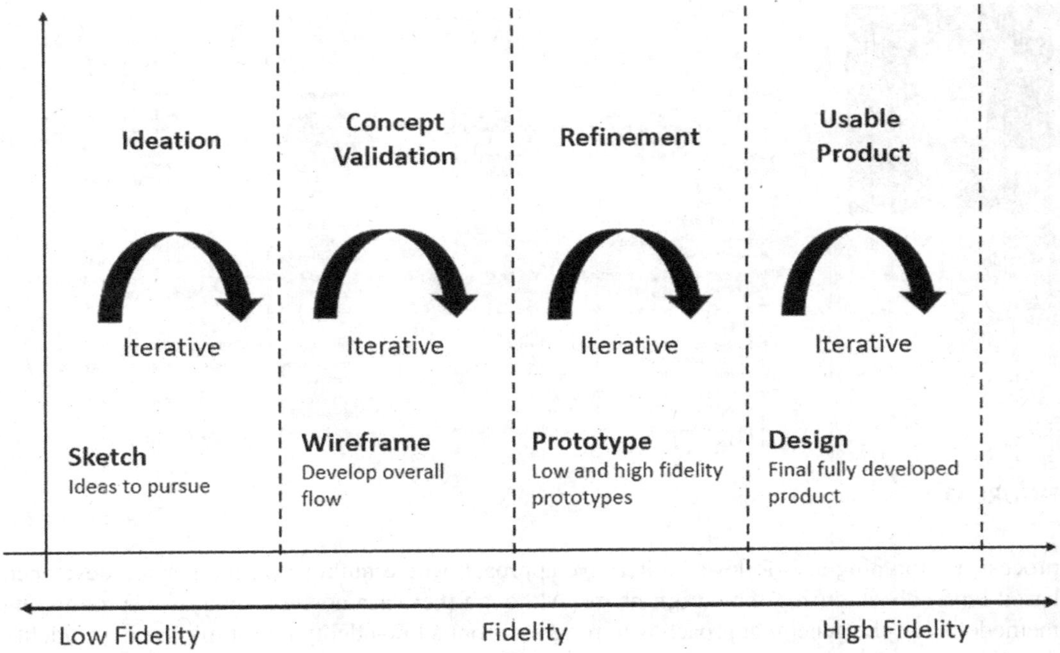

FIGURE 33.8 General Prototyping Guidelines.

developed using prototype development software such as Adobe XD, Figma, etc., or using computer programming languages. The advantages of digital prototypes are that they are interactive, dynamic, animated, and offer smooth transitions allowing the users to click through and have a close-to-real interface experience. However, digital prototypes are developed only in the later stages of the design process when the design team agrees on most design considerations and interface flow and is ready for expert and user evaluations. This is primarily because developing digital prototypes is time-consuming and expensive. Figure 33.8 represents a general guideline used by design teams for prototype development (Lepore 2010).

Figure 33.8 provides a detailed guideline that can be adopted for the interface design process. For example, it can be seen that Sketches are low-fidelity prototypes and communicate ideas, possibilities, and directions to pursue. Further, it can also be observed that sketches are used during the ideation phase, and as discussed throughout this chapter, the entire design process and prototype development follow an iterative approach. Similar details for wireframes, high-fidelity digital prototypes can be identified from the figure. After developing the high-fidelity prototypes, the next step is interface evaluation, where various validated tools are used to evaluate the usability of the developed system. A few of these interface evaluation tools and the iterative design process are discussed in the next section.

33.5 EVALUATING AN HSI INTERFACE DESIGN

One of the critical steps in developing and implementing functional and usable interfaces that meet user needs is evaluating the interface. In this step of the iterative design process, the primary goal is to evaluate the proposed interface design against the usability standards and task goals before finalizing the design. Two broad approaches for evaluating a user interface are (1) heuristic evaluation and (2) empirical evaluation. The main difference is that heuristic evaluation is conducted by subject matter experts following a validated set of steps, whereas empirical testing involves testing and collecting data using end-users.

33.5.1 Heuristic Evaluation

Heuristics evaluation is an interface evaluation method for identifying the usability issues associated with an interface design. As mentioned earlier, the heuristic evaluation is performed by a small set of subject matter experts where the interface under consideration is evaluated against a set of rules (heuristics). The first step in heuristic evaluation is identifying the number of evaluators required for testing. Empirical evidence suggests that including five evaluators could identify 75% of the usability problems, and the minimum recommended evaluators are three. The next step is a briefing session where each evaluator is familiarized with the interface, and details regarding the task that should be performed are provided. To ensure consistency, the briefing session should be the same across all the evaluators. The number of tasks under consideration can vary between heuristic evaluations, and usually, evaluations can last up to 1–2 hours. Also, it is recommended that evaluators perform a minimum of two interface passes as it is difficult to identify problems through a single inspection. After defining the task, each evaluator would perform the task independently, and during each step of the task, the interface will be compared against the ten usability heuristics developed by (Nielsen 1994). The ten usability heuristics are as follows:

1. **Visibility of System Status**: The user interface should provide feedback to the users about the system status without delay. For example, the spinning wheel representing the loading action provides feedback to the user that an action is being completed.
2. **Match between Interface and the Real World**: The user interface and design actions, including logo, fonts, wording, etc., should match the real world. For example, "trash can" represents a place where we remove old files.
3. **User Control and Freedom**: This usually represents the "emergency exit" mechanism that allows users to undo an action or back out of the process without a lot of delays. For example, undo and redo buttons for actions.
4. **Consistency and Standards**: Interface design, including wordings and action, should be consistent throughout the interface. Moreover, consistency should be maintained across the family of products.
5. **Error Prevention**: Interface design should be designed such that slips and mistakes are prevented. The design should eliminate all error-prone conditions, and in case of errors that cannot be prevented, the interface should allow for correcting the error.
6. **Recognition Rather than Recall**: The interface should be designed such that it minimizes the user's memory load by creating actions and options visible. For example, the "help" button allows for assisting users during an action rather than remembering the process.
7. **Flexibility and Efficiency of Use**: The interface should be designed considering a wide variety of users. For example, shortcuts for expert users allow for efficient communication.
8. **Aesthetic and Minimalist Design**: Irrelevant information should be avoided from the interface. Contents and features should be prioritized, and only the essentials should be included on the interface.
9. **Help Users Recognize, Diagnose, and Recover from an Error**: Error messages in the interface should be represented in plain language, and constructive advice that allows users to recover from the error should be provided.
10. **Help and Documentation**: Although an ideal interface should not need any additional information and explanation, help and documentation should be readily available for users. For example, these include frequently asked questions (FAQs), tutorials, how-to questions, etc.

The evaluator should note down any heuristic violations during the comparison, and the process should be continued until the task is completed. On completing the task, the violations identified by each evaluator are aggregated by an independent observer (moderator). The next step is to circulate the list with identified violations along with a brief description to each evaluator for assigning a severity rating. Since each evaluator would usually identify only a subset of violations, it is critical that a brief description is included. Each evaluator would then assign a severity rating for the identified violation on a 0–4 scale where 0 represents "I don't agree that this is a usability problem at all" and 4 represents "Usability catastrophe," meaning it is imperative that this violation is fixed before developing the final interface. After assigning the severity scores, the findings are shared with the design and development team to brainstorm and address the critical issues. Although it is not mandatory to share a solution recommendation, evaluators can provide recommendations on how the violations can be addressed, and in a few cases, these recommendations are intuitive when the violations are defined. The heuristic evaluation process can be repeated until an interface with no major usability problems is designed. A few advantages of conducting a heuristic evaluation are that it helps identify and fix usability issues during the early phase of designing and prototyping without end-users, relatively inexpensive and quick as evaluators can be in-house experts. However, a key disadvantage is that it fails to capture the user preferences and additional insights from a user standpoint.

33.5.2 Usability Testing

Usability testing is a systematic approach for observing end-users performing specific tasks using the proposed interface. The primary goals of usability testing are to identify problems in the interface design, understand user preferences and behavior while performing a task. It can also lead to uncovering new opportunities and identifying easier ways of completing tasks. There are two types of usability testing: (1) moderated and (2) unmoderated user testing. A moderated usability testing includes a facilitator or moderator who asks the user to perform a specific task and feedback, whereas an unmoderated usability testing participant completes the tasks and provides feedback at their own pace without a moderator. Both approaches have their advantages and disadvantages, and the application of each would depend on the specific project. The main advantage of moderated usability testing is that the facilitator can observe body language, pick up on subtle behaviors, and probe the user for more information. However, the cost of moderated user testing is higher compared to the unmoderated and takes more time to collect data for the same number of users because of logistical limitations. Additionally, both moderated and unmoderated usability testing can be conducted remotely or on-site.

Although there are different types of usability testing, the core elements in usability tests are tasks and participants. A few considerations while selecting the participants are that they represent the intended user community, at least five users are considered (identifies 85% of issues), and testing is performed until no new insights are gained. While conducting usability testing, one of the critical aspects is identifying the right tasks that users will perform in real life and crafting the task since errors in phrasing can bias the opinion and performance. A realistic task that can be used for user testing if you are developing a new online shopping website can be "*Please visit abc.com and select any product on the screen and leave a customer feedback.*" After crafting the task, the next step is to set goals and success criteria for the task to evaluate the user performance. In this example, the success criteria would be successfully leaving customer feedback on the product.

After setting the success criteria and goals, there are generally two approaches for data gathering and evaluation: (1) quantitative methods and (2) qualitative methods. Quantitative methods are those where the observations can be represented as numbers and provide an indirect assessment of the usability of an interface. Since these metrics are numbers, they are hard to interpret without a reference point. Hence, it is critical to have a predefined standard value or competing product interface for benchmarking or comparison. Qualitative metrics can be used to represent specific task performance or the perceived usability of the interface. Metrics that represent a task performance are known as task analysis metrics, and a few of the most common quantitative metrics used to evaluate a task performance are success rates, task completion times, and the number of errors. Success rates represent the probability of successfully completing the task, task completion times represent the time it takes to successfully complete the task under consideration, and the number of errors represents slip, mistake, or omission a user makes while attempting the task. The quantitative metric representing the user's perception of the ease of using the interface is a satisfaction survey. Although custom satisfaction surveys can be used, using validated questionnaires such as System Usability Scale (SUS), IBM Computer Usability Satisfaction Questionnaires (CSUQ), and Post-Study System Usability Questionnaire (PSSUQ) allows for comparing the reported satisfaction score to a standard value (Brooke 1996; Lewis 2002; 1995). For example, the average SUS Score for cell phone apps is 67. Hence while developing a new cell phone app interface, the average user satisfaction survey score collected using SUS should be higher than 67 to be considered as a good app. Further, development in sensor technology allows for collecting eye-tracking data while interacting with an interface. Commonly used eye-tracking metrics include gaze patterns and heatmaps which help in identifying the area of interest on the interface and user scanning patterns while interacting with the interface. One of the main advantages of quantitative data compared to qualitative data is that it can be analyzed and represented in terms of statistical significance. Specifically, the ability to perform hypothesis testing with this data and reporting the confidence intervals allow for decision-making based on the likelihood of scenarios.

In contrast to quantitative methods, which provide an indirect assessment of the usability of an interface, qualitative methods provide a direct evaluation of the usability of an interface by identifying the user's thoughts, feelings and behavior while interacting with the interface. Some of the most common qualitative methods include "think-aloud" techniques, probing, interviews, and observations. Many researchers consider the think-aloud technique one of the most important usability testing methods because of the raw feedback and input collected from the participants. There are two common think-aloud approaches used by the facilitators: (1) concurrent think aloud and (2) retrospective think aloud. In a concurrent think-aloud approach, after defining the task, the participant is asked to verbalize their thoughts including, why they are making certain decisions, problems faced, etc., as they are working through the task at hand. The goal here is to get real-time feedback on user thoughts and feelings while working through the task. However, the downside of this approach is that the quantitative metrics, such as task completion time, etc., can be affected as the user is providing continuous feedback. An alternative approach is to use a retrospective think-aloud where the participant is asked to retrace their actions and thoughts after completing the task. Although this approach does not affect the quantitative usability metrics, there is a possibility that the participant misses out on key issues, decisions, and actions they performed while completing the task. Advantages of the think-aloud approach include cheap data collection, getting users' unfiltered thoughts, robust findings from a small sample, and easy implementation.

Probing is another critical tool that is used concurrently with other qualitative techniques, including the think-aloud approach to get user feedback. However, unlike the think-aloud approach here, the facilitator will probe the users when making certain decisions or actions in the interface to understand their underlying thoughts. Similar to the think-aloud approach, there are two probing techniques: (1) concurrent probing and (2) retrospective probing. In concurrent probing, the facilitator asks questions while the user makes unique decisions or certain actions, whereas, in retrospective probing, the facilitator notes down user actions and probes the user after task completion. Interviews performed during usability testing are similar to probing; however, the facilitator will have a structure about questions they plan to ask before the usability testing. The type of interview question can vary based on the type of information the facilitator is trying to gather. In the case of gathering specific data related to an action or task, the facilitator can use a structured or semi-structured interview. Unstructured interview questions or open-ended questions are not commonly utilized during the usability testing as the participants can diverge from the topic, but they are used in scenarios where participants' opinions and suggestions about specific design considerations are required.

Apart from collecting data by interacting with the user, another critical qualitative data gathering approach is observing the user. This approach aims to closely observe the user while performing the task to understand their behaviors, body language, and emotions. Additionally, observational studies help understand the user's mental model for troubleshooting while facing issues when completing a task. Based on the type of usability study, the observation can be in-person or using video recordings in case of remote or unmoderated usability study.

Both quantitative and qualitative data gathering methods are an essential part of the iterative design approach for developing a functional and usable interface. Qualitative methods help improve the design process by identifying the poor design considerations and why the interface is not usable. In contrast, quantitative methods help compare the interface to an existing standard or competing products and provide insights on how much it varies from the benchmark. Hence it is critical to consider quantitative and qualitative methods during usability testing to develop a user-centered interface.

33.6 CONCLUSIONS

This chapter discussed the interface design process from systems thinking perspective. The advantages and importance of the systems thinking approach in designing and developing a user interface is discussed with the human-tech ladder, which helps identify the potential failure of the system. Further, various aspects and steps for developing a functional interface, including guidelines, limited human cognitive resources, information processing, decision-making, UCD approach, and interface evaluation methods, were discussed in depth. Although the theories, methodologies, and tools discussed in the chapter are not exhaustive, these can be used as foundational guidelines to develop a usable and functional user interface.

REFERENCES

Anderson, J. A., Silverstein, J. W., Ritz, S. A., & Jones, R. S. (1977). Distinctive features, categorical perception, and probability learning: Some applications of a neural model. *Psychological Review* 84(5), 413–451. doi: 10.1037/0033-295X.84.5.413.

Booher, H. R. (2003). *Handbook of Human Systems Integration. Hoboken*, NJ: Wiley.

Brooke, J. (1996). SUS: A 'quick and dirty' usability scale. In P. W. Jordan, B. Thomas, B. A. Weerdmeester, & I. L. McClelland (eds.), *Usability Evaluation in Industry*, (pp. 189–94). London: Taylor & Francis.

Checkland, P. (1999). Systems thinking. In W. Currie, & B. Galliers (eds.), *Rethinking Management Information Systems: An Interdisciplinary Perspective* (pp. 45–55). Oxford Academic. doi: 10.1002/9780470926963.ch2.

Fitts, P. M., & Jones, R. E. (1947). Psychological aspects of instrument display. I: Analysis of 270 'pilot-error' experiences in reading and interpreting aircraft instruments. Memorandum Report, vol. TSEAA-694-12A, Dayton. https://apps.dtic.mil/sti/pdfs/ADA800143.pdf.

Girishan Prabhu, V., Stanley, L., Morgan, R., & Shirley, B. (2023). Designing and developing a nature-based virtual reality with heart rate variability biofeedback for surgical anxiety and pain management: Evidence from total knee arthroplasty patients. *Aging & Mental Health*, 28(5), 738–753. doi: 10.1080/13607863.2023.2270442.

Gray, D. (2017). *Empathy Map*. gamestorming, xplaner.com. https://gamestorming.com/empathy-mapping/.

Handbook of Human-Computer Interaction (1997). *Handbook of Human-Computer Interaction*. Elsevier. doi: 10.1016/b978-0-444-81862-1.x5065-1.

Harley, A. (2015). *Personas Make Users Memorable for Product Team Members*. Nielsen Norman Group (NN/G). https://www.nngroup.com/articles/persona/.

Hoffman, P. F., Kaufman, A. J., Halverson, G. P., and Schrag, D. P. (1998). A neoproterozoic snowball earth. *Science*, 281(5381), 1342–1346. doi: 10.1126/science.281.5381.1342.

Lepore, T. (2010). *Sketches and Wireframes and Prototypes! Oh My! Creating Your Own Magical Wizard Experience*. UXmatters. https://www.uxmatters.com/mt/archives/2010/05/sketches-and-wireframes-and-prototypes-oh-my-creating-your-own-magical-wizard-experience.php.

Lewis, J. R. (1995). IBM computer usability satisfaction questionnaires: Psychometric evaluation and instructions for use. *International Journal of Human-Computer Interaction*, 7(1), 57–78. doi: 10.1080/10447319509526110.

Lewis, J. R. (2002). Psychometric evaluation of the PSSUQ using data from five years of usability studies. *International Journal of Human-Computer Interaction*, 14(3–4), 463–488. doi: 10.1080/10447318.2002.9669130.

Mayer, R. E. (2014). Cognitive theory of multimedia learning. In *The Cambridge Handbook of Multimedia Learning*, 2nd Edition (pp. 43–71). doi: 10.1017/CBO9781139547369.005.

Miller, G. A. (1956). The magical number seven, plus or minus two: Some limits on our capacity for processing information. *Psychological Review*, 63(2), 81–97. doi: 10.1037/H0043158.

Morville, P. (2004). *User Experience Honeycomb*. Semantic Studios. https://semanticstudios.com/user_experience_design/.

Nielsen, J. (1994). Usability inspection methods. In *Conference on Human Factors in Computing Systems – Proceedings* (pp. 413–414). doi: 10.1145/259963.260531.

Prabhu, V. G., Linder, C., Stanley, L. M., & Morgan, R. (2019). An affective computing in virtual reality environments for managing surgical pain and anxiety. In *Proceedings -2019 IEEE International Conference on Artificial Intelligence and Virtual Reality, AIVR 2019*. doi: 10.1109/AIVR46125.2019.00049.

Rasmussen, J. (1983). Skills, rules, and knowledge; Signals, signs, and symbols, and other distinctions in human performance models. *IEEE Transactions on Systems, Man and Cybernetics*, SMC-13(3), 257–266. doi: 10.1109/TSMC.1983.6313160.

Shneiderman, B. (1987). Designing the user interface strategies for effective human-computer interaction. *ACM SIGBIO Newsletter*, 9(1), 6. doi: 10.1145/25065.950626.

Svenson, O., & Maule, A. J. (1993). *Time Pressure and Stress in Human Judgment and Decision Making*. doi: 10.1007/978-1-4757-6846-6.

Sweller, J. (2011). Cognitive load theory. In *Psychology of Learning and Motivation - Advances in Research and Theory* (vol. 55, pp. 37–76). Academic Press. doi: 10.1016/B978-0-12-387691-1.00002-8.

Vicente, K. J. (2004). *The Human Factor: Revolutionizing the Way People Live with Technology. The Human Factor: Revolutionizing the Way People Live with Technology*. Canada: Penguin Random House. doi: 10.4324/9780203944479.

Wickens, C. D. (1984). Processing resource in attention. In R. Parasuraman, & D. R. Davies (eds.), *Varieties of Attention* (pp. 63–101). New York: Academic Press.

Wickens, C. D., Hollands, J. G., Banbury, S., & Parasuraman, R. (2012). *Engineering Psychology and Human Performance. Engineering Psychology and Human Performance*, 4th Edition. Psychology Press, Taylor and Francis Group LLC (Books) US. doi: 10.4324/9781315665177.

Wickens, C. D., McCarley, J. S., Alexander, A. L., Thomas, L. C., Ambinder, M., & Zheng, S. (2007). Attention-situation awareness (A-SA) model of pilot error. In *Human Performance Modeling in Aviation* (pp. 213–242). doi: 10.1201/9781420062984.ch9. https://www.taylorfrancis.com/chapters/mono/10.1201/9781420062984-17/attention-situation-awareness-sa-model-pilot-error-david-foyle-becky-hooey?context=ubx&refId=d8f594b8-4795-4171-b002-8f5a6b5f708c

Wiegmann, D. A., Goh, J., & O'Hare, D. (2002). The role of situation assessment and flight experience in pilots' decisions to continue visual flight rules flight into adverse weather. *Human Factors*, 44(2), 189–197. doi: 10.1518/0018720024497871.

34 Human and Privacy Rights

Razieh Nokhbeh Zaeem and K. Suzanne Barber

34.1 PRIVACY IS A HUMAN RIGHT

Privacy is important. Individuals should have the right to control the disclosure of personal data that describes them, identifies them, and reveals information about them. Individuals should not be subjected to invasions of their privacy, family, home, or correspondence, nor to assaults upon their reputation. These rights to privacy must be protected by law to guard against such interference, invasions, or attacks.

Many, many different types of personal data serve to describe, identify, and reveal information about a person, their devices, and their organizations. This information can be organized into four categories:

- **What you KNOW**: memorized assets such as name, address, or mother's maiden name.
- **What you HAVE**: held assets such as driving license and employee credential badges.
- **What you ARE**: biometric assets describing physical characteristics.
- **What you DO**: assets describing behavioral patterns such as Internet browsing or location patterns.

This personal data serves as identifiers, the "Little i," that describe and imprint the "Big I"—individuals. Consequently, this "Little i" describes, imprints, points to and ultimately offers benefits or threatens the "Big I" in the physical world.

Because the "Little i," is directly related to "Big I," theft, fraud and abuse of personal data can be used to expose, exploit, threaten or attack an individual. In a world of increasing physical and digital surveillance where an enormous amount of information is collected about individuals, privacy is challenged in every aspect of an individual's life. Privacy achieved though anonymity is extremely difficult if not impossible. Consequently, privacy rights must encompass the right to control one's personal data, "Little i" to best protect the "Big I" in the physical world.

Agency and control are prerequisites for controllable and measurable privacy. For example, the U.S. Constitution gives individuals the expectation for certain rights and protections. These constitutional rights empower individuals with agency (power and control) and give individuals well-defined authority over their assets and promise consequences when violated. For example, individuals own assets in the physical world such as a home, car, or filing cabinet and have expectations based on their constitutional rights that protect ownership of those assets and guard against unauthorized forfeiture, search or theft.

An individual's personal data, their "identity assets," are constantly being collected and, in fact, individuals knowingly and unknowingly share their identity assets in exchange for goods, services and access on a daily basis. This personal data is often referred to by the authors as "identity assets" due to the importance of this information in constructing identities and the value of this information as it serves as a means to access a wide range of goods, services, and systems. In fact, identity has often been called "the new currency" or "the new oil" of the Internet.

Whether used to grant benefits, gain access or pose threats, these identity assets are inseparable from us as human beings. Thus, the human rights offered to us as individuals must be extended to the rights of privacy and control of information that identifies us. Without the rights to control one's identity assets, our privacy rights and thereby our human rights are at risk.

DOI: 10.1201/9781003188377-41

34.2 DEFINING PRIVACY

"Privacy is a plurality of different things" (Solove, 2008). "Currently, privacy is a sweeping concept, encompassing (among other things) freedom of thought, control over one's body, solitude in one's home, control over personal information, freedom from surveillance, protection of one's reputation, and protection from searches and interrogations" (Solove, 2008). Privacy has been connected to the creation of knowledge, to dignity, and to freedom (Post, 2000). Philosophers, legal theorists, jurists, psychologists, and computer scientists have endeavored to understand what it means and how it should be protected.

The Merriam-Webster dictionary defines privacy as "the quality or state of being apart from company or observation", "freedom from unauthorized intrusion", and "secrecy".[1] The European General Data Protection Regulation (GDPR) defines data protection as "keeping data safe from unauthorized access" and data privacy as "empowering users to make their own decisions about who can process their data and for what purpose".[2] According to the GDPR, all "natural persons" have a right to data privacy under the European Union law. The California Consumer Privacy Act (CCPA) secures new privacy rights for California consumers.[3]

34.3 PRIVACY IN THE DIGITAL SOCIETY

Toward the end of the twentieth century, the development of new information technologies–in particular, the rise of the computer and the Internet–has made privacy erupt into the front-line (Kang, 1997; Moor, 1997).

34.3.1 Privacy Policies: The Reality of Privacy Online

Many websites collect, share, and use their users' Personally Identifiable Information (PII)—"any information relating to an identified or identifiable natural person" (European Parliament Regulation, 2016). Online privacy policies are legal documents that explain how an organization collects, handles, shares, uses, and discloses user data. Privacy policies have grown into the de facto method of communicating such data practices for organizations, and particularly their websites.

The ever-growing use of the Internet and the collection of PII over it has raised concerns for over two decades (Culnan, 1999; FTC, 1998). In particular, the problem of how companies handle users' PII collected over the Internet involves three main players: companies, regulators, and users.

Companies, across industries, currently are faced with tough decisions when constructing their privacy policies. On the one hand, many business models are built on collecting, using, sharing, and selling personal information. Such information can be profitable for the company, and can be leveraged to improve their product offerings and consumer-facing services. On the other hand, collecting and storing personal information about consumers carries considerable risk, as evidenced by the financial and public relations fallout from high-profile hacks. Data breaches are occurring at alarming rates, and the fallout from such hacks can be massive. Companies must assess the balance of these risk/value propositions as they construct their privacy policies.

In response to high-profile data breaches, *regulators* and policy makers—the second important player—have employed two lines of strategy: (1) holding corporations liable for breaches, imposing fines and sanctions on organizations that handled consumer data inappropriately and (2) attempting to increase the transparency of privacy and data management practices in privacy policies. The regulators, however, must constantly assess the current state of privacy policies across industries and evaluate the effects of the regulations they establish (Romanosky et al., 2011). Many regulatory bodies around the globe have long enforced requirements on posting privacy policies online. Over the past two decades, as the concerns regarding PII use and misuse were growing, newer laws have gone into effect to protect user privacy. Prominent examples of such laws are the General Data Protection Regulation (GDPR)[4] in the European Union and the California Consumer Privacy Act (CCPA)[5] in the United States.

Finally, in the face of companies' carefully constructed privacy policies and regulators' endeavors to encourage transparency in privacy policies, *users* have neither the time (Kohavi, 2001; McDonald and Cranor, 2008; Meinert et al., 2006; Milne and Culnan, 2004) nor the inclination (Graber et al., 2002; Milne et al., 2006) to read privacy policies thoroughly, choosing instead to agree absent-mindedly to the various privacy policies. More than ever, consumers need information to help them compare what a privacy policy offers with the status quo (e.g., average privacy practices among the companies that provide similar services).

34.3.1.1 Prevalence of Privacy Policies

The Federal Trade Commission (FTC) has provided several reports on online privacy practices since 1995. Its 1998 report (FTC, 1998) on US commercial websites' privacy disclosures revealed that while 92% of websites were collecting PII, only 14% disclosed any privacy policies. In its 2000 report (FTC, 2000), the FTC investigated a group of 335 websites chosen randomly and another group of 100 most busiest websites, both groups from the US market. The FTC noted that a vast majority of the websites studied collected some PII, e.g., 97% of the random sample and 99% of the busiest websites asked for email addresses. The same study found that, in the year 2000, 88% of the random group and all of the 100 busiest websites disclosed some form of their privacy policy. In the same time frame, a survey of 100 most heavily used websites (Culnan, 1999) focused on Notice, Choice, Access, and Security—the four Fair Information Practice Principles (FIPPs).

Since the first round of studies performed by the FTC, many researchers have analyzed the content of privacy policies in various ways. Many have investigated privacy policies with respect to a set of factors. Prior to the GDPR, the majority of such investigations were focused on the FTC's four Fair Information Privacy Practices. These investigations have considered different sample sets of privacy policies (Cha, 2011; Li et al., 2012; Rains and Bosch, 2009; Ryker et al., 2002; Shalhoub, 2006; Zhang et al., 2007).

It appears that privacy policies are becoming more and more common. In Liu and Arnett (2002). examined websites of the Global 500 and showed that only 61% of companies in the US had posted privacy policies. They extended their search effort for companies' privacy policies by emailing the companies and asking for their policies, when one could not be found online. Even with that extra effort, they showed that only 24% of the websites without posted privacy policy that they contacted indeed did have a policy elsewhere. In Nokhbeh Zaeem and Barber (2017), an evaluation of 10% of all listings on NYSE, Nasdaq, and AMEX stock markets revealed that 69% had their privacy policies posted online.

34.3.1.2 Who Reads Privacy Policies?!

The topic of user interest, or lack thereof, in privacy policies has been on the minds of many researchers. McDonald and Cranor (2008) noted that if users were to read–just once a year–the privacy policy for each site they visit, they would need to spend over 200 hours doing so. In fact, less than half of website users claim to have *ever* read a privacy policy (Meinert et al., 2006). Studies that used self-reported data from users found that only 4.5% claim to always read them (Milne and Culnan, 2004) and more reliable server side observation of websites revealed that only 1% or less of users click on a website's privacy policy (Kohavi, 2001). Studies using advanced eye-tracking techniques come to the same conclusion: users barely make the effort to read privacy policies thoroughly (Steinfeld, 2016). More recent work (Obar and Oeldorf-Hirsch, 2020) demonstrated that three out of four users completely ignore privacy policies. Other users skim through policies that take 29–32 minutes to read in less than 2 minutes.

Researchers have also long criticized (Graber et al., 2002; Milne et al., 2006; Ermakova et al., 2014; Fabian et al., 2017) poor readability of online privacy policies. An average privacy policy was previously measured to need 12 to 14 years of schooling to comprehend (Graber et al., 2002; Milne et al., 2006; Amos et al., 2020; Srinath et al., 2020). Many researchers have investigated privacy policy readability in general, and inside one sector in particular (Graber et al., 2002; Antón et al.,

2004; Bowers et al., 2017; Das et al., 2018; Ermakova et al., 2015; Sunyaev et al., 2015; Vallina et al., 2019). Prior work has also considered privacy policy readability across market sectors (Milne et al., 2006; Li et al., 2012; Fabian et al., 2017). In addition, multiple studies of online privacy policies found that privacy policies are getting longer and harder to read, with their readability score decreasing over time (Milne et al., 2006; Jensen and Potts, 2004).

Therefore, it should not be surprising that researchers have attributed (Ermakova et al., 2014) the fact that users rarely read privacy policies to the lack of readability in these lengthy documents. After two decades of research, readability of privacy policies is still an important and relevant issue. First, the investigation of privacy policy readability across categories enlightens users and policy makers about how users read (or ignore) privacy policies. Second, the poor readability of these documents makes the case for the thriving field of automatic analysis of privacy policies (e.g., Polisis (Harkous et al., 2018), Pribots (Harkous et al., 2016), PrivacyCheck (Nokhbeh Zaeem and Barber, 2020; Nokhbeh Zaeem et al., 2020; Zaeem et al., 2018, 2020), MAPS (Zimmeck et al., 2019), PolicyLint (Andow et al., 2019), and PrivacyGuide (Tesfay et al., 2018b)), as we cover in Section 34.4. Informing users, for example by clearly displaying privacy policies (Pan and Zinkhan, 2006; Tsai et al., 2011), motivates them to incorporate privacy into their online decisions. Therefore, it is vital to educate users through tools, information, and statistics about privacy policies.

34.4 PRIVACY ENHANCING TECHNOLOGIES TO DIGEST PRIVACY POLICIES

Privacy policies are lengthy and hard to read, yet are profoundly important as they communicate the practices of an organization pertaining to user data privacy. To address the poor readability of privacy policies, an emerging field of research focuses on privacy enhancing technologies (PETs) that summarize and visualize online privacy policies (e.g., Polisis (Harkous et al., 2018), Pribots (Harkous et al., 2016), PolicyLint (Andow et al., 2019), Privee (Zimmeck and Bellovin, 2014), PrivacyGuide (Tesfay et al., 2018b), tools from the Usable Privacy project (Sadeh et al., 2013), other similar tools and research (Zimmeck et al., 2019; Bracamonte et al., 2020; Fawaz et al., 2019), and our own publicly available PrivacyCheck (Nokhbeh Zaeem and Barber, 2020; Nokhbeh Zaeem et al., 2020; Zaeem et al., 2018)). Rich PET tools not only inform users about details of privacy policies, but also empower them to understand privacy policies at a higher level, make informed decisions, and even select competitors with better privacy policies.

The flourishing field of PETs development has resulted in research and (sometimes publicly accessible) tools that digest long privacy policies and automatically answer questions about them. These tools utilize Machine Learning (ML), Natural Language Processing (NLP), crowd sourcing, etc. In this section, we review the most related PET tools and research.

Privee (Zimmeck and Bellovin, 2014) was the first automatic privacy policy analysis tool to utilize machine learning. Building on the crowd sourcing privacy analysis framework ToS;DR (2024), Privee combines crowd sourcing with rule and machine learning classifiers to classify privacy policies that are not already rated in the crowd-sourcing repository. Privee, however, does not go beyond this basic analysis.

Polisis, available as a web page[6] and a browser extension, utilizes deep learning to summarize what user data privacy policies collect and share. At its core, Polisis is a neural network classifier trained on privacy policies retrieved from the Google Play store. In addition to providing the summary, Polisis visualizes user data collection/sharing, mapping types of data the policy collects/shares to the collection/sharing reasons outlined therein. Furthermore, Polisis displays user choices, security, data retention, etc. as graphs, making it easier for the user to comprehend what is covered in the *current* privacy policy. Notably, Polisis particularly extracts statements about how the policy claims to handle changes in its content. None of these capabilities, nonetheless, go beyond the analysis of the current policy at hand. Even the "policy change" is limited to information extraction from the current privacy policy. Pribots (Harkous et al., 2016) is from the authors of Polisis and is a chatbot that answers free form questions about a given privacy policy.

The Usable Privacy Project[7] (Sadeh et al., 2013) takes advantage of machine learning and crowd sourcing to semi-automatically annotate privacy policies. This project annotates (Wilson et al., 2016a, b) a corpus of 115 policies with attributes and data practices, the same corpus that Polisis and Pribots use to extract coarseand fine-grained classes.

PolicyLint (Andow et al., 2019) is a natural language processing tool that identifies potential contradictions that may arise inside the same privacy policy. PrivacyGuide (Tesfay et al., 2018b) (not publicly available) is a machine learning and natural language processing tool inspired by the GDPR. PolicyLint, PrivacyGuide, and many other recently developed tools (Bracamonte et al., 2020; Bui et al., 2021) are solely focused on the automatic extraction of information from *one* privacy policy.

Researchers have also investigated the consistency, or lack thereof, between privacy policies of mobile applications and how their actual code treats user data (Zimmeck et al., 2016, 2019). As one prominent example, MAPS (Zimmeck et al., 2019) analyzes privacy policies of more than one million mobile applications.

At the Center for Identity at the University of Texas at Austin[8] we target many aspects of identity management and privacy (Rana et al., 2019; Zaeem et al., 2016a, b; Zaeem et al., 2017; Zaiss et al., 2019). We developed PrivacyCheck v1 (Zaeem et al., 2018), v2 (Nokhbeh Zaeem and Barber, 2020; Nokhbeh Zaeem et al., 2020), and v3 (Nokhbeh Zaeem et al., 2021) as detailed in the next section.

34.4.1 PrivacyCheck

PrivacyCheck is a publicly available browser extension that summarizes privacy policies with machine learning. It automatically answers 20 questions, rooted in the FIPPs (Fair Information Practice Principles) and GDPR (European General Data Protection Regulation). Our previous work covered how we chose these questions and trained LightGBM machine learning models for them (FIPPs questions (Zaeem et al., 2018) and GDPR questions (Nokhbeh Zaeem and Barber, 2020; Nokhbeh Zaeem et al., 2020)). We have also applied PrivacyCheck in a variety of applications: e.g., to study the effect of the GDPR on the landscape of privacy policies (Nokhbeh Zaeem and Barber, 2020), to compare privacy policies in the public and private sectors (Zaeem and Barber, 2021), to study privacy policies across industries (Nokhbeh Zaeem and Barber, 2017), and to study PET usage patterns (Zaeem et al., 2020).

To use PrivacyCheck, the user navigates to a web page using the Chrome browser and then opens and runs the PrivacyCheck Chrome extension. PrivacyCheck's machine learning models digest the privacy policy and assign two scores to it, one for the (FIPPs-based) User Control and one for the GDPR standards. Clicking on each of the scores takes the user to score breakdowns explaining why the privacy policy received this score, based on the questions and their corresponding answers according to this privacy policy.

PrivacyCheck is free and publicly available online[9] and can also be found by searching for "PrivacyCheck" on the Google Chrome Web Store.[10] PrivacyCheck currently has over 800 users around the globe.

The first version of PrivacyCheck (Zaeem et al., 2020) was used to summarize privacy policies based on ten *User Control* privacy questions that were rooted in the work of the Organization for Economic Cooperation and Development (Regard, 1980), and the Federal Trade Commission Fair Information Practices (FTC, 2000). The second version of PrivacyCheck added ten new *GDPR* questions (Nokhbeh Zaeem and Barber, 2020; Nokhbeh Zaeem et al., 2020). PrivacyCheck v2 also added (1) a consumer-facing tool with higher performance, (2) new interface, and (3) the ability to find the top three competitors with better privacy policies (according to the User Control or GDPR standards) in the same market sector as of the privacy policy under evaluation. Finally, new capabilities were introduced in PrivacyCheck v3. In particular, the third version added the ability to

(1) find the competitors of an organization with Alexa traffic analysis and compare policies across them, (2) follow privacy policies the user has agreed to and notify the user when policies change, (3) track policies over time and report how often policies change and their trends, (4) automatically find privacy policies in domains, and (5) provide a bird's-eye view of privacy policies the user has agreed to.

34.4.2 Corpora of Privacy Policies

The studies that investigate privacy policies (e.g., their readability, or the effect of regulation on them) and the more recent body of PETs that utilizes machine learning and natural language processing to automatically summarize privacy policies greatly benefit, if not rely on, corpora of privacy policies collected from the web or mobile app stores. Such corpora of privacy policies are valuable tools at the researchers' disposal to investigate privacy policies. For example, they facilitate comparison among different methods of privacy policy summarization by providing benchmarks, and can be used in unsupervised machine learning to summarize privacy policies. Any tool or research technique that addresses the length and poor readability of these policies, such as those that apply machine learning, natural language processing, and crowd sourcing to automatically summarize privacy policies (e.g., (Harkous et al., 2018; Zaeem et al., 2018; Zimmeck et al., 2019; Andow et al., 2020; Bannihatti Kumar et al., 2020; Tesfay et al., 2018a)), would require (or benefit from) large corpora of privacy policies.

Many studies have privately gathered corpora of privacy policies and some have publicly shared them with the research community. Some researchers have dedicated their attention to manually annotating privacy policy corpora. The fact that these corpora are manually annotated by researchers or crowd-sourced workers prohibits them from including more than a couple hundred privacy policies. Two of the most widely used privacy policy corpora are OPP-115 (Wilson et al., 2016a, b) and APP-350 (Zimmeck et al., 2019), containing 115 and 350 privacy policies respectively. There exist other manually labeled corpora containing less than 1K policies (e.g., 236 policies (Bannihatti Kumar et al., 2020), 400 policies (Nokhbeh Zaeem and Barber, 2017, 2020; Nokhbeh Zaeem et al., 2020; Zaeem et al., 2018), 45 policies (Tesfay et al., 2018b), and 64 policies (Costante et al., 2012)). While valuable for supervised machine learning, these corpora fall short when it comes to unsupervised machine learning, natural language processing, and testing/validation because of their limited size.

Seeking to put together larger corpora of privacy policies, one could collect privacy policies from (1) the web or (2) mobile app stores. Interestingly, there are multiple corpora of *mobile app* (particularly Google Play) privacy policies that are available. For example, Kumar et al. used 150K (Kumar et al., 2019) policies from Google Play and Sunyaev et al. (2015) considered the privacy policies of 183 health iOS and Android apps. Notable is the MAPS framework (Zimmeck et al., 2019), which evaluated the privacy policies of over one million Android apps and released 441,626 app privacy policies with their app categories. Mobile app privacy policies have received a lot of attention, arguably, among other reasons, because of the research that analyzes a mobile app's code alongside its privacy policy (Andow et al., 2019, 2020; Arzt et al., 2014; Enck et al., 2014; Kales et al., 2019). Privacy policies of websites, nonetheless, are equally important. There are meaningful differences between the contents of web privacy policies and mobile app privacy policies. For instance, the use of cookies is more applicable to web privacy policies or, generally speaking, mobile apps can obtain finer grade location information when compared to websites and should address how they deal with this location information in their privacy policies.

There have been recent efforts to curate very large corpora of *web* privacy policies (Amos et al., 2020; Srinath et al., 2020; Nokhbeh Zaeem and Barber, 2021). Among corpora gathered from the web, some are larger but not available to the public—for instance, corpora of 9,295 policies (Zimmeck et al., 2017) and 130K policies (Harkous et al., 2018). Some smaller corpora of web privacy policies are made publicly available, for example, a corpus of 1,010 policies (Ramanath et al., 2014). Srinath and his colleagues (Srinath et al., 2020) created a corpus of one million privacy policies. They crawled the web for links with the words "privacy" or "data protection" in the URL. We (Nokhbeh Zaeem and Barber, 2021) recently published a dataset of over 100K English website privacy policies across 15 categories. These categories come from the DMOZ (now known as Curlie[11]) project—a community of volunteers who created, manually categorized, and maintained a collection of over 1.5 million links according to a hierarchical ontology. Our use of Curlie/DMOZ is particularly advantageous—we are reusing a manual categorization already done by volunteers.

34.4.3 PET Usage Patterns

There are a variety of PET tools to summarize privacy policies, but how do actual users take advantage of these tools? We were the first to monitor the usage patterns of about a thousand actual PrivacyCheck users, the first work to track the usage and traffic of an ML-based privacy analysis tool. Results show: (1) there is a good number of privacy policy URLs checked repeatedly by the user base; (2) the users are particularly interested in privacy policies of software services; and (3) PrivacyCheck increased the number of times a user consults privacy policies by 80%. Our work demonstrates the potential of ML-based privacy analysis tools and also sheds light on how these tools are used in practice to give users actionable knowledge they can use to proactively protect their privacy.

We observe that PrivacyCheck users have a tendency to check the same privacy policies, presumably the most commonly used services or the most important sources of privacy concerns. There is, generally, a good number of URLs investigated by the PrivacyCheck user base more than once: among 534 calls to PrivacyCheck in a given period of time, only 366 (68%) were on unique URLs. We find it fascinating that the user base of PrivacyCheck was disproportionately interested in running it on a variety of software service privacy policies (online social networks, large software companies, and predominantly smaller software companies) versus any other market sector/category. We think that the reason is the sheer amount of information such services can collect from their users.

Finally, we found that a typical PrivacyCheck user investigates 1.8% of all the new websites he/she visits in a year. Previous literature estimates a typical (non PrivacyCheck) web user investigates 1% of the privacy policies of new websites visited. We concluded that PrivacyCheck has the potential to increase the number of times a user consults a privacy policy and, consequently, increase their knowledge about an organization's privacy commitments.

34.4.4 Other Related PET Tools

In this section, we review tools, services, and PETs that help users protect their privacy, but do not focus on summarizing *privacy policy text*, in the following categories:

1. Privacy seals require web page operators to enroll in order to evaluate their privacy policies.
2. New formats encourage web page operators to adopt machine-readable notation of privacy policies to be automatically interpreted.
3. Crowd-sourced services have an online community that reads and rates privacy policies.
4. Tracking monitors observe web pages in action, instead of investigating their privacy policies.

34.4.4.1 Privacy Seals

Privacy seals are logos of organizations or agencies that evaluate and rate privacy policies. For example, TRUSTe (TRUSTArc, 2024) (now TrustArc) is a data privacy management company that examines privacy policies and helps businesses align their privacy policies with legal requirements. While the information provided by TRUSTe is manually extracted, TRUSTe and other similar services suffer from two major drawbacks: (1) they significantly lack comprehensive web coverage; even though TRUSTe owns the majority of the market share among all similar services it covers only roughly 55,000 in one million web pages,[12] and (2) web page operators have to sign up and potentially pay for such services which hinders their universal adoption.

Another privacy seal is provided by Better Business Bureau (BBB) (Better Business Bureau, 2024), a non-profit organization that provides free business reviews. However, BBB accredited businesses pay a fee for accreditation review.

Overall, researchers have expressed concerns with privacy seals in general: insufficient scrutiny of privacy seal organizations, negative self-selection of websites that participate in a seal, and users' ignorance regarding privacy seals (Kobsa, 2007).

34.4.4.2 New Formats

The Platform for Privacy Preferences Project (P3P) (W3 C, 2006) is a standard for websites to express their privacy policies in a both human and machine-readable format. Following such standards enables automatic interpretation of privacy policies. P3P and similar standards require web page operators to adopt new formats. As a result, P3P has always suffered from lack of industry participation. Consequently, the P3P working group was closed in 2006 and P3P 1.1 was never finalized (Cranor, 2012).

Currently, P3P is used in only 70,000 in one million web pages.[13] The failure of P3P in attracting industry participation, however, is not limited to the number of websites that do not support it. A large fraction of the websites that support P3P chose to include a minimal and mis-representative version of their privacy policy just to prevent Internet Explorer, the major web browser supporting P3P, from blocking their cookies. In fact, thousands of websites were found to use an identical erroneous policy recommended by a Microsoft support website to avoid cookie blocking by Internet Explorer (Cranor, 2012).

Researchers have attempted visualizing privacy policies represented in P3P (e.g., Privacy Bird extension for Internet Explorer (Cranor et al., 2006)), some with only little success in improving the comprehension (Reeder et al., 2008). Internet Explorer 6.0 included a feature to textually present privacy policies formatted as P3P.

Nutrition Label (Cranor, 2012; Kelley et al., 2009; 2010) introduced another new format that asks website operators to consolidate their privacy policies in a one-page standardized format inspired by the nutrition facts panel found on food and drug packages. Complying with this new format also places an additional, unwanted burden on website operators. In 2020, Apple, Inc. moved to adopt the Nutrition Label for privacy policies of its mobile apps.[14]

34.4.4.3 Crowd-Sourced Services

Terms of Service; Didn't Read (ToS;DR, 2024) is a free software project that started in 2012 to address the problem that very few users actually read the terms of service for websites they use. In this project, an online community of volunteers read, discuss, and rate privacy policies. The ratings and discussions are available online and as free software in the form of browser extensions for Mozilla Firefox, Google Chrome, Apple Safari, and Opera. Even though privacy policies addressed in this project are read and rated by humans and discussed thoroughly, the in-depth and occasionally long discussions pose a new challenge to the usefulness of the ratings: one might as well read, selectively, the original privacy policy itself. Furthermore, the coverage of ToS;DR is even more limited than privacy seals and new formats (ToS;DR, 2024).

34.4.4.4 Tracking Monitors

Ghostery (2021) is a software available as free browser extensions for Mozilla Firefox, Google Chrome, Internet Explorer, Opera, and Apple Safari. Ghostery tracks cookies, tags, web bugs, pixels, and beacons and then notifies the user of their presence as well as the companies that operate them, giving the user the choice to make informed decisions about blocking them. Similarly,

AdblockPlus (2024) is a free extension that blocks ads and disables tracking. Adblock Plus is available for Android, Google Chrome, Mozilla Firefox, Internet Explorer, Opera, and Apple Safari among other browsers. Ghostery, Adblock Plus, and other similar services are fundamentally different from PrivacyCheck and ML-based PET tools, in that they focus on the actions a website takes, instead of the legal privacy policy it posts. Hopefully, such actions are aligned with the privacy policy, but that is not guaranteed. In addition, tracking monitors do not indicate the usage of the information gathered from users.

34.5 PRIVACY REGULATIONS

As we mentioned earlier, regulatory bodies around the world have long enforced requirements on online privacy policies. Over the past few years, newer laws have gone into effect to protect user privacy. The most important examples of such laws are the General Data Protection Regulation (GDPR)[15] in the European Union and the California Consumer Privacy Act (CCPA)[16] in the United States.

34.5.1 Regulations Governing Privacy Policies

In this section, we list some recent and old privacy policy regulations, with an emphasis on the United States (US) and the European Union (EU).

- The GDPR is the newest regulation in the EU law on data protection and privacy. The key principles of the GDPR are (1) Lawfulness, fairness, and transparency, (2) Purpose limitation, (3) Data minimization, (4) Accuracy, (5) Storage limitation, (6) Security, and (7) Accountability.
- The California Consumer Privacy Act (CCPA) is a state statute to enhance privacy rights and consumer protection for residents of California.
- The International Safe Harbor Privacy Principles were principles (overturned in 2015) developed to prevent private organizations within the European Union or United States that store customer data from accidentally disclosing or losing personal information (Brown and Blevins, 2002).
- The Federal Trade Commission (FTC) Fair Information Practice Principles (FIPP) are recommendations, though not legally enforced, for maintaining privacy-friendly, consumer-oriented data collection practices and include Notice, Choice, Access, and Integrity.
- The Children's Online Privacy Protection Act (COPPA) protects the personal information of children under 13.
- The Gramm-Leach-Bliley Act (GLBA) requires financial institutions (i.e. companies) to explain their information-sharing practices to their customers and to safeguard sensitive data.
- The Health Insurance Portability and Accountability Act (HIPPA) applies to healthcare providers, suppliers, and vendors (business associates).
- The Telephone Consumer Protection Act (TCPA) regulates the collection of information by telephone service providers.
- The Privacy Act of 1974 is the primary law in the US that governs government collection, maintenance, use, and dissemination of PII by federal agencies.
- The Freedom of Information Act (FOIA) governs the collection, maintenance, use, and dissemination of PII that is maintained in systems of records by federal agencies.
- Federal Information Security Management Act (FISMA) mandates that each federal agency implements an information security program for the information and information systems that support the operations and assets of the agency.
- The Electronic Communications Privacy Act (ECPA) restricts the government's access and disclosure of electronic communication.

34.5.2 The Effect of Privacy Regulations such as the GDPR on Online Privacy Policies

The research community has looked into quantifying the effect of privacy policy regulations on the landscape of online privacy. Interestingly, the PET tools we discussed earlier have found a new use in addition to assisting final consumers in understanding privacy policies: researchers have leveraged these tools to automatically analyze huge corpora of privacy policies and study their statistics. For instance, both Polisis and PrivacyCheck have been utilized to study the effect of the GDPR on the privacy landscape (Nokhbeh Zaeem and Barber, 2020; Linden et al., 2018; Nokhbeh Zaeem and Barber, 2021).

The GDPR is considered by some to be the most important change in data privacy regulation in 20 years. The European Union GDPR privacy law applies to any organization that collects and processes the personal information of EU citizens within or outside the EU. While the GDPR is a European Union law, it covers any organization that collects or processes EU citizen data independent of the organization's location. Due to the global nature of commerce and people's movements, the GDPR drove businesses around the world to make important decisions and changes regarding how they collect and process their employee's and customers' PII.

The GDPR went into effect on May 25, 2018. As a result, many companies that do business completely or partially in the EU or handle EU citizens' data updated their online privacy policies around the same time in order to comply with the GDPR (Degeling et al., 2018). It is also important to note that the GDPR has already motivated other advancements in privacy regulation and continues to do so around the globe as consumers demand their data rights.[17] Evaluating the actual effect of the GDPR on online privacy policies of companies is a significant research question, which has not received the attention it deserves.

Degeling and colleagues are among the few who studied a sizable set of privacy policies (from more than 6,500 websites) across the EU (Degeling et al., 2018) with the help of automation, and compared policies before and after the GDPR. That work, however, focused exclusively on the consent to use cookies in privacy policies. Libert et al. (2018) studied privacy policies of seven countries of the EU before and after the GDPR went into effect, but also focused on cookies only.

A closely related work comes from Linden et al. (2020) which examines policies from both inside and outside the EU. While they use over 6,000 privacy policies, with the exception of the visual representation evaluation that is done by Amazon MTurk users, the rest of the compliance assessment is done automatically. The automation is in turn based on Polisis (Harkous et al., 2018), a deep learning tool developed by the authors. They concluded that, even though the GDPR has prompted a general overhaul in privacy policies, many policies still do not meet several GDPR requirements. The authors of Claudette (Contissa et al., 2018) compare pre- and post-GDPR privacy policies with a data mining tool, but their work is limited to only few policies (from 14 (Contissa et al., 2018) to 50 policies (Lippi et al., 2019)).

In a recent work (Nokhbeh Zaeem and Barber, 2020), we quantified the progress the GDPR has made in improving privacy policies around the globe. We leveraged our data mining tool, PrivacyCheck, to automatically compare three corpora (totaling 550) of privacy policies, pre- and post-GDPR. In addition, to evaluate the current level of compliance with the GDPR around the globe, we manually studied the policies within two corpora (450 policies).

We applied PrivacyCheck to compare the privacy policies before and after the GDPR through PrivacyCheck's ten questions to assess the impact of the GDPR. We manually studied these privacy policies after the GDPR went into effect to measure how close the policies are to full compliance with the GDPR. We distill another ten questions, previously not supported by PrivacyCheck, directly from the GDPR.

In this work, we examined the landscape of online privacy policies, to evaluate the effect of the GDPR as well as to paint a clearer picture of data privacy and best practices regarding privacy policies and their level of compliance with the GDPR. The verdict is that, with modest changes, most of the privacy policies were able to satisfy many but not all GDPR requirements. The most notable non-compliance of the investigated privacy policies was found when policies fail to indicate compliance with a GDPR requirement, either affirmatively or negatively. Consequently, the most notable non-compliance with the GDPR results when an organization lacks transparency and explicit disclosure of their processing and protection of consumer personal information.

Based on the manual examination of policies, when non-compliance does appear, it is often in failing to explicitly indicate compliance. We identify the following areas for improvement in privacy policies:

- Many policies fail to mention whether they encrypt data while at rest.
- Many do not mention if and when they notify the supervisory authority in case of a data breach.
- Some US-based policies protect children under 13 as per US regulation and hence do not necessarily protect children between 13 and 16 years of age.
- Privacy policies lack a 100% consensus to require informed consent.

We reported that the GDPR has made progress in protecting user data, but more progress is necessary—particularly in the area of giving users the right to edit and delete their information—to entirely fulfill the GDPR's promise. We also observed that the GDPR encourages sharing user data with law enforcement, and, as a result, many policies have facilitated such sharing after the GDPR. Finally, we saw that when there is non-compliance with the GDPR, it is often in the form of failing to explicitly indicate compliance, which in turn speaks to an organization's lack of transparency and disclosure regarding the processing and protection of personal information.

We found that websites have modestly changed their privacy policies after the GDPR. Changes have often been geared toward compliance. This is not surprising since the penalties for non-compliance far exceed prior regulations. Overall, the landscape of privacy policies after the GDPR is promising when considering all the privacy factors measured by PrivacyCheck. For many factors, over 90% of the privacy policies we considered were in compliance with the GDPR.

34.6 CONCLUSIONS AND FUTURE WORK

Personal data or PII–the "little i"–is directly tied to an individual human–the "big I"– giving individuals benefits or posing threats. Thus, we must protect personal data like we protect humans.

In today's world, where a huge amount of information is collected about individuals, achieving privacy through anonymity is extremely difficult if not impossible, and achieving such anonymity will involve losing access to many goods, systems, and services. As anonymity is nearly impossible and interferes with full participation in society, privacy rights over one's personal data are essential. The rights of agency and control are central to these protections.

To protect *individual users'* privacy rights, in particular when *companies* collect user data, privacy laws put in place by *regulators* are critical. In fact, users, companies, and regulators are the three main stakeholders required to protect privacy rights.

PETs, such as PrivacyCheck, are critical to ensuring individuals can implement privacy controls and defend their privacy rights. These PETs can also serve both companies and regulators, to gauge adherence to privacy laws.

Laws are changing to offer more privacy protections. The two most recent examples are the GDPR in the European Union and the CCPA in the United States. As these privacy laws change, it is paramount to study their tangible effects to guide further regulation, ensure compliance, and inform and empower users.

All in all, the PETs and privacy regulations have made progress in educating users and enforcing data protection regulations. More work is necessary to enable and empower users to make informed decisions with respect to privacy policies. For future work, we envision answering multiple sets of questions with PrivacyCheck to better inform users to understand their privacy risks and provide actionable knowledge to use their usage and buying power to select products and services offering by privacy-respecting organizations. Privacy enhancing tools like PrivacyCheck aim to give users the agency and control necessary to protect and defend their privacy rights. Finally, pertaining to the privacy regulations like the GDPR, more work is necessary, particularly in the area of granting users the right to edit, update, and delete their data, to entirely fulfill the GDPRs promise.

NOTES

1 https://www.merriam-webster.com/dictionary/privacy?utm_campaign=sd&utm_medium=serp&utm_source=jsonld
2 https://gdpr.eu/data-privacy
3 https://oag.ca.gov/privacy/ccpa
4 https://gdpr-info.eu
5 https://oag.ca.gov/privacy/ccpa
6 https://pribot.org/polisis
7 https://usableprivacy.org
8 https://identity.utexas.edu
9 https://tinyurl.com/ydf7h7dr
10 https://chrome.google.com/webstore
11 https://curlie.org
12 https://www.datanyze.com/market-share/security/truste-market-share
13 http://trends.builtwith.com/docinfo/P3P-Policy
14 https://developer.apple.com/app-store/app-privacy-details
15 https://gdpr-info.eu
16 https://oag.ca.gov/privacy/ccpa
17 California Consumer Privacy Act (CCPA), https://www.caprivacy.org

REFERENCES

AdblockPlus (2024). Adblock plus surf the web without annoying ads! https://getadblock.com/en/ (Accessed October 3, 2024).

Amos, R., Acar, G., Lucherini, E., Kshirsagar, M., Narayanan, A., & Mayer, J. (2020). Privacy policies over time: Curation and analysis of a million-document dataset. *arXiv preprint arXiv:2008.09159.*

Andow, B., Mahmud, S. Y., Wang, W., Whitaker, J., Enck, W., Reaves, B., … & Xie, T. (2019). {PolicyLint}: Investigating internal privacy policy contradictions on google play. In *28th USENIX Security Symposium* (pp. 585–602).

Andow, B., Mahmud, S. Y., Whitaker, J., Enck, W., Reaves, B., Singh, K., & Egelman, S. (2020). Actions speak louder than words:{Entity-Sensitive} privacy policy and data flow analysis with {PoliCheck}. In *29th {USENIX} Security Symposium ({USENIX} Security 20)* (pp. 985–1002).

Antón, A. I., Earp, J. B., He, Q., Stufflebeam, W., Bolchini, D., & Jensen, C. (2004). Financial privacy policies and the need for standardization. *IEEE Security & Privacy*, 2(2), 36–45.

Arzt, S., Rasthofer, S., Fritz, C., Bodden, E., Bartel, A., Klein, J., … & McDaniel, P. (2014). Flowdroid: Precise context, flow, field, object-sensitive and lifecycle-aware taint analysis for android apps. *ACM Sigplan Notices*, 49(6), 259–269.

Bannihatti Kumar, V., Iyengar, R., Nisal, N., Feng, Y., Habib, H., Story, P., … & Sadeh, N. (2020). Finding a choice in a haystack: Automatic extraction of opt-out statements from privacy policy text. In *Proceedings of the Web Conference 2020* (pp. 1943–1954).

Better Business Bureau (2024). BBB Research: BBB Seal is TRUSTED When It Comes to Online Retail. https://www.bbb.org/all/trust-hub/deep-dive-into-trust/bbb-seal-trusted-online-retail. (Accessed October 3, 2024).

Bowers, J., Reaves, B., Sherman, I. N., Traynor, P., & Butler, K. (2017). Regulators, mount up! Analysis of privacy policies for mobile money services. In *Thirteenth Symposium on Usable Privacy and Security (SOUPS 2017)* (pp. 97–114).

Bracamonte, V., Hidano, S., Tesfay, W. B., & Kiyomoto, S. (2020). Evaluating the effect of justification and confidence information on user perception of a privacy policy summarization tool. In *ICISSP* (pp. 142–151).

Brown, D. H., & Blevins, J. L. (2002). The safe-harbor agreement between the United States and Europe: A missed opportunity to balance the interests of e-commerce and privacy online? *Journal of Broadcasting & Electronic Media*, 46(4), 549–564.

Bui, D., Shin, K. G., Choi, J. M., & Shin, J. (2021). Automated extraction and presentation of data practices in privacy policies. *Proceedings on Privacy Enhancing Technologies*, 2021(2), 88–110.

Cha, J. (2011). Information privacy: A comprehensive analysis of information request and privacy policies of most-visited web sites. *Asian Journal of Communication*, 21(6), 613–631.

Contissa, G., Docter, K., Lagioia, F., Lippi, M., Micklitz, H. W., Pałka, P., … & Torroni, P. (2018). Claudette meets GDPR: Automating the evaluation of privacy policies using artificial intelligence. Available at SSRN 3208596.

Costante, E., Sun, Y., Petković, M., & Den Hartog, J. (2012). A machine learning solution to assess privacy policy completeness: (Short paper). In *Proceedings of the 2012 ACM Workshop on Privacy in the Electronic Society* (pp. 91–96).

Cranor, L. F. (2012). Necessary but not sufficient: Standardized mechanisms for privacy notice and choice. *Journal on Telecommunications and High Technology Law*, 10, 273.

Cranor, L. F., Guduru, P., & Arjula, M. (2006). User interfaces for privacy agents. *ACM Transactions on Computer-Human Interaction (TOCHI)*, 13(2), 135–178.

Culnan, M. J. (1999). *Georgetown Internet Privacy Policy Survey: Report to the Federal Trade Commission.* Washington, DC: Georgetown University, The McDonough School of Business.

Das, G., Cheung, C., Nebeker, C., Bietz, M., & Bloss, C. (2018). Privacy policies for apps targeted toward youth: Descriptive analysis of readability. *JMIR mHealth and uHealth*, 6(1), e3.

Degeling, M., Utz, C., Lentzsch, C., Hosseini, H., Schaub, F., & Holz, T. (2018). We value your privacy... now take some cookies-measuring the GDPR's impact on web privacy. *Informatik Spektrum*, 42(5), 345–346, 2019.

Enck, W., Gilbert, P., Han, S., Tendulkar, V., Chun, B. G., Cox, L. P., … & Sheth, A. N. (2014). Taintdroid: An information-flow tracking system for realtime privacy monitoring on smartphones. *ACM Transactions on Computer Systems (TOCS)*, 32(2), 1–29.

Ermakova, T., Baumann, A., Fabian, B., & Krasnova, H. (2014). Privacy policies and users' trust: Does readability matter? In *Proceedings of the Twentieth Americas Conference on Information Systems* (p. 12).

Ermakova, T., Fabian, B., & Babina, E. (2015). Readability of privacy policies of healthcare websites. In *Wirtschaftsinformatik*.

European Parliament Regulation. (2016). General data protection regulation. *Official Journal of the European Parliament, Legislation series, OJ L 119*, 4.5, 1–88, 2016/679.

Fabian, B., Ermakova, T., & Lentz, T. (2017). Large-scale readability analysis of privacy policies. In *Proceedings of the International Conference on Web Intelligence* (pp. 18–25).

Fawaz, K., Linden, T., & Harkous, H. (2019). The applications of machine learning in privacy notice and choice. In *2019 11th International Conference on Communication Systems & Networks (COMSNETS)* (pp. 118–124). IEEE.

FTC (1998). Privacy Online: A Report to Congress.

FTC (2000). Privacy Online: Fair Information Practices in the Electronic Marketplace: A Federal Trade Commission Report to Congress.

Ghostery (2021). Join over 40 million Ghostery Users and Download the Web's Most Popular Privacy Tool.

Graber, M. A., D Alessandro, D. M., & Johnson-West, J. (2002). Reading level of privacy policies on internet health web sites. *Journal of Family Practice*, 51(7), 642–642.

Harkous, H., Fawaz, K., Lebret, R., Schaub, F., Shin, K. G., & Aberer, K. (2018). Polisis: Automated analysis and presentation of privacy policies using deep learning. In *27th USENIX Security Symposium* (pp. 531–548).

Harkous, H., Fawaz, K., Shin, K. G., & Aberer, K. (2016). Pribots: Conversational privacy with chatbots. In *Twelfth Symposium on Usable Privacy and Security (SOUPS) 2016.*

Jensen, C. & Potts, C. (2004). Privacy policies as decision-making tools: An evaluation of online privacy notices. In *Proceedings of the SIGCHI Conference on Human Factors in Computing Systems* (pp. 471–478).

Kales, D., Rechberger, C., Schneider, T., Senker, M., & Weinert, C. (2019). Mobile private contact discovery at scale. In *28th USENIX Security Symposium (USENIX Security 19)* (pp. 1447–1464).

Kang, J. (1997). Information privacy in cyberspace transactions. *Stanford Law Review*, 50, 1193.

Kelley, P. G., Bresee, J., Cranor, L. F., & Reeder, R. W. (2009). A nutrition label for privacy. In *Proceedings of the 5th Symposium on Usable Privacy and Security* (p. 4). ACM.

Kelley, P. G., Cesca, L., Bresee, J., & Cranor, L. F. (2010). Standardizing privacy notices: An online study of the nutrition label approach. In *Proceedings of the SIGCHI Conference on Human Factors in Computing Systems* (pp. 1573–1582). ACM.

Kobsa, A. (2007). Privacy-enhanced web personalization. In P. Brusilovsky, A. Kobsa, & W. Nejdl (eds.), *The Adaptive Web* (pp. 628–670). Heidelberg, Berlin: Springer.

Kohavi, R. (2001). Mining e-commerce data: The good, the bad, and the ugly. In *International Conference on Knowledge Discovery and Data Mining* (pp. 8–13). ACM.

Kumar, V. B., Ravichander, A., Story, P., & Sadeh, N. (2019). Quantifying the effect of in-domain distributed word representations: A study of privacy policies. In *AAAI Spring Symposium on Privacy-Enhancing Artificial Intelligence and Language Technologies.*

Li, Y., Stewart, W., Zhu, J., Ni, A., & Rohm Jr, C. E. (2012). Online privacy policy of the thirty Dow Jones corporations: Compliance with FTC fair information practice principles and readability assessment. *Communications of the IIMA*, 12(3), 5.

Libert, T., Graves, L., & Nielsen, R. (2018). Changes in Third-Party Content on European News Websites after GDPR. Reuters Institute for the Study of Journalism Reports: Factsheet.

Linden, T., Harkous, H., & Fawaz, K. (2018). The privacy policy landscape after the GDPR. *arXiv preprint arXiv:1809.08396.*

Linden, T., Khandelwal, R., Harkous, H., & Fawaz, K. (2020). The privacy policy landscape after the GDPR. *Proceedings on Privacy Enhancing Technologies*, 2020(1), 47–64.

Lippi, M., Pałka, P., Contissa, G., Lagioia, F., Micklitz, H. W., Sartor, G., & Torroni, P. (2019). Claudette: An automated detector of potentially unfair clauses in online terms of service. *Artificial Intelligence and Law*, 27(2), 117–139.

Liu, C., & Arnett, K. P. (2002). Raising a red flag on global www privacy policies. *Journal of Computer Information Systems*, 43(1), 117–127.

McDonald, A. M., & Cranor, L. F. (2008). The cost of reading privacy policies. *I/S: A Journal of Law and Policy for the Information Society*, 4, 543.

Meinert, D. B., Peterson, D. K., Criswell, J. R., & Crossland, M. D. (2006). Privacy policy statements and consumer willingness to provide personal information. *Journal of Electronic Commerce in Organizations*, 4(1), 1.

Milne, G. R., & Culnan, M. J. (2004). Strategies for reducing online privacy risks: Why consumers read (or don't read) online privacy notices. *Journal of Interactive Marketing*, 18(3), 15–29.

Milne, G. R., Culnan, M. J., & Greene, H. (2006). A longitudinal assessment of online privacy notice readability. *Journal of Public Policy & Marketing*, 25(2), 238–249.

Moor, J. H. (1997). Towards a theory of privacy in the information age. *ACM Sigcas Computers and Society*, 27(3), 27–32.

Nokhbeh Zaeem, R., & Barber, K. S. (2017). A study of web privacy policies across industries. *Journal of Information Privacy and Security*, 13(4), 169–185.

Nokhbeh Zaeem, R., & Barber, K. S. (2020). The effect of the GDPR on privacy policies: Recent progress and future promise. *ACM Transactions on Management of Information Systems*, 12(1), 1–20.

Nokhbeh Zaeem, R., & Barber, K. S. (2021). A large publicly available corpus of website privacy policies based on dmoz. In *the 11th ACM Conference on Data and Application Security and Privacy*. To Appear. https://tinyurl.com/ybsq9qc8.

Nokhbeh Zaeem, R., & Barber, K. S. (2021). Comparing privacy policies of government agencies and companies: A study using machine-learning- based privacy policy analysis tools. In *ICAART (2)* (pp. 29–40).

Nokhbeh Zaeem, R., Ahbab, A., Bestor, J., Djadi, H. H., Kharel, S., Lai, V., … & Barber, K. S. (2021). PrivacyCheck v3: Empowering users with higher-level understanding of privacy policies. In *20th Workshop on Privacy in the Electronic Society (WPES 21)*. Under Submission.

Nokhbeh Zaeem, R., Anya, S., Issa, A., Nimergood, J., Rogers, I., Shah, V., … & Barber, K. S. (2020). PrivacyCheck v2: A tool that recaps privacy policies for you. In *Proceedings of the 29th ACM International Conference on Information & Knowledge Management* (pp. 3441–3444).

Obar, J. A., & Oeldorf-Hirsch, A. (2020). The biggest lie on the internet: Ignoring the privacy policies and terms of service policies of social networking services. *Information, Communication & Society*, 23(1), 128–147.

Pan, Y., & Zinkhan, G. M. (2006). Exploring the impact of online privacy disclosures on consumer trust. *Journal of Retailing*, 82(4), 331–338.

Post, R. C. (2000). Three concepts of privacy. *The Georgetown Law Journal*, 89, 2087.

Rains, S. A., & Bosch, L. A. (2009). Privacy and health in the information age: A content analysis of health web site privacy policy statements. *Health Communication*, 24(5), 435–446.

Ramanath, R., Liu, F., Sadeh, N., & Smith, N. A. (2014). Unsupervised alignment of privacy policies using hidden Markov models. In *Proceedings of the 52nd Annual Meeting of the Association for Computational Linguistics (Volume 2: Short Papers)* (pp. 605–610).

Rana, R., Zaeem, R. N., & Barber, K. S. (2019). An assessment of blockchain identity solutions: Minimizing risk and liability of authentication. In *2019 IEEE/WIC/ACM International Conference on Web Intelligence (WI)* (pp. 26–33).

Reeder, R. W., Kelley, P. G., McDonald, A. M., & Cranor, L. F. (2008). A user study of the expandable grid applied to P3P privacy policy visualization. In *Proceedings of the 7th ACM Workshop on Privacy in the Electronic Society* (pp. 45–54). ACM.

Regard, H. (1980). Recommendation of the Council Concerning Guidelines Governing the Protection of Privacy and Transborder Flows of Personal Data.

Romanosky, S., Telang, R., & Acquisti, A. (2011). Do data breach disclosure laws reduce identity theft? *Journal of Policy Analysis and Management*, 30(2), 256–286.

Ryker, R., Lafleur, E., McManis, B., & Cox, K. C. (2002). Online privacy policies: An assessment of the fortune E-50. *Journal of Computer Information Systems*, 42(4), 15–20.

Sadeh, N., Acquisti, A., Breaux, T. D., Cranor, L. F., McDonald, A. M., Reidenberg, J. R., ... & Wilson, S. (2013). The Usable Privacy Policy Project, Technical Report, CMU-ISR-13-119. Carnegie Mellon University.

Shalhoub, Z. K. (2006). Content analysis of web privacy policies in the gcc countries. *Information Systems Security*, 15(3), 36–45.

Solove, D. J. (2008). *Understanding Privacy*. Cambridge, MA: Harvard University Press.

Srinath, M., Wilson, S., & Giles, C. L. (2020). Privacy at scale: Introducing the privaseer corpus of web privacy policies. *arXiv preprint arXiv:2004.11131*.

Steinfeld, N. (2016). I agree to the terms and conditions: (How) do users read privacy policies online? An eye-tracking experiment. *Computers in human Behavior*, 55, 992–1000.

Sunyaev, A., Dehling, T., Taylor, P. L., & Mandl, K. D. (2015). Availability and quality of mobile health app privacy policies. *Journal of the American Medical Informatics Association*, 22(e1), e28–e33.

Tesfay, W. B., Hofmann, P., Nakamura, T., Kiyomoto, S., & Serna, J. (2018). PrivacyGuide: Towards an implementation of the EU GDPR on internet privacy policy evaluation. In *Proceedings of the Fourth ACM International Workshop on Security and Privacy Analytics* (pp. 15–21). ACM.

Tesfay, W. B., Hofmann, P., Nakamura, T., Kiyomoto, S., & Serna, J. (2018). I read but don't agree: Privacy policy benchmarking using machine learning and the EU GDPR. In *Companion Proceedings of the Web Conference 2018* (pp. 163–166). International World Wide Web Conferences Steering Committee.

ToS;DR (2024). Terms of service; didn't read. https://tosdr.org/about#:~:text=%E2%80%9CTerms%20of%20Service%3B%20Didn',agree%20to%20all%20the%20time (Updated October 3, 2024).

TRUSTArc (2024). HomePage: Privacy and Governance. https://trustarc.com/?utm_source=google&utm_medium=cpc&utm_campaign=brand-phrase&utm_term=trustarc&gclid=CjwKCAjwgfm-3BhBeEiwAFfxrG7R6rAVpiFY48aTxWyyZKGTWFP2lQqW7RoG0B6yw8LHAEUKbb3r3xxoCRMwQAvD_BwE&gad_source=1 (Updated October 3, 2024).

Tsai, J. Y., Egelman, S., Cranor, L., & Acquisti, A. (2011). The effect of online privacy information on purchasing behavior: An experimental study. *Information Systems Research*, 22(2), 254–268.

Vallina, P., Feal, Á., Gamba, J., Vallina-Rodriguez, N., & Anta, A. F. (2019). Tales from the porn: A comprehensive privacy analysis of the web porn ecosystem. In *Proceedings of the Internet Measurement Conference* (pp. 245–258).

W3 C (2006). The Platform for Privacy Preferences 1.1 (P3P1.1) Specification.

Wilson, S., Schaub, F., Dara, A. A., Liu, F., Cherivirala, S., Leon, P. G., ... & Sadeh, N. (2016a). The creation and analysis of a website privacy policy corpus. In *Annual Meeting of the Association for Computational Linguistics* (pp. 1330–13340).

Wilson, S., Schaub, F., Ramanath, R., Sadeh, N., Liu, F., Smith, N. A., & Liu, F. (2016b). Crowdsourcing annotations for websites' privacy policies: Can it really work? In *Proceedings of the 25th International Conference on World Wide Web* (pp. 133–143).

Zaeem, R. N., Anya, S., Issa, A., Nimergood, J., Rogers, I., Shah, V., ... & Barber, K. S. (2020). Privacychecks machine learning to digest privacy policies: Competitor analysis and usage patterns. In *2020 IEEE/WIC/ACM International Joint Conference on Web Intelligence and Intelligent Agent Technology (WI-IAT)* (pp. 291–298). IEEE.

Zaeem, R. N., Budalakoti, S., Barber, K. S., Rasheed, M., & Bajaj, C. (2016a). Predicting and explaining identity risk, exposure and cost using the ecosystem of identity attributes. In *2016 IEEE International Carnahan Conference on Security Technology (ICCST)* (pp. 1–8). IEEE.

Zaeem, R. N., German, R. L., & Barber, K. S. (2018). Privacycheck: Automatic summarization of privacy policies using data mining. *ACM Transactions on Internet Technology (TOIT)*, 18(4), 53.

Zaeem, R. N., Manoharan, M., & Barber, K. S. (2016b). Risk kit: Highlighting vulnerable identity assets for specific age groups. In *2016 European Intelligence and Security Informatics Conference (EISIC)* (pp. 32–38). IEEE.

Zaeem, R. N., Manoharan, M., Yang, Y., & Barber, K. S. (2017). Modeling and analysis of identity threat behaviors through text mining of identity theft stories. *Computers & Security*, 65, 50–63.

Zaeem, R., & Barber, K. (2021). Comparing privacy policies of government agencies and companies: A study using machine-learning-based privacy policy analysis tools. In *Proceedings of the 13th International Conference on Agents and Artificial Intelligence - Volume 2: ICAART* (pp. 29–40). INSTICC, SciTePress.

Zaiss, J., Nokhbeh Zaeem, R., & Barber, K. S. (2019). Identity threat assessment and prediction. *Journal of Consumer Affairs*, 53(1), 58–70.

Zhang, X., Toru, S., & Kennedy, M. (2007). A cross-cultural analysis of privacy notices of the global 2000. *Journal of Information Privacy and Security*, 3(2), 18–36.

Zimmeck, S., & Bellovin, S. M. (2014). Privee: An architecture for automatically analyzing web privacy policies. In *23rd USENIX Security Symposium* (pp. 1–16). San Diego, CA: USENIX Association.

Zimmeck, S., Story, P., Smullen, D., Ravichander, A., Wang, Z., Reidenberg, J., … & Sadeh, N. (2019). Maps: Scaling privacy compliance analysis to a million apps. *Proceedings on Privacy Enhancing Technologies*, 2019(3), 66–86.

Zimmeck, S., Wang, Z., Zou, L., Iyengar, R., Liu, B., Schaub, F., … & Reidenberg, J. (2016). Automated analysis of privacy requirements for mobile apps. In *2016 AAAI Fall Symposium Series*.

Zimmeck, S., Wang, Z., Zou, L., Iyengar, R., Liu, B., Schaub, F., … & Reidenberg, J. (2017). Automated analysis of privacy requirements for mobile apps. In *NDSS*.

35 Humanitarian Well-Being

Kristen Intemann

35.1 INTRODUCTION

The mission of the IEEE's Global Initiative on Ethics of Autonomous and Intelligent (A/I) Systems is, "to ensure every stakeholder involved in the design and development of autonomous and intelligent systems is educated, trained, and empowered to prioritize ethical considerations so that these technologies *are advanced for the benefit of humanity*" (IEEE 2017a, emphasis added). Moreover, the IEEE's principles for "ethically aligned design" include a commitment to increasing human well-being as a *primary success criterion* for A/I systems (IEEE 2017b, 11). What exactly is "human well-being"? How can human well-being be measured (particularly if it is subjective and/or the well-being of some humans may conflict with the well-being of others)? Finally, even if we are committed to the value of "human well-being," which in the abstract seems uncontroversial, how can it practically guide the design of A/I systems? These are the central questions of this chapter.

35.2 WHAT IS WELL-BEING?

Although there has been a longstanding push among funding agencies and professional organizations to produce research and innovation for "social or human benefit," it is not always clear what this means or what would count as "beneficial." Historically, such benefits have been interpreted in rather narrow ways with a focus on economic benefit. As the IEEE notes, the indicators that are frequently used to measure success in complex systems include increasing productivity for customers and society, as measured by gross domestic product (GDP), profit, or consumption levels (IEEE 2017b, 21). Yet these sorts of measures would likely not be identified as those things most important to many humans throughout the globe. Technology can be legal, profitable, safe, and create jobs, and yet not contribute at all to human well-being. Consider, for example, recent revelations about social media platforms and the algorithms they employ that are widely used and highly profitable but may lead to various mental health challenges in young people (such as addiction, shorter attention spans, low self-esteem, poor body-images, feelings of isolation and depression). In short, such platforms may have negative impacts on human well-being.

Well-being is also distinct from the protection of human rights, another core principle of ethically aligned design (IEEE 2017b, 17). While the protection of human rights may also be important to promoting well-being, it is not always sufficient. Internationally recognized human rights intend to provide a baseline that ought to be protected (such as the right not to be killed) unless it is outweighed by some other competing right (such as the right to self-defense). Rights essentially provide us with limits on the treatment of others. The concept of well-being, however, is more complicated. Well-being may come in degrees and is multifaceted, such that some aspects of well-being might be promoted even if other aspects are neglected. As the social media example discussed previously shows, there may be technologies that do not, strictly speaking, violate individual rights, but that negatively impact psychological well-being in certain ways. Of course, social media platforms may also contribute to well-being in *other ways,* such as connecting loved ones who are geographically distant or increasing access to information. Similarly, there may be things that are not captured by individual rights but that are nonetheless vital to well-being, such as having free time, or living in ways that facilitate our relationships with others.

DOI: 10.1201/9781003188377-42

Well-being is, thus, a broad concept that is closely tied to judgments about what we think is necessary for living a good life or achieving a certain quality of life. This is something that may differ from culture to culture, or even person to person. In this section, we will consider the conceptual groundwork for approaches to well-being, which has implications for how well-being might be measured.

The notion of well-being is often attributed to the ancient Greek philosopher Aristotle, who argued that the ultimate goal of humans is "eudaimonia," which has been translated as "happiness," or, perhaps more accurately, "flourishing" (Crisp 2014). For Aristotle, eudaimonia is not just the subjective state of satisfaction of one's desires, but the achievement of flourishing as a human in relation to human needs and abilities (Fowers 2016). A similar concept also arose within Eastern philosophical traditions, African ethics, as well as Indigenous philosophies. For example, the Buddhist term "sukha," has been interpreted as a state of flourishing that arises from mental balance and insight into the nature of reality (Ekman et al. 2005).

There are important differences in conceptions of well-being that reflect different beliefs about our nature as humans. While some Western traditions have tended to focus on humans as individuals and thus provide accounts of individual well-being, some non-Western traditions view humans as creatures that are, by nature, interdependent on other beings and the world of nature they live in. For example, sub-Saharan African Ubuntu ethics focuses on what promotes the flourishing of communities and the relationships within them (Lutz 2009). Yusufu Turaki (2006) states: "People are not individuals, living in a state of independence, but part of a community, living in relationships and interdependence." These relationships and the nature of interdependence we have with each other and with the environment around us are frequently viewed as central to understanding both our nature as humans and our well-being in a variety of ethical traditions including feminist care ethics (Noddings 2013), and various indigenous views (Panelli and Tipa 2007; Taylor 2008; Watene and Merino 2018).

While there are cultural, historical, and other contextual differences in how well-being is conceived, a common thread running through all of these philosophical views is the notion that there is some state of flourishing or good for the kinds of beings we are (even when there is disagreement about whether we are best understood as independent individuals or interdependent communities, or some combination of both). If complex systems are to be evaluated with respect to how they might best promote human well-being, we need ways to identify more specifically what counts as well-being in a particular context, as well as metrics or ways to measure it.

One prominent approach used by many development theorists and policymakers is to focus on what humans can do and be – that is, their capability to function (Alkire and Deneulin 2009). Amartya Sen (1997, 2009) and Martha Nussbaum (2001, 2006) have both advocated that we think of well-being (and progress toward well-being) in terms of the extent to which people are able to live the kinds of lives that they value. Such an approach recognizes that a range of different factors (physical, emotional, cultural, social, economic, political, and geographical) can impact what people are able to do and be (Robeyns 2005, 2009). This approach has also partly informed the IEEE's *Ethically Aligned Design* principles, which defines well-being as "an evaluation of the general quality of life of an individual and the state of external circumstances… encompassing the full spectrum of personal, social, and environmental factors that enhance human life and on which human life can depend" (IEEE 2017b, 71). While not intended to be a definitive or exhaustive list, the IEEE provides the following general categories (in alphabetical order) as a starting point for identifying important aspects of well-being:

Community: The ability to have a sense of belonging, inclusion, participation, non-discrimination, and social support.

Culture: The ability to have and participate in a cultural identity, practices, language, and live in accordance with cultural values.

Economy: The ability to participate in innovation, conduct meaningful and well-compensated work, and live in ways that are sustainable, equitable, and at a certain standard of living.

Education: The ability to have formal education, life-long learning, and access to quality teachers.

Environment: The ability to sustainably thrive with clean air, water, soil, and biodiversity.

Government: The ability to participate in governance and have confidence in governmental institutions and their protection of human rights.

Human Settlements (Infrastructure): The ability to access energy, healthy food, housing, modes of transportation, and information and communication technology.

Physical Health: The ability to be physically healthy and have physical needs met that are necessary for human life, including having access to healthcare.

Psychological Health: The ability to be mentally healthy and satisfied with one's life.

Work: The ability to do the sort of work one wants to do and finds fulfilling, to have a supportive and non-hostile workplace environment, and to balance work with other needs and desires.

There are several important things to note about this list. First, as mentioned, the list is intended to provide a starting point, rather than an exhaustive list of considerations that might be relevant to well-being. There may be vital dimensions of human life not *explicitly* mentioned above, such as the ability to have and enjoy a family, engage in relationships, or engage in creative or leisure activities (Nussbaum 2001). It is possible that this list could or should be expanded or that we might think of these things as *implicitly* connected to one or more of the categories above. For example, these sorts of things may turn out to be important to psychological well-being. Second, the dimensions of well-being listed here are, to a large extent, interconnected. For example, our ability to live in safe community conditions have a healthy environment or achieve work-life balance may be important for physical and psychological health. Third, what exactly each of these elements might look like, as well as how they may or may not be promoted, may depend very much on the specific geographical, socioeconomic, and cultural context. While we might recognize the importance of well-being very generally as the ability to live the sort of life one values, the kind of life that *is* valued may be different across cultures, religious traditions, geographical locations, or even individuals. Fourth, while professional organizations and non-profits have called attention to the importance of taking human well-being into account in designing and implementing technologies, this may also be interconnected with other principles of ethically aligned design. For example, the IEEE's principle of data agency – or empowering individuals with the ability to access and securely share their data (IEEE 2017b, 17), or maintain control over their identity, may be necessary to promote human well-being. Moreover, how we think of well-being may have implications for what kind of control and security individuals or communities need. Thus, it is important for engineers and managers who are aiming to incorporate ethically aligned design principles to think about these principles wholistically, recognizing the ways in which they may interact.

Yet, even if we use this list of considerations as a starting point, it may start to feel overwhelming. If agreement on a framework for thinking about well-being is reached, how can such factors be measured and incorporated into design decisions? In the next section, we will consider metrics for evaluating well-being.

35.3 METRICS FOR WELL-BEING

To promote well-being in the design, implementation, and operation of complex systems, we need practical ways of incorporating well-being metrics to guide, measure, and monitor such systems. The capability approach to well-being evaluates technologies (including complex systems) according to their impact on people's capabilities to live the kind of lives that they value, including those things that are necessary for human life, given the sorts of beings that we are. It asks whether a complex system, at each stage of its lifecycle, will either enhance or hinder any of the above listed capabilities for individuals or communities likely to be affected. This approach requires identifying not just direct impacts (e.g., whether a particular technology will hinder physical health), but also whether the technology might enhance or hinder the means or resources necessary for a particular capability. For example, will it increase or decrease access to clean water, protection from infections and diseases, or access to healthcare? Will it increase or decrease the ability to produce healthy food, or engage in cultural activities? Some of the indicators we might use to measure this are objective insofar as there are objective states, such as having enough food necessary for sustaining human life. Yet, objective indicators may not capture some of the aspects of well-being listed above. For example, some aspects of well-being are tied to human *satisfaction,* such as being able to *feel* a sense of belonging as part of a cultural community, or *perceiving* work to be fulfilling, or being able to engage in the kinds of activities one *enjoys*). Feelings, perceptions, and evaluations of enjoyment or satisfaction are subjective states that may vary from person to person. Measuring these requires evaluating the extent to which individuals feel satisfied, fulfilled, supported, or protected. Thus, evaluating the extent to which some technology might enhance or hinder human well-being may require examining both objective and subjective indicators.

35.3.1 Objective Indicators

Objective indicators are primary measures of states of affairs or circumstances that relate to some of the dimensions of well-being discussed above. Data such as unemployment rates, income levels relative to a cost of living, crime rates, statistics on access to affordable housing, healthcare, childcare or transportation, levels of pollution, literacy rates, graduation rates, voter participation rates, and the existence of legal protections, can all provide some indicators for whether people have access to the resources necessary for exercising their capabilities and meeting their needs. There can also be objective indicators of psychological well-being. For example, the prevalence of mental illness, addiction, or suicide might be an indicator of threats to well-being within a population. What makes an indicator objective is that it attempts to measure some state of affairs that we take to be relevant to human well-being generally.

35.3.2 Subjective Indicators

Subjective indicators are primarily survey-based well-being indicators that are used to understand and measure perceptions of well-being and satisfaction with respect to certain domains of life for a particular population (Kahneman and Krueger 2006). Such indicators are subjective in the sense that they intend to measure individuals' perceptions, feelings, and evaluations of their own lives. There may be a temptation to think that subjective indicators are not reliable. After all, there may be people (especially in wealthy countries) who are not satisfied despite the fact that they have lives that provide them with many opportunities, a higher standard of living, and greater ability to exercise their capabilities. However, subjective indicators may be important given the differences between individuals and communities about *what makes them happy* and satisfied. For example, in some rural areas, there may be less access to transportation, but in some cases, that (relative) isolation may be part of what people living in those areas value about their lives. Similarly, communities may have access to sufficient food to avoid hunger or malnutrition, but they may rely on others to produce that food, or they may have a lack of access to culturally significant foods that causes dissatisfaction. Thus, subjective indicators not only measure perceptions and feelings within a population, but they can also alert us to considerations that may be relevant to well-being that we previously neglected or give us a better sense of which particular aspects of well-being are most valued within different populations. Subjective indicators of well-being are widely used by governments, such as those employed in the European Social Survey (ESS 2015), Bhutan's Gross National

Happiness Indicators (Royal Government of Bhutan 2012), and the well-being surveys created by the UK Office of National Statistics (see Hicks et al, 2013).

35.3.3 Composite Indicators

Composite indicators combine both objective and subjective metrics. Both objective and subjective indicators of well-being are important as they reveal different kinds of information. Subjective indicators tell us how people feel about their lives and the extent to which they are satisfied with certain circumstances. Objective indicators tell us about the conditions under which people are living that may thwart or promote their well-being, given the kinds of beings we are as humans. It is not the case that one is "better" than the other – rather they both reveal data that can help us understand different aspects of well-being. Subjective indicators can also be helpful in that they may call attention to aspects of well-being that are not currently well captured by objective metrics. Some composite indicators aim to combine objective and subjective indicators to generate overall well-being assessments that reflect the totality of the data. Both the OECD's Better Life Index (OECD 2014) and The UN's Human Development Index (HDI) ranking in their Human Development Report are attempts to do this (UN's Human Development Index 2020). The OECD's Better Life Index does this by allowing users to give more or less weight to aspects of well-being that are most important to them.

35.4 HOW CAN HUMAN WELL-BEING METRICS GUIDE THE DESIGN OF COMPLEX SYSTEMS?

While there are a variety of metrics that are used to measure the multifaceted concept of human well-being, how can these be integrated into the design of complex systems? The first step is to clearly define the type of system that is being developed (and the subsystems that may be involved), including identifying its typical intended uses, potential applications, and the ways in which the system could be misused. While it may be difficult to predict every possible use, many are reasonably foreseeable. The next step is to identify the potential stakeholders or those likely to be impacted by the technology. These obviously include users, but it may also include third parties that are impacted by the consequences of others using the technology. Having a clear vision of what the system is intended to do, how it might be used (or abused), and who might be impacted helps make visible the different elements of well-being that may be relevant to the particular complex system and how different groups might be differently impacted (for good or ill).

The third step is to work with social scientists and others with expertise in well-being concepts and metrics to identify aspects of well-being that may be relevant to a particular project, or to those stakeholders effected. Well-being researchers will be able to assess what existing data may be relevant, what new data might need to be collected with respect to a particular population (consistent with consent, confidentiality, and ethical use of this data) (IEEE 2017b, 74).

Because well-being is partly subjective and different groups in different contexts value different things about their lives, it is important to have some engagement or feedback mechanisms with both internal stakeholders (such as industry partners or clients) and external stakeholders (those likely to be impacted by the use of the systems). This is important not only to make sure that relevant considerations of well-being are assessed but also to respect the autonomy of stakeholders in allowing them to articulate what they believe is relevant to their own well-being. Social scientists typically start with some commonly used indicators and metrics for well-being but will change or fine-tune them in response to stakeholder input.

This engagement with stakeholders is often accomplished through stakeholder advisory boards. There are obviously some practical limitations and challenges to doing this. External stakeholders may be a large and diverse group, and there is a need to ensure that the group is representative, but small enough so as to get feedback and incorporate it in a timely manner. Moreover, certain groups (often those disproportionately impacted by technology) may not have the ability or inclination to volunteer their time or participate. In some cases, members of advocacy groups or non-profit organizations or governmental agencies serve as representatives (Sedlacek 2013). In any case, such participation needs to be incentivized in ways that ensure the engagement is mutually beneficial. Moreover, in some instances, there are limits to what stakeholders can know or be told in sensitive or proprietary contexts. Yet even in these contexts, there are successful models for how such advisory boards can be done (Boiko et al. 1996, Malcolm 2008, Murray 2017).

As with assessing the social impacts of complex systems, well-being evaluations should be used throughout the design and testing of such systems to inform design, and potentially make course corrections if needed. For example, work on assistive systems for those with cognitive or physical disabilities such as "smart" wheelchairs (Viswanathan et al. 2008) personal robot assistants (Sarkar 2018), and wayfinding navigation systems (Delnevo et al. 2018) has benefited from having input on well-being by intended users. Such technology has been designed to enhance the well-being of users by enabling greater mobility, having assisted reminders, and enhancing independence. At the same time, developers also learned that for some, such technologies (insofar as they eliminated human caregivers) also increased a sense of isolation and prevented social contact and the development of caring relationships that often exist with caregivers (Scherer et al. 2011). Indeed, some neurotechnologies designed to "enhance" abilities for those with cognitive "impairment" have been criticized as viewing such conditions as medical conditions that require a technological "fix" as opposed to a problem created by social conditions that make it difficult for those who are cognitively diverse to function (Klein et al. 2015) Because well-being involves many different dimensions, it is often the case that systems may enhance well-being in some ways while hindering it in others, and the evaluations about whether benefits are worth the costs may vary between users. Knowing this can help us understand the limits of certain solutions as well as reveal other needs. In addition, trust and transparency can be increased if designers and managers communicate clearly and honestly about the ways in which a complex system may advance or hinder aspects of their lives they care about. Operationalizing well-being metrics provided by stakeholders and providing them with reports that show how various aspects are advanced or hindered in the expected or actual outcomes of a project can facilitate avenues for improvement, autonomous choice by users, and an alignment of user expectations about what a particular system is likely to do.

Another, perhaps more proactive strategy that designers can use, is to consult existing scientific data on well-being, in order to anticipate which aspects of well-being might be impacted by particular kinds of systems and then build those metrics of well-being into the objectives of autonomous or intelligent systems. This can ensure that the ethical aims are among those goals that are being optimized within the system. For example, multi-objective algorithms are being used in precision agriculture to reduce fertilizer and water use to optimal levels that will reduce environmental damage and harms to health, while also maximizing crop yield (Kropp et al. 2019).

Regardless of which particular strategies are used, the effects of complex systems on well-being should be monitored continuously throughout their lifecycles and engineers, managers, and some stakeholders need to be prepared to significantly modify or even cease using technology that is shown to reduce well-being (IEEE 2017b, 75). Having an organizational project leader assigned to issues of well-being and stakeholder engagement can help facilitate these efforts (IEEE 2017b, 80).

35.5 CONCLUSIONS

The IEEE is urging engineers and managers to use human well-being as a primary criterion for the success of complex systems. Well-being is a fairly broad concept but can be understood to refer to the capability of individuals or society to live the kind of lives they want to live, or the sorts of lives that are likely to allow individuals and communities to flourish. While this may seem like a very broad goal, there is a significant body of research on indicators for well-being that are common across individuals and cultures, even though there are also important differences at a finer-grained level. There is also a growing body of scientific research on how to measure both objective and subjective aspects of well-being. Engineers and managers should be familiar with common well-being indicators and should, to the extent possible, involve those with appropriate expertise to ensure that systems being developed do more to enhance (rather than hinder) human well-being (broadly construed to include the environment in which human beings live and depend on to flourish). It is also highly recommended that design and development projects include a stakeholder advisory board that can provide feedback on both well-being metrics as well as how those aspects of well-being might be impacted by particular kinds of systems.

Understanding how complex systems or the technologies that comprise them impact well-being is important for several reasons. It is important insofar as engineers have the responsibilities to avoid harms with their innovations. But, just as importantly, engineers have the opportunity to design technologies and complex systems in ways that can genuinely *enhance* well-being. For example, much attention has been given to the ways in which autonomous vehicles might present threats to safety under certain conditions, but it is equally important to consider the ways in which such technology might be designed in ways that make transportation more accessible (e.g., to those who are disabled) or help address the stress and time wasted in commuting (insofar as it might allow drivers to multitask), or to make transportation systems more sustainable. Designing to enhance human well-being would not only lead to increased social benefits but may also be more economically viable. If complex systems are designed and operate in ways that enhance well-being, there is much more likelihood that they will be widely accepted.

REFERENCES

Alkire, S., & Deneulin, S. (2009). *An Introduction to the Human Development and Capability Approach.* London: Earthscan.

Boiko, P. E., Morrill, R. L., Flynn, J., Faustman, E. M., van Belle, G., & Omenn, G. S. (1996). Who holds the stakes? A case study of stakeholder identification at two nuclear weapons production sites. *Risk Analysis*, 16(2), 237–249.

Crisp, R. (ed.) (2014). Aristotle*: Nicomachean Ethics*. Cambridge University Press.

Delnevo, G., Monti, L., Vignola, F., Salomoni, P., & Mirri, S. (2018, January). AlmaWhere: A prototype of accessible indoor wayfinding and navigation system. In *2018 15th IEEE Annual Consumer Communications & Networking Conference (CCNC)* (pp. 1–6). IEEE. doi: 10.1109/CCNC.2018.8319242.

Ekman, P., Davidson, R. J., Ricard, M., & Wallace, B. A. (2005). Buddhist and psychological perspectives on emotions and well-being. *Current Directions in Psychological Science*, 14(2), 59–63.

European Social Survey (ESS) (2015). https://www.europeansocialsurvey.org/findings/europeans-wellbeing/europeans-wellbeing.

Fowers, B. J. (2016). Aristotle on eudaimonia: On the virtue of returning to the source. In J. Vittersø (ed.), *Handbook of Eudaimonic Well-Being* (pp. 67–83). Cham: Springer.

Hicks, S., Tinkler, L., & Allin, P. (2013). Measuring subjective well-being and its potential role in policy: Perspectives from the UK office for national statistics. *Social Indicators Research*, 114(1), 73–86.

IEEE (2017a). The Global Initiative on Autonomous and Intelligent Systems. https://standards.ieee.org/industry-connections/ec/autonomous-systems.html.

IEEE (2017b). Ethically Aligned Design—A Vision for Priori Tizing Human Well-Being with Autonomous and Intelligent Systems. https://standards.ieee.org/wp-content/uploads/import/documents/other/ead1e.pdf

Kahneman, D., & Krueger, A. B. (2006). Developments in the measurement of subjective well-being. *Journal of Economic Perspectives*, 20(1), 3–24.

Klein, E., Brown, T., Sample, M., Truitt, A. R., & Goering, S. (2015). Engineering the brain: Ethical issues and the introduction of neural devices. *Hastings Center Report*, 45(6), 26–35.

Kropp, I., Nejadhashemi, A. P., Deb, K., Abouali, M., Roy, P. C., Adhikari, U., & Hoogenboom, G. (2019). A multi-objective approach to water and nutrient efficiency for sustainable agricultural intensification. *Agricultural Systems*, 173, 289–302.

Lutz, D. W. (2009). African Ubuntu philosophy and global management. *Journal of Business Ethics*, 84(3), 313–328.

Malcolm, J. (2008). *Multi-Stakeholder Governance and the Internet Governance Forum*. Wembley, WA: Terminus Press.

Murray, J. (2017). Adopting stakeholder advisory boards. *American Business Law Journal*, 54, 61.

Noddings, N. (2013). *Caring: A Relational Approach to Ethics and Moral Education*. Oakland, CA: University of California Press.

Nussbaum, M. C. (2006). Poverty and human functioning: Capabilities as fundamental entitlements. In D. B. Grusky, & R. Kanbur (eds.), *Poverty and Inequality* (pp. 47–75). Redwood City: Stanford University Press.

Nussbaum, M. C. (2001). *Women and Human Development: The Capabilities Approach*, No. 3. Cambridge: Cambridge University Press.

Organization for Economic Cooperation and Development (OECD) (2014). Better Life Index. https://www.oecdbetterlifeindex.org/.

Panelli, R., & Tipa, G. (2007). Placing well-being: A Maori case study of cultural and environmental specificity. *EcoHealth*, 4(4), 445–460.

Robeyns, I. (2009). Capability approach. In J. Peil, & I. van Staveren (eds.), *Handbook of Economics and Ethics*. North Hampton, MA: Edward Elgar Publishing.

Robeyns, I. (2005). The capability approach: A theoretical survey. *Journal of Human Development*, 6(1), 93–117.

Royal Government of Bhutan (2012). *The Report of the High-Level Meeting on Wellbeing and Happiness: Defining a New Economic Paradigm*. New York: The Permanent Mission of the Kingdom of Bhutan to the United Nations.

Sarkar, D. P. (2018). A nurse bot for elderly people. In *Proceedings of the 2018 ACM International Joint Conference and 2018 International Symposium on Pervasive and Ubiquitous Computing and Wearable Computers* (pp. 1798–1799).

Scherer, M. J., Craddock, G., & Mackeogh, T. (2011). The relationship of personal factors and subjective well-being to the use of assistive technology devices. *Disability and Rehabilitation* 33(10), 811–817.

Sedlacek, S. (2013). The role of universities in fostering sustainable development at the regional level. *Journal of Cleaner Production*, 48, 74–84.

Sen, A. (1997). *Resources, Values, and Development*. Harvard University Press.

Sen, A. (2009). *The Idea of Justice*. Harvard University Press.

Taylor, J. (2008). Indigenous peoples and indicators of well-being: Australian perspectives on United Nations global frameworks. *Social Indicators Research*, 87(1), 111–126.

Turaki, Y. (2006). *Foundations of African Traditional Religion and Worldview*. WordAlive Publishers.

UN's Human Development Index (2020). https://hdr.undp.org/en/content/latest-human-development-index-ranking.

Viswanathan, P., Mackworth, A. K., Little, J., Hoey, J., & Mihailidis, A. (2008). NOAH for wheelchair users with cognitive impairment: Navigation and obstacle avoidance help. In *AAAI Fall Symposium: AI in Eldercare: New Solutions to Old Problems* (pp. 150–152).

Watene, K., & Merino, R. (2018). Indigenous peoples: Self-determination, decolonization, and indigenous philosophies. In *Routledge Handbook of Development Ethics* (pp. 134–147). Routledge.

36 The Social Impacts of Complex Systems

Kristen Intemann

36.1 INTRODUCTION

Complex systems are not just comprised of their technical subsystems but the interactions between them. Such systems also interact with various users and aspects of the particular social, cultural, and environmental contexts in which they are embedded and used. Consider the Internet of Things (IoT), for example. The IoT can be thought of as a complex network of technical components including sensors, informational processors, and actuators, but it operates in particular cultural, social, political, and economic environments in ways that can significantly impact humans, our relationships with each other, and our physical environment (Allhoff and Henschke 2018). For example, the IoT can make it possible to accomplish many things at a distance through voice commands that can lock doors, close blinds, or adjust heat. It can allow certain things to be done more efficiently, such as aiding law enforcement in identifying and locating suspects or helping us in everyday chores such as generating grocery lists or paying for parking without waiting in a line. It may also enable us to do things that would have been previously challenging to do, such as locate children or pets, monitor our health, or conduct work or schooling remotely. Many of the ways that the IoT can transform our lives can be beneficial, but its power to change the way the world operates can also have negative (unintended) consequences. The ways things are networked create new vulnerabilities that can be exploited by hackers, the move toward autonomation may have significant implications for work and educational opportunities, and the kinds of interactions the IoT allows with others may change the nature of our relationships. The IoT has also allowed multiple parties to collect data about users, data that may be used in ways users do not fully understand and has the potential to infringe on privacy rights without genuine consent. Such technologies can also create inequalities as we become dependent on technologies that may not be equally accessible, particularly in poor or rural areas. Thus, it is important to understand, and anticipate as much as possible, the impacts that complex systems can have in order to develop these systems in responsible ways. This chapter has four main aims: (1) distinguish the different kinds of impacts that may arise in relation to complex systems; (2) identify how such impacts may arise; (3) consider what ethical obligations (if any) managers and engineers may have in minimizing negative impacts; and (4) provide some guidance on how development and management of complex systems can be handled responsibly.

36.2 TYPES OF SOCIAL IMPACTS FROM COMPLEX SYSTEMS

The "social impacts" of complex systems, broadly speaking, refers to the effects (either positive or negative) on societies and the individuals who comprise them. Complex systems and their subsystems are comprised of materials, some of which may be scarce and some of which may have long lifecycles. Moreover, they interact with users, and other existing systems, and can even have impacts on those who do not or cannot use them. There are many kinds of impacts that might be considered "social" in this sense, which will be identified in this section. Most if not all engineers (in academia and in industry) design complex systems with the hope of bringing about positive impacts, and

 DOI: 10.1201/9781003188377-43

usually with the aim of solving or addressing a particular problem or need. Yet unintended negative impacts can occur, directly or indirectly, making them challenging to identify. This will be our focus here.

36.2.1 Environmental Impacts

Complex systems can have consequences for the environment and those humans that depend on the environment to live. Consider the complex communication system of mobile phone technologies. Global manufacturing of mobile phones requires large-scale extraction of a variety of minerals, and mining operations can generate air and water pollution, particularly in areas that have relaxed environmental regulations (He 2020). Moreover, the typical amount of time before a user has a cell phone before replacing it is 9 months to 2 years generating a significant amount of electronic waste, which itself can negatively impact the environment (Wilhelm et al. 2015). The short life first users give to cell phones is itself a consequence of the ways in which software and new capabilities are quickly developed and designed to be incompatible with older phones, so users must upgrade in order to keep using programs and applications efficiently.

36.2.2 Health Impacts

The life cycle of complex systems from manufacturing, transporting, operating, using, and disposing of the material components of complex systems can also have impacts on human health. Consider again all of the materials mined for the manufacturing of electronics. Harmful levels of heavy metals have impacted miners and those living near mining activities (Banza et al. 2009, Castro and Sánchez, 2003, Li et al., 2008) through wastewater and soil contamination and air pollution. Metal exposure has been linked to cancer, organ damage, DNA damage, and degeneration of the nervous system (resulting in diseases such as Alzheimer's and Parkinson's) (Csavina et al. 2012). The manufacturing of circuit boards has been linked to various forms of cancer and other health problems from exposure to toxic solvents, chemicals, and vapors (LaDou 2006, Morose et al. 2011). The processing of primary palladium and platinum for mobile phone networks results in sulfur dioxide emissions into air, which poses a risk to human respiratory organs (Scharnost et al. 2005). There are also health concerns about mobile phone use. Mobile phone use has been found to impact drivers' concentration, limiting their ability to react quickly, even when they are using a hand-free device (Ferlazzo 2008). There are even concerns about whether mobile phone use might be linked to certain forms of cancer, primarily because of the radiofrequency radiation emitted (Khurana 2009; Luo et al. 2020). Social media applications, such as Facebook and Instagram, on mobile phones have been linked to stress, depression, anxiety, and other health problems, particularly in adolescents (Calancie et al. 2017, Marino et. al. 2018).

36.2.3 Political Impacts

Often complex systems are engineered with the goal of increasing our national security or related political and economic interests. Yet the manufacturing of this technology can also have negative political consequences. High-tech minerals that are often needed are not dispersed uniformly throughout the world. For example, 97% of the world's supply of certain minerals needed in cellphone production comes from China, which makes us dependent on them (Humphries 2015). Complex systems that rely on drones and sensors to attack targets from a distance may make us more likely to enter into conflicts (because we can do so without as much risk to troops). Communication, financial, and surveillance systems that network personal information and data can become vulnerable to cyber espionage by foreign actors, being hacked or ransomed (Wangen 2015).

36.2.4 Economic Impacts

Complex systems such as transportation systems, communication systems, agricultural systems, and defense systems are often expensive, which raises concerns about who will be able to have or use such systems. The ability to access and use such systems results in power and those with less access and resources may have even less. In the case of smartphones, we have become incredibly dependent upon them, such that ways of life (education, financial transactions, smart homes, and smart cities) are being built upon having these devices. Yet such devices are expensive, and not everyone has access. The result is a deepening of economic and educational inequalities, where those who cannot afford a smartphone are also excluded from certain kinds of activities. Moreover, when complex systems are unintentionally designed to exclude certain users (such as the integration of facial recognition software that does not accurately recognize certain races, or transportation systems that are not accessible to those who are disabled), social and economic inequalities are also exacerbated.

36.2.5 Cultural Impacts

Our use and reliance on complex systems can also radically reshape our cultures. Transportation systems can shape who we are able to interact with and how we do so. They can isolate some communities and connect others. They can impact the ways in which cities and towns organize around those systems. Or in the case of networked computers and phones and applications, the nature of our social interactions can change. Instead of having in-person interactions, we are seeing an increase in people interacting and communicating over virtual technologies such as Zoom or FaceTime calls and meetings, online classes, email, text, and social media. Indeed, the rise of social media applications has potentially even changed the way we meet people and our very notion of "friendship." (Sandler 2014). The ways in which we use email, texting, and instant messaging have also changed the ways we interact and, in some cases, the nature of work. They have created an expectation of constant accessibility to people, which often generates more work, particularly outside the traditional norms of the workday and work week.

36.2.6 Ethical Impacts

All of the above social impacts end up impacting certain widely held values or ethical principles. Environmental, health, political, economic, and cultural impacts can result in harm, restrict rights or opportunities, or impact the well-being of individuals or communities. Moreover, complex systems have the potential to either alleviate or reinforce existing social inequalities. For example, systems or subsystems could disproportionately negatively impact certain groups. Transportation systems, for example, can make freedom of movement or the ability to participate in society less accessible to some (or more accessible). Systems that involve unintentional social biases may result in discrimination on the basis of features we think are morally irrelevant (such as gender, ethnicity, class, or physical abilities). They also have the power to potentially make the *value* of certain rights either stronger or weaker. For example, while citizens may enjoy a right to freedom of movement, the value of this liberty can decrease if, for example, one's community is cut off from new transportation systems.

36.3 HOW CAN NEGATIVE IMPACTS OCCUR?

Of course, with all of the ways in which complex systems may have social impacts (either directly or indirectly), it may not be reasonable to think that engineers can either anticipate or address all of them. Yet understanding how negative impacts can occur may be important to preventing or mitigating them to the greatest extent possible. This section examines different ways that negative impacts can arise.

36.3.1 Impacts that Occur from Manufacturing Complex Systems

Complex systems require many different parts, all of which come from somewhere. Materials must be extracted, which can have both environmental and health impacts, as well as political impacts because such materials are not uniformly distributed throughout the world. The raw materials must also undergo processing which can have health impacts on workers and those who live near the places where such processing takes place.

36.3.2 Impacts that Occur from Use or Misuse of Complex Systems

We tend to think of the negative impacts of complex systems as coming from the misuse of such systems by bad actors. Cyber espionage, hacking, or malware impacts, for example, occur when bad-faith actors exploit vulnerabilities within complex systems. Yet as Section 36.2 illustrates, negative impacts can also happen when complex systems operate as they are intended and function well. Even well-functioning complex systems can have unintended impacts, such as increased use of certain technologies that change our interactions with other humans, impact our notion of friendship, or give rise to the expectation that individuals always be accessible.

36.3.3 Impacts that Occur from Complex Systems Failures

Impacts can also arise from catastrophic failures in complex systems. Complex systems are characterized by a large number of interacting components (often in complex nonlinear ways) that produce emergent properties difficult to predict from the knowledge of single components (Santos and Zhou 2017). Large systems typically involve many safeguards (redundancies, emergency shut down devices, physical barriers, and regulations for operating and maintaining equipment). Thus, when they fail it is due to much more than a single element or mistake. While complex systems are designed to absorb some random disruptions or failures, they can be highly vulnerable to failure in adverse conditions (Dueñas-Osorio and Vemuru 2009). This can occur because of unexpected interactions between subsystems or unexpected interactions that are triggered by external forces –such as extreme weather events. These "cascading failures" are usually the result of failures that begin in one component of the system, which can put stresses on other parts of the system that can then also fail (Baldick et al. 2009). Components within the system particularly when stressed can fail due to a multitude of factors including maloperation, poor maintenance, design flaws, or poor decision-making by humans.

36.3.4 Impacts that Occur from the Disposal or End of Use of Complex Systems

Disposal of electronic waste is an increasing problem from a variety of complex systems. Impacts can arise not from the design, production, or use of complex systems, but also because the materials involved may have very long lifecycles that outlast their use. Disposal of this waste can cause a variety of environmental and public health problems, particularly in developing countries (Needhiasan et al. 2014). Even the recycling of some electronic waste is causing health problems for workers (Decharat and Kiddee 2020).

36.4 ADDRESSING IMPACTS: WHAT ARE THE ETHICAL OBLIGATIONS OF ENGINEERS?

All human beings have some ethical obligations to consider how their actions or decisions may impact others and try to prevent harm (Douglas 2009). Yet engineers also have additional professional obligations, which arise from their expertise and the power that their actions and decisions have to impact and shape societies in significant ways (which is far more than the average individual). In other words, engineers have special obligations in virtue of their professional role.

This is widely recognized by a variety of professional organizations. The Institute of Electrical and Electronics Engineers (IEEE) code of ethics, for example, requires engineers "to uphold the highest standards of integrity, responsible behavior, and ethical conduct in professional activities" (IEEE 2020, 1). This includes holding "paramount the safety, health, and welfare of the public, to strive to comply with ethical design and sustainable development practices, to protect the privacy of others, and to disclose promptly factors that endanger the public or the environment" (IEEE 2020, 1). Indeed, this code not only requires engineers to engage in ethical design, practices, and decision-making but also to "improve understanding by individuals and society of the capabilities and societal implications of conventional and emerging technologies" (IEEE 2020, 1). This is significant, as it suggests that it is not enough merely to act in ways to minimize obvious harm, but also to engage with individuals and communities to better understand the potential implications of technologies in order to prevent or minimize less obvious harm.

Given the different kinds of social impacts that can arise from complex systems and the complicated ways in which they can arise from design, to manufacturing, to use (or misuse), and disposal, such obligations may seem quite daunting. In particular, it is unreasonable to hold engineers responsible for impacts that they fail to predict, consequences that may be beyond their control, circumstances that may arise from many disparate decision-makers or unforeseeable consequences. Engineers are, after all, also human beings. Moreover, complex systems can involve significant uncertainties. There are uncertainties about how subsystems may interact or impact each other, uncertainties about the environmental conditions they may be subject to (e.g., increasing extreme weather events in a particular location), uncertainties about how complex systems might be used or adopted, and uncertainties about the environmental or social circumstances that might arise (such as impacts from climate change or political conflict). While it would be unreasonable to hold individual engineers legally or even ethically responsible for every direct or indirect negative social impact that arises from complex systems, especially given so many uncertainties, we might conclude that engineers, teams of engineers, managers, and professional engineering organizations have duty to proceed responsibly by attempting to anticipate and minimize negative impacts in the design, manufacturing, management, and disposal of complex systems, at least insofar as possible.

What would such a duty involve and how can engineers, teams of engineers, managers, and professional engineering organizations fulfill this duty? "Ethical technology assessments" (ETAs) can be a helpful tool in the design and implementation of emerging technologies (Palm and Hansson 2006). The goal of these assessments is to identify or anticipate potentially serious negative impacts early in the design process to guide design, policies, or safety procedures in ways that can avoid, minimize, or mitigate such impacts. A variety of frameworks for conducting assessments of the ethical impacts of technology have been proposed (Palm and Hansson 2006, Johnson 2007, Boenink et al. 2010, Stahl et. al 2010, Johnson 2011, Brey 2012). All of these frameworks share certain common elements. First, they all attempt to *anticipate* or predict (to different degrees) the social impacts of technologies and their applications. Second, these frameworks agree that these assessments should begin early in technological development for the purposes of influencing that development in directions and should address the whole life cycle of any technological artifacts produced. Third, they agree that to be successful, an "ethical assessment" should not be understood *merely* as a "checklist" to be completed but as involving processes that ensure continual dialogue and feedback between developers, designers, managers, ethicists, and stakeholders from research and development, to design, to implementation and use.

36.5 CONSIDERATIONS FOR ETHICAL TECHNOLOGY ASSESSMENTS OF COMPLEX SYSTEMS

Conducting an ethical assessment requires four interactive repeating stages. First, *identifying the technological features* of the complex system under development, its subsystems, its intended domains of use, and the conditions under which it is likely to operate. This is knowledge that needs to be generated by engineers to a large extent, or those who are most familiar with the technological and scientific aspects of the systems. In some cases, it may require consultation with other project stakeholders, so that engineers understand the wider context in which the system is likely to operate. Second, *identifying any potential social impacts* of the system being developed by cross-referencing those technological features with a variety of ethical considerations. The goal at this stage is to consider how the operation of the system or subsystems may have environmental, health, political, economic, and cultural impacts that would be ethically problematic (or beneficial). To identify ethical impacts, we must assess how those features are likely to affect key ethical considerations and values that are widely held, particularly among stakeholders. Which value considerations would be relevant might depend on the particular context and groups likely to be impacted, but Brey (2012) has developed an approach built around four categories of ethical considerations that are widely recognized in ethics and policy documents such as the United Nations's *Universal Declaration of Human Rights*. Negative impacts are identified by asking whether the system, its subsystems, interactions between subsystems, or potential applications have the potential to:

1. **Produce harms** (including harms to mental or physical health, bodily harm, environmental harms, economic harms, or other social harms).
2. **Restrict individual rights** (such as freedom, autonomy, human dignity, privacy, and property).
3. **Create, reinforce, or widen injustices** (such as inequitable distribution of primary goods, capabilities, risks, and hazards, or discrimination against certain groups or populations).
4. **Hinder well-being or the common good** (such as happiness, knowledge, friendship, trust, democracy, or cultural traditions).

In considering the development of, for example, communication systems, we must identify whether there are health risks, environmental risks, or other social risks. We must also examine what the potential risks are for individuals. Are there features of the communication system that might make personal information of users vulnerable in ways that violate privacy? Are there features that give users less control over their information or decision-making in ways that might violate individual autonomy? We must also identify the extent to which use and dependence on a complex system might create or reinforce inequities. For example, do the risks of the system fall disproportionately to certain groups and, if so, are those the same groups that will receive greater benefits from its use? Finally, we must consider how the complex system may impact the well-being or the common good for communities. Is it a system that might, for example, facilitate the spread of disinformation in ways that might undermine democracy? Does the technology have the potential to transform our relationships with others in ways that may weaken those relationships? Might the systems involved have vulnerabilities that might undermine trust in some vital relationships?

The fact that some complex system has the *potential* to have one or more negative impacts is not ethically "disqualifying." After identifying potential ethical impacts, the third step is to assess and *evaluate* the significance of those potential impacts (Brey 2012). The goal here is to determine how problematic those potential consequences really are and whether they are outweighed by other considerations. In evaluating ethical impacts, one must consider:

1. **The Likelihood of the Potential Impacts:** We must assess the likelihood of each impact, to the extent this is possible. Is the impact only likely to occur under certain highly improbable or infrequent circumstances, or is it highly likely given the nature of the complex system, its subsystems, or its intended use?
2. **The Extent of the Potential Impacts:** That is, how broadly are the impacts likely to be within societies? Is the complex system likely to impact many people, are those impacts likely to be serious and/or irreversible?

3. **The Extent to Which Potential Negative Impacts are Outweighed by Potential Positive Impacts:** There may be concerns, for example, about privacy, but are there other conflicting values involved? When values conflict, are there some that are more important than others to stakeholders? For example, relatively minor restrictions to freedom of movement or privacy might be tolerated in order to ensure security or avoid the risk of bodily harm.

After evaluating which ethical impacts are potentially significant or serious, this information needs to be relayed to engineers so as to try to assess the extent to which negative impacts can be prevented or mitigated in the design of the complex system. There may be concerns that cannot be adequately addressed through design, which may require mitigation through either the development of operation procedures, regulatory policies, or limits to their use. This fourth stage needs to involve engineers, ethicists, and policymakers, in dialogue with each other about how to address these issues. This process may need to be repeated, as, for example, design changes may result in new or different impacts that would need to be addressed, and stakeholders may need opportunities for providing feedback on how their concerns are being met. Figure 36.1 aims to capture this interactive process.

What may now be clear is that conducting ETAs (and particularly doing them well) is not easy. They can take time, particularly since they require collaboration with others. Moreover, they can be logistically difficult, insofar as they require the participation of representative stakeholders, who may have their own challenges influencing their ability or willingness to participate. Engaging with stakeholders also raises questions about what counts as a "representative group" of stakeholders, particularly since large numbers of participants would make this process unwieldy. All of these issues can also result in increased project costs. At the same time, as ETAs are conducted with emerging technologies and particular kinds of complex systems, greater awareness of the issues involved can also help inform other projects of the kinds of considerations they should be attentive to. So, managers can also examine ETAs that have been done on similar sorts of technologies or systems to help inform design and development. Thus, it may be more useful to think of this as an aspirational model, as we think about how to do our best to avoid or minimize negative social impacts in developing complex systems.

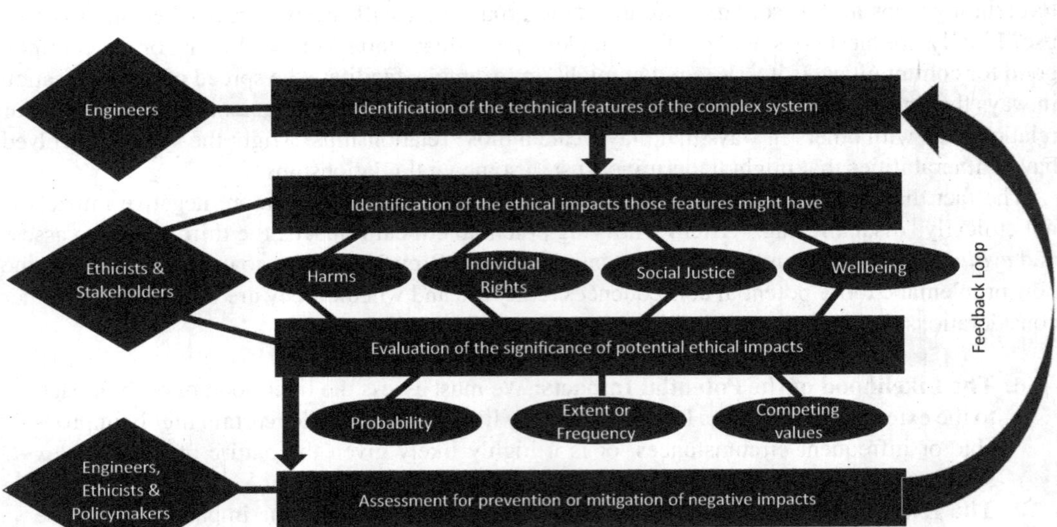

FIGURE 36.1 Ethical Technology Assessment.

36.6 CONCLUSIONS AND FINAL CONSIDERATIONS

While engineers have a professional duty to hold "paramount the safety, health, and welfare of the public, to strive to comply with ethical design and sustainable development practices, and to protect the privacy of others," (IEEE 2020, 1), it is not always clear what this involves or how it can be accomplished. This is particularly the case in the development of complex systems that may involve several different technologies and uncertainties about how they will interact, how they might be used or misused, or how they might impact the socio-environmental context in which they operate. Yet engineers should acknowledge the tremendous power that their decisions and actions have for impacting others, and society as a whole. It is important to try to identify and anticipate negative impacts to the extent possible and ETAs are one tool that can help in identifying and evaluating such impacts in ways that can guide design, implementation, and use of complex systems. When negative impacts cannot be avoided through the design of complex systems, engineers may have obligations to disclose risks and uncertainties to the public and policymakers. It is likely not possible to anticipate every social or ethical impact that will result from a complex system, and there will be uncertainties that are simply beyond the control of engineers and managers. Yet engineers have a duty to try to design and innovate responsibly as much as humanly possible.

REFERENCES

Allhoff, F., & Henschke, A. (2018). The internet of things: Foundational ethical issues. *Internet of Things*, 1, 55–66.

Baldick, R., Chowdhury, B., Dobson, I., Dong, Z., Gou, B., Hawkins, D., … & Zhang, X. (2009). Vulnerability assessment for cascading failures in electric power systems." In *2009 IEEE/PES Power Systems Conference and Exposition* (pp. 1–9). IEEE.

Banza, C. L. N., Nawrot, T. S., Haufroid, V., Decrée, S., De Putter, T., Smolders, E., … & Nemery, B. (2009). "High human exposure to cobalt and other metals in Katanga, a mining area of the Democratic Republic of Congo. *Environmental Research*, 109(6), 745–752.

Boenink, M., Swierstra, T., & Stemerding, D. (2010). Anticipating the interaction between technology and morality: A scenario study of experimenting with humans in bionanotechnology. *Studies in Ethics, Law, and Technology*, 4(2), doi: 10.2202/1941-6008.1098.

Brey, P. A.E. (2012). Anticipatory ethics for emerging technologies. *NanoEthics*, 6(1), 1–13.

Calancie, O., Ewing, L., Narducci, L. D., Horgan, S., & Khalid-Khan, S. (2017). Exploring how social networking sites impact youth with anxiety: A qualitative study of Facebook stressors among adolescents with an anxiety disorder diagnosis. *Cyberpsychology: Journal of Psychosocial Research on Cyberspace*, 11(4), 2. doi: 10.5817/CP2017-4-2.

Castro, S. H., & Sánchez, M. (2003). Environmental viewpoint on small-scale copper, gold and silver mining in Chile. *Journal of Cleaner Production*, 11(2), 207–213.

Csavina, J., Field, J., Taylor, M. P., Gao, S., Landázuri, A., Betterton, E. A., & Sáez, A. E. (2012). A review on the importance of metals and metalloids in atmospheric dust and aerosol from mining operations. *Science of the Total Environment*, 433, 58–73.

Decharat, S., & Kiddee, P. (2020). Health problems among workers who recycle electronic waste in southern Thailand. *Osong Public Health and Research Perspectives*, 11(1), 34.

Dueñas-Osorio, L., & Vemuru, S. M. (2009). Abstract for "Cascading failures in complex infrastructure systems." *Structural Safety*, 31(2), 157–167.

Ferlazzo, F., Fagioli, S., Di Nocera, F., & Sdoia, S. (2008). Shifting attention across near and far spaces: Implications for the use of hands-free cell phones while driving. *Accident Analysis & Prevention*, 40(6), 1859–1864.

He, P., Feng, H., Hu, G., Hewage, K., Achari, G., Wang, C., & Sadiq, R. (2020). Life cycle cost analysis for recycling high-tech minerals from waste mobile phones in China. *Journal of Cleaner Production*, 251, 119498.

Humphries, M. (2015). China's mineral industry and US access to strategic and critical minerals: Issues for congress. In *Congressional Research Services Report* (pp. 1–26).

Johnson, D. G. (2007). Ethics and technology 'in the making': An essay on the challenge of nanoethics. *NanoEthics*, 1(1), 21–30.

Johnson, D. G. (2011). Software agents, anticipatory ethics, and accountability. In G. E. Marchant, B. R. Allenby, & J. R. Herkert (Eds.), *The Growing Gap between Emerging Technologies and Legal-Ethical Oversight* (pp. 61–76). Dordrecht: Springer.

Khurana, V. G., Teo, C., Kundi, M., Hardell, L., & Carlberg, M. (2009). Cell phones and brain tumors: A review including the long-term epidemiologic data. *Surgical Neurology*, 72(3), 205–214.

LaDou, J. (2006). Printed circuit board industry. *International Journal of Hygiene and Environmental Health*, 209(3), 211–219.

Li, P., Li, P., Feng, X., Qiu, G., Li, Z., Fu, X., Sakamoto, M., … & Wang, D. (2008). Mercury exposures and symptoms in smelting workers of artisanal mercury mines in Wuchuan, Guizhou, China. *Environmental Research*, 107(1), 108–114.

Luo, J., Li, H., Deziel, N. C., Huang, H., Zhao, N., Ma, S., ... & Zhang, Y. (2020). Genetic susceptibility may modify the association between cell phone use and thyroid cancer: A population-based case-control study in Connecticut. *Environmental Research*, 182, 109013.

Marino, C., Gini, G., Vieno, A., & Spada, M. M. (2018). The associations between problematic Facebook use, psychological distress and well-being among adolescents and young adults: A systematic review and meta-analysis. *Journal of Affective Disorders*, 226, 274–281.

Morose, G., Shina, S., & Farrell, R. (2011). Supply chain collaboration to achieve toxics use reduction. *Journal of Cleaner Production*, 19(5), 397–407.

Needhidasan, S., Samuel, M., & Chidambaram, R. (2014). Electronic waste–an emerging threat to the environment of urban India. *Journal of Environmental Health Science and Engineering*, 12(1), 1–9.

Palm, E., & Hansson, S. O. (2006). The case for ethical technology assessment (eTA). *Technological Forecasting and Social Change*, 73(5), 543–558.

Petavratzi, E., Kingman, S., & Lowndes, I. (2005). Particulates from mining operations: A review of sources, effects and regulations. *Minerals Engineering*, 18(12), 1183–1199.

Sandler, R. L. (2014). Introduction: Technology and ethics. In R. L. Sandler (Ed.), *Ethics and Emerging Technologies* (pp. 1–23). London: Palgrave Macmillan.

Scharnhorst, W., Althaus, H. J., Classen, M., Jolliet, O., & Hilty, L. M. (2005). The end of life treatment of second generation mobile phone networks: Strategies to reduce the environmental impact. *Envionmental Impact Assessment Review*, 25(5), 540–566.

Shilton, K. (2015). Anticipatory ethics for a future Internet: Analyzing values during the design of an internet infrastructure. *Science and Engineering Ethics*, 21(1), 1–18.

Stahl, B. C., Heersmink, R., Goujon, P., Flick, C., Van Den Hoven, J., Wakunuma, K., … & Rader, M. (2010). Identifying the ethics of emerging information and communication technologies: An essay on issues, concepts and method. *International Journal of Technoethics*, 1(4), 20–38.

Wangen, G (2015). The role of malware in reported cyber espionage: A review of the impact and mechanism. *Information*, 6(2), 183–211.

Wilhelm, M., Hutchins, M., Mars, C., & Benoit-Norris, C. (2015). An overview of social impacts and their corresponding improvement implications: A mobile phone case study. *Journal of Cleaner Production*, 102, 302–315.

Index

Note: **Bold** page numbers refer to tables; italic page numbers refer to figures and page numbers followed by "n" denote endnotes.